AF756145

Near Eastern Helmets of the Iron Age

Near Eastern Helmets of the Iron Age

Tamás Dezsö

UNIV DUNELM
0 1 FEB 2002

BAR International Series 992
2001

~~059.13~~
930
BRI

(992)

British Archaeological Reports are published by John and Erica Hedges Ltd and by Archaeopress

This volume has been published by:
John and Erica Hedges Ltd
British Archaeological Reports
7 Longworth Road
Oxford OX2 6RA
England
Tel/Fax +44(0)1865 511560
E-mail: bar.hedges@lineone.net
www.barhedges.com

Enquiries regarding the submission of manuscripts for future publication may be sent to the above address

BAR S992

Near Eastern Helmets of the Iron Age

© Tamás Dezsö 2001.

Volume Editor: Erica Hedges

Printed in England by Biddles Ltd

ISBN 1 84171 193 4

All BAR titles available from:

Hadrian Books
122 Banbury Road
Oxford OX2 7BP
England

The current BAR catalogue with details of all titles in print, prices and means of payment, is available free from Hadrian Books

All volumes are distributed by Hadrian Book Ltd

TABLE OF CONTENTS

Foreword

Forty years after publication Yigael Yadin's *The Art of Warfare in Biblical Lands* (Jerusalem 1963 remains an isolated introduction to the subject. Its author, who combined the field experience of an archaeologist and a soldier with the expertise and learning of a professional scholar, was uniquely qualified to write such a panoramic survey. The formidable task of assembling, critically assessing and interpreting the basic evidence, which is needed to carry the subject decisively forward, has even now barely been started. This applies particularly to the material remains. Relevant art-historical and textual sources are rather better published and consequently more widely known. Amongst arms and amour the former have attracted more attention from archaeologists than the latter, as sheet metal is more vulnerable to the ravages of passing time. Dr. Dezső's contribution to this famous series of volumes marks an important step towards redressing the balance. The helmets, the most accessible element amongst surviving body armour, are various enough to provide a basis for the study of Iron Age sheet metal armour in the Near East.

Research of this kind requires an unusual combination of skills. They include the ability to handle ancient written sources, to present systematic typologies based on the visual arts, both monumental (reliefs and statuary) and minor (cylinder seals), and finally, but by no means least, an ability to deal with surviving objects. Here the skills of a draughtsman are a particular asset to a scholar, not only for the obvious reasons. They develop an attention to detail greater than is common in those who only look and write when describing corroded and damaged objects, but do not themselves draw. Dr. Dezső displays each of these skills here.

The student of ancient Near Eastern metalwork is well acquainted with problems provided by a widely scattered database of objects often not well preserved or well published. Now there is a new hazard which is represented in this catalogue. The present body of evidence for ancient Near Eastern Iron Age helmets is contaminated at times not by total forgeries, but by what may be ingenious *pastiches*, potentially involving both the forms and decorations of the helmets. In the absence of professional scientific reports (and sometimes even with them) this new problem requires the exercise of balanced judgement, if the bad is not to drive out the good gratuitously. The limitations of modern knowledge of ancient craft skills and of iconography have always to be borne in mind when reaching an opinion on authenticity. Absence of evidence is not necessarily evidence of absence when the known range of Assyrian or Urartian imagery, particularly in a special context like this one, is still restricted.

The contexts of retrieval and some aspects of the technology of the helmets catalogued here are distinctive enough to merit initial comment. It is striking how many of the Iron Age helmets from the Near East at present come from the highland zone. Surviving material evidence for Bronze Age helmets is at present very rare but, as Dr. Dezső demonstrates at the beginning of his monograph, pictures and texts may be combined to outline their history. Mudbrick architecture and the processes of *tell* formation have rarely favoured the survival of sheet metal objects in good condition. Moreover, when found, they demand skilled excavation and conservation on site. The two helmets which open the catalogue are indeed from the lowlands, but significantly from a well and a harbour respectively.

In the highlands ruined stone-built structures and rockcut or stone-lined graves increase the potential for the survival of sheet metal objects. Regrettably, for much of the last century, such sites were especially accessible and vulnerable to illicit excavations, and the objects found to dismemberment by their plunderers to satisfy multiple claims on them. Consequently, research of the kind undertaken here is too often rescue archaeology in the most explicit sense. In retrospect, it is painfully clear that pioneer excavators of Assyrian palaces and temples in northern Iraq in the nineteenth century found arsenals and armouries, but were unable to retrieve much of the contents. Opportunity and improved techniques of excavation have more recently allowed Urartian temple store-rooms and palace arsenals to contribute more than half the total number of helmets catalogued here.

As ancient Near Eastern sheet metal has rarely been the subject of technological study the contribution made by some of the catalogued helmets is unusually instructive. The significant role of military workshops and the demands of warfare in technological innovation is a commonplace throughout history; but it may rarely be observed so readily in the ancient world as in some of the Assyrian and Urartian helmets published here. The earliest iron helmets were constructed from pieces rivetted together. Within a century they were hammered out of a single iron sheet. It is also clear from the bi-metallic examples (copper alloy/iron) that the helmet makers were exploiting colour contrasts, long since lost to corrosion and decay, indicating that appearance as well as defensive qualities were appreciated in this armour, as they have been through the ages. Particular interest attaches to the incised and *repoussé* decoration. This appears to have been used, both in Assyria and Urartu, in meaningful ways to distinguish votive helmets from those worn by various contingents within the army.

Catalogues and typologies are the means to an end and that end is best served, as in this case, when the commentaries are rich in detail, offering a summary of the present state of knowledge. It is remarkable what Dr. Dezső has managed to contribute to general understanding of the war machines of Assyria and Urartu, particularly, by concentrating his focus on a single aspect of their armoury.

Ashmolean Museum, Oxford — P.R.S. Moorey

Acknowledgements

This work is the updated English version of my PhD dissertation (Loránd Eötvös University, Budapest, 1992). I started the work in Oxford, Linacre College (1989—1990), as a Hungarian Soros Fellow. So I am very much indebted to the College, and the Hungarian Soros Foundation.

I am greatly in the debt of Dr. P.R.S. Moorey, my former supervisor, whose guidance and advice I have sought. He carefully read the drafts of this text and made important comments. I am especially grateful to Dr. J.E. Curtis for his help and inspiration. As I did not always take their advice, I alone am responsible for what survived.

I must also thank to the Institutions and Foundations, who helped my work, especially my department, the Department of Assyriology and Hebrew, Eötvös Loránd University, Budapest, its professor, Dr. Géza Komoróczy; the W.F. Albright Institute of Archaeological Research, its director Dr. S. Gitin, and the Andrew Mellon Foundation. I am very much indebted for the financial help of my project to the Hungarian Ministry of Education (FKFP-2000 Project), and the Bolyai Fellowship of the Hungarian Academy of Sciences.

Budapest, 2001. 10. 01. — Tamás Dezső

INTRODUCTION

3rd millennium B.C. Mesopotamian helmets

In the 1927—1928 excavation season at Ur, in the dromos of Tomb 789, Sir Leonard Woolley found six copper helmets belonging to six skeletons of soldiers. The soldiers might be members of a "royal" bodyguard, because their spearheads were decorated with the same symbol, an incised bull's leg[1]. All of the helmets were badly crushed by the weight of the earth, so their shape can hardly be reconstructed. Owing to the inconsistencies of the publication[2] it is impossible to identify the helmets with the excavation numbers (U.10809, U.10825, U.10826, U.10827, U.10828), inventory numbers[3], photographs[4] or drawings[5]. A similar copper helmet of presumably the same age was found by Gaston Cros at Telloh[6]. These seven helmets represent the following helmet types:

1. Hemispherical helmet without earflaps (Ur): **helmet no. 1**
2. Hemispherical helmet with rounded earflaps (Ur): **helmet no. 2**
3. Pointed helmet with rounded earflaps (Ur): **helmet no. 3**
4. Pointed helmet with rounded earflaps and neckguard (Ur, Mari): **helmet nos. 3-6**, *Chart 1, nos. 3-6*
5. Pointed helmet with pointed earflaps and neckguard (Telloh [Lagash]): **helmet no. 7**, *Chart 1, nos. 1-2*

Comparing these types with the Early Dynastic III representational evidence the first three types are unknown from art. The fourth type appears on the famous "Standard of Ur" found in PG/779 at the Royal Cemetery of Ur (*Chart 1, no. 3*). The shell inlay decoration of one of the sides of the standard (probably the sound-box of a lyre) shows Sumerian soldiers on the march and in battle wearing this type of pointed helmet with rounded earflaps, neckguard and chin-strap[7]. The same type of helmet is known from shell inlay representations of soldiers from Mari[8]. These representations show that this Sumerian type of copper helmet (presumably with the hoplite tactic connected to the Sumerian type of weaponry) was used widely in other parts of Mesopotamia. The fifth type – conical helmet with pointed earflaps and neckguard – might be a local variant of type no. 4. This type is known exclusively from Telloh [Lagash] (**helmet no. 7**). This type of helmet can be seen on the "Stele of Vultures" of Eannatum, king of Lagash found at Telloh (*Chart 1, no. 2*). On the stele the king wears a strange variant of helmet type 5: a ceremonial helmet with *repoussé* ears on the earflaps and probably a knot of his own hair at the back, outside of the helmet (*Chart 1, no. 1*). This knot of hair was probably a sacral manner of wearing the *LUGAL*'s (king's) hair. This sacral manner of wearing the hair is known from the famous electrum helmet of Meskalamdug (**helmet no. 8**), and from the shell inlay figure of a king from the Temple of Ishtar at Mari[9].

The helmet tradition of the first known Semitic dynasty of Mesopotamia, the Dynasty of Agade, is different from the earlier Sumerian helmet tradition. Unfortunately there are no actual helmet finds from this period but the "Stele of Naram-Sin"[10] recovered from Susa shows higher, conical helmets with neckguard (*Chart 1, nos. 7-8, 10*). The shape of these high conical helmets furnished with

---

1 Woolley 1934, 64, pl. 189, U.10825-10828, types 3, 6.

2 Woolley 1934, 66-67, 349, 562, pls. 29, 148, 149, 218, skeleton nos. 45-50. Only a single helmet (p. 66, cat. no. 18) can be connected with a drawing (pl. 218, PG/789b) but the excavation number of this helmet is unknown. I am much indebted to Dr. P.R.S. Moorey, who at my request checked Woolley's original excavation cards in the British Museum. He wrote: "There appear to be seven in all (helmets): the one with burial no. 18, for which only the twisted silver chinstrap is recorded on the card for U.10809; and six with burials 45-50. Three of these, "in poor condition", share the same field number: U.10828. Only one BM number, BM 121414 = U.10825 (with burial no. 46) turned out to be a helmet." The field cards give the following details: U.10809. Tomb Group (UE II, pl. 218: PG/789b): B. Fragment of twisted silver-wire chin strap from a copper helmet (this must be the source of the more detailed published description). PG/789, burial no. 18. U.10825. Tomb Group: A. Copper helmet on head = BM 121414 (not illustrated), PG 789, burial no. 33 (original in pencil, changed to 46 in ink). U.10826. Tomb Group: A. Copper helmet on skull (UE II, pls. 148a, 218: PG/789a), PG/789, burial no. 34 (in pencil, changed to 50 in ink). U.10827. Tomb Group: A. Copper helmet (UE II, pl. 218: PG/789c), PG/789, burial no. 35 (in pencil, changed to 48 in ink). U.10828. Tomb Group: A. Copper helmets; three; in poor condition; PG/789 "from the soldiers of the dromos", 49 (added in ink). (Presumably also 45 and 47).

3 Woolley 1934, 562.

4 Woolley 1934, pls. 148, 149.

5 Woolley 1934, fig. 5, pl. 218: PG/789a-c.

6 Parrot 1948, 106, fig. 26d.

7 Woolley 1934, 61, 238-239, 243, 255, 258, 266, pls. 91-92.

8 "L'Étandard de Mari", Mari, Temple of Ishtar. Paris, Louvre, AO 18215, AO 19820: Parrot 1956, pls. LV, LVI; 2. Mari, Presargonic Palace. Damascus, Archaeological Museum: Parrot 1983, pl. 169.

9 "L'Étandard de Mari": Parrot 1956, pl. 77 (*Chart 1, no. 4*).

10 Paris, Louvre, Sb 4. Strommenger 1962, pls. 122-123; Börker-Klähn 1982, 134-136, no. 26; Parrot 1983, pl. 192.

earflaps and neckguard is different from the earlier, squat pointed or hemispherical shape of the Sumerian helmets of the ED Period. Similar helmets are shown on a stele from the same period found in Telloh[11]. This change might be connected to new people but it could simply represent a change in the representational tradition. The helmet of Naram-Sin (*Chart 1, no. 7*) is unique. The shape is conical, there are the embossed ribs running around the rim of the helmet and there is a long neckguard (metal or leather) attached to the back of the helmet. Two pairs of horns are the most anomalous feature of the helmet. In Mesopotamian iconography the pairs of horns belong exclusively to the well-known headdresses or crowns of deities. Such a representation of Naram-Sin (2254—2218 B.C.) suggests a certain level of divinity attached to the person or the status of the king.

Written sources of the 3rd millennium B.C. mention helmets made of different materials. Copper[12] helmets (*sagšu*[13] *urudu*) are known from Fara Period (Early Dynastic II-IIIa)[14], from the period of the Dynasty of Agade[15] and from the time of the IIIrd Dynasty of Ur[16]:

2 sag-$šu_4$ urudu kù gar-ra ("2 copper helmets decorated or plated with silver")

On the representations it is impossible to distinguish the shape of copper helmets from the shape of bronze helmets. It is interesting that "copper helmets" are known as late as the time of the IIIrd Dynasty of Ur and "bronze helmets" appear as early as the first "copper helmets". The first documents which mention "bronze helmets" are two parallel inscriptions of Uru-KA-gi-na, king of Lagash (Telloh) (*sag-šu₄ zabár*)[17]. The most interesting documents are those from the period of the Dynasty of Agade mentioning bronze and leather helmets and "eagle-headdresses" (probably helmets to which eagle figures were attached) decorated with silver. One of the tablets is a weapon list[18]: bronze helmets and copper spears were assigned to different persons presumably from the central arsenal. An other tablet from the same corpus is, however, a different kind of inventory list: raw materials for the manufacture of weapons were listed[19]:

vs. II: 11. 3 sagšu zabar
12. zabar-*śu-nu* 5 ma-na
13. *in* 1 $^2/_3$ *ša* ma-na-ta
("3 bronze helmets, (weight of the) bronze of which is 5 *mana*, 1 $^2/_3$ *mana* each")

vs. III: 18. 8 sagšu z[abar]
19. zabar-*śu-nu* [13 $^1/_3$ ma-na]
20. *in* [1 $^2/_3$ *ša* ma-na-ta]
("8 bronze helmets, (weight of the) bronze of which is [13 $^1/_3$ *mana*, 1 $^2/_3$ *mana* each")

rs. II: 12. 7 sagšu zabar
13. zabar-*śu-nu* 11 $^2/_3$ ma-na
14. *in* 1 $^2/_3$ *ša* ma-na-ta
("7 bronze helmets, (weight of the) bronze of which is 11 $^2/_3$ *mana*,1 $^2/_3$ *mana* each")

It is very important to point out that the weight of these bronze helmets or rather the quantity of bronze assigned to the manufacture of a helmet is the same (let us say standardized): all of the helmets weighed probably 1 $^2/_3$ mana = 833 g[20]. It has already been mentioned that the shape of the helmets in each city was probably standardized (*see* the "Standard of Ur" or the "Stele of Vultures"). But it is clear that not only the shape, but the weight of the helmets of the professional soldiers equipped from the arsenal by the state was standardized as well. The bad condition of the Ur helmets (**helmet nos. 1-6**) makes it impossible to weight them, but the weight of later helmets made of the same material with the same technique shows considerable similarity. The weight of the 15th century B.C. bronze helmet

[11] Paris, Louvre, AO 2678, 2679. Strommenger 1962, pl. 117; Börker-Klähn 1982, 130, no. 21.
[12] Sumerian word *urudu* means probably copper or comparing with the word *zabar* ("bronze") an inferior quality bronze alloy (tin bronze *versus* arsenic bronze).
[13] Sumerian SAGŠU = Akkadian *kubšu*. Landsberger 1955, 251 (Syllabary B): sag-šu U+SAG = *ku-ub-šu*.
[14] Unpublished economic documents from Isin. *See* Wilcke 1975, 312.
[15] Thureau-Dangin 1910, 7, no. 1125; de Genouillac 1911, no. 5723.
[16] Limet 1960, 288, no. 17, rs. 1.
[17] Sollberger 1956, Ukg. 4 V 11 = 5 V 8.
[18] Scheil 1913, no. 85.
[19] Scheil 1913, no. 86.
[20] 1 gín (*šiqlum*) = 8.33 g. 1 mana = 60 gín (*šiqlum*) = 0.5 kg. 1 $^2/_3$ mana = 100 gín (*šiqlum*) = 833 g. Wilcke 1975, 312: 673 g.

found at Knossos[21] is 695 g, while the weight of a 8th century B.C. Urartian bronze helmet (**Cat. no. 86**)[22] is 767 g. The weight of the helmet increases considerably only in the second half of the 2nd millennium B.C. with the appearance of helmets made of scale armour plates. These scale armour helmets were however worn by chariot warriors (*maryannu*) who – fighting on the platform of the chariot – were not hindered by the greater weight of the helmet. With the help of finds of actual scale armour plate finds and texts mentioning helmets made of scale armour plates T. Kendall reconstructed the weights of helmets made of 140 *usque* 200 armour plates. The result was 2008 g *usque* 3560 g[23]. This weight is comparable with the 3049 g weight of the corroded Nimrud Assyrian iron helmet (**Cat. no. 10**) of the late 8th – early 7th century B.C. The scale armour helmets were furnished with earflaps, neck-guards and probably a protective apron covering the shoulders as well. But in the antiquity the weight of bronze helmets hammered out from a sheet of bronze remained always between 700 and 1300 g[24].

The same cuneiform text quoted above gives us details regarding not only the copper or bronze helmets, but helmets made of different materials, for example leather as well. The identification of leather helmets on the 3rd millennium B.C. representations is difficult. The distinction between the non-metal or leather headdresses and non-metal or leather helmets is similarly difficult. Some stele fragments[25] found "near Nasiriya" and dated to the time of the Dynasty of Agade show leather helmets furnished with nose-guard, earflaps and neck-guard and made of horizontal leather bands sewn together (*Chart 1, no. 9*). The same helmet type is shown on a stele-fragment of the same age found at Susa[26]. The structure of the helmet is curious. It was presumably made of different types of non-metallic materials: layers and bands of different types of leather. As the same written source quoted above shows, this type of helmet lacked any inner metal structure:

vs. II:
17. 4 sagšu kuš-gu$_4$
18. kuš-gu$_4$-*śu-nu* 4 kuš-[máš-*śu*-]
19. *-nu*
20. *in* 1-ta
21. siki-*śu-nu* 1 $^{1}/_{3}$ *ša* [ma-na]
22. *in* $^{1}/_{3}$ *ša* [ma-na-ta]
("4 oxhide helmets. 4 (pieces) of oxhide, 4 (pieces) of goat-skin, 1 piece (of leather) to each helmet. Their wool in all is 1 $^{1}/_{3}$ *mana*, $^{1}/_{3}$ *mana* each.")

rs. I:
4. 16 sag[šu kuš-gu$_4$]
5. kuš-gu$_4$-*śu-nu* 1[6 kuš-máš-]
6. *-śu-nu* 16
7. *in* 1-[ta]
8. siki-*śu-nu* 5[$^{1}/_{3}$ ma-na]
9. *in* $^{1}/_{3}$ *ša* [ma-na-ta]
("16 oxhide helmets. 16 (pieces) of oxhide, 16 (pieces) of goat-skin, 1 piece (of leather) to each helmet. Their wool in all is 5$^{1}/_{3}$ *mana*, $^{1}/_{3}$ *mana* each.")

rs. II:
15. 41 sagšu kuš-gu$_4$
16. kuš-gu$_4$-*śu-nu* 41 kuš-máš-*śu-nu* 41
17. *in* 1-ta
18. siki-*śu-nu* 13 $^{2}/_{3}$ *ša* ma-na
19. *in* $^{1}/_{3}$ *ša* [ma-na]-ta
"41 oxhide helmets. 41 (pieces) of oxhide, 41 (pieces) of goat-skin, 1 piece (of leather) to each helmet. Their wool in all is 13 $^{2}/_{3}$ *mana*, $^{1}/_{3}$ *mana* each."

---

[21] Knossos, Hospital Site, Tomb V (Archaeological Museum, Iraklion). H.: 17 cm; diam.: 21 cm. Earflaps: H.: 16.5 cm; W.: 9 cm. Hood – de Jong 1952, 256ff., pls. 50-52; Hencken 1971, 20, fig. 3e-g (LM II, 1450-1400 B.C.); Borchhardt 1972, 55-60, pl. 37:1, 4, Kat. 11 I.

[22] Boston, Museum of Fine Arts, Inv. no. 1981.25; Kendall 1981, 213, note 45.

[23] Kendall 1981, 212.

[24] For example *see* four ancient Greek helmets in the Museum of Fine Arts, Budapest: 1. Geometric helmet made of five pieces of bronze sheet, Inv. no. 8442: 1020 g; 2. Corinthian helmet (the "Myros helmet"), Inv. no. 8444: 1303 g; 3. Illyrian helmet, Inv. no. 8440: 1041 g; 4. Chalkidian helmet, Inv. no. 8441: 1167 g.

[25] Baghdad, Iraq Museum, IM 55639, IM 59205; Boston, Museum of Fine Arts, Inv. 66893; Strommenger 1962, pl. 119; Börker-Klähn 1982, 131-133, no. 22; Parrot 1983, pl. 197.

[26] Paris, Louvre, Sb. 6641, 6641*bis*; Börker-Klähn 1982, 133, nos. 23, 24.

It is clear that in this central distribution system, similar to that for copper helmets, the weight of the raw material supplied by the state, the weight of leather and wool of which the helmet was made was standardized. The manufacture of a leather helmet (ordered by the state) needed 1 piece of oxhide, 1 piece of goatskin and $^1/_3$ mana[27] wool. The leather bands were sewn together horizontally and the lining was made of wool. The ratio of cheaper leather helmets (61 pieces are mentioned in the document) and more expensive bronze helmets (18 pieces) is probably commensurate with the ratio of auxiliary troops conscripted from the ranks of different nomadic peoples and citizens of tributary cities (levies) and professional soldiers of the Akkadian/Sumerian *élite* troops.

An interesting third group of 3rd millennium B.C. helmets are those decorated with precious metals. There were probably several types of helmets made of or decorated with precious metals. The ceremonial golden helmet of Meskalamdug (**helmet no. 8**) has already been discussed. As has been mentioned earlier the helmet of Eannatum depicted on the "Stele of Vultures" (*Chart 1, no. 1*), which featured the same "royal" hairstyle as the helmet of Meskalamdug, was presumably made of a precious metal, or gilded bronze. There are cuneiform sources referring to helmets or headdresses decorated with precious metals. There is a cuneiform text[28] dated to the period of the Dynasty of Agade which mentions three helmets gilded or inlaid with gold:

3 sagšu kù-sig$_{17}$ gar-ra
"3 gilded helmets (to the king)"

The same grammatical form was used in another text[29] to describe copper helmets plated with silver:

2 sag-šu urudu kù gar-ra
"2 copper helmets plated with silver"

These two loci only prove that metal helmets were decorated (gilded and plated?) with gold and silver. The manner of decoration however remains obscure. But in one of the cuneiform texts[30] (an armoury list) mentioned earlier appears the name of a mysterious type of helmet, the "bird", probably "eagle (*ti*$_8^{\text{mušen}}$, *A*^{mušen}) headdress":

vs. II: 14. 3 sagšu [ti$_8$]^{mušen}
15. [kù]-babbar-*śu-nu* $^1/_2$ [ma-na]
16. *in* 10 gín-[ta]
"3 'eagle' helmets, (weight of) their silver is $^1/_2$ *mana*, 10 *gín* each"

rs. I: 2. 1 sagšu [ti$_8$ ^{mušen}]
3. kù-babbar-*śu* 10 g[ín]
"1 'eagle' helmet, (weight of) its silver is 10 *gín*"

It is obvious that the amount of silver used to decorate this type of helmet or headdress is standard: 10 gín (*šiqlum*)[31]. So these headdresses were probably made not to individual order, but were "mass produced" as the copper/bronze helmets. There are no known representations of helmets or headdresses decorated with birds in Mesopotamian art. There are however a few helmets decorated with birds (eagles) known from Northern Iran from the end of the 2nd millennium B.C. (**Cat. nos. 122-124**). These helmets were decorated with relief sculptures: figures of deities and a raptorial bird with wings outstretched and head rendered in the round above the human figures. Each of the figures was sculpted from bitumen, overlaid with silver and gold foil, and then fastened to a bronze plate riveted to the helmet (An alternative technique might be: the figures were hammered up in *repoussé* in sheet gold or silver then reinforced with bitumen [poured in hot] before being attached to the helmet). The same technique was used in Mesopotamia as well, but most of the objects decorated with this technique (for example disks or roundels[32]) came from Elam (Susa) and the Safid River Region. The iconography of the deities is Mesopotamian. The "eagle helmet or headdress" (*sagšu ti*$_8$*-mušen*) might fall in the same category as the Iranian helmets mentioned above, gilded and plated figurative decoration which might have originated in an unfortunately unknown cultic or apotropaic tradition, possibly Elamite.

---

[27] 166 g (*see* note 20); Wilcke 1975, 313: 134.5 g.
[28] Thureau-Dangin 1903, no. 222, II:3; Limet 1960, 154.
[29] Paris, Louvre, AO 3606, Limet 1960, 288, no. 17, rs. 1.
[30] Scheil 1913, no. 86.
[31] 10 gín (*šiqlum*) = 83 g.
[32] New York, Metropolitan Museum of Art, MMA 62.115 (Wilkinson 1965, fig. 10; Muscarella 1988, fig. 16); MMA 66.31.1, MMA 64.63 (Muscarella 1988, figs. 17, 19); Norbert Schimmel Collection (Wilkinson 1965, fig. 11; Muscarella 1988, fig. 18).

In the mid-3rd millennium B.C. Ebla almost the same types of helmet were in use. In the Ebla texts the word for the helmet is *nì-sagšu*[33]. As it is known from the Palace archive, the Ebla helmets were mass-produced, and the army of Ebla used almost the same equipment as the contemporary Sumerian and/or Akkadian armies. One of the texts mentions altogether 156 bronze helmets[34]. Some of the Ebla helmets were furnished with what was probably a face guard (*IGI.DUB*), the weight of which was 40 *šiqlum* (Ebla: 316 g)[35].

The Ebla texts mention helmets made of gold and silver, the weight of which was between 73 and 240 *šiqlum* (Ebla: 576.7—1896 g).

vs. V: 4. 1 níg-sagšu
5. 1 ma-na TAR kù-sig$_{17}$
("1 helmet, gold of which is $1^1/_2$ *mana*") [36]

As Waetzold has shown[37], these ceremonial helmets were probably worn by officers or the members of the Eblaite *élite*. It is unfortunately unknown, whether these helmets were entirely made of gold and silver, or were simply gilded or silverplated[38]. The price of the most expensive, probably golden helmet was around 266 *šiqlum* silver:

vs. I: 3. 13 ma-na «GUR$_{8}$»
vs. II: 1. níg-sa$_{10}$ 3 níg-sagšu
("3 helmets, price of which is 13 *mana* 20 gín")[39]

Taking the close connections of the Eblaite and Mesopotamian civilizations into account, it is not surprising, that there were "eagle helmets" or headdresses at Ebla as well.

vs. I: 1. 15 gín Dilmun kù-sig$_{17}$
2. šir-za [a]n-dùl [
vs. II: 1. 1 Á^{mušen} ŠA.PI
(„1 eagle: 15 *gín* Dilmunite gold")[40]

vs. VI: 2. šir-za
3. Á^{mušen} 1 níg-sagšu
4. *ìr-^{d}da-mu*
5. lú *i-rí-ik-gú-nu*
6. 5 gín Dilmun kù-sig$_{17}$
(„eagle: 1 helmet for Ir-Damu, the man of Irik-Gunu, 5 *gín* Dilmunite gold")[41]

The Ebla texts mention more than 500 pieces of such "eagle helmets" (*nì-sagšu ti$_{8}$-mušen*)[42]. The weight of these helmets or the "eagle appliques" differs considerably[43], so these pieces were made probably for private order. However, there are eagle helmets made of 30-34 *šiqlum* bronze or 40 *šiqlum* silver, which means, that these helmets or headdresses were probably the helmets of officers, officials (even priests) and the ceremonial helmets of a special service (bodyguard?). This idea seems to be corroborated by the fact, that other objects were also decorated in the same manner with this eagle badge(?)[44]. These eagle helmets were offered to deities as *ex votos* as well[45].

---

[33] Waetzold 1990, 29-30.
[34] Pettinato 1979, 1080 rs. II:6; Waetzold 1990, 29.
[35] Edzard 1981, 11 I:1 - II:2, II:3 - rs. III:4; Waetzold 1990, 29, note 158.
[36] Edzard 1981, 4 V:4-5; Waetzold 1990, 29, note 159. At Ebla 1 gín (*šiqlum*) = 7.9 g. 1 mana = 60 gín (*šiqlum*); $1^1/_2$ mana (90 *šiqlum*) = 711 g.
[37] Waetzold 1990, 29.
[38] For the prices of the different helmets *see* Waetzold 1990, 29.
[39] Edzard 1981, 41 I:3-II:1; 13 mana 20 gín (*šiqlum*) = 800 gín (*šiqlum*). So the price of a helmet is 266 *šiqlum* (2101 g) silver.
[40] Archi – Biga 1982, 431 II:1. 15 gín (*šiqlum*) = 118.5 g gold.
[41] Pettinato 1980, 12 rs. VI:1-6. 5 gín (*šiqlum*) = 39.5 g gold.
[42] Waetzold 1990, 29.
[43] Waetzold 1990, 29: one eagle helmet was made of 230 *šiqlum* (1817 g) gold.
[44] Waetzold 1990, 30, note 163.
[45] Waetzold 1990, 30, note 164.

3rd millennium B.C. cuneiform sources mention "helmets" or headdresses made of different materials that can be excluded from the category of helmets. Such helmets are the headdresses made of wool (siki, sík [*šipātum*])[46] and the very interesting reed matting (^{ú}ninni$_5$ [*ašlum*])[47].

1. Ur, Royal Cemetery, PG/789 (? U.10826, burial no. 50) – Hemispherical copper helmet without earflaps. There is a horizontal rib running around the rim of the helmet. Crushed. – ED Period (26th–25th century B.C.) – London, British Museum – Woolley 1934, pls. 148a, 218: PG/789a.

2. Ur, Royal Cemetery, PG/789 (? U.10825, burial no. 46) – Hemispherical copper helmet with rounded earflaps. The back of the helmet is damaged. Crushed. There was probably a silver-wire chin strap attached to the helmet. – ED Period (26th–25th century B.C.) – London, British Museum – Woolley 1934, pl. 218: PG/789b.

3. Ur, Royal Cemetery, PG/789 (? U.10827, burial no. 48) – Slightly conical copper helmet with rounded earflaps. Crushed. – ED Period (26th–25th century B.C.) – London, British Museum – Woolley 1934, pl. 218: PG/789c.

4. Ur, Royal Cemetery, PG/789 (U.10828, burial no. 49?) – H. without point: 21.5 cm; H. of the point bent back: 9 cm; Diam. (crushed): 34.4 cm; W. of the sheet 2 mm. – Conical copper helmet with rounded earflaps and neckguard. The point of the helmet is bent back and crushed. – ED Period (26th–25th century B.C.) – London, British Museum, WA 121414 – Woolley 1934, 66-67.

5. Ur, Royal Cemetery, PG/789 (U.10828, burial no. 45) – Copper helmet of unidentified shape (probably similar to helmet **4**). "*In poor condition*", badly damaged. – ED Period (26th–25th century B.C.) – London, British Museum – Woolley 1934, 66-67.

6. Ur, Royal Cemetery, PG/789 (U.10828, burial no. 47) – Copper helmet of unidentified shape (probably similar to helmet **4**). "*In poor condition*", badly damaged. – ED Period (26th–25th century B.C.) – London, British Museum – Woolley 1934, 66-67.

7. Telloh – Hemispherical copper helmet with a small point. The helmet is damaged at the front, above the face cutout and at the back. The helmet has combined earflaps and neck-guard. There are a series of holes around the rim of the helmet for the fastening of a lining. On the pointed earflaps there is a further hole for the fastening of a chinstrap. – ED Period (26th–25th century B.C. – Paris, Louvre, AO 4119 – Parrot 1948, 106, fig. 26d.

8. Ur, Royal Cemetery, PG/755 (U.10000) – H.: 23 cm, Diam.: 26 cm. Ceremonial electrum helmet of Meskalamdug. The overall shape of the helmet forms a probably sacral manner of wearing the very long hair braided in a plait twisted around the head and fastened in a knot of hair at the back of the head. The helmet covered the ears and the back of the head as well. "Hammered up from a single sheet of metal, with features in relief and details engraved. Into the interior was fitted a quilted cap, fragments of whose cloth and wool stuffing were found; the discoloration of the metal showed that the cloth was brought up on to the outside of the helmet, so as to protect the wearer's skin from the sharp edges of the metal, and secured by laces passed through the small holes round the rim. The ears are pierced to allow of hearing. When found the helmet was intact except the right cheek-piece had been crushed in and the right side of the band above the forehead was bent inwards; This minor damage has been made good." – ED Period (26th–25th century B.C.) – Baghdad, Iraq Museum, IM.8269 – Woolley 1934, 244, 292, 294, 552, colour plate on frontispiece, pl. 150.

---

[46] In Wilcke's reconstruction (Wilcke 1975, 312) the helmets listed in one of our cuneiform tablet (Scheil 1913, no. 86, vs. col. II, 21-22; rs. col. I, 8-9; rs. col. II, 18-19) discussed above were made of wool. In our reconstruction the wool was used only for the lining, and the helmets were made of leather. For woolen "helmets" *see*: Wilcke 1969, 106, Lugalbanda II, 150: ^{túg}*sagšu pirig-mè* ("woolen 'helmet', the lion of the battle"); Legrain 1937, no. 1588, 6: ^{túg}*sagšu*. Waetzold (Waetzold 1990, 29 and note 156) mentioned Eblaite woolen caps which were probably the lining of leather or metal helmets.

[47] Crawford 1954, no. 197, 1-2: 6 kuš-gu$_4$ A.GAR ^{kuš}sagšu-^{ú}ninni$_5$-šè ("6 pieces of oxhide for helmets/headdresses made of matting"); Thureau-Dangin 1910, 7, no. 1125.

2nd millennium B.C. Near Eastern helmets

The 2nd millennium B.C. continuation of 3rd millennium B.C. helmet tradition is unknown. The most important section of the 3rd millennium B.C. Sumerian armies was the heavy infantry using the phalanx tactic. The heavy, four-wheeled and equid-drawn chariots were probably manned by the same infantrymen, who used infantry weapons (spears, axes) and wore the same helmets (and heavy woollen cloaks) as the foot soldiers (*see* for example the "Stele of Vultures" or the "Standard of Ur", *Chart 1, nos. 1-3*). These slowly moving heavy chariots probably used similar tactics to the infantry. The early 2nd millennium B.C. Mesopotamian armies probably consisted of the same skilled or semi-skilled masses of infantrymen. It is known that the army of Hammurapi, king of Babylon (1792—1750 B.C.) consisted mainly of foot soldiers of various types[48]. Unfortunately only a few helmet representations of the Old Babylonian period are known. The linen headdresses shown on a Mari wall painting[49] and a head made of alabaster[50] can be excluded from the category of helmets. There is a helmet, however, on an Old Babylonian shell-inlay plaque found at Mari. The shell-inlay scene shows a crested helmet on the top of a standard (*tropaion*) (*Chart 1, no. 11*)[51]. The helmet is conical with rounded top and an omphalos on the side. The crest was fastened to the helmet probably diagonally, starting at the front and ending at the back. There is not a single similar helmet known from this period. Only a 600 hundred years younger Submycenaean (11th century B.C.) bronze helmet found at Tiryns shows a similar form[52]. It consists of two symmetrical side plates, a long bronze band fastening the two sides together, and two earflaps. The long joining band was probably the base of a diagonal crest placed similarly as the crest of the Mari helmet. In spite of the openwork triangles forming a zigzag motive and the punched decoration, there is an *omphalos* on the side plates of the helmet.

However, the helmet of the Mari standard is the only known archetype of a later well-known helmet type: the helmet decorated with lengthwise, diagonally fastened crest. This type of crested helmet was widespread from the 14th century B.C. on the northern peripheries of Mesopotamia: Eastern Anatolia (*Chart 1, no. 38*), Transcaucasia (Ltshashen: *Chart 1, nos. 46-49*; Lori-Berd: *Chart 1, nos. 50-51*; Asthi-Blur: *Chart 1, no. 54*) and Northern Iran (Khurvin: *Chart 1, no. 53*). This type of helmet was worn by Urartian soldiers (*Chart 3, nos. 107-110*) on the Balawat Gates of Shalmaneser III (858—824 B.C.) and on a late 9th century B.C. ivory plaque (*Chart 3, no. 106*). This type of helmet is shown on a series of bronze statuettes found in Urartian sites (Toprak Kale: *Chart 3, no. 113*; Maraş: *Chart 3, no. 114*) and on a horse's bronze pectoral reported as from "Northern Iran" (*Chart 3, nos. 111-112*). The latest and only known examples of this type of helmet are the Hasanlu hemispherical crested helmets (**Cat. nos. 52-56**) from about 800 B.C. (Hasanlu IV, Iron II Period)[53].

The crucial change in the warfare of the ancient Near East was the appearance of the two wheeled, light, horse drawn chariot in the 18th–17th centuries B.C. These chariots deployed in large numbers[54] with their speed and manoeuvrebility easily put the "centre of gravity" of a battle from one flank to the other. The infantry of the enemy was usually unable to follow this quick move or to react to it. The new type of chariot warrior (the best known name is the Hurrian *maryannu*, the Mitannian chariot warrior[55]) was equipped with the weapons of close range fighting, but his most important weapon was the bow. The primary role of the chariot in the battle was probably to serve as a moving firing platform[56]. Consequently the chariot warrior became the most important target of enemy fire. The chariot, which lost its warrior, became ineffective. The chariot crew in the 2nd millennium B.C. consisted of two, later three men (warrior and driver, later the shield bearer). Not only the driver but also the warrior (bowman) needed both hands in the battle, so they could not hold a shield to defend themselves. Consequently the armour became essential. The best type of armour, scale armour ensured the highest level of personal defense. Its main advantage was its flexibility, which enabled the chariot warrior to shoot with his bow, to move his arms and elbows freely. The Akkadian word for the scale of the scale armour (*kurṣimtu*) expressed this flexibility and high level of defense: it derives probably

---

48 Voth 1981, *passim*.

49 Parrot 1958, pl. XX:2; Borchhardt 1972, 96, pl. 21:2, Kat. 21 I 3; Parrot 1983, pl. 254.

50 Aleppo, Archaeological Museum, Inv. no. 1648; Parrot 1959, pls. VII-VIII; Borchhardt 1972, 96, pl. 21:3, Kat. 21 I 2; Parrot 1983, pls. 250-251.

51 Damascus, Archaeological Museum, Inv. no. 1915; Borchhardt 1972, 96, fig. 8, Kat. 21 I 4; Calmeyer 1975, 314 questions Borchhardt's dating.

52 Tiryns, Tomb XXVIII. H.: 17 cm, Diam.: 22.5 cm, L. of the central bronze band: c. 75 cm, H. of the earflaps: 16 cm, W. of the earflaps: 9 cm. Daux 1958, 707, fig. 26; Verdelis 1963, 10-24, figs. 2, 4-7, 9, pls. 6-7; Snodgrass 1964, 4; Hencken 1971, 23, 26, fig. 8; Borchhardt 1972, 43-44, pl. 8:4-5; Guida 1973, pl. 30:3; Borchhardt 1977, E68, fig. 9E, pl. EIVa.

53 There is only a single later representation of this type from a Zinçirli relief showing a chariot warrior: Maxwell-Hyslop 1959, pl. III:A. However the 8th century B.C. dating of the relief is questionnable.

54 For example in the battle of Qadesh the Hittite army put 3500 chariots in action.

55 Albright 1930—1931, 217-221.

56 In chariot battles and against the enemy infantry fighting in close order.

from the Akkadian word *kurṣindu* ("snake", "scale of a snake"). In the second half of the 2nd millennium B.C. this type of armour became widespread from Anatolia to Egypt, from Phoenicia to Iran together with the light chariot and the new chariot tactic. The original Hurrian word for scale armour (*sariam*) can be found in all the ancient Near Eastern languages as a loanword[57]. Actual finds of armour scales are known from Mesopotamia[58], Iran[59], Urartu[60], Anatolia[61], Syria[62], Phoenicia[63], Palestine[64], Egypt[65] and Cyprus[66]. Unfortunately there is not a single complete scale armour known. But the 184 scales and 157 fragments found during the excavations of Kāmid el-Lōz[67] made reconstruction of the structure of a suit of scale armour possible. The second group of evidence, the representational evidence is known from the wall paintings of Egyptian tombs of the New Kingdom. The most important representation showing two, long, multicoloured[68] suits of scale armour with high neckguard made of bronze plate[69] is on the walls of the Theban tomb of Ken-Amun[70]. These suits of scale armour were probably imports or gifts from Syria. They show the most developed type of the period: the armour covered almost the whole body of the warrior down to his ankles. This armour was made of about 450 scales. A later and more detailed representation, on a wall-painting of the tomb of Ramesses III (1194—1163 B.C.), shows a long suit of scale armour made of probably 470-500 scales (on the painting there are 237 scales on the front of the armour). This scale armour was made of two different metals represented in alternating bands with two colours. This armour covered the upper arm, the neck and the body down to the ankles. The third group of evidence consists of written sources. The most important written sources are the cuneiform tablets of the famous Nuzi library dated to the late 15th – early 14th century B.C. Altogether 16 types of armour can be reconstructed from the Nuzi tablets:

1. *sariam maški* (leather armour: Lacheman 1950, 236:9)
2. *sariam ša maški* (armour made of leather scales: Lachemam 1955, 7:6)
3. x *kurṣimētu ša siparri* x *kurṣimētu ša maški* (armour made of bronze scales and big leather scales: Lacheman 1955, 11:1-7)
4. *sariam maški ša ramān awēli ša aḫišunu ša siparri* (leather body armour with bronze sleeves: Chiera 1934, 527:7)
5. *sariam ša ramāni ša maški ša aḫišunu ša siparri* (body armour made of leather scales with bronze scales on the sleeves: Pfeiffer – Lacheman 1942, 195:17-18)

---

[57] Akkadian "*sariam*", "*siriam*" (Reiner, *et al.* 1984, 313-315; Oppenheim 1950, 191-192.); Egyptian "*tryn*" (*ti-ir(ra)-ya-na*: Albright 1934, 36, no. 15; Kendall 1979, 263); Hittite "*saryanni*", "*siryanni*" (Oppenheim 1950, 192); Ugaritic "*tryn*" (Gordon 1965, 506, no. 1123:5,6; Kendall 1979, 264); Hebrew "*širyōn*", "*širyān*", (שִׁרְיוֹנָא) "*siryōn*" (סִרְיוֹן) (Speiser 1950, 47). The Akkadian word for helmet: "*ḫuliam*" is also a Hurrian loanword. For further details on the words for parts of the armour *see* Salonen 1966, 100-108.

[58] Nuzi (Yorgan Tepe, 15th–14th century B.C., Starr 1939, 475-477, pls. 126: A_1-A_3, B, D, E, F, G, H, J, K, L, O); Tell al-Fakhar (14th century B.C., Mahmoud 1970, 122, pl. 24); Kar Tukulti-Ninurta (13th century B.C., Bittel 1958, 35); Kalhu (Nimrud, 9th century B.C., Mallowan 1966, 404, 409-411, figs. 336a-e; Muscarella 1988, 317-318, nos. 451a-c, 452, 453, 454a-b, 455, 456, 457a-f, 458a-b, 459, figs. 451-459; Stronach 1958, 172-173, pls. XXXIV:1-10).

[59] Dur-Untash (Choga Zanbil, 13th century B.C., Ghirshman 1966, 71, 101, pls. XLVIII:3, LV:4a-c, LXXVI, XCIII); Hasanlu (Muscarella 1988, 54, nos. 62-63).

[60] Karmir Blur (Argishti I, *ca.* 789—766 B.C., Piotrovsky 1955, 31f., fig. 21, pl. 14; Barnett 1959, 15, fig. 14).

[61] Hattushash (Boğazköy, 18th (?) century B.C., Bittel *et al.* 1958, 35; Boehmer 1972, 102-104, pl. XXV, nos. 803-808.A); Troy (15th–14th century B.C., Blegen 1953, 297-298, pl. 297).

[62] Ugarit (Rās Shamra, 15th–14th century B.C., Margueron 1977, 172, fig. 11:2-8; Schaeffer 1937, 144, fig. 9; Schaeffer 1951, 13, fig. 6; Schaeffer 1962, 74, fig. 61:C, I); Alalakh (Tell Atchana, 14th–13th century B.C., Woolley 1955, 278, pl. LXXI); Hama (11th–9th century B.C., Riis 1948, 124, fig. 143).

[63] Kāmid el-Lōz (15th century B.C., Ventzke 1983, 94-100, Cat. no. 77, figs. 45-50, 54, colour pl. on p. 117; Miron 1990, 65-86, Cat. nos. 144-345, figs. 12-15, pls. 15-18, colour pl. = fig. 46).

[64] Megiddo (Tell el-Mutesellim, 15th–7th century B.C., Lamon – Shipton 1939, pls. 85:1-10; Loud 1948, pls. 177:6-8); Beth-Shean (Amenhotep III, 1391—1353 B.C., Rowe 1940, 74, pls. XXXI:18, 20); Tell el-Far'ah (North, Tirzah?, 11th–10th century B.C., de Vaux 1951, pls. XIX.3:1-3; Chambon 1984, 89, pls. 69:2-5, 83); Deir Alla (13th–12th century B.C., Franken 1961, 365ff., pl. 12); Lachish (10th–8th century B.C., Tufnell 1953, 215, 253, 386-387, pls. 39:7-9, 56:11-12, 57:16-17, 58:4-14; Aharoni 1975, 76, pls. 37:3-5, Ussishkin 1982, 55, fig. 48; Shanks 1984, fig. on p. 53); Gaza (Tell el-Ajjul, 16th–13th century B.C., Petrie 1933, 9, pls. XXII:81-83; Petrie 1934, 11, pl. XXXV:551-553).

[65] Beth-Pelet (Tell Fāra, 12th–11th century B.C., Petrie 1930, pl. 32, no. 106); Thebes (15th–14th century B.C., Petrie 1917, 38, pls. XLII:105-108; Winlock 1947, 162-163, pl. 29); without provenance (New York, MMA 09.183.2, MMA 11.151.183-5, 192; MMA 22.3.46; MMA 34.1.72-73; Winlock 1947, 163, pl. 29; New York, Brooklyn Museum, Sheshonk I, 945—924 B.C., Petrie 1917, 38, pls. XLIII:129-130; Yadin 1963, fig. on p. 354); Petrie 1917, 38, pls. XLII:103-104; British Museum, BM 67522-67528 (Amarna?, *ca.* 750 B.C.); Oxford, Ashmolean Museum, Inv. no. 1969.753; London, British Museum, WA 135236 (82 pieces).

[66] Catling 1970, 441-449; Karageorghis – Masson 1975, 209-222; Catling 1977, 74-118.

[67] Kāmid el-Lōz, "Schatzhaus", Raum T-O, Ventzke 1983, 94-100, figs. 45-50, 54, Cat. no. 77, colour pl. on p. 117; Ventzke 1986, 162-166, pls. 21-22; Miron 1990, 65-86, Cat. nos. 144-345, figs. 12-15, pls. 15-18, colour pl. = fig. 46.

[68] Probably made of different metals.

[69] Similar to the neck-guard of the famous Dendra corselet. Snodgrass 1971, 33-50; Catling 1977, 74-118; Åström 1977, *passim*; Åström 1983, *passim*.

[70] Thebes, TT 93, Amenhotep II (1427—1401 B.C.), de G. Davies 1930, 27, pls. XVI, XXIV; Yadin 1963, fig. on p. 197.

6. *sariam maški ša ramān awēli 4 kalku u ša aḫišunu ša siparri* (leather body armour with bronze sleeves and 4 *kalkus* made of bronze (*kalku* means probably armour plates hanging on the waist)[71]
7. *sariam maški ša ramān awēli* [x] *kalku kī liwissunuma u ša aḫišunu ša siparri* (leather body armour with [x] *kalkus* like belts and with bronze sleeves: Chiera 1934, 527:1-2, 3-4)
8. *sariam ša ramāni ša irti 1 mana ša siparri* (body armour with a bronze plate on the front weighting 1 *mana*: Pfeiffer – Lacheman 1942, 195:15-16)
9. [*sariam kurṣi*]*mātu ša ramān awēli ašar irtišu ša siparri* (body armour made of scales with bronze on the front: Lacheman 1955, 8:2-3, 13-14)
10. *sariam ša ramānišu ša maški ašar ṣērišu siparri* (leather body armour with bronze on its back: Lacheman 1955, 15 4:15)
11. *sariam ša ramānišu ša tutiwašunu u gurpisišunu siparri* (body armour with its *kardio phylakes* and its helmet made of bronze: Lacheman 1950, 616:12-14)[72]
12. *tutiwa ša awēl rākib narkabti ina warkišu nadî* ([armour with] *tutiwa* hanging on the back of the chariot driver: Lacheman 1955, 15 12:6-7)
13. *tikkišu ana warkišu nadî* ([armour with] collar hanging on the back: Lacheman 1955, 12:16, 19, 27-28, 29, 30)
14. *sariam siparri* (bronze armour: Lacheman 1955, 15 4:44, 50)
15. *sariam ša siparri qadu gurpisišunuma ša ramānišu ša awēli* (bronze armour with helmets for the body of the man: Lacheman 1950, 616:9-10)
16. *sariam siparri ša awēli 7 kalku ša liwissunu* (bronze armour with 7 *kalku* like a belt: Chiera 1934, 533:6)

Several types of armour for horses and chariots can be reconstructed from the same tablets:

1. *2 tapālu sariam ša sīsî ištēnnūtu paraššannu ša aḫišu* (2 sets of horse armour and 1 set of *paraššannu* armour for the side: Lacheman 1955, 15 16:20)
2. *sariam ša awēli u narkabti* (armour for the man and chariot: Lacheman 1955, 15:19-20)
3. *paraššannu ša narkabtišu* (*paraššannu* armour for the chariot: Lacheman 1955, 15:45)[73]

Furthermore several types of bronze scale armour can be reconstructed from the numbers of bronze scales allocated for the body, the sleeves and for the helmet (*Diagram 1*). The weight of the different bronze scales differs considerably. Consequently it is hard to determine the weight of a complete body armour. T. Kendall used the weights of the armour scales found at Nuzi (for example: HSM 1012: 9.4g, HSM 1282: 12 g, HSM 1006: 14 g) to calculate the average weight of the different types of scale armours[74]. His results ranged from 37.3 lb (16.92 kg) to 54.8 lb (24.86 kg). C. Zaccagnini using the same values got similar results[75]. It has to be mentioned however that there were also much heavier scales found in Nuzi[76]. It means that the scale armours mentioned in the Nuzi tablets might have been much heavier as well[77].

Calculation of the weight of helmets made of armour scales (*Diagram 2*) will probably be more reliable, than the weights of other parts of the armour, since the helmets were only made of "small" armour scales.

TEXT	NUMBER OF ARMOUR SCALES				WEIGHT	TYPE
	Body	Sleeves	Helmet	Total		
HSS 15 5:1-3	500 (big)	500 (small)	200 (small)	1200	22.5 kg	Hanigalbat
HSS 15 5:5-8	500 (big)	500 (small)	200 (small)	1200	22.5 kg	Hanigalbat
HSS 15 5:10-13	560 (big)	160 (small)	140 (small)	860	18.88 kg	Arrapha
HSS 15 5:15-18	[420] (big)	[1]30 (small)	[1]70 (small)	720	15.06 kg	Arrapha
HSS 15 3:2-6	400 (big)	280 (small)	190	(870)	16.56 kg	Nuzi militia
HSS 15 3:10-13	598 (big)	440 (small)	–	(1038)	21.6 kg	Nuzi militia
HSS 15 3:17-21	500 (big)	360 (small)	190	(1050)	20.25 kg	Nuzi militia
HSS 15 3:27-31	435 (big)	312 (small)	190	(937)	17.89 kg	Nuzi militia

[71] Kendall 1981, 209, note 33.
[72] For *tutiwa* = *kardio phylax see* Kendall 1979, 183; Kendall 1981, 202, note 9.
[73] For further types *see*: Lacheman 1955, 15:12, 26, 43, 48, 52, 54.
[74] Kendall 1979, 277; Kendall 1981, 212-213.
[75] Zaccagnini 1979, 26-27.
[76] HSM 1201, 1213, 1229: 25 g, HSM 1202: 27.3 g, HSM 1021: 31 g, HSM 1081: 52 g, Kendall 1981, 213, note 44.
[77] For the following calculation we used mean values as follows: "big" scale: HSM 1202 = 27.3 g; "small" scale: HSM 1282 = 12 g. The weight calculations of *Diagram 1* refer only to the weights of the body armours without the helmets.

HSS 15 9	400 (big)	279 (small)	–	(679)	14,26 kg	Nuzi militia
HSS 15 9b	–	–	3 x 193 or 2 x 190, 1 x 200	–	–	Nuzi militia
HSS 15 11	[x]79	[x]00	246 (leather)	–	–	Nuzi militia

Diagram 1: Reconstruction of the different types of scale armour used by the different Nuzi chariot units.

One of the earliest known representations of a suit of scale armour and its helmet is a graffito on a potsherd found at Boğazköy (*Chart 1, no. 16*). This 15th century B.C. graffito depicts a warrior with a sword. He wears scale armour reaching to his elbow on his arm. On his body (probably over the top of his suit of scale armour) there is a waistcoat strengthened with metal disks. His distinctive helmet is a type of helmet so far known only in the Aegean[78]: the skull-piece, the earflaps and the neck-guard of his helmet were made of small scales fixed between horizontal metal bands (hoops). There are 4 such bands on the skull-piece, 4 on the earflaps and 3 on the neck-guard. There are two horns on the helmet: a small horn projects from the front, and a big horn projects from the top of the helmet. There is a long plume made of horsehair hanging on the back of the helmet. In the Aegean this type of helmet was made from scales made of boar's tusks[79]. However there are no such boar's tusk plates known from the Near East. Since the plates fixed between bands on the helmet are depicted on the graffito in the same way (zigzag) as the scales of the scale armour, it can be assumed that the helmet was made not of boar's tusk plates, but of bronze plates or scales. This assumption might be correct as there are scale armour plates earlier then the 15th century B.C. known from Boğazköy[80], and there are several late 15th century B.C. representations of scale armour helmets known from Egypt, from a bronze relief of the chariot of Thutmose IV (1401—1391 B.C.) (*Chart 1, nos. 27-32*). Probably the same type of helmet is shown on a horseman depicted on a 18th–17th century B.C. Anatolian cylinder seal. On this seal there are two warriors one of which wears a hemispherical helmet (*Chart 1, no. 12*) while the other wears a helmet with a high horn projecting from the top (*Chart 1, no. 13*) similar to the helmet of the Boğazköy potsherd (*Chart 1, no. 16*). This cylinder seal is contemporary with the earliest known representations of the Aegean boar's tusk helmet, so the question, whether this type of helmet originated in Western Anatolia or in the Aegean (for example in Crete), remains unsettled. It seems likely that this type of helmet spread quickly. Throughout the Middle and Late Helladic and Minoan periods the boar's tusk helmet (*Eberzahnhelm*) remained the most important helmet type of the Aegean[81] – there are actual pieces of this type known from Dendra (the helmet of the Dendra panoply)[82], and from Zapher Papoura[83] and even Homer mentions it[84] – in Anatolia it was replaced by bronze scale armour helmets (*Chart 1, nos. 27-32*) and bronze helmets (*Chart 1, nos. 37-39*). In other parts of the Near East this type was unknown. In North Syria the development of bronze scale armour helmets (*Chart 1, nos. 27-32*) and hammered bronze helmets (*Chart 1, nos. 18-26, 40, 41*) was continuous.

From the middle of the 2nd millennium B.C. there were three types of helmets used in the Near East:

1. Helmets made of bronze scale armour plates (*Chart 1, nos. 27-32*)
2. Helmets made of leather and strengthened with metal plates and disks (*Chart 1, nos. 33-36*)
3. Helmets hammered out from bronze plate
    - I. Conical helmets (with rounded point)
        - a. Helmets without crest (*Chart 1, nos. 18, 23-26*)
        - b. Crested helmets (*Chart 1, no. 22*)
        - c. Plumed helmets (*Chart 1, no. 19*)
        - d. Crested and plumed helmets (*Chart 1, nos. 20-21, 37*)

[78] Borchhardt 1972, 27 and map; Borchhardt 1977, E64, fig. 12.
[79] Xenaki-Sakellariou 1953, *passim*; Deonna 1954, 253-257; Hencken 1971, 18-20, fig. 4; Borchhardt 1972, Greek Mainland: 16-37, Kat. 2 I 1, pl. 4:1; Kat. 5 I 1, pl. 4:3; Kat. 5 I 3, pl. 4:8; Kat. 5 IV, pl. 4:2, (MH III-) LH II-III (ca. 1500-1400 B.C.); Crete: 47-52, Kat. 9 I, pl. 4:9; Kat.9 II, pl. 4:10; Kat. 9 III 1, pl. 4:5-6; Kat. 9 III 2, pl. 4:7; Kat. 9 IV, pl. 4:4, (MM III-) LM I-III, LH I-III (16th–15th century B.C.); Borchhardt 1977, E62-E65, figs. 11-12.
[80] Boehmer 1972, 102-104, pl. XXV, nos. 803-808.A. 18th century B.C.(?).
[81] There is only a single bronze helmet known from the middle of the 2nd millennium B.C. Aegean (LM II, ca. 1450—1400 B.C.): Knossos, Hospital Site, Tomb V (Archaeological Museum, Iraklion). H.: 17 cm; diam.: 21 cm; h. of earflaps: 16.5 cm; w.: 9 cm; h. of the knob: 3.9 cm; w.: 5 cm; weight: 695 g. Hood – de Jong 1952, 256-260, pls. 50-52; Snodgrass 1964, 3-4; Hencken 1971, 20, fig. 3e-g (Hencken assumed that this helmet is a bronze imitation of the boar's tusk helmet, because the boar was a relatively rare animal on Crete, and consequently a bronze helmet was cheaper than a boar's tusk helmet); Borchhardt 1972, 55-60, pl. 37:1, 4, Kat. 11 I; Guida 1973, pl. 30; Borchhardt 1977, E65-E66, pl. EIIId.
[82] Dendra, Tomb XII: boar's tusk plates and a pair of bronze earflaps. Verdelis 1967, 42-44, pls. 20-21; Daux 1961, 671-675, fig. 1; Verdelis 1977, 28-65, pl. XV:2; Borchhardt 1972, 30-32, pl. 6, Kat. 4 IV 3, LH IIB-IIIA1 (1450—1400 B.C.).
[83] Archaeological Museum, Iraklion. Daux 1968, 994, 996, fig. 5.
[84] *Iliad*, X:260. For the map of the "journey" of this helmet *see* Borchhardt 1972, 81, fig. 7.

II. Pointed helmets (forerunners of Assyrian pointed helmets), (*Chart 1, nos. 38, 40-41*)
III. Hemispherical crested helmets (*Chart 1, nos. 39, 46-54*)

1. The most important pictorial evidence for helmets made of bronze scales are the scenes on the sides of the chariot of Thutmose IV (1401—1391 B.C.)[85] made of bronze, where a battle is depicted between the pharaoh and bearded, close cropped North Syrian (Mitannian?) charioteers. Several of the types of armour mentioned above are shown: scale armour (*sariam kurṣimētu*), (leather) body armour with a bronze plate on the front weighting 1 *mana* (0.5 kg) (*sariam ša ramāni ša irti 1 mana ša siparri*)[86], leather body armour with bronze on its back (*sariam ša ramānišu ša maški ašar ṣērišu siparri*)[87], body armour with its *kardio phylakes* and its helmet made of bronze (*sariam ša ramānišu ša tutiwašunu u gurpisišunu siparri*)[88], leather body armour with bronze sleeves and 4 *kalkus* (*sariam maški ša ramān awēli 4 kalku u ša aḫišunu ša siparri*)[89], horse armour (*sariam ša sīsî, paraššannu ša sīsî*) and even chariot armour (*sariam ša narkabti, paraššannu ša narkabti*). The ten helmets depicted on the bronze parapet fall into two groups: scale armour helmets (*Chart 1, nos. 27-32*) and leather helmets strengthened with metal plates and disks (*Chart 1, nos. 33-36*). There are two types of scale armour helmets shown on the bronze relief: 1. Helmets made of one row of long bronze scale plates (*Chart 1, nos. 27-30*); 2. Helmets made of two rows of shorter bronze scale plates (*Chart 1, nos. 31-32*). All of these scale armour helmets are furnished with a long plume made of leather or horsehair. One of the helmets has a bronze or leather neck-guard covering the warrior's ears and throat (*Chart 1, no. 27*).

There are *ca.* 80 tablets (of the total four thousand) found in the Nuzi archive which deal with the question of issuing the members of the local militia with armour and weapons from the central arsenal of the town. These tablets describe all the weapons, armour (at least 16 types), horse-harness and chariot equipment, but do not mention a single known word – neither the Hurrian *ḫuliam* nor the Akkadian *elû*[90] – for "helmet". It would be surprising and incomprehensible if one of the most important pieces of the armour the "helmet" was not mentioned at all.

Scholars for long time have been interested in defining the exact meaning of a word, the *gurpisu*, which is part of the armour relating to both the head[91] and the body[92]. Lacheman, who published most of the Nuzi tablets, considered it as a piece of armour protecting the head (but not definitely a helmet)[93]. Von Brandenstein basing his theory on the Nuzi material assumed that it is a kind of breastplate (*Halsberge*)[94]. Speiser supposed that the *gurpisu* made of scale plates is a piece of armour attached to the body armour[95], but for the "*gurpisu* of the head" he thought possible that it might even be a helmet[96]. Oppenheim argued that it was a kind of neck-guard made of leather and covered with bronze scales, which was attached to the rim of the helmet and protected the neck, the ears, the chin and the throat, and left only the nose and eyes free[97]. A. Salonen referred to the Sanskrit and Iranian etymology of the word ("neck and body protection")[98], and accepted the helmet, collar, and pectoral interpretations as well[99]. E. Salonen examining the cuneiform sources considered it a kind of body armour or helmet and translated it as armour[100]. Volume "G" of *The Assyrian Dictionary of the University of Chicago* (CAD) translates it scale armour hauberk[101], but the later volumes translate it "helmet"[102]. Von Soden[103] regarded it as a kind of armour furnished with a protective apron for soldiers

[85] Carter – Newberry 1904, *passim*; Yadin 1963, 192-193; Kendall 1981, fig. 8.
[86] Pfeiffer – Lacheman 1942, 195:15-16.
[87] Lacheman 1955, 4:15.
[88] Lacheman 1950, 616:12-14.
[89] Chiera 1934, 533:6; Kendall 1981, 209, note 33.
[90] Thompson 1904, 19 (K.9983), Rs. 7: [*e*]-*lu-ú* = *ḫu-li-*[*am*]; von Soden 1965—1981, *s.v.* "*ḫuli(j)am*", 353-354; Oppenheim *et al.* 1956b, *s.v.* "*ḫuliam*", 228-229.
[91] Lacheman 1955, 9b:2: *gurpisu ša qaqqadi* ("*gurpisu* of the head").
[92] Chiera 1934, 527:15: *gurpisu siparri ša ramān awēli* ("bronze *gurpisu* for the bady of the man").
[93] Lacheman 1939, 541: "I am thus inclined to think that kurpisu was the head part of the protective armour".
[94] von Brandenstein 1940, 104-105.
[95] Speiser 1950, 47-49.
[96] Speiser 1950, 48: "a visor, a beaver, or perhaps even the helmet itself."
[97] Oppenheim 1950, 192-193, note 17.
[98] Salonen 1951, 130, note 2: "sanskrit *kūrpāsa* (später auch *kurpāsa*, prakrt. *kuppāsa*), "Jacke als Schutzwaffe"; pers. *gryvp'n*: *grivβan* "Halsschutz".
[99] Salonen 1955, 141-142: "eine Art 'Helm' für Pferde und Soldaten"; "einen mit Metallplättchen überzogenen ledernen Brustschutz".
[100] Salonen 1966, 101-102: "eine Art Panzer".
[101] Oppenheim *et al.* 1956a, 139, *s.v.* "*gurpisu*": "leather hauberk covered with metal scales (as part of the armour of soldiers and horses)".
[102] Reiner *et al.* 1984, 313-315, *s.v.* "*siriam*".
[103] von Soden 1965—1981, 929, *s.v.* "*qurpi(s)su(m)*": "eine Panzerüberwurf mit einem Nackenschutz für Menschen und Pferde".

and horses. Kendall has recently shown that *gurpisu* most probably means "helmet"[104]. This result may solve the problem of the almost total lack of words meaning "helmet" in Akkadian[105].

The earliest known occurrence of the word *gurpisu* is an Old Babylonian cuneiform text[106] found at Eshnunna:

1. 4 *šu-ši gur-pí-su ša* SAR-*ki-im*
2. 7 *šu-ši* 38 *gur-pí-su ša aš-li-im*

This locus has been a *hapax legomenon* for a long time[107]. Nobody could really explain it. In his article Kendall proposed the *ašlum* = "matgrass", "reed" meaning[108] and found a representation of this type of "corded helmet"[109], but he neglected the interpretation of word *sarkum*. The *gurpisu ša ašlim* = corded helmet interpretation might have been verified by the term sagšu ^{ú}ninni$_5$ (*ašlum*) = "reed matting headdress" mentioned above[110], and in line 2 of an Old Babylonian cuneiform text found at Mari[111]:

2. 2 ^{gi}*gur-si-pu* UŠ

But the lack of the interpretation of word *sarkum* makes further examination necessary. Another reading of *sar-ki-im* is *šar-ki-im*, which makes another interpretation possible. Namely that *šarkum* means "group of soldiers"[112]. And the word *ašlum* had a similar meaning instead of "reed matting" in the Old Babylonian period: "group of 8-12 men"[113]. It can be concluded that this locus mentions probably helmets for two (unidentifiable) groups of soldiers (*šarkum* and *ašlum* men).

There is a further cuneiform text of the Old Babylonian period from Mari[114] mentioning the *gurpisu*, but most of its occurrences are known from the middle of the 2nd millennium B.C. Cuneiform texts in the Nuzi archive of the same period have already been discussed, but there is another Mitannian document, the letter of Tushratta, king of Mitanni mentioning armour pieces from the same period. This letter[115], written by Tushratta to Amenhotep III, king of Egypt (1391—1353 B.C.), found at el-Amarna, is a list of dowry resembling the Nuzi inventory lists[116]:

vs. III:

37. 1-*šu*(*ištēnšu*) *sa-ri-am* UD.KA.BAR(*siparri*) 1 *gur-sí-*[*ip* UD.KA.BAR(*siparri*) *ša*] LÚ(*awēli*)
38. 1-*šu*(*ištēnšu*) *sa-ri-am ša* KUŠ(*maški*) 1 *gur-sí-*[*ip* UD.K]A.BAR(*siparri*)
39. *ša* ^{lú}*sà-ar*-GU-*ti* 1-*nu-tu*$_4$ (*ištēnnūtu*) *sa-ri-am ša* KUŠ(*maški*)
40. ANŠE.KUR.RA.MEŠ(*sīsî*) *ku-la-a-na ša* UD.KA.BAR(*siparri*) *mu-uḫ-ḫu-zu*
41. 2 *gur-si-ip ša* UD.KA.BAR(*siparri*) *ša* ANŠE.KUR.RA.MEŠ(*sīsî*)

The meaning of *gurpisu* mentioned in the text together with the word *sariam* (armour) is almost certainly "helmet". From the contemporary Egyptian representations[117] it is clear that there was no additional piece of armour, cuirass and collar – identified by so many scholars with the *gurpisu* – worn by the chariot warriors. The frequent reference to *gurpisu* would have been unnecessary if it was part of

---

104 Kendall 1981, *passim*.

105 The best known word for helmet "*ḫuliam*" occurs altogether three times in Akkadian (Assyrian and Babylonian dialects). Oppenheim 1956b, 228, *s.v.* "*ḫuliam*"; von Soden 1965–1981, 353-354, *s.v.* "*ḫuli(j)am*".

106 Baghdad, IM 31309, 1-2; Salonen 1966, 101; Oppenheim *et al.* 1956a, 139, *s.v.* "*gurpisu*"; Kendall 1981, 205, note 24.

107 Oppenheim *et al.* 1956a, 139, *s.v.* "*gurpisu*": "240 hauberks with ..., 458 hauberks with ...".

108 Kendall 1981, 205, note 24: "240 helmets of ..., 458 corded helmets"; *gurpisu ša ašlim* (lit. "helmet of rope").

109 On an Old Babylonian terracotta. Kendall 1981, 205, note 24.

110 *See* p. 5, note 33.

111 Bottéro 1956, 255:2; Oppenheim *et al.* 1956a, 139, *s.v.* "*gurpisu*": "two hauberks (reaching to) the genitals"; Kendall 1981, 205, note 24: "two helmets of (woven) reed, ordinary quality", or perhaps "two codpieces of woven reed".

112 von Soden 1965—1981, 1029, *s.v.* "*šarku*" (<heth. *šarikuwa-*) "eine Gattung Soldaten"; Boğazköy: Weidner 1922, 51 Rs. 6: 1-*en*(*iltēn*) ^{lú}*sà-ar*-GU; El-Amarna: Schroeder 1915, 199 (VAT 395), Rs. col. III, 38-39:
38. 1-*šu*(*ištēnšu*) *sa-ri-am ša* KUŠ(*maški*) 1 *gur-sí-*[*ip* UD.K]A.BAR(*siparri*)
39. *ša* ^{lú}*sà-ar*-GU-*ti* ...
Ugarit: Nougayrol 1968, 98:12, 16: ^{I}*A-mu-ta-ri* ^{lú}ZA-*ar*-GU/KU IGI-*ia*(*pānija*); CAD, "Z", p. 69, *s.v.* "*zarku*" (or *zargu*, *zarqu*, *sark/qu*).

113 von Soden 1965—1981, 81, *s.v.* "*ašlu(m)*" 2c., aB "Messtrupp (zu 8-12 Mann)".

114 Bottéro 1956, 240:3-4.

115 Winckler 1896, no. 296; Winckler – Abel 1889-1890, no. 26; Schroeder 1915, no. 199; Knudtzon 1915, no. 22.

116 "37) 1 bronze armour and 1 bronze helmet of man, 38) 1 leather armour and 1 bronze helmet 39) of a *sarku*-soldier, 1 leather armour 40) of horses, decorated with bronze applique, 41) 2 bronze helmets of horses".

117 The bronze sides of the chariot of Thutmose IV (1401—1391 B.C.) shows the complete armament of chariot warriors. Yadin 1963, 192-193.

the armour and not the helmet, and similarly the total lack of other words for helmet[118] makes it probable that the meaning of *gurpisu* is "helmet".

The only locus which makes the *gurpisu* = "helmet" identification questionable is a reference to the *gurpisu* of the body[119]:

14. 6 *ta-pa-la* [*sa*]-*ri-am* KUŠ(*maški*) *ša* NÍ(*ramān*) [LÚ(*awēli*)]
15. 1 [*gur*]-*pí-sú* UD.KA.BAR(*siparri*) [*ša*] NÍ(*ramān*) L[Ú(*awēli*)]
16. 1 [*gur*]-*pí-sú* UD.KA.BAR(*siparri*) [*ša*] LÚ(*awēli*)

But this locus is probably an error by the scribe. The long inventory list mentions several sets of "armour of the body of man" (*sariam ša ramān awēli*) and several "helmets of the man" (*gurpisu ša awēli*). This single line of the text mistakenly affixes the attributes of the *sariam* ("armour") to *gurpisu* ("helmet"). Consequently all the theories which regarded the *gurpisu* as an additional (but unknown) part of the body armour can be rejected.

There are altogether 7 types of helmet (*gurpisu*) and further 5 types of "horse-helmet" that can be reconstructed from the Nuzi inventory lists. In several cases the word helmet (*gurpisu*) occurs without attributes[120] or with a neutral attribute (*gurpisu ša qaqqadi*, "helmet of the head")[121]. The seven types of helmets are as follows:

1. **Leather helmet** (*gurpisu ša maški*)[122]. This type of helmet occurs only in a single case, in an inventory list, in which armour, helmets, horse-armour, horse-helmets, and weapons were issued from the central arsenal to chariot warriors, members of the local militia named in the text. Consequently it is clear that charioteers of the Nuzi militia wore mainly bronze helmets, while the infantry of the same militia, who may have been hindered by the heavy equipment on the battlefield, probably wore the lighter leather helmets.

There are however three tablets in the Nuzi archive mentioning that a *Ḫutip-tilla* leather-worker (*aškāpu*) had to make leather helmets and woollen linings as communal work (*iškaru*) for the palace or arsenal. The first of these texts[123] reports that a helmet weighting 3 *mana* 40 *šiqlu* (1.833 kg) was issued to him as a sample. He was expected to manufacture six similar helmets. However the weight of the sample helmet is so great, that it is certain that this helmet was not a simple leather helmet. Its weight is too great for bronze helmets hammered out from bronze sheet as well[124]. Accordingly it is probable that *Ḫutip-tilla* leather-worker had to manufacture the leather base for scale armour helmets and even had to sew the bronze armour scales onto this leather base. The second tablet reports that he got 7 pieces of goatskin to manufacture 3 leather helmets (*gurpisu*.MEŠ *maški*)[125]:

12. 7 KUŠ.MEŠ(*maški*) *ša en-zi a-na* 3 *gur-pí-is-sú*.MEŠ KUŠ(*maški*) *ša* DÙ(*epēši*)

Consequently each helmet was made from 2 $^1/_3$ goatskins. Unfortunately it is not known what the weight of these helmets was and how *Ḫutip-tilla* used the goatskins. The third tablet reports that *Ḫutip-tilla* received 3 *mana* of wool to manufacture 2 helmets[126]:

9. 3 *ma-na* SÍK.MEŠ(*šipāti*)
10. *a-na* 2 *gur-pí-su-ú*
11. *ša* SAG-*du*(*qaqqadu*) *iš-ka*$_4$-*ri*

It is unlikely that *Ḫutip-tilla* would have made helmets from wool. It is most probable that the wool was the raw material for the lining of different types (leather, bronze or scale armour) of helmets as has

[118] The only known Hurrian/Akkadian word for helmet, the *ḫuliam* is completly missing from the detailed armour descriptions of the Nuzi tablets! It occurs three times in the letter of Tushratta mentioned above, but in each context means "vessel": Schroeder 1915, no. 199, col. II:62: 1 *ḫu-li-am ša* UD.KA.BAR(*siparri*) *ša* IZI(*išāti*) ("1 bronze *ḫuliam* vessel of the fire [for cooking]"). The word *ḫuliam* meaning helmet only occurs twice, in the royal inscriptions of Sennacherib and Asarhaddon (Oppenheim *et al.* 1956b, 228-229, *s.v.* "*ḫuliam*"; von Soden 1965—1981, 353-354, *s.v.* "*ḫuli(j)am*").

[119] Chiera 1934, 527:14-16: "14) 6 sets of leather armour for the body of the man, 15) 1 bronze *gurpisu* for the {body} of man, 16) 1 bronze helmet for man".

[120] Pfeiffer – Lacheman 1942, 195:19; Lacheman 1950, 616:2, 6, 10; Lacheman 1955, 4:2, 22 (*gurpisu jānu* = "there is no helmet"), 28, 30, 50, 55, 60; 7:8, 12, 22, 27, 33; 12:9; 23:1, 4, Rs. 4, 5.

[121] Lacheman 1955, 9b:2, 3; 208:9-10.

[122] Lacheman 1955, 4:38.

[123] Lacheman 1955, 24.

[124] As has earlier been discussed (*see* pp. 3-4) their weight fluctuated between 700 g and 1300 g.

[125] Lacheman 1955, 196:12.

[126] Lacheman 1955, 208:9-11: "9) 3 mana wool 10) for two helmets 11) of the head as *iškaru* work".

been discussed in case of the 3rd millennium B.C. leather helmets[127]. While in the 3rd millennium B.C. texts the wool/helmet ratio was $^1/_3$ *mana*, in the Nuzi texts the same ratio was 1.5 *mana* (750 g). This weight is too heavy for a simple wool helmet, and it is possible that this type of "wool helmet" served as base onto which the bronze scale armour plates were fixed or it was simply the lining of such a scale armour helmet.

Unfortunately not a single leather helmet can be identified on contemporary representations. The only evidence is the four leather helmets reinforced with metal disks represented on the chariot sides of Thutmose IV (*Chart 1, nos. 33-36*). These helmets were worn by Mitannian or Syrian charioteers. It is possible that the Nuzi leather helmets were reinforced with metal disks as well.

2. **Bronze helmet** (*gurpisu siparri*[128], *gurpisu ša siparri*[129], *gurpisu siparri ša awēli*[130]). The "bronze helmet" is the most common type of helmet mentioned in the Nuzi tablets. It is likely that this is a general category to which several types of bronze helmets identified with different attributes (*see* later) could belong. From the Nuzi inventory lists it is clear that the bronze helmet was the most common type of helmet for charioteers. Since the Nuzi lists mentioning scale armour helmets always give the number of scale plates needed to manufacture the helmet, we are perhaps right if we identify this general category with bronze helmets hammered out from a single piece of bronze sheet. Furthermore, since the scale armour helmet was the most expensive type of helmet (*see* later), it is likely that the charioteers of the Nuzi militia wore the cheaper sheet bronze helmets. Several representations of this type of helmet are known from the wall paintings of Egyptian tombs (Mitannian or Canaanite gifts sent to the pharaoh, *Chart 1, nos. 18-26*) and from the sculpture of the King's Gate at Boğazköy (*Chart 1, no. 37*). But some of these representations can be connected to different types of Nuzi bronze helmets identified in the Nuzi tablets.

3. **Decorated bronze helmets** (*gurpisu siparri tegipu*[131], *gurpisu ša siparri tegipu*[132]). The identification of this type of helmet is uncertain, because the exact meaning of the word *tegipu* is unknown. Kendall collecting all the occurrences of this word in Nuzi tablets concluded that its meaning might be "finely elaborated", "polished" or "decorated"[133]. There is only a single locus, where it's meaning is uncertain[134]

> 22. *sa-ri-am* KUŠ.MEŠ(*maški*) *a-na* ANŠE.KUR.RA-*šu*(*sīsêšu*) TI(*ilqê*)
> 23. *gur-pí-is-sú-nu ša* UD.KA.BAR(*siparri*) 2 *ma-ti* 42 *kur-ṣí-me-tu₄-šu-nu*
> 24. *ka-qa-ni-aš-wa-na ša* UD.KA.BAR(*siparri*) *te-gi-pu*

Here the word *tegipu* is an attribute of a horse's bronze scale armour helmet. In this case it is hard to decide whether the bronze armour plates were "polished" or "decorated". Since the word *tegipu* is always an epithet of the bronze, it might mean a kind of alloy or might denote the quality of the bronze.

4. **"Crested" bronze helmets** (*gurpisu si[parri ṣu]ppuru*[135], *gurpisu siparri ša awēli ṣuppuru*[136]). If the identification is correct, this is the best-documented type of helmet. Its representations are known from wall paintings in Egyptian tombs depicting Syro-Mitannian gift helmets (*Chart 1, nos. 19-22*)[137]. Three variants of this type can be reconstructed from these wall paintings:

1. Plumed helmet (*Chart 1, no. 19*)
2. Crested and plumed helmet (*Chart 1, nos. 20-21*)
3. Crested helmet (*Chart 1, no. 22*)

Unfortunately it is impossible to separate such details in the Nuzi texts. Furthermore it has to be mentioned that – as shown on the chariot sides of Thutmose IV (*Chart 1, nos. 27-36*) – the scale armour helmets might have been plumed as well.

---

127 *See* pp. 3-4, Scheil 1913, 86, vs. col. II, 17-22; rs. col. I, 4-9; rs. col. II, 15-19.
128 Lacheman 1950, 236:10; 616:41; Lacheman 1955, 4:44; 8:7.
129 Lacheman 1950, 616:8; Lacheman 1955, 4:5-6, 34.
130 Chiera 1934, 527:16; Lacheman 1955, 106:6-7.
131 Lacheman 1955, 3, rs. 35.
132 Lacheman 1955, 3, vs. 16.
133 Kendall 1981, 208, note 31.
134 Lacheman 1955, 3:23-24: "22) leather armour for his horse received. 23) Their helmets (made) of 242 24) *kaqaniašwana*(-type?) polished? bronze armour scales."
135 Lacheman 1955, 8:15.
136 Chiera 1934, 527:9.
137 Thebes, Tomb of Menkheperre-Seneb (TT 86, Thutmose III (1479—1425 B.C.)), de G. Davies 1933, pls. 5, 8; Tomb of Amenmose (TT 112, Amenhotep II (1427—1401 B.C.)), de G. Davies 1933, pls. 34, 35: the helmets were painted yellow (shining bronze), the crest blue, the plumes red; Kendall 1981, 215-224.

5. **Bronze helmet with "earflaps, neckguard, and leather plume"** (*gurpisu siparri ša awēli mašku ṣuppuru 3 kalku ištu pānišu ša siparri*[138]. This type of helmet is a well-identified one. There are only a few representations depicting bronze helmets furnished with earflaps, neckguard, and plume. It has to be mentioned that all the helmets depicted on the wall paintings of Egyptian tombs are furnished with earflaps, but there is not a single helmet among them, which was furnished with a neckguard. However, on the relief sculpture of the King's Gate at Boğazköy there is a helmet with detachable(?) earflaps, neckguard, and a plumed crest (*Chart 1, no. 37*), which might be an almost exact depiction of our helmet with "earflaps, neckguard, and leather plume" reconstructed from a Nuzi text. Helmets furnished with earflaps, neckguard and crest or plume might be made not only from sheet metal but from scale armour plates as well. This type of scale armour helmet is shown on the well-known chariot sides of Thutmose IV (*Chart 1, no. 27*). But in this case the scale armour cap of the helmet, which covered the ears, was furnished with a leather neck- and throat-guard as well.

6. **Bronze scale armour helmet** (*gurpisu kurṣimētu*[139]). Certainly the scale armour helmets were the most developed and most expensive helmets of the period. The manufacture of this type of helmet might be easier than the manufacture (hammering out) of sheet bronze helmets, but the hammering or casting of bronze scales and sewing them onto the leather base was a process, which undoubtedly had to be done with meticulous care. As will be discussed the scale armour helmets were made from 140 to 200 scales. The flexibility of the scale armour helmets however provided better defense against the enemy blows and cuts than the sheet bronze helmets, which could easily be dented or broken.

It is clear that because of its expensiveness only the elite troops (chariot warriors, *maryannu*) were equipped with scale armour and scale armour helmets. In the battle of Megiddo for example Thutmose III (1479—1425 B.C.) seized one suit of scale armour from the reigning prince of Megiddo and 200 suits of leather armour from his defeated troops[140]. This locus is however questionable. Kendall suggested[141] that the 200 captured suits of armour were not made of leather but of bronze scales, because in this broken passage the word "garment" (*mss*) was preceded not by the word for "leather", but by the word for "bronze" (*ḥsmn*). Kendall concluded that from the nearly 1800 crewmembers of the captured 924 chariots only 10 % were equipped with bronze scale armour. However, from the battle scene depicted on the bronze chariot sides of Thutmose IV (1401—1391) it is clear that in a chariot only one man wore scale armour. The same tendency can be observed in the Nuzi texts as well: for one chariot only one suit of armour was issued from the palace or arsenal. Consequently, if Kendall is right, and the 200 suits of armour seized at Megiddo were bronze scale armour, nearly 20 % of the chariot warriors of the Megiddo army were equipped with bronze scale armour. So it can be assumed that on the whole only 20-25 % of the chariotry, the elite troops of the period, were equipped with the expensive scale armour. It looks a relatively small percentage, but if we take it into consideration that such a small, provincial town like Nuzi was able to issue chariots, scale armour, helmets, horses, horse armour, and weapons to the members of its local militia in large quantities, it can easily be appreciated that in the army of the whole Mitannian kingdom thousands of chariots and suits of scale armour were arrayed.

The different types of scale armour helmets can be identified with the help of the number of armour scales used. It can be assumed that in Nuzi only a limited number of scale types were used. On the chariot sides of Thutmose IV there are representations of scale armour helmets made of a single row of armour scales (*Chart 1, nos. 27-30*) and two rows of armours scales (*Chart 1, nos. 31-32*). However, the number of scales given in the Nuzi texts is much higher (*see* below) than the number that can be reconstructed from these representations. It can be assumed that the Nuzi helmets were made of much more (5-6) rows of smaller armour scales than the helmets depicted on the chariot sides of Thutmose IV. There is only a single long armour scale known from Nuzi[142], the size of which does not correspond with the number of scales (140-200) given for scale armour helmets. For the reconstruction of the weight[143] of different scale armour helmets *see Diagram 2*.

---

[138] Chiera 1934, 527:10-11: "10) bronze helmet for man, (with) leather plume, 11) three plates (are hanging) on it (on the front of it).")

[139] Lacheman 1955, 3:6, 20-21, 30-31; 5:3, 7, 12, 17; 9b:1-3.

[140] Sethe 1907, IV, 664.

[141] Kendall 1981, 222, note 63.

[142] Nuzi, House of Šilwatešup, Room 23; Baghdad, Iraq Museum; L.: 13.2 cm, W.: 3.4 cm. Kendall 1979, 275, Type VI; Kendall 1981, fig. 9.

[143] Weight of the two smallest armour scales found at Nuzi: HSM 1012 = 9.4 g; HSM 1282 = 12 g. Kendall 1981, 213. Total weight: with the weight of the lining (Lacheman 1955, 208:9-11: 3 manas of wool for two helmets. The lining of one helmet weights 1.5 mana = 750 g) added to the calculated weight of the scales.

TEXT	NUMBER OF ARMOUR SCALES	WEIGHT		TOTAL WEIGHT		TYPE
		min.	max.	min.	max.	
HSS 15 5:3	200	1880 g	2400 g	2630 g	3150 g	Hanigalbat
HSS 15 5:7	200	1880 g	2400 g	2630 g	3150 g	Hanigalbat
HSS 15 5:12	140	1316 g	1680 g	2066 g	2430 g	Arrapha
HSS 15 5:17	170	1598 g	2040 g	2348 g	2790 g	Arrapha
HSS 15 3:6	190	1786 g	2280 g	2536 g	3030 g	Nuzi militia
HSS 15 3:21	190	1786 g	2280 g	2536 g	3030 g	Nuzi militia
HSS 15 3:30-31	190	1786 g	2280 g	2536 g	3030 g	Nuzi militia
HSS 15 9b:1-3	190	1786 g	2280 g	2536 g	3030 g	Nuzi militia
HSS 15 9b:1-3	190	1786 g	2280 g	2536 g	3030 g	Nuzi militia
HSS 15 9b:1-3	200	1880 g	2400 g	2630 g	3150 g	Hanigalbat

Diagram 2: Reconstruction of the different types of scale armour helmet used by the Nuzi chariot units.

As has been cited earlier *Ḫutip-tilla* "leather-worker" (*aškāpu*) got a helmet from the palace as a proof sample, the weight of which was 3 *mana* 40 *šiqlu* (1.833 kg). He had to manufacture six similar helmets. Although the text does not mention the type of the helmet, its weight suggests that this might be a scale armour helmet[144]. It is obvious that the scale armour helmets – like bronze scale armour – were extremely heavy[145]. Their weights were often double or triple the weight of contemporary sheet bronze helmets (*gurpisu siparri*). Consequently it is obvious that the scale armour and scale armour helmets were worn only by charioteers (mainly chariot warriors[146]) and not infantrymen.

The standard number of armour scales could refer to the identical structure and shape of the helmets. *Diagram 2* shows that – if the number of armour scales of the separate types were standard – at least four types of scale armour helmets can be reconstructed from the Nuzi texts:

140 armour scales	Arrapha type of scale armour helmets
170 armour scales	Arrapha type of scale armour helmets
190 armour scales	Nuzi type of scale armour helmets[147]
200 armour scales	Hanigalbat type of scale armour helmets

7. **"Big" helmet** (*gurpisu arābu*[148]). This idiom most probably referred to the biggest and heaviest type of helmets, the scale armour helmet.

Horse-"helmets":
1. Crested leather helmet for horse (*gurpisu maški ṣuppuru ša sīsî*[149])
2. (Leather) helmet for horse with bronze on the sides ([*gurpi*]*su ašar* [*ṣērišu si*]*parri*[150])
3. Bronze helmet for horse (*gurpisu ša sīsî siparri*[151], *gurpisu ša siparri*[152], *gurpisu siparri*[153])
4. Crested bronze helmet for horse (*gurpisu siparri ṣuppuru ša sīsî*[154])
5. Bronze scale armour helmet for horse (*gurpisu ša siparri 242 kurṣimētušunu kaqaniašwana ša siparri tegipu*[155])

Unfortunately there is not a single horse-helmet representation known from the middle of the 2nd millennium B.C. Depictions of contemporary horse armour are shown on the battle scene of the bronze chariot sides of Thutmose IV[156] and on the Enkomi ivory game-board[157]. Later representations of horse

[144] As has been mentioned, the weight of ancient sheet bronze helmets was between 700 g and 1300 g.
[145] For comparison *see* the weight calculations of C. Zaccagnini (Zaccagnini 1979, 27) and T. Kendall (Kendall 1981, 213).
[146] *See* the battle scene on the chariot parapet of Thotmes IV; Yadin 1963, 192-193.
[147] Lacheman 1955, 9b: "1) 580 armour plates 2) for three helmets 3) of the head, 4) to (the hand of) Ḫutip-tilla 5) were given."
[148] Lacheman 1955, 167:21: *gur-pí-sú a-ra-bu.*
[149] Chiera 1934, 527:25.
[150] Lacheman 1955, 4:46, 52.
[151] Lacheman 1955, 4:56.
[152] Lacheman 1955, 4:8, 18, 36, 40.
[153] Lacheman 1955, 4:36.
[154] Chiera 1934, 527:24.
[155] Lacheman 1955, 3:23-24: "22) leather armours for his horse (he) received. 23) Their helmets (made) of 242 24) *kaqaniašwana*(-type?) polished? bronze armour scales."
[156] Yadin 1963, 192-193.
[157] de Mertzenfeld 1954, pl. LXIX, no. 788b.

armour are known from a Sakçegözü basalt-relief[158], from a Nimrud bronze bowl[159], and from Neo-Assyrian ivories found at Nimrud[160]. The horse-helmet might have covered the forehead or even the whole head and neck of the horse. From the 1st millennium B.C. however there are several representations of horse headdresses with crests (*gurpisu siparri ṣuppuru ša sīsî*) similar to the crests of the soldiers' helmets[161]. Unfortunately the further reconstruction of the Nuzi horse-helmets is impossible.

In the middle of the 2nd millennium B.C. simultaneously with helmet types discussed above another type of helmet, the pointed bronze helmet was developed (Anatolia: *Chart 1, no. 38*, Syria: *Chart 1, no. 40* and Assyria: *Chart 1, no. 41*). In the first half of the 1st millennium B.C. the significance of this type became greater than all other helmet types. The 18th century B.C. Mari crested helmet (*Chart 1, no. 11*) might have been the archetype of the hemispherical crested helmets of the 14th–13th centuries (Anatolia: *Chart 1, no. 39*, Transcaucasia: *Chart 1, nos. 46-54*), from which the 9th–8th centuries B.C. hemispherical crested helmets[162] developed. The mid 2nd millennium B.C. North Syrian and Anatolian conical crested helmets (*Chart 1, nos. 20-22, 37*) became the archetype[163] of the 1st millennium B.C. conical crested helmets. This type of helmet became the predecessor of the famous "Assyrian" pointed helmets of the first half of 1st millennium B.C. Nearly 90 % of the surviving Near Eastern helmet corpus consists of this type of helmet.

[158] Madhloom 1970, pl. XIII:2.
[159] Madhloom 1970, pl. V:4.
[160] Madhloom 1970, pl. XIII:1; 2. ND 10316, BM 132939: Herrmann 1986, vol. I, 156, no. 657, vol. II, pl. 161; 3. de Mertzenfeld 1954, pl. CXVII, no. 1115.
[161] *See* for example Assyrian palace reliefs, *passim*.
[162] For detailed discussion *see* Group H.2: Hasanlu hemispherical crested helmets.
[163] The intermediary might be the conical crested helmet tradition of the North Syrian Neo-Hittite states (*Chart 3, nos. 101-105*).

ASSYRIAN HELMETS

Group A.1: Assyrian conical bronze helmets (Cat. nos. 1-8)

At present the earliest representation[164] of the Assyrian conical helmet is on a cylinder seal impression (*Chart 1, no. 41*)[165] on a cuneiform tablet from Ashur[166]. The tablet dates to the end of the 15th century B.C., to the reign of Ashur-nirari II (1424—1418 B.C.). There are heroes wearing pointed helmets and fighting with wild animals on the impression. The caps of the sphinxes around them are decorated with pairs of horns (horned crown) and lotus buds.

This type of helmet appears on the bronze statuettes of the second half of the 2nd millennium B.C. However in consequence of the frequent superficial shaping of the statuettes of warriors and gods, we can hardly establish whether they wear this („Assyrian") type of conical helmet or variants of caps made of either metal or perishable material[167]. One of the rare exceptions might be the conical helmet of a bronze statuette which may be Syrian in origin (*Chart 1, no. 40*)[168] dated to the second half of the 2nd millennium B.C.[169], and the single combat scene on an Anatolian cylinder seal[170], on which a warrior wearing a conical helmet (*Chart 1, no. 38*) defeats a warrior wearing a crested helmet (*Chart 1, no. 39*). Alexander suggested that the Tyskiewicz stamp cylinder "stems from the period immediately following the Colony Age"[171]. This early date and the probably Old-Assyrian (Anatolian) origin of this type of helmet can be reinforced by two metal skull caps found in cyst graves in the Karum of Old-Assyrian Kanish (Kültepe). Their shape is conical. One of them was made of gold[172], the other of silver[173]. A third gold object in the shape of a pointed cap with a gold rivet remaining in the tip was found in the palace of Warshama (contemporary with Karum level Ib)[174]. Furthermore, two Anatolian sealings show the same type of conical headdresses – possibly helmets from the same period (*Chart 1, nos. 14-15*)[175].

Owing to the bad condition of the stone there are no signs of decoration on the helmets of Assyrian warriors on the White Obelisk (*Chart 2, no. 75*)[176]. The first appearance of the typical decoration of Assyrian conical helmets which characterized the 9th–7th century B.C. Near Eastern helmet tradition is on the palace reliefs of Ashur-nasir-pal II (883—859 B.C.): 1-3 pairs of *repoussé* curves bending towards the middle ("*Krummstab*") with a vertical rib running from the point of the helmet between them (*Fig. 1*). As it is widely used on the conical helmets of the Iron Age in the Near East a number of explanations have been proposed for the origin of this device. Calmeyer[177] derived the motif from the Zagros region, whence he argued the Urartians adopted it in the early 8th century B.C.[178]. But the earliest known examples in Western Iran are on conical helmets[179] from Hasanlu IVB (Iron II Period), *ca.* 800 B.C.[180], where there is considerable evidence for Assyrian influence. As this motif appears in Assyria by at least the reign of Ashur-nasir-pal II, it is perhaps more likely to be of Assyrian origin, although it is already shown in a developed form at this time.

---

164 An earlier point of view (Hrouda 1965, 89) regarded the White Obelisk (British Museum, WA 118807, reign of Ashur-nasir-pal I (1050–1032 B.C.)) as the first appearance of Assyrian conical helmets (*Chart 2, no. 75*). The dating of the obelisk is doubtful (Ashur-nasir-pal I or Ashur-nasir-pal II (883—859 B.C.)). *See* Börker-Klähn 1982, 179-180, no. 132, pls. 132a-d. Hrouda 1965, 181 mentioned a fragment of a glazed brick from Ashur with a representation of a helmeted Assyrian soldier on it (Ass. 10756, Andrae 1925, pl. 9, no. 9e) as from the reign of Tukulti-Ninurta I (1243–1207 B.C.). But the bronze helmet (painted yellow), the iron scale armour (painted blue) attached to the rim of the helmet and the characteristic spiked shield dates the scene to the reigns of Ashur-nasir-pal II or Shalmaneser III (858–824 B.C.).

165 Beran 1957, fig. 1.

166 Berlin, VAT 8951 = Ass. 14446n, Ebeling 1927, no. 177.

167 Seeden 1980, nos. 1738, 1743, 1744, 1776-1779, 1784, 1786, 1793, 1809.

168 Damascus, National Museum, S.O. 112; Joundi 1980, 36, fig. 14 ("Hittite style"); Seeden 1980, 115, no. 1750, pl. 107.

169 Joundi 1980, 36. Seeden 1980, 115 dates it to the early 1st millennium B.C.!

170 Boston, Tyskiewicz Collection; Terrace 1962, fig. 12; Alexander 1973-1976, 148.

171 Alexander 1973-1976, 178, note 124.

172 Özgüç 1986, 28, fig. 28, pl. 66:13a-b. Kt. f/K 7; 19213; H.: 11 cm, Diam.: 18 cm.

173 Özgüç 1986, 28, fig. 24, pl. 64:1a-c; Kt. f/K 13; 18755.

174 Özgüç 1986, 35, pl. 73:1; Kt. v/T 30; Kayseri Archaeological Museum; L.: 2.1 cm; W.: 1.9 cm.

175 Özgüç 1965, pl. VIII, no. 24: Kültepe, Karum level II, Anatolian sealing (Littauer – Crouwel 1979, 50-51, fig. 28, 19th–18th century B.C.); Louvre, AO 8306 ("*tablette cappadocienne*"), Anatolian sealing, Littauer – Crouwel 1979, 50-51, fig. 29.

176 Börker-Klähn 1982, 179-180, no. 132, pls. 132a-d.

177 Calmeyer 1969, 90-91.

178 On Urartian helmets with royal inscriptions: Argishti I (789—766 B.C.): **Cat. nos. 57-58**; Sarduri II (765—733 B.C.): **Cat. nos. 91-94**.

179 **Cat. nos. 49-51**.

180 For the problem of the connection of an Urartian attack and the destruction level of Hasanlu IVB *see* recently: Medvedskaya 1988, 1-15; Dyson – Muscarella 1989, 1-27; Medvedskaya 1991, 149-161.

The basis of this misconjecture was the distinction between two types of helmet decoration: the 9th century B.C. pairs of *repoussé* curves bending inwards (on the sculptures of Ashur-nasir-pal II (883—859 B.C.), *Chart 2, nos. 76-86*) and the continuous arcs (framing a "stele") replacing the curves on the front of the Assyrian helmets from the middle of the 8th century B.C. (reign of Tiglath-Pileser III (745—727 B.C.)), (*Fig. 1*). Calmeyer's theory did not allow for the fact that these two motifs are two stages in the linear evolution of Assyrian helmets, not the result of two different developments[181].

So long as only the standard forms of helmets shown on Assyrian reliefs were studied[182] the chronology of the development of Assyrian helmets was distorted. The subject has to be studied on a broader Near Eastern basis. New helmet finds of the last 20 years, and the recently published Assyrian iron helmet fragments from Nimrud now in the British Museum (**Cat. nos. 16-31**, *Pls. 25-43*)[183], with the help of helmet representations on Neo-Assyrian reliefs, allow for the chronological and typological classification of Iron Age Near Eastern helmets. Three periods may be distinguished in the evolution of Assyrian conical/pointed helmets (*Fig. 1*): 1. The characteristic design on the conical helmets of the sculptures of Ashur-nasir-pal II (883—859 B.C.)[184] consists of two (*Chart 2, nos. 76-77, 81-82*) or more (*Chart 2, nos. 78-79, 83-86*) horizontal ribs around the base of the helmet, one (*Chart 2, nos. 76-78, 81-84, 86*) or more (*Chart 2, nos. 79-80, 85*) pairs of curves (semicircular arcs) at the front of the helmet. On some helmets there is a vertical rib running from the point in between the curves (*Chart 2, nos. 81-86*). These horizontal and vertical ribs and curves undoubtedly represent the *repoussé* ribs strengthening the bronze sheet of the actual conical helmets. This decorative system can be exclusively connected to the early 9th century B.C. Assyrian conical helmets, and could not have appeared in any other context. The curves on the sculptures have rounded extremities[185]. But on the 9th century B.C. Assyrian conical helmets (**Cat. nos. 3-7, 8**, *Pls. 3-6, 7-8, see* below) the curves terminate in animal heads (ram's head)[186]. Overlaet assumed that the curves terminating in animal heads appear only after the reign of Ashur-nasir-pal II[187]. However, it may simply be a question of what the sculptors did or did not choose to represent. The sculptors of the sculptures of Ashur-nasir-pal II, as on the reliefs of other Assyrian kings, were satisfied with carving only the outlines and main motifs of the helmets, and did not represent the detailed incised decoration (*e.g.* on **Cat. nos. 3-7, 8** prepared with meticulous care), even on the surface of helmets of Assyrian high ranking military officials. So the lack of animal headed extremities should not be taken as a distinctive feature in the chronology of 9th century B.C. Assyrian conical helmets[188]. On the other hand the curves without animal heads may have been meaningless. There are at least three possibilities for the significance of the decoration (curves or crooks, interpreted often as "snakes") on the front of Iron Age Near Eastern helmets. They may have originated: 1. in the use of animals (or animal headed snakes) in an unknown cultic tradition as apotropaic symbols, which would be very appropriate on the front of a helmet; 2. in the representation of eyebrows on the front of a helmet, which may also have been apotropaic[189]; 3. in frames for figural scenes on the front of helmets (**Cat. nos. 3-7, 8**), as on the later Assyrian conical helmets (a new type of frame reminiscent of a stele, *see* below and *Fig. 1*). The explanation may owe something to all three possibilities.

The same design appears on the helmets shown on the Balawat gates[190] of Shalmaneser III (858—824 B.C.) (*Chart 2, nos. 92-94*), and on the helmet of a bronze statuette from Beqaa, Libanon (*Chart 2, no. 90*)[191]. The helmet worn by one of the two figures in an undated bronze chariot model from "Tartus", Syria (*Chart 2, no. 91*)[192] is also of interest in this context. It shows significant

---

181 S. Salvatori (Salvatori 1975, 260-261) and B.J. Overlaet (Overlaet 1979, 58-61) have pointed out the insufficiency of Calmeyer's theory of deriving the motif (pairs of curves) from N.-W. Iran.

182 Hrouda 1965, 89-90, pl. 23; Madhloom 1970, 37-40, pl. 18.

183 Dezső – Curtis 1991, 105-122, figs. 1-20.

184 Or some of them under the reign of Tukulti-Ninurta II (890—884 B.C.); Hrouda 1965, 89.

185 But on the 9th century B.C. Assyrian conical bronze helmets (**Cat. nos. 3-7, 8**) there are curves ending in animal heads.

186 The curves on the ancient Near Eastern helmets usually end in bovid heads, rarely in snakeheads; on Urartian helmets in lion heads.

187 Overlaet 1979, 59.

188 It has often been suggested that the small details of the helmets represented on sculptures – as with the details of helmets on the Til-Barsip frescoes – were painted on the reliefs.

189 In Argos (Perseus street 41), in a warrior grave of Geometric Age (third quarter of 8th century B.C.) the excavators found a conical helmet, the greek "paraphrase" of a Near Eastern conical helmet, on which the oriental features were easily observable: squat conical shape, vertical rib running from the point and two curves. But in Greece, in an another cultural complex, the wearers did not understand the function (presumably well known in the contemporary Near East) of the curves, and incised two eyes under the "eyebrows", which was undoubtedly an apotropaic feature in Greece. Protonotariou-Deilaki 1973, 97-99, pl. 95e; Gröschel 1986, 71-72, fig. 4. From the Near East we know only a single example of incising eyes under the curves: **Cat. no. 53** – a Hasanlu crested helmet.

190 King 1915, *passim*.

191 Beirut, National Museum; Seeden 1980, 109, no. 1723, pl. 102.

192 Paris, Louvre, AO 22265; Seeden 1980, 109-110, no. 1725, pls. N, 103.

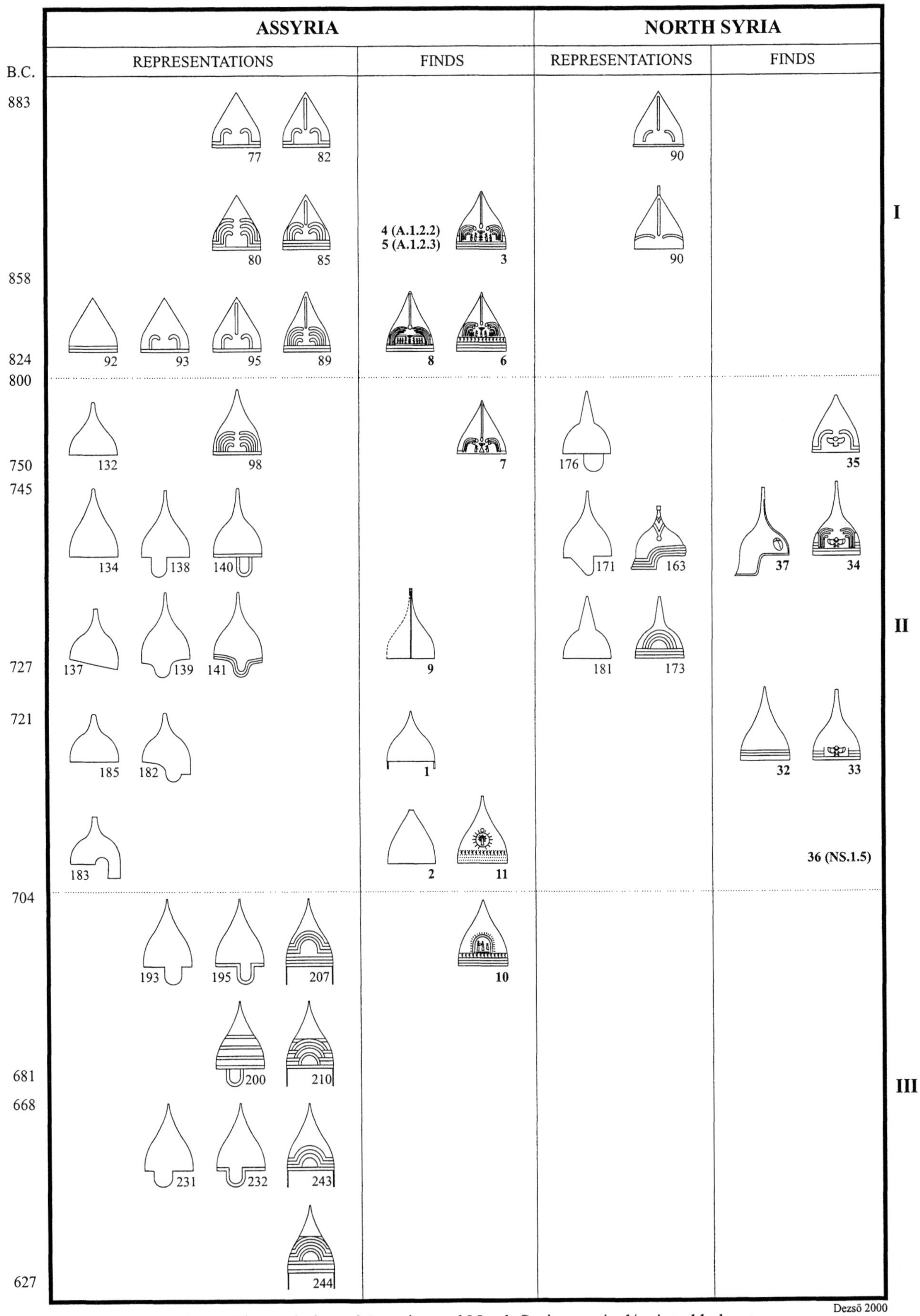

Fig. 1: The evolution of Assyrian and North Syrian conical/pointed helmets

similarity to that of the Beqaa statuette. Seeden did not consider them in relation to the Assyrian conical helmet tradition, but in the case of the first statuette (*Chart 2, no. 90*) she regarded the curves on the front of the helmet as horns. She dated the statuette around the late 2nd – early 1st millennium. The helmet of the second statuette (*Chart 2, no. 91*) is similar: there is a pair of curves (mentioned by Seeden as horns[193]) and a vertical rib between them on the front of the helmet. But a crest-shape vertical fin at the point makes this helmet curious. There are no published examples of fins of similar form[194]. The characteristic features of these two helmets (*Chart 2, nos. 90-91*) exclude the possibility that they are Levantine helmets decorated with horns. So their suggested 11th–10th century B.C. dating seems to be a little early, unless we consider that the Assyrian conical helmets with curves on their front appeared in the early 9th century B.C. (on the sculptures of Ashur-nasir-pal II (883—859 B.C.)) in their developed form. So the emergence of this motif can be dated to the 10th century B.C. (or even to the 11th century B.C., but there is no sign of it on the White Obelisk). The statuettes might be either the first examples of this early tradition or can be dated to the 9th or 8th centuries B.C.[195]. Since there are no helmet representations in Assyrian art from the period between the White Obelisk (Ashur-nasir-pal I (1050—1032 B.C.)) and the palace reliefs of Ashur-nasir-pal II (883—859 B.C.), the emergence of this motif can not be dated exactly.

The Assyrian conical helmet decorated with curves and a vertical rib (*Fig. 1*) discussed above spread widely in the late 9th century – early 8th century B.C. in the Near East[196], and remained in use on the periphery of the Assyrian Empire during the 8th–7th centuries B.C., while in Assyria the front design of helmets was modified to form a new type of frame reminiscent of a stele.

2. The main characteristic of the new decorative system was that the curves terminating in animal heads or rounded extremities were replaced by one or more arches framing a figural scene ("stele") on the front of the helmet (*Fig. 1*). At the same time the vertical rib running down from the point of the helmet to the curves disappeared from Assyrian helmets. The earliest appearance of the new decorative system is on the Arslan-Tash (Hadatu) stele (*Chart 2, no. 173*)[197] and on the helmets of the Til-Barsip wall-paintings (*Chart 2, nos. 163-167*)[198], both from the reign of Tiglath-Pileser III (745—727 B.C.). This kind of decoration is entirely missing from the palace reliefs of the Assyrian homeland from the reign of Tiglath-Pileser III (745—727 B.C.)[199] and Sargon II (721—705 B.C.)[200] and reappears only on the sculptures of Sennacherib (705—681 B.C.)[201] and Ashurbanipal[202] (669—627 B.C.)[203]. This characteristic design appears only on Assyrian helmets (the best examples are the Nimrud iron helmets [**Cat. nos. 10, 16-17**] discussed below). This new decorative system originates in the monumental arts of the second half of the 8th century B.C. The decorative system of the Nimrud conical iron helmet (**Cat. no. 10**, *Pls. 9-11*) parallels the wall paintings of the throne-room (Palace K) of Sargon II in Khorsabad (*Fig. 2*)[204], where there are three horizontal friezes (genii, bulls, genii); on the helmet there is only one figural frieze of processing officials in the upper register.

In the throne-room at Khorsabad, set into the upper border, is an arched panel within which Sargon II with an attendant behind him, to the right, is shown before a statue of a deity set on a podium, to the left. The whole scene originates in Assyrian monumental sculpture and the arched border around the figural scene imitates the outline of a stele. So the continuous arches and the scenes

---

[193] Seeden 1980, 110.

[194] The best parallels for this feature appear on incised helmet representations of a horse's bronze breastplate depicting the combat scene of chariot warriors wearing this type of helmet (*Chart 2, nos. 116-117*) and fleeing horsemen wearing "Hasanlu-type" crested helmet (*Chart 2, nos. 111-112*). The only examples of this type are **Cat. nos. 59-61**, three conical bronze helmets with vertical fin probably from Urartu. Both the breastplate and these helmets can be dated to the late 9th – early 8th century B.C.

[195] Amiet 1980, 434, no. 727, J. Crouwel and M.A. Littauer (Collon – Crouwel – Littauer 1976, 71-79; Littauer – Crouwel 1979, 146-147) dated the statuettes on the basis of chariot typology to the 6th–5th centuries B.C., while D. Collon (Collon – Crouwel – Littauer 1976, 79-81) examining the figures (charioteer and "smiting god") proposed a 14th century B.C. date.

[196] The earliest dated representation of conical helmets in use outside Assyria is shown on the throne basis (panel f) of Shalmaneser III (858—827 B.C.) from Nimrud: *Chart 2, no. 89* on which the tribute bearers of Mushallim-Marduk son of Ukani and of Adini, son of Dakuri, Chaldaean rulers from South-Babylonia, bring a conical helmet of Assyrian type (Mallowan 1966, 446-447, fig. f on p. 499). The earliest dated appearance of conical helmets outside Assyria is a series of helmets from Urartu, with the inscription of king Ishpuini (830—810 B.C.), (**Cat. nos. 62-64**). These helmets are decorated with a characteristic Urartian motif, the "lightning fork of Teisheba". The curves terminating in lion's heads appear only in the reign of Argishti I (789—766 B.C.), (**Cat. nos. 57-58**).

[197] Thureau-Dangin 1931, pls. IX:1, XI:1-2.

[198] Thureau-Dangin – Dunand 1936, pls. XLIX-LI.

[199] *Chart 2, nos. 134-142.*

[200] *Chart 2, nos. 182-185.*

[201] *Chart 2, nos. 200-211.*

[202] *Chart 2, nos. 233-235.* On the palace reliefs of Ashurbanipal the arcs appear not only on the front of conical (and crested) helmets, but on their backs as well: *Chart 2, nos. 236* (WA 124956), *240* (WA 124957).

[203] We can not exclude the possibility that the now missing decoration of the helmets on the palace reliefs of Tiglath-Pileser III and Sargon II was originally painted.

[204] Moortgat 1959, fig. 6; Heinrich 1984, fig. 101; Lloyd – Müller 1986, 43, fig. 81.

depicted on the front of the Assyrian conical helmets follow the same pattern inspired by the form and imagery of Assyrian monumental stelae. This is proved by an iron helmet-fragment from Nimrud (**Cat. no. 16**) which is decorated with the figure of Ishtar of Arba'il standing on the back of her lion inlaid in bronze (*Pl. 25*) within a stele-like curved border. An exact parallel to this scene is known on a stele from Til-Barsip dedicated to the goddess Ishtar of Arba'il[205].

At the latest in this period appeared a new motif, which modified the decorative system of the front of Assyrian helmets. The earliest and only example of the new system is helmet **Cat. no. 11**, *Pls. 12-16*), a fragmentary iron helmet without provenance[206]. The punched and engraved decoration of the helmet may be more archaic than the technically more developed bronze inlay, which might represent a higher standard in manufacturing iron helmets. Its decoration diverges from the standard system for Assyrian helmets in the 9th century B.C. by having an "aura" (*melammu* forming a shining circle) above the figurative frieze (officials in procession)[207] as the border for the central figurative scene rather than a "stele-field". There is the figure of Ishtar in the middle of the "aura"-border. This new motif did not destroy the decorative system of Assyrian helmets, because of the main elements of the system (parallel lines with figurative frieze of officials between the two upper lines; central figurative scene above the lines on the front of the helmet) remained canonical. The only divergence is the framing motif: "stele" or "aura". This new border motif ("aura") has no parallels or representational evidence on the front of helmets in the Assyrian art, and the "stele border" with figurative scene on parallel lines remained the most important (if not the only) decorative element on the front of Assyrian helmets (*Chart 2*). Without doubt the front decoration of helmet **Cat. no. 11** (*Pl. 16*) (Ishtar within a shining circle or "aura") can be derived similarly from the Assyrian monumental art as have been proved in the case of helmets **Cat. nos. 10, 16** (Khorsabad wall painting [*Fig. 2*] and stele of Ishtar of Arba'il from Til-Barsip [*Fig. 4*]).

The other change in the 8th century B.C. on conical helmets was the appearance of the earflaps. Earflaps are entirely missing from 9th century B.C. Assyrian helmet-representations, and appear only about the end of the 9th century B.C. (*Chart 2, nos. 116-117*). Because of the 80 years lack of Assyrian helmet-representations between the age of the Balawat gates (Shalmaneser III, 858—824 B.C.) and the palace-reliefs of Tiglath-Pileser III (745—727 B.C.) the first earflaps appear only on the sculptures of Tiglath-Pileser III[208].

The shape of helmets changed as well. On the palace-reliefs of Tiglath-Pileser III (*Chart 2, nos. 134-142*) there are representations of pointed helmets with curving profile and long, tube-shaped point. This shape is also depicted on the sculptures of Sargon II (721—705 B.C.) from Khorsabad (*Chart 2, nos. 182-185*) and in the art of the second half of 8th century B.C. in North Syria: on the wall-paintings of Til-Barsip (*Chart 2, nos. 163-171*)[209]; on the sculptures of Tell Ta'yinat (*Chart 2, no. 181*); and on the orthostat-slabs of Arslan-Tash (*Chart 2, no. 173*) and Karatepe (*Chart 2, nos. 174-176*).

The reason for the changes in the decoration on the front of Assyrian conical helmets in the middle of the 8th B.C. ("stele-border" replacing the curves bending to the centre) might be that the traditional (canonical) but stereotyped figurative scenes of the 9th century B.C. Assyrian conical bronze helmets (king with his attendant behind him adoring a "tree-of-life") was replaced by an other type of central scenes (which was in fact canonical). The historical background for these changes in the representations of Assyrian helmets (in fact for the changes in the uniform of the Assyrian troops) might be the army and military reform of Tiglath-Pileser III. While at the end of the 9th century – first half of 8th century B.C. the crested helmets indicated the nationality of its wearer (Syrian, Anatolian or Phrygian allies or mercenaries, or troops of these territories), later, from the third quarter of the 8th century B.C. (in fact from the reign of Tiglath-Pileser III), after the final conquest of most of these territories by the Assyrian army, they lost their national or ethnic character, and simply demonstrated the wearer's membership of a special service of the "new model Assyrian imperial army", the light infantry. The light infantry of the imperial army was always equipped with spear, round shield, sword,

---

205 Paris, Louvre, AO 11503; Thureau-Dangin – Dunand 1936, 156-157, pl. 14:1; Börker-Klähn 1982, 226, no. 252, Dezső – Curtis 1991, 107-108, pl. XV.

206 Said to come from Eastern Turkey. Kellner 1993, 325.

207 **Cat. nos. 3-4**, *Pls. 3-5*, pp. 26-29.

208 There is no known helmet with earflaps hammered out from the same piece of metal as the helmet itself (as shown on the sculptures of Tiglath-Pileser III and Sargon II (721—705 B.C.)). Surviving helmets have only the holes of the missing detachable earflaps on the sides.

209 The first appearance of the neck-guards bending behind the ears is on the wall paintings of Til-Barsip (reign of Tiglath-Pileser III, 745—727 B.C.). There are neck-guards on the helmets of the palace-reliefs of Sargon II (721—705 B.C.) from Khorsabad. Under Assyrian influence this kind of neck-guard is widespread in the art of the Eastern Mediterranean in the second half of the 8th century B.C.: on the shields of the Zeus cave, Ida mountain, Crete (*Chart 2, no. M.2.21*); on Phoenician bronze bowls from Delphi (*Chart 2, nos. M.2.22-23*) and Olympia (*Chart 2, no. M.2.24*). Only one helmet is known (**Cat. no. 34**, *Pl. 46*) with broad neck-guard and long tube-shaped point similar to those on the Til-Barsip wall paintings.

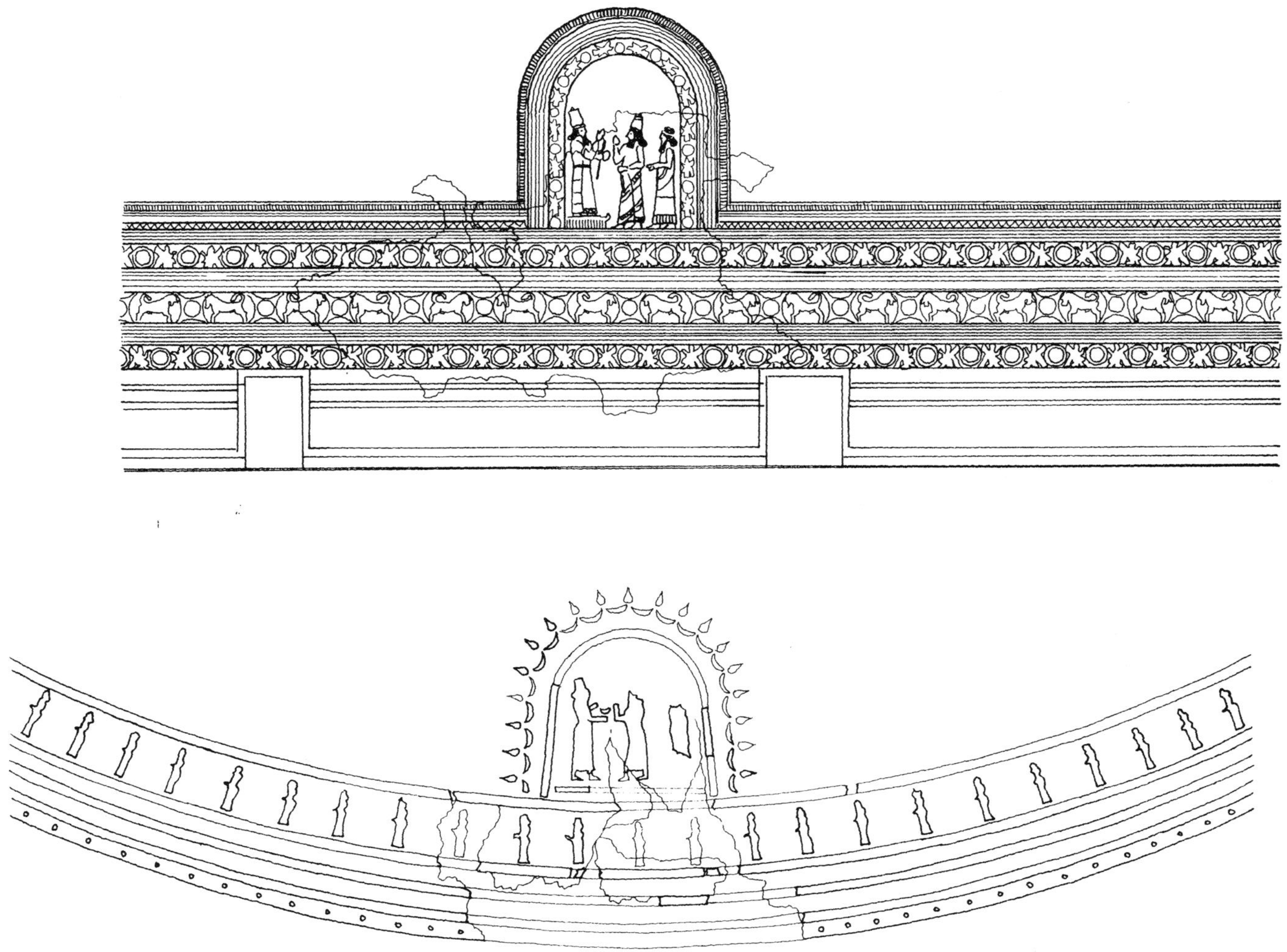

Fig. 2: *Above:* Mural painting, Khorsabad, Palace K, Throne-room, Sargon II; *Below:* Bronze inlay decoration of helmet **Cat. no. 10**

breast and back plates (*kardio phylakes*)[210]. At the same time (mid 8th century B.C., reign of Tiglath-Pileser III) appeared the new type of front decoration on Assyrian conical helmets mentioned above. This new type of decoration (images of gods or goddesses: Ishtar, **Cat. no. 11**; Ishtar of Arba'il, **Cat. no. 16**; or scene of the cultic activity of the king himself: **Cat. no. 10**) might indicate that the wearer of the helmet belonged to a special (territorial or functional) unit of the new model army[211]. The helmet decoration might have separated the different units of the same service (i.e. the units of the heavy infantry, chariotry or cavalry).

3. There were two important changes in the third phase of the evolution of Assyrian conical helmets. 1. From the last quarter of the 8th century B.C. the decoration gradually spread on the helmets. On the palace reliefs of Sennacherib (705—681 B.C.) horizontal ribs and framing arches in superimposed registers frequently covered the whole surface of the helmet (*Chart 2, nos. 202-212*)[212]. Perhaps a better example of the new system of decoration is observable on the cylinder seal of Ashur-nadin-shumi (*Chart 2, no. 226*), the short-lived Assyrian king of Babylon (700—695 B.C.), on which the helmet of the ruler is decorated entirely not only with horizontal ribs, but with rows of rosettes between them[213]. We know this linear and dotted design well from the iron helmet-fragments from Nimrud now in the British Museum (**Cat. nos. 16-31**, *Pls. 25-43*)[214] dated to the last third of the 8th century B.C. We know an earlier example of the new lavish decorative system from 46 fragments of an iron helmet inlaid in bronze from the destruction level of Zinçirli associated with the siege of the city by Sargon II in 720 B.C. (**Cat. no. 38**, *Pl. 57*). 2. The second important change was the increasing use of iron helmets on a large scale. The first iron helmets must have already appeared at least in the 2nd phase of development of Assyrian conical helmets, possibly in the early 8th century B.C. We can date the earliest written sources mentioning iron helmets (*gurpisi parzilli*)[215] to the same period. The first representations undoubtedly depicting iron helmets are the Til-Barsip wall paintings possibly from the reign of Tiglath-Pileser III (745—727 B.C.). On the Til-Barsip wall paintings (*Chart 2, nos. 163-165*)[216] there are helmets coloured light blue, as are the iron blades of the swords, certainly depicting iron helmets[217]. The yellow decoration on two of them (*Chart 2, nos. 163-164*) might indicate the same bronze inlay decoration as on the well known Zinçirli (**Cat. no. 38**) and Nimrud iron helmets (**Cat. nos. 10, 16-31**). The Til-Barsip helmet representations are the earliest evidence both for the Near Eastern iron helmets and iron helmets decorated with bronze inlay. By the end of the 8th century B.C. there can be no doubt that iron was a common material and widely used[218]. Place mentioned in his work[219] the discovery of a hoard of very well preserved ironwork: grappling irons and chains, ploughshares, hammers, pick-axes and several hundred iron ingots, the total weight of which was estimated by Place in the region of 160.000,000 kg.

The decoration of the Til-Barsip helmets – similar to the Zinçirli iron helmet fragments (**Cat. no. 38**, *Pl. 57*) – differs from the canonical Assyrian decoration system ("stele-field" on the front, *Fig. 1*), and represents the local North Syrian tradition of helmet decoration on Assyrian helmets. From the middle of the 8th century B.C. the North Syrian metalworking centres[220] might have been at least as important as the Assyrian. The only difference between the two productions is the difference in decoration (*Fig. 1*). However, we cannot assume that the North Syrian weapon industry and metalworking were independent from Assyrian metalworking centres. The products of the North Syrian weapon industry ("North Syrian" bronze and iron helmets) were worn by soldiers of the Assyrian

---

[210] *See* p. 39.

[211] *See* p. 50 on Ishtar of Arba'il.

[212] Hrouda 1965, 89 suggested, that the number of horizontal ribs and arches on the helmets indicated the military rank of its wearer. After systematically assembling the representations of Assyrian helmets, there is no a clear indication that this was the case. It remains hypothetical, but there is a considerable possibility, that the level of decoration (*i.e.* the number of lines and arcs combined with the figurative decoration of the helmet) together with additional elements of the soldiers equipment: the material (*i.e.* colour) of their garment, the staff and mace sometimes hold by Assyrian soldiers identified as military commanders, and of course the extent of the use of precious metals in the decoration of their garment and equipment might indicate the military rank.

[213] Collon 1988, 129-130, no. 555.

[214] Dezső – Curtis 1991, 105-122, figs. 1-20.

[215] *Gurpisu parzillu*: Friedrich *et al.* 1940, nos. 49:1, 52:10, Tell Halaf. Unfortunately Assyrian written sources mention helmets (*ḫuliam, elû* or *gurpisu*) only a few times. The royal inscriptions of Sennacherib and Esarhaddon mention *ḫuliam*, and an administrative tablet from Tell-Baqqaq (dated by its *limu* (*ᴵIa-ḫa-lu*) to 833/824/821 B.C.) lists 20 gurpisus made of copper/bronze (20 *gur-pis erî*). Ismail 1989, 61-62.

[216] Thureau-Dangin – Dunand 1936, pls. XLXI-LI.

[217] A 9th century B.C. (Ashur-nasir-pal II or Shalmaneser III) example of the same colour system is a fragment of a glazed brick from Ashur (Ass. 10756, Andrae 1925, pl. 9e) depicting an Assyrian warrior with spiked shield. The helmet of the soldier is yellow (bronze), but the scale armour attached to the rim of the helmet is painted blue (iron). The piece was dated mistakenly by Andrae to the 13th century B.C.

[218] Curtis *et al.* 1979, 382.

[219] Place 1867—1870, vol. I, 84-88.

[220] The helmets illustrated at Til-Barsip (*Chart 2, nos. 163-172*) were presumably made locally, as the Zinçirli helmets (**Cat. nos. 32-33, 38**) perhaps were made in a North Syrian province, even if worn by Assyrian soldiers.

imperial army[221]. So the metalworking of North Syria, under continuous Assyrian rule from the middle of the 8th century B.C., may only be interpreted as part of the "Assyrian imperial production".

The helmets classed among this group are with the exception of three pieces[222] without provenance, or are associated with "Luristan". Therefore the definition of this group and the classification of helmets are supported only by formal iconographical analysis. There is no agreement on the identification of Assyrian helmets. Curtis, discussing the Nimrud iron helmets decorated with bronze inlay[223], assumed that no single helmet found outside Assyria can be identified as certainly of Assyrian manufacture. Assyrian helmets seem to have been copied all over the Near East. Ligabue and Salvatori[224] listed the same four bronze helmets and the Nimrud iron helmet (**Cat. no. 10**) in their group of Assyrian conical helmets. The present writer included altogether 11 conical helmets in this group. Three of them (**Cat. nos. 1, 2, 10**) are from excavations, the rest of them (**Cat. nos. 3-7, 8-9, 11**) are unprovenanced helmets from the art market.

Group A.1.1: Undecorated Assyrian conical bronze helmets (Cat. nos. 1-2)

Helmets of this group can be dated to the second half of the Neo-Assyrian period (second half of 8th century – 7th century B.C.). The origin and dating of helmet **Cat. no. 1** is uncertain. Petrie found it together with a bronze phiale and iron tools. In Assyria, pointed helmets with slightly curved contour were used in the second half of the 8th century B.C. (second phase of development, *Fig. 1*). On Assyrian palace-reliefs this type of helmet is usually depicted as furnished with earflaps, but our helmet (**Cat. no. 1**) like several Urartian helmets and a Median helmet found in Olympia (**Cat. no. 141**) has two loops made of bronze wire for the chin-strap. However, it cannot be considered as a distinctive feature, since – in spite of the frequent representations (*Chart 2*) of pointed helmets with earflaps – we do not know a single surviving ancient Near Eastern conical/pointed helmet equipped with earflaps. There are holes for fastening the detachable earflaps on only a few helmets[225]. The lack of decoration is not a distinctive feature. On contemporary Assyrian palace-reliefs (Tiglath-Pileser III and Sargon II) the helmets are undecorated without exception. Unfortunately there are no Assyrian bronze helmets of that age (second half of the 8th century B.C.) from excavations. The Khorsabad bronze helmets excavated by Place in the palace of Sargon II (721—705 B.C.)[226] unfortunately no longer exist(?). So the identification and dating of the Theban helmet remains doubtful. It might reach Thebes in the last third of the 8th century B.C. as booty, taken from the Assyrians.

The origin of the Atlit conical bronze helmet (**Cat. no. 2**, *Pl. 2*) is doubtful as well. It was in the Phoenician harbour of Atlit, together with various bronze objects including "a leg of a throne in Assyrian style"[227]. Without any further indications (*e.g.* decoration, *etc.*) there is no possibility to date it exactly, thus as G. Radan assumed, its date may range from the Neo-Assyrian period (from the middle of the 8th century B.C.) through the Neo-Babylonian period, till the Persian period[228]. The conical/pointed form of Assyrian helmets undoubtedly survived for centuries after the fall of Assyria (*see* the chapter of Persian Helmets).

1. Thebes, North of Tausert's temple (from a well). Group no. **A.1.1.1** – H.: 21 cm. It has a pointed shape with a slight curve. There is no decoration on the helmet. There are two loops made of bronze wire on the sides of the helmet to fasten the chinstrap. Complete (*Pl. 1*). Last quarter of 8th century – early 7th century B.C. – Manchester, University of Manchester Museum, Inv. no.:

---

[221] The wall paintings of Til-Barsip (*Chart 2, nos. 163-172*); helmets from Zinçirli (**Cat. nos. 32-33, 38**).

[222] **Cat. no. 1**, Manchester, University of Manchester Museum, Inv. no. 1618, from Thebes, north of Tausert's temple. Petrie 1897, 18, pl. 21; Meissner 1920, pl. 58; von Luschan – Andrae 1943, 75; Borchhardt 1972, 100, Cat. no. 22 B II; Williams – Maxwell Hyslop 1976, 283-286, fig. 1; Curtis 1979, vol. I, 184; Özgen 1982, 20, 53, note 10; **Cat no. 2**, Haifa, National Maritime Museum, from Atlit, Phoenician harbour, Radan 1981, 49-57, pl. VIII:3; **Cat no. 10**, London, British Museum, BM 22496, from Nimrud, Northwest Palace, Barnett 1953, 101-102, pls. 31-32; Calmeyer 1969, 90, note 295; Borchhardt 1972, 99, Cat. no. 22 A I; Pleiner – Bjorkman 1974, 291, fig. 5; Calmeyer 1975, 316; Ligabue – Salvatori 1977, 8; Gamber 1978, fig. 196; Overlaet 1979, 54, note 6; Curtis 1979, vol. I, 178-181, vol. II, 32, pl. 27; Curtis, *et al.* 1979, 385, fig. 34; Horn – Rüger 1979, 313, fig. 187; Özgen 1982, 20, 51, note 3; Gröschel 1986, 73, note 225; Dezső – Curtis 1991, 105-107, 122-126, fig. 21, pl. 19.

[223] Curtis 1979, vol. I, 184-185.

[224] Ligabue – Salvatori 1977, 7-8.

[225] The detachable earflaps appear only on the palace-reliefs of Sennacherib (705—681 B.C.), *Chart 2, nos. 197, 200-215.*

[226] Chambers 16, 17, Palace of Sargon II, Khorsabad (Dur-Sharrukin); Place 1867-1870, vol. I, 64-65: "*La chambre 16 contenait des casques en cuivre, fortement déprimés sous la terre éboulée; à part quelques fêlures, ils avaient conservé leur forme, encore très-apparente; il était facile d'y reconnaître le casque si simple, qui était la coiffure habituelle des soldats assyriens dans les bas-reliefs; la forme de cette armoure défensive, terminée au sommet en pointe arrondie, est exactement la même dans les deux cas.*" Ibid., 65: "*Dans la chambre 17 étaient encore des casques identiques aux précédents; plus, quelques plats ou assiettes, également en cuivre, déprimés et brisés en partie.*"

[227] Radan 1981, 49.

[228] Radan 1981, 54-56.

1618. – Petrie 1897, 18, pl. 21; Meissner 1920, pl. 58; von Luschan – Andrae 1943, 75; Borchhardt 1972, 100, Cat. no. 22 B II; Williams – Maxwell-Hyslop 1976, 283-286, fig. 1; Curtis 1979, vol. I, 184; Özgen 1982, 20, 53, note 10.

2. Atlit, Phoenician harbour (together with an Assyrian style bronze throne leg). Group no. **A.1.1.2** – H.: 20.5 cm, Diam.: 20 cm. Plain conical bronze helmet. Near the rim of the helmet there are holes, 5.4 cm wide, for the attachment of either movable earflaps or cheek pieces. Higher up towards the top, cracks have developed from the persistent action of the of saltwater. There is a tubular hole on top of the dome of the helmet (*Pl. 2*). Second half of 8th century B.C. – Haifa, National Maritime Museum. – Radan 1981, 49-57, pl. VIII:3.

Group A.1.2: Decorated Assyrian conical bronze helmets (Cat. nos. 3-7)

Helmets of this group can be dated to the first half of the Neo-Assyrian period (9th century – first half of 8th century B.C.). With their arcs curving towards the centre and with their traced figural decoration these five helmets undoubtedly represent the 9th century B.C. Assyrian helmet tradition, the first phase of the evolution of Assyrian helmets (*Fig. 1*). Both the conical form and the figural decoration are characteristically 9th century B.C. features. The figural decoration on the front of the helmets (sun-disk, tree-of-life, king and attendants) may only be dated stylistically. Unfortunately we do not know anything of the decoration of helmets depicted in the 9th century B.C. Assyrian palace reliefs (Ashur-nasir-pal II, 883—859 B.C.), because there are no carved details representing traced figural decoration on the helmets on the sculptures. Examining the three helmets mentioned above G. Ligabue and S. Salvatori assumed that the repertory of the 9th century B.C. Assyrian helmet-decoration was limited and consisted of only a few stereotyped motifs[229]. The composition of the winged sun-disk is characteristic in the 9th century B.C. Its best parallels are on a stele from the fifth regnal year of Ashur-nasir-pal II (879 B.C.)[230] and on the Black Obelisk[231] of Shalmaneser III (858—824 B.C.). Representations of "trees-of-life" similar to those on the helmets discussed above (**Cat. nos. 3-5**) occur also in 9th century B.C. Assyrian art. One of the best example occurs on a palace relief of Ashur-nasir-pal II (883—859 B.C.), on which the whole central scene known from helmets **Cat. nos. 3-5** (king on either sides of a tree-of-life under a winged sun-disk) appears on the king's carved pectoral[232]. There are however, two unique helmets in the Axel Guttmann Collection, decorative system of which differ in some details from the decoration of these canonical 9th century B.C. helmets: **Cat. no. 6**: with a "coronation scene" in the centre field and **Cat. no. 7**: with a "praying king" in the centre field under the arches. The decoration of only one, probably 9th century B.C. Assyrian helmet, the Karlsruhe conical crested helmet (**Cat. no. 14**, *Pls. 20-23*) differs entirely from this 9th century B.C. decorative system. However the decorative system of crested helmets differs generally from the decoration of conical/pointed helmets: with a single exception[233] the figural scene bordered by curves bending towards the centre, or continuous arcs (semicircles) is missing[234] on them. Nevertheless, on the front of this helmet, in the centre of the figural frieze around the base, there is the scene of the king with his attendant behind him worshipping beside a "tree-of-life". Above the tree, in a winged sun-disk, there is the figure of a god (probably of Ashur).

The frieze of officials in procession visible on two of the helmets mentioned above (**Cat. nos. 3-4**, and partly on **Cat. no. 6**) is a characteristically Assyrian motif. The elaborate scabbards (with unique double projections), and garments of the officials are typically 9th century B.C. Assyrian features[235]. This motif (officials in procession) remained in use through 8th (and possibly 7th) century B.C. Assyrian art, and as a canonical scene appears on the Nimrud conical iron helmet (**Cat. no. 10**, *Pls. 9-11*) and on the Munich fragmentary iron helmet (**Cat. no. 11**, *Pls. 12-16*), both from the second half of the 8th century B.C. As a result of Assyrian influence this motif spread into the art of regions around Assyria[236].

In his recently published monograph O.W. Muscarella[237] questioned the originality of several Near Eastern (Assyrian[238] and Urartian[239] helmets). His minimalist view is based on the otherwise

[229] Ligabue – Salvatori 1977, 7.
[230] Baghdad, Iraq Museum; Nimrud, NW-Palace, ND 1104; Börker-Klähn 1982, 182-183, no. 137, pls. 137a-b.
[231] Nimrud; London, British Museum, BM 118885; Börker-Klähn 1982, 190-191, no. 152, pls. 152a-d, A1-D5.
[232] Layard 1853, pls. 5-6. For the representations of trees-of-life and winged sun-disks *see* Layard 1853, *passim*.
[233] **Cat. no. 13**, *Pls. 18-19*.
[234] Continuous arcs on the front of crested helmets appear only on the palace reliefs of Ashurbanipal (668—627 B.C.), *Chart 2, no. 229*.
[235] The details of the garments (*e.g.* the incised design of the trimming and fringes: stripes, transversal stripes and quadrants) are similar to the incised decoration of the garments of officials in procession on Nimrud ivories. Mallowan – Davies 1970, nos. 13-14, pl. VI; nos. 16-17, pl. VIII; no. 29, pl. X, nos. 52, 53a-l, 54a-f, pls. XIV-XV.
[236] *E.g.* on an Urartian assyrianizing quiver; Vanden Berghe 1982, 145-157, pls. 16-19.
[237] Muscarella 2000.

logical viewpoint, that only those artifacts can be considered original, which came from controlled excavations. In case of Near Eastern helmets his method of criticism is based on the following standpoints. He discredited helmets because of:

1. Their motifs do not have authentic iconographical parallels.
2. Their composition is irregular and superficial.
3. Their execution is „crude and unskilled".

Muscarella concludes that the originality of these pieces can only be proved by a metallurgical analysis, the examination of their corrosion history[240]: if the incisions of the different scenes cut through the corrosion of the helmet, the piece is a forgery. At this point it is interesting to mention, that he based his criticism only on questioning figurative elements of the decorated helmets. He does not question the originality of the helmet itself. He supposed, that the helmets were original and were later (nowadays) decorated with incised figurative scenes by forgers. It means that he considered most of the questionned helmets *pastiches*. However, only a smaller part of helmets of uncertain origin are decorated.

Muscarella questioned the originality of helmet **Cat. no. 3**[241]. From the iconographical point of view he did not raise an objection, but he argued, that the execution of the central scene is „especially crude"[242]. He wrote, that „no artisan could have executed this work, for he would have been deported". The question is, however not so simple. I think it is obvious that those tens of thousands of Assyrian helmets which were made during the three hundred years of the Neo-Assyrian period simply should not have been without exception masterpieces. But much more interesting is the subsidiary scene, which was termed by Muscarella as showing „Assyrian (?!) dignitaries wandering with no place to go"[243]. This scene is an integral part of several Assyrian conical helmets: **Cat. nos. 3-4, 10-11** one of which (**Cat. no. 10**) was excavated by Layard in Chamber I, North-West Palace, Nimrud[244]. If helmet **Cat. no. 3** is a forgery, where is this scene from? Salvatori published this helmet in 1975[245], and two years later published a similar fragment[246] (**Cat. no. 4**). At that time this scene (frieze showing officials and running around the base of the helmet) was unknown from other examples. Though Barnett published the Nimrud iron helmet (**Cat. no. 10**) as early as 1953[247], he did not recognize the same scene running around the base of the Nimrud piece. It was only in 1990, that during a new X-ray examination the present author and J. Curtis noticed the scene[248]! So an excavated piece corroborated the authenticity of the Ligabue helmet (**Cat. no. 3**), which was published some 15 years earlier. If the Ligabue helmet (**Cat. no. 3**) is a forgery as argued by Muscarella, the forger should have copied an original, which was decorated similarly with this authentic scene. If so, where is the original?

In case of helmet **Cat. no. 6** the problem is much more serious. As has been mentioned earlier the decorative system of helmets **Cat. nos. 6-7** differs in significant details from the decorative system of the three 9th century B.C. canonical Assyrian pointed helmets (**Cat. nos. 3-5**). As U. Seidl's detailed description shows[249], these helmets added new iconographical elements to the canonical pattern of the 9th century B.C. Assyrian conical bronze helmets. However, these helmets added new iconographical elements not only to the canonical pattern of the 9th century B.C. Assyrian conical bronze helmets, but to the Assyrian art as a whole as well. Some of the new motifs are virtually unknown in the canonical Assyrian art (palace-reliefs, wall-paintings, glazed tiles, carved ivories, decorated metalwork, *etc.*).

---

238 Actually all the decorated Assyrian bronze helmets: **Cat. no. 3**: Muscarella 2000, 186-187, no. 44, pl. on p. 507; **Cat. no. 6**: Muscarella 2000, 184-186, no. 42, pls. on pp. 505-506; **Cat. no. 7**: Muscarella 2000, 186, no. 43. He does not question the originality of fragments **Cat. nos. 4-5**, or he had never seen them.

239 **Cat. no. 98**: Muscarella 2000, 149-150, no. 17, pl. on p. 460; **Cat. no. 8**: Muscarella 2000, 150, no. 18; **Cat. no. 14**: Muscarella 2000, 150, no. 19. The last two pieces are Assyrian helmets. It is unambiguous from the corpus of excavated Urartian helmets, that the Urartians never accepted prominent and therefore obvious elements of their rival's Assyrian royal iconography. Muscarella should have listed these two pieces between the Assyrian forgeries, and would not have accepted their alleged Urartian origin. It is obvious that whoever made these pieces intended to represent Assyrian scenes.

240 Muscarella 2000, 149, 186.

241 Muscarella 2000, 186-187, no. 44, pl. on p. 507.

242 Muscarella 2000, 186 wrote that the execution of the central scene is especially crude: there are no feet. This phenomenon is not necessarily a sign of a forgery. There are several examples known, by way of illustration see an unprovenanced Urartian belt (Muscarella 1981, 193, no. 150).

243 Muscarella 2000, 186. Furthermore he wrote that some of the dignitaries "wave to each other". I think the disproportionate distance between the figures is not an obvious sign of a forgery. Muscarella himself published three fragments from a bronze receptacle (from Ziwiye?, Muscarella 1988, no. 473) on which the same phenomenon can be detected: some of the tribute bearers wave to each other (figs. 35-36). In spite of this he wrote that the scene "is classically Assyrian with regard to both style of execution and iconography" (Muscarella 1988, 343).

244 Dezső – Curtis 1991, 105-107.

245 Salvatori 1975, 255-264, figs. 1-2, pls. 36-38.

246 Ligabue – Salvatori 1977, 7-8, fig. 1.

247 Barnett 1953, 101-102, pls. 31-32.

248 Dezső – Curtis 1991, 105-107, 122-126, fig. 21, pl. 19.

249 Born – Seidl 1995, 11-47: **Cat. no. 6**, 48-50: **Cat. no. 7**.

Muscarella not only questioned but completely rejected the originality of these helmets on the basis of the 1) complete lack of iconographical parallels, 2) the irregular and superficial composition and 3) the unskilled execution[250]. He undoubtedly raised serious questions. The present author does not want to repeat Muscarella's detailed argumentation, but wants to add some further details. On helmet **Cat. no. 6** the central scene, "coronation of the Assyrian king" is unknown in the iconographical repertory of the Assyrian (and Mesopotamian) culture, where the most important act of the inauguration of a new king's era would probably be an anointment ceremony. However, as the Middle Assyrian text cited by U. Seidl[251] shows, the king was actually crowned by a crown or diadem. Furthermore, Sargon II during his 8th campaign (714 B.C., against Urartu) desribes the plunder of the city of Musasir and the Haldi Temple. He mentions that this temple is where the Urartian kings (in this case Rusa) were crowned (received the crown and the sceptre)[252]:

> IGI(*pān*) ^d^*Ḫal-di-a* DINGIR(*ili*)-*šu* AGA(*agû*) *be-lu-ti* *ip-pi-ru-šu-ma* *ú-šá-áš-šu-šu* ^giš^GIDRU(*ḫaṭṭu*) LUGAL(*šarrū*)-*ti* ^kur^*Ur-ar-ti* *ù* UN.MEŠ(*nīšē*)-*šu* *i-nam-bu-ú* MU(*šum*)-*šu*
>
> "In front of Haldi, his god, the crown of the rule was put on (his head) and the sceptre of the kingship of Urartu was given to him. And his nation declared his name."

This scene is strikingly similar to the scene depicted on the front of helmet **Cat. no. 6** (putting on the crown and receiving the sceptre(?) in front of the god)[253]. In spite of this fact, the coronation and/or anointment ceremonies are (surprisingly?) not part of the Assyrian iconographical repertory, and are not represented for example on Assyrian palace-reliefs. The representation of these ceremonies was not part of the Assyrian royal/imperial propaganda. However, written sources do not legitimate automatically such a scene. Muscarella's reason, that authentic iconographical parallels are needed is a very important point. However, the present author is sure that there were alternative iconographical traditions, which are unknown from the official or canonical Assyrian iconographical repertory of the palace-reliefs and minor arts. One of the missing subjects is the king's activity as a high-priest[254]. Such a striking discovery was, when the present author in 1993 in a London conference (The Furniture of Western Asia) heard the lecture of M. Mahmoud, one of the Iraqi excavators of the Nimrud tombs of Assyrian queens. One of the lecturer's colour slides showed a series of large terracotta plaques, *in situ*, resting against the wall of one of the tombs. The first terracotta plaque showed an Assyrian queen standing in her chariot. This uncanonical scene – without parallels – on the art market would easily be considered as a forgery. But its excavation context corroborates a theory that there were iconographical traditions unknown to us. In this case at least two possibilities can be considered: 1) There was an independent iconographical tradition for Assyrian queens unknown to us. 2) There was a "funerary iconography" unknown to us. The second possibility would explain the existence of "curious scenes".

Interestingly, the present author has many more objections to the subsidiary decoration. In addition to Muscarella's long list of "mistakes" the present author would like to point to some iconographical details in the lower frieze of the helmet, which are very strange. For example the small flag(?) in the hand of one of the courtiers behind the king and the figure of the other courtier coming out from the royal tent(!) raising a chalice (with vertical handle) and not a bowl. Both the chalice and the phenomenon of a flag are unknown or at least unusual in Assyrian art, and the present author has no idea how to explain them[255]. The representation of the city is eclectic as well, but this may be the result of a kind of "artistic freedom" obvious in details of minor importance. This composite frieze running around the base of the helmet mixes two independent iconographical elements of the Assyrian helmet art: procession of Assyrian officials[256] (**Cat. nos. 3, 4, 10, 11**) and tribute bearers[257] (**Cat. nos. 8, 14**).

---

[250] Muscarella 2000, 184-186, no. 42, pls. on pp. 505-506.

[251] Born – Seidl 1995, 30. "May Aššur (and) Ninlil, the lords of your 'turban' (*kulūlu*), put the 'turban' on your head for a hundred years", Oppenheim et al. 1971, 527; further Middle Assyrian sources mention that the king was crowned by deities: "when Aššur put the lordly crown (*agû*) on my head (Tukulti-Ninurta I.)", Oppenheim et al. 1964, 155; Weidner 1959, no. 1, I:28; "whom (Tiglath-pileser I) thou (the gods) crowned with the majestic tiara (*agû*)", Oppenheim et al. 1964, 155; Budge – King 1902, 30, I:21; "Aššur, who gives the royal sceptre and crown (*agû*)" (Tiglath-pileser I.), Budge – King 1902, 28, I:2; *etc.*; One of the later sources, the 'Coronation Hymn of Ashurbanipal' (VAT 13831, r. 5; Livingstone 1989, 27) mentions only that the king received his crown from Anu. One of the inscriptions of Esarhaddon mentions the same: "Anu had bestowed on me his tiara (*agû*)", Oppenheim et al. 1964, 155; Borger 1956, 81, r. 1.

[252] 8th campaign, line 342; Mayer 1983, 102-103, line 342.

[253] If this helmet is a forgery, the forger probably knew this text, or his work was based on this passage. U. Seidl did not mention it.

[254] This is the reason, why the corroded central scene of helmet **Cat. no. 7** is so unfamiliar for us.

[255] Similar features are known from Urartian art (sic!). For the chalice see the Urartian belt mentioned above

[256] *See* pp. 35, 43.

[257] *See* pp. 31, 43.

The incised figurative scene of helmet **Cat. no. 7** is unique as well. Independent elements of the scene ("praying" king in *ubāna tarāṣu* position[258], triangle as a mountain, winged sun-disk) are known from Assyrian and Near Eastern art, but this combination is unique. The composition is a little bit superficial: the position of the mountain-triangle is asymmetrical, secondary motifs (punched rosettes and plants) fill the place of main figures. The whole scene looks secondary and fills a space which was significantly reduced by the loss of the lower part of the helmet (at least two horizontal ribs were cut down with the base of the helmet). It is possible that this figurative scene was incised after the mutilation of the helmet. Muscarella expressed his objections, and termed the scene "as a modern creation"[259].

For the origin of helmets **Cat. nos. 6-7** – if they are originals – one probable possibility is the provincial origin. It is known that similar, unofficial elements might appear in the provincial or local tradition, where the rules of representation are not so strict, new, local elements may appear together with well known motifs of the official art of the Assyrian palaces, and the canonical order of iconographical system may loosen. The Til-Barsip frescoes[260] represent for example such an alternative, North Syrian (Syro-Assyrian) artistic tradition.

However it may be, the originality of helmets **Cat. nos. 6-7** can only be proved convincingly by a metallographic examination, the examination of their incision and corrosion history. The arguments and counter-arguments concerning their iconography, technique and quality of their execution are not conclusive.

3. Provenance unknown (said to come from Luristan), Group no. **A.1.2.1** – H.: 29.5 cm, Diam.: 22.5 cm. Conical helmet with repoussé and traced decoration. There are three embossed ribs around the base above which, on the front of the helmet there are three pairs of "crooks" and a vertical rib. A fourth, uppermost horizontal rib is interrupted by the "crooks". The innermost pair of "crooks" has rounded extremities, both the other pairs and the vertical rib end in animal (ram) heads. Just behind the heads the fleece is depicted. Between the embossed "crooks" there are 21-22 punched rosettes. There is an incised/chased figurative scene on the front and the side of the helmet. Between the "crooks" there is a symmetrical scene of a tree-of-life under a winged sun-disk, and on either side of the tree the king with his attendant behind him. The king wears the Assyrian royal crown (without tassels) and a court dress (*Schalgewand* Nr. 1) and raises his right hand in *ubāna tarāṣu* position. The attendant behind him holds a mace, a bow and a quiver. On the sides of the helmet – between the two upper ribs – there is a symmetrical scene of 2 x 8 armed and unarmed attendants/officials in procession. Complete (*Pls. 3-4*). 9th century B.C. – Venice, Maria and Giancarlo Ligabue Collection. – Salvatori 1975, 255-264, figs. 1-2, pls. 36-38; Ligabue – Salvatori 1977, 7-8; Overlaet 1979, 56; Curtis 1979, vol. I, 184; Gröschel 1986, 73, note 226; Fales 1989, 218, colour plate on p. 219; Dezső – Curtis 1991, 122-124; Born – Seidl 1995, 48, Abb. 40; Dezső 1998, ills. 10-11.

4. Provenance unknown (said to come from Luristan), Group no. **A.1.2.2** – H.: 12 cm, W.: 18 cm. "Fragment of the frontal part of a conical helmet with part of the border preserved. The border decoration consists of a series (2-3 + 3) of embossed horizontal annular lines demitting a strip 2.3 cm high which is taken up by six figures of dignitaries armed with long swords proceeding to the left. In the upper part, in the middle of the scene (of which we have only the right half) stands a tree of life surmounted by a winged sun-disk. To the left of the tree of life there are still traces of the figure of a king. To the right of the tree there is again the figure of a king with right hand raised (this is the well-known scene of the "worship of the tree"). The king is followed by two officials armed with swords. To the right and above them there is preserved, though only partially, the customary "snake" decoration, i.e. three embossed cords which rise up parallel to each other then curve down inwards. Of the three cords only one is fully intact, namely the inner one, which terminates in an animal protome, perhaps a horse. The neck of the animal is adorned with a series of vertical lines, of which the first is formed of short horizontal strokes, the second, third and fourth by columns of dots; after these come two pairs of lines, the second pair bordered by a saw-tooth pattern. On the outer cord there is a discontinued narrow band worked in relief on the upper side only. The snakes are outlined with a row of dots made by a very finely pointed tool. The ornamentation is worked exclusively by chasing. The fragment in question, like other similar works such as the splendid helmet in the Ligabue collection, has no traces of engraving." (Ligabue – Salvatori 1977, 7-8). The king wears a ritual dress (*Schalgewand* Nr. 2), raises his right hand in *ubāna tarāṣu* position, and with his left holds a mace. (*Pl. 5*). 9th century B.C. – Venice, Private collection. – Ligabue – Salvatori 1977, 7-8, fig. 1; Gröschel 1986, 73, no. 3, note 227; Dezső – Curtis 1991, 122-124; Dezső 1998, ill. 12.

5. Provenance unknown (said to come from Luristan), Group no. **A.1.2.3** – H.: 6.1 cm. Small fragment from the front of a helmet decorated in *repoussé*. There is an arc ending in a snakehead. Under the arc, there is a figure of an Assyrian king with his attendant behind him The king wears the Assyrian crown (without tassels) and the court dress (*Schalgewand* Nr. 1), and raises his right hand in *ubāna tarāṣu* position The figure of the king is armed with a sword. (*Pl. 6*). 9th century B.C. – Tehran, Private collection. – Calmeyer 1975, 316, fig. 2; Ligabue – Salvatori 1977, 8; Overlaet 1979, 56, note 16; Gröschel 1986, 73, note 228; Dezső – Curtis 1991, 122-124; Dezső 1998, ill. 13.

6. Provenance unknown, Group no. **A.1.2.4** – H.: 31 cm, Diam.: 21.5-22 cm, W.: 875 g. Conical helmet with pointed top. There are three horizontal embossed ribs around the base, and the rim of the helmet is hammered out to form a fourth rib. There are two pairs of holes for the attachment of probably the chin-strap loops. 2.4 cm above the upper horizontal rib, there is a depression in the contour of the helmet forming a cylindrical band, which is a separate base for the inset conical upper part. This cylindrical

---

[258] The *ubāna tarāṣu* position ("stretching the forefinger") together with *appa labānu* position ("touching the nose" with an aromatic plant) is the canonical gesture of the representation of Assyrian kings on stelae, rock reliefs and palace reliefs. Börker-Klähn 1982, *passim*.

[259] Muscarella 2000, 186, no. 43.

[260] *See* the weaponry of Assyrian soldiers (for example the curved swords) and courtiers/bodyguards, and further details on the Til-Barsip frescoes. Thureau-Dangin – Dunand 1936, *passim*.

band is decorated by an incised figural frieze running around the helmet. On the front of the helmet above the ribs and the cylindrical band, there are three pairs of symmetrical *repoussé* "crooks" bending inward. The crooks end in animal (bovid) heads. The fleece of the animals is indicated by incised lines cutting each other diagonally. There is a vertical embossed rib running from the point of the helmet and ending between the "crooks". This rib is terminating in a lion's head, mane of which is indicated by incised or chased lines. The outer "crooks" and the upper edge of the cylindrical band is bordered by the characteristic incised Assyrian rope motif. The same motif runs around the base of the helmet between the horizontal ribs. On the front of the helmet, between the "crooks" there is an incised figurative scene showing the "coronation" ceremony of the Assyrian king. In the middle of the five figures scene between two deities stands the king. Above the figure of the king there is a winged sun-disk. The king wears the ritual dress (*Schalgewand* Nr. 2). In his raised right hand he holds an unknown object. In front of him there is a male deity, probably Ashur himself, who, with his raised right hand blesses(?) the new king. In his left hand he holds an axe decorated with a ribbon(?). The deity wears a double-horned *polos* with a *fleur-de-lys* emblem on its top. In his left hand holds a bow, while with his raised right hand greets the procession of high-ranking Assyrian officials. Behind the figure of the king stands a female deity, perhaps the goddess Ishtar[261], who with her right hand places the Assyrian crown onto the head of the king. The crown has no ribbons. Ishtar wears a pair of quivers crossing each other on her back and terminating in stars. On her head there is a simple-horned *polos* with a star emblem on its top. The deities wear the same, traditional long dress opened at the front. The scene is bordered by two antithetic, four-winged griffin demons holding purificatory buckets and cones. The figurative scene on the cylindrical band consists of a continuous frieze running around the base of the helmet: the Assyrian king receives the submission and tribute of a western, perhaps maritime city. In the centre of the frieze, on the front of the helmet stands the Assyrian king wearing court dress (*Schalgewand* Nr. 1). In his left hand holds a bow, while with his raised right hand greets the procession of high-ranking Assyrian officials. Behind him there are two courtiers armed with sword. One of them holds a flag(?!). Behind them there is the royal tent with the tent-cloth rolled up at the entrance. A third courtier is coming out from the tent. In his raised left hand holds a chalice(?) or bowl, in his right hand carries a mace. Behind the tent there are three Assyrian soldiers: the first of them is equipped with conical helmet, rounded shield hanging on his shoulder, sword, bow, and quiver; the second is equipped with conical helmet, rounded shield hanging on his shoulder, sword, spear, and quiver; the third is equipped with conical helmet, rounded shield hanging on his shoulder, sword, bow, quiver, and spear. In front of the king there is an Assyrian soldier fallen on his knees before the king and kissing his feet. This soldier (high ranking officer?) wears a conical helmet and is armed with a sword. Behind him, in front of the king there is a high ranking Assyrian official leading the procession of Assyrian dignitaries and tribute bearers. His position, at the head of the column, is usually taken by the crown-prince, but in this case the characteristic tiara of the crown-prince (with two ribbons) is missing, and the beardless figure is probably a eunuch. He is armed with sword, and in his raised right hand holds a bunch of palm leaves(?) and blesses the king. Behind him there is a human head (probably the head of the defeated ruler of the conquered city) on the ground. Behind the head there is a procession of 9 Assyrian dignitaries armed with sword. Two of them are bearded, seven of them are beardless. Behind them there is a further beardless Assyrian official leading the procession of western tribute bearers wearing the characteristic headdresses, boots, and long robes. The first of them is probably their leader, the second carries a bow and a quiver; the third carries a situla and a mace-form object; the fourth carries a basket and a duck or goose; the fifth and sixth carry cauldrons; the seventh and eighth lead bulls; the ninth carries two tusks of elephant; the tenth carries a sack on his sholuder. Behind them there is the conquered city with a parapeted outer wall and gardens indicated by two palm trees. The strongly fortified inner city or acropolis has four high towers with windows/loop-holes, crenellated battlements projecting from the wall on their top. The high parapeted curtain-walls have an eclectic decoration imitating different types of parapets at half of walls height. Between the two left side towers there is a high vaulted gate with two wings of door. Complete. – 9th century B.C. – Berlin, Axel Guttmann Collection, AG 504 – Born – Seidl 1995, 11-47, 124-134, Abbs. 1-22, 96-102, Taf. II-III.

7. **Provenance unknown**, Group no. **A.1.2.5** – H.: 26.5 cm, Diam.: 21.5-22 cm, W.: 995 g. Conical helmet with pointed top decorated with *repoussé* and traced decoration. The base of the helmet for some reason (technical defect?) was mutilated in the antiquity: from the original three(?) embossed horizontal ribs running around the base of the helmet two were cut down, and there is only a single horizontal rib remained. On the front of the helmet two pairs of embossed "crooks" bending inward interrupt the horizontal rib. There is a vertical embossed rib running from the point of the helmet and ending between the "crooks". The vertical rib ends in a *repoussé* lion's head, while the "crooks" end in *repoussé* ram's head with traced details. The mane of the lion and the fleece of the rams are indicated by incised lines behind the heads. There is an incised figurative scene on the front of the helmet, between the "crooks". The symmetrical scene shows the Assyrian king on the sides of an asymmetrically placed triangle, which – according to the traced scale decoration – symbolizes a mountain. Above the triangle there is a winged sun-disk. The king wears the Assyrian crown and the ritual dress (*Schalgewand* Nr. 2) and raises his right hand in *ubāna tarāṣu* position. Behind the figures of the king there is a punched plant design (a bush or a small tree) with additional punched rosettes. The surface of the helmet is worn, which might be the result of frequent polishing. In consequence of this abrasion, some of the incised details of the figurative scene are probably missing. On the sides of the helmet there are three holes for the fastening of earflaps. 9th – early 8th century B.C. – Berlin, Axel Guttmann Collection, AG 424 – Born – Seidl 1995, 48-50, Taf. IV-V.

Group A.1.3: "Assyro-Babylonian" conical bronze helmet (Cat. no. 8)

Besides the Nimrud iron helmet (**Cat. no. 10**) mentioned above and a bronze crested helmet (**Cat. no. 14**) catalogued in a separate group by the present writer – S.-G. Gröschel listed an Assyrianizing, unprovenanced conical bronze helmet (**Cat. no. 8**) in the group of Assyrian conical bronze helmets.

The decorative system of this helmet is similar to the decoration of 9th century B.C. Assyrian conical bronze helmets (**Cat. nos. 3-7**). The most important characteristic of the helmet's decoration is the 9th century B.C. Assyrian figurative scene around the base of the helmet. Parallels for this incised frieze (the king receives tribute from Western Syria or Phoenicia[262]) are well known from the palace-

[261] For the detailed discussion of the role of goddess Ishtar in Assyrian art, especially in helmet art *see* the chapter of Assyrian conical crested iron helmets, pp. 47-50.

[262] Wäfler 1975, 72ff., figs. 35-46: Israel; 83ff., fig. 57: Phoenicia; 121ff., fig. 64: Damascus; 163ff., fig. 93: Hattina; 200ff. figs. 120-121, 127: Bit-Adini; 225ff., fig. 143: Carchemish.

reliefs of Ashur-nasir-pal II (883—859 B.C.)[263], from the Rassam Obelisk[264], from the bronze reliefs of Ashur-nasir-pal II's Balawat Gates[265] and from the bronze reliefs of the gates of the Northwest Palace, Nimrud[266]. This scene remained in frequent use under his successor, Shalmaneser III (858—824 B.C.) and is shown on the bronze reliefs of the Temple of Adad in Ashur[267] and of the Balawat Gates[268], on the Black Obelisk[269], on the throne-base of Shalmaneser III[270] and on ivory plaques from Nimrud[271]. A similar scene appears on an Assyrian conical bronze helmet (**Cat. no. 6**), and on an Assyrian conical crested helmet (**Cat. no. 14**, *Pls. 20-23*). The central scene (king adoring the tree-of-life under a winged sun-disk) is also well known from Assyrian art (*see* on p. 26). It is sure that this scene was the canonical central scene of 9th century B.C. Assyrian helmets: **Cat. no. 3** (*Pls. 3-4*), **Cat. no. 4** (*Pl. 5*), **Cat. no. 5** (*Pl. 6*); **Cat. no. 11** (*Pls. 12-16*).

On the basis of the technique of its decoration (*repoussé* outlines and incised details) some scholars attributed this helmet to Assyria[272], while other scholars emphasize the provincial execution of its figurative scenes and identify it as Urartian manufacture[273]. S. Eichler set the genii and demons, which can be seen on the helmet into the Urartian iconographical tradition[274]. But the goat-headed, winged, fish-bodied demon ("goat-fish") of the helmet with hoofed foreleg, forked fishtail and dorsal fin differs from that of Urartian iconography[275]. A similar demon of Urartian origin has a goat-head as well, but it has a body of a bird (and not of a fish), lion legs and shoots with bow and arrow[276]. Exact parallels for the demon on the helmet (considered to be a generally magically protective type[277]) can be found in Mesopotamian art: on an Old-Babylonian cylinder seal[278], on a Kassite *kudurru*[279], on Neo-Babylonian seals[280] and cylinder seals[281], and in Neo-Assyrian sun-dried clay figurines of the demon[282]. So there is no reason to attribute the iconographic details of the figures of the helmet to the Urartian tradition. A similar bird-demon but with lion's head and legs carrying a situla appears on a horse's breastplate[283], which has already been discussed. This breastplate can be connected to the helmet not only by the demon but by the rare scale-and-tassel/fringe border motif, which appears on the helmet and on the breastplate as well. The horse's breastplate, also unprovenanced, can be dated to the late 9th century B.C. with the help of its representations of hemispherical crested and conical helmets with vertical fin and "Hasanlu earflaps". Both objects may have been in the same group of workshops.

A similar high relief lion's head is known from other territories of the Near East (*see* for example the Hasanlu bronze belt buckles[284]). It of course does not mean that this helmet was made in Hasanlu. However, it emphasizes the "international" character of the different details of the helmet. Further *repoussé* lion heads are known from the vertical ribs of the two Axel Guttmann helmets (**Cat. nos. 6-7**) as well.

As can be seen, the decorative system of the helmet dates it to the 9th century B.C. But the separate motifs of its decoration can be derived not only from the Assyrian, but from the Babylonian tradition as well. As has been mentioned, the figure of the "goat-fish"[285] for example can be derived from the Old Babylonian and Kassite Periods. Furthermore there are representations, which prove the existence of a Babylonian helmet tradition – probably derivative from the Assyrian helmet tradition – at least from the reign of Shalmaneser III (858—824 B.C.). As shown on the throne base (panel f) of

---

263 Budge 1914, pl. XXVIII; Barnett – Falkner 1962, pl. CXXIV.
264 London, British Museum, WA 90925 + 118800 + 132013 + 136897-8; Börker-Klähn 1982, 183-184, no. 138, pls. 138a-t.
265 London, British Museum, WA 124687-124700: 1st, 2nd and 6th registers, WA 124685-124698: 1st, 4th registers.
266 London, British Museum, WA N 2063-2065, upper register.
267 Andrae 1977², fig. 195.
268 King 1915, pls. XIII, XIV, XXII, XXV-XXVIII, XXXI-XXXIV, XL, XLII-XLV.
269 London, British Museum, WA 118885, 31st year of Shalmaneser III (827 B.C.); Börker-Klähn 1982, 190-191, no. 152, pls. 152a: 2nd and 5th registers; 152b: 2nd and 5th registers; 152c: 2nd registers; 152d: 1st, 2nd and 5th registers.
270 Baghdad, Iraq Museum, ND 11000; Orthmann 1975, fig. 209b.
271 Mallowan – Davies 1970, nos. 67-69, 71-78, 81-85, 96d-e, 100-102.
272 Gröschel 1986, 74; Calmeyer 1986, 81. P. Calmeyer later included it in the Jerusalem Urartian exhibition catalogue (Calmeyer 1991a, no. 11) but did not exclude its Assyrian origin and emphasized its Assyrian style decoration.
273 Vanden Berghe – De Meyer 1982, no. 22, fig. 33; Maass 1987, 69.
274 Eichler 1984, 45, 2.4.1.2, 47, 2.7.1.
275 Eichler 1984, 47, 2.7.1.
276 Eichler 1984, 47, 2.7.1.
277 Black – Green 1992, 93, *s.v.* "goat-fish".
278 Porada 1948, nos. 366, 367E.
279 Moortgat 1969, fig. 229. Kudurru of Melishishu II from Susa.
280 Porada 1948, no. 803E.
281 Porada 1948, nos. 783, 788.
282 Black – Green 1992, fig. 70.
283 Seidl 1986, 229-236, figs. 1-4.
284 Winter 1980, figs. 70a-b (HAS 58-450), 71a-b (HAS 58-244).
285 For the discussion of the "goat-fish" *see* Black – Green 1992, 93, figs. 7, 70, 90.

Shalmaneser III from Nimrud[286], the tribute bearers of Mushallim-Marduk son of Ukani and of Adini, son of Dakuri, Chaldaean chiefs from Babylonia, bring a conical helmet of Assyrian type (*Chart 2, no. 89*). The high conical shape of this helmet is similar to that of helmet **Cat. no. 8**, but the iconographical details of its decoration ("goat-fish") mentioned above could indicate the possible Babylonian origin of this helmet. Unfortunately there is no pictorial or written evidence available on the helmet tradition of 8th century B.C. Babylonia under Assyrian rule. But on a cylinder seal of Ashur-nadin-shumi, the short-lived Assyrian king of Babylon (700—695 B.C.) the ruler wears a conical helmet with separately added high point and decorated entirely not only with horizontal ribs, but also with rows of rosettes between them (*Chart 2, no. 226*)[287]. It seems that a helmet with a hammered conical body and a solid point riveted to it on the top was the typical helmet or royal headgear in Babylonia in the 7th and 6th centuries B.C. It can be seen on the rock relief of Nebuchadrezzar II in Lebanon (586 B.C.)[288], on the stelae of Nabonidus[289], on two stelae from Harran[290], from Tema oasis[291] and the rock relief of an unknown Neo-Babylonian king: Gebel Akrum (Lebanon)[292].

In Muscarella's opinion helmet **Cat no. 8** is a forgery[293]. He based his critique on the unskillful execution of mixed motifs. However, he questiones the originality of only the incised scene and not of the helmet again. It looks that the incised figurative frieze is an integral part of the decorative system, and without this decoration the helmet would be a „halfmade piece". The construction of the *repoussé* decoration differs from the other conical bronze helmets. The three pairs of thin ribs curving inward are placed on three horizontal strengthening ribs. The elaborate *repoussé* (high relief) lion on the front of the helmet bears wittness to artistic ambitions. This lion without incised figurative design would be a superabundantly luxurious feature. It would be similarly luxurious for a forgery as well. In this case the incised details of the *repoussé* decoration (curving bands terminating in animal's heads and the lion) – the originality of which was not questioned by Muscarella – would help us solve the problem of its corrosion history.

8. **Provenance unknown** (said to come from "Urartu, East Turkey"), Group no. **A.1.3.1** – H.: 30.5 cm, Diam.: 20.8-23.2 cm. Conical helmet with pointed top. There are three horizontal ribs around the base of the helmet. There are three pairs of curves (arcs) ending in animals' heads (there is an incised decoration of the animals' skin or mane behind the heads) on the front of the helmet. There is a vertical rib on the front of the helmet ending between the curves in a "human-handed lion head" in *repoussé*. There is a continuous frieze (interrupted only by the curves) around the helmet bordered by scale-and-tassel motifs (above and below). The frieze contains figural decoration. There is a symmetrical scene on the front of the helmet, in three registers between the arcs. There are a "column-headed" tree-of-life and a winged sun-disk with a rosette above the tree, in the middle of the scene. On either side of the tree there is an adoring king with his two attendants, an eagle-headed winged genius and a bearded, winged, helmeted genius behind him. The king wears the ritual dress (*Schalgewand* Nr. 2), while the genii wear "*Schalgewand* Nr. 1". The king raises his right hand in *ubāna tarāṣu* position, and with his left hand holds the ribbon(?) of the winged sun-disk. The attendants are armed with maces, bows and quivers. One of them wears a conical helmet. Above them (in the second register) there are four birds on either side of the winged sun-disk. Above them, below the curves, there are two animals with goat's head and forelegs and fish-body. There are two continuous procession scenes around the helmet, moving from the front (from the arcs) to the back of the helmet, meeting in the middle (approx. ?). Each procession is led by the king himself with his attendants and, on the left side of the helmet, with tribute-bearers as well behind him. The seven left side officials following the king are armed with maces, bows and swords (5), with helmet and rounded shield (1) and one of them (the last in the row) wears a kind of cultic dress. On the right side, behind the king and three of his officials, there are 9 (probably western) tribute-bearers, bringing tribute (two cauldrons, one mixing vessel?, and four oxhide-shaped moulds?). There is a zigzag motif above the arcs. Complete (*Pls. 7-8*). (9th century – early 8th century B.C. – Schaffhausen, Museum Allerheiligen (formerly: Les Arcs, Ebnöther Collection; Galvano-Copy: Munich, Archäologische Staatssammlung, Neg no. 26-27). – Vanden Berghe – De Meyer 1982, 78-79, 130, no. 22 (fig.), fig. 33; Eichler 1984, 19, 39, 59, 61, pls. 9:2, 10; Gröschel 1986, 74, notes 234-237; Calmeyer 1986, 81, note 14; Esaian 1986, 27, pls. 13:5, 17; Maass 1987, 69, note 23; Magen 1986, pl. 25:6; Egg – Waurick 1990, pl. 6:2; Calmeyer 1991a, no. 11.

Group A.1.4: Assyrian conical composite bronze and iron helmet (Cat. no. 9)

This group consists of a single, very curious helmet. The helmet was made of two identical part[294] – and iron and a bronze half – which were riveted together[295]. The technique – riveting together two

[286] Mallowan 1966, 446-449.
[287] Collon 1988, 129-130, no. 555.
[288] Börker-Klähn 1982, 228, nos. 259-260, pls. 259-260.
[289] From Babylon, BM 90837; Börker-Klähn 1982, 230-231, no. 266, pl. 266.
[290] Urfa Museum; Börker-Klähn 1982, 229-230, nos. 263-264, pls. 263a-b, 264a-b.
[291] Louvre, AO 1509; Börker-Klähn 1982, 230, no. 265, pl. 265.
[292] Börker-Klähn 1982, 231-232, no. 269, pl. 269.
[293] Muscarella 2000, 150, no. 18. However, on this helmet there is not a single Urartian element, so Muscarella should have discussed this piece in the chapter of Assyrian forgeries.
[294] For the perfect join both halves were probably hammered into the same negative mould or were hammered onto a wooden model; Born – Seidl 1995, 144.
[295] For the very detailed study of the technique of this helmet *see* H. Born's article: Born – Seidl 1995, 134-149, Abbs. 103-115, Taf. XIII-XV.

halves – is well known in the second half of 8th century Assyrian (**Cat. no. 11**), Urartian (**Cat. nos 104-106** and **107**) and Northwest Iranian (**Cat. no. 48**) iron helmet tradition[296]:

Its shape – terminating in a high, tubular point – is characteristic of the middle and second half of 8th century B.C. Assyrian and North Syrian helmets: **Cat. nos. 11, 32-34, 37** and **48**. Its best Assyrian parallels are shown on the palace reliefs of Tiglath-Pileser III (745—727 B.C.): *Chart 2, nos. 134-142.* However, its earliest parallels date to the gap, in which Assyrian helmet representations are scarce: similar helmet shapes with high tubular point are shown on Karatepe sculptures (*Chart 2, nos 175-176*) which may be contemporary with the sculptures of Tiglath-Pileser III or slightly earlier. Further North-Syrian parallels can be dated to the reign of Tiglath-Pileser III: a Hadatu (Arslan-Tash) relief (*Chart 2, no. 173*); a Tell Ta'yinat slab (*Chart 2, no. 181*) and the Til-Barsip frescoes (*Chart 2, nos. 163-172*).

This bimetallism (combination of bronze and iron) has probably no technical reason or structural necessity. The background might be purely artistic or ceremonial. The combination of bronze (its colour if it is polished is shining yellow or reddish yellow) and iron (its colour if it is simply polished and not coloured with technical methods[297] is shining silver) has a good tradition in Near Eastern (mainly Assyrian and North Syrian) helmet industry and art[298]. As has been mentioned, the Til-Barsip frescoes (*Chart 2, nos. 163-165*)[299] show helmets coloured light blue, as are the iron blades of the swords. These wall-paintings certainly depict iron helmets. The yellow decoration on two of them (*Chart 2, nos. 163-164*) might indicate the same bronze inlay decoration as on the Zinçirli (**Cat. no. 38**) and Nimrud iron helmets (**Cat. nos. 10, 16-31**). This bimetallism might originate from the ancient Mesopotamian colour harmony tradition of yellow – reddish yellow (gold, bronze, copper) and white – light blue (silver, polished iron)[300].

Unfortunately we do not even know how this helmet was worn. Whether the two colours were divided between the front and the back of the helmet (crosswise division), or the helmet was worn with the join at the front and the back of the helmet, forming a left and right side made of different metals (lengthwise division). In case of the few decorated iron helmets made of two halves riveted together (**Cat. nos. 11, 48**), the front view of the decorative system located obviously on one of the halves, so the join divided the helmets crosswise.

9. Provenance unknown, Group no. **A.1.4.1** – H.: 35.3 cm, Diam.: 20 cm, W.: 716 g (bronze half). Bronze half of a pointed helmet made of bronze and iron ("*Zweifarbenhelm*"). The helmet was a high pointed helmet with tubular point. The helmet was made of two similar parts, which were riveted together. This riveted joint was strengthened by two separately added iron bands on the outer side of the helmet. The long tubular point (L.: *ca.* 11.5 cm) was closed by an iron ring (Diam.: *ca.* 1.5 cm) and a similar bronze cap (Diam.: *ca.* 1.5 cm). From the helmet only the bronze half and corroded remains of the iron half (corroded to the bronze) exist. On the bronze half there are 18 holes around the base probably for the fastening of a lining. There are 15—15 rivet-holes along the edges of the bronze half, which were used – together with the parallel holes on the one-time overlapping part of the iron half – to fasten the two helmet halves together. This riveted joint was strengthened on the outer surface of the helmet by two, 0.7 cm wide iron bands (one on each side of the helmet) mentioned above. These iron bands were riveted to the overlapping bronze and iron halves by 5 rivets. These five rivet-holes are visible on the edge of the bronze half as half-holes (*see* H. Born's detailed description and reconstruction: Born – Seidl 1995, 134-149, Taf. XV). A sixth rivet fastened the upper end of this iron band to the closing iron ring at the top of the tubular point. Second half of 8th century B.C. – Berlin, Axel Guttmann Collection, AG 204 – Born – Seidl 1995, 50, 134-149, Abbs. 103-115, Taf. XIII-XV.

Group A.2: Assyrian conical iron helmets (Cat. nos. 10-11)

There are two helmets, which belong to this group. As J. Curtis showed in a recent article[301] helmet **Cat. no. 10** was mistakenly attributed to Nineveh by R.D. Barnett. When he published[302] the piece he identified it as one of the finds sent to the British Museum from Quyunjik in 1880 by Rassam. Barnett cited Samuel Birch, who referred to the arrival of a consignment of antiquities from Nineveh in July 1880 that included an "Assyrian bronze (*sic*!) helmet"[303]. Almost certainly however, Birch was referring to one of the three bronze and iron helmets ("Spangenhelme") of Parthian and Sassanian date

[296] For detailed study of the production techniques of Near Eastern iron helmets *see* the next chapter: Group A.2: Assyrian conical iron helmets, especially pp. 35-37.
[297] "*Ob die assyrischen Eisenhelme auch eisensichtig, d.h. silbrig auspoliert waren, oder aber tatsächlich 'bebläut' bzw. 'brüniert', d.h. durch Erhitzen (im Feuer bis ca. 400 °C) und nachfolgendem Abschrecken in Wasser oder Öl gefärbt wurden, muß allerdings wohl für immer ein altorientalisches Geheimnis bleiben.*" Born – Seidl 1995, 140.
[298] Dezső – Curtis 1991; Born – Seidl 1995, 134-149; Moorey 1999, 278-291, especially 285.
[299] Thureau-Dangin – Dunand 1936, pls. XLXI-LI.
[300] For the parallel gold and silver decoration *see* pp. 4-5 (3rd millennium B.C. Mesopotamian helmets) and pp. 99-101 (Group I.1: Iranian hemispherical bronze helmets).
[301] Dezső – Curtis 1991, 105-107.
[302] Barnett 1953, 101-102, pls. 31-32.
[303] Trustees of the British Museum, Original Papers 22 July 1880, p. 3228.

that arrived with this consignment, now in the British Museum[304]. The two other examples were mentioned by Birch in his annual report for 1877, when he says that "two helmets from Kouyunjik" were mounted in that year[305].

So it is very unlikely that our helmet (**Cat. no. 10**) comes from Nineveh. It is much more probable that this piece, together with 16 iron helmet fragments decorated with bronze inlay (**Cat. nos. 16-31**)[306], came from Layard's excavations in the Northwest Palace at Nimrud in 1845—1847. This is confirmed by Layard's own account. The find was made during his first campaign of excavations (1845—1847) in the autumn of 1846. Unfortunately Layard's original excavation reports for this period are missing, or at least are not amongst the Layard Papers in the British Library. However he mentioned his armour finds in one of his letters[307] and in his published account, *Nineveh and its Remains*. He describes his armour finds in Chamber I, Northwest Palace at Nimrud as follows:

"As the earth was removed, other portions of armour were found; some of copper, others of iron, and others of iron inlaid with copper. At length a perfect helmet (resembling in shape and in the ornaments the pointed helmet represented in the bas-reliefs) was discovered. When first separated from the earth it was perfect, but immediately fell to pieces. I carefully collected and preserved the fragments, which were sent to England. The lines, which are seen round the lower part of the pointed helmets in the sculptures, are thin strips of copper, inlaid in the iron.Several helmets of other shapes, some with the arched crest, were also uncovered; but they fell to pieces as soon as exposed; and I was able, with the greatest care, to gather up a few of the fragments which still held together; for the iron was in so complete a state of decomposition that it crumbled away on being touched.

Portions of armour in copper, and embossed, were also found, with small holes for nails round the edges."[308]

The conical iron helmet inlaid with bronze and sent to England as mentioned by Layard is without a doubt our helmet **Cat. no. 10**, while the other fragments mentioned above are those now in the British Museum under the registration numbers 48-11-4, 113-115 (**Cat. nos. 16-31**).

In Chamber I of the Northwest Palace, apart from the helmets and other pieces of armour, there were alabaster vessels (some of them with cuneiform inscriptions) and a glass vessel with an inscription of Sargon II (721—705 B.C.)[309]. Although Nimrud was not sacked until 614/612 B.C., the date of the debris in which these helmets were found, it ceased to be the capital of Assyria from Sargon II's reign so many 8th century B.C. objects survived there through the 7th century B.C.

The dating of the helmet to the reign of Sargon II (721—705 B.C.) is proved by the decorative system of the helmet, which is significantly similar to the wall-paintings of the throne-room of Sargon II in Palace K at Khorsabad[310] (*Fig. 2*). On the wall-painting between four parallel lines there are three figural friezes: winged genii, bulls, winged genii. On the uppermost painted line there is figurative scene with three figures inside an arched, stele-shaped border. In the middle there is the king (Sargon II), with one of his attendants behind him (right), adoring in front of a god (probably Ashur). The inner line of the border is decorated with the figures of winged genii, the central line is decorated with geometric design, the outer parallel line is decorated with "bud-and-garland" border similar to the border decoration of the helmet **Cat. no. 10**. The rim of the Zinçirli iron helmet (**Cat. no. 38**) was decorated with this well-known design inlaid with bronze. This motif appears on the front of the other Assyrian iron helmet (**Cat. no. 11**), as the *melammu* ("aura") border around the figure of Ishtar. The decoration running around the base of the Nimrud iron helmet (**Cat. no. 10**) and on the wall-painting, consists of four parallel lines, but on the helmet there is only one figural scene between the two upper (3rd and 4th) lines: the frieze of 23 officials in procession (*Fig. 2, Pl. 11*). On the front of the Nimrud iron helmet (**Cat. no. 10**) in a stele-like field there is a libation scene of three figures slightly different from those on the wall-painting: the figure of the king facing to the right is on the left side of the "stele"; the crown-prince facing the king is in the middle with one of his attendants behind him (right). The king holds a cup in his right hand, his left hand is raised. This scene is well-known from Assyrian art of the 9th century B.C.[311]. Since this scene was a canonical representation, it should have had without a doubt an exact parallel on one of the walls of Assyrian royal palaces from which the decorative system of the Nimrud iron helmet (**Cat. no. 10**) originated. Its appearance on this helmet

---

304 BM 22495, BM 22497, BM 22498.
305 Trustees of the British Museum, Original Papers 19 February 1878, p. 1724.
306 London, British Museum, 48-11-4, 113-115.
307 To Rev. Josiah Forshall, on 2nd November 1846; Original Letters and Papers, vol. XXXVI (Nov. 1846 – Feb. 1847).
308 Layard 1850, vol. I, 341-342.
309 "Sargon Vase", London, British Museum, BM 90952; Barag 1985, no. 26, fig. 2, pl. 3, col. pl. B.
310 Moortgat 1959, 16, fig. 6; Heinrich 1984, fig. 101; Lloyd – Müller 1986, 43, fig. 81.
311 *See* for example on ivory carvings: Mallowan – Davies 1970, no. 1 (ND 1082, IM 56344), no. 2 (ND 4251, BM 132155); or instead of the cup with a bow in the hand of the king: Barnett – Falkner 1962, pls. LXXXIV-LXXXV; Rassam Obelisk (BM 90925 + 118800 + 132013 + 136897-98): Börker-Klähn 1982, pl. 138e.

shows how their decorative system was not random or casual, it was based upon strict rules and observed basic principles of composition with a specific purpose (*i.e.* to indicate a unit or special service of the Assyrian army; military rank, *etc.*; *see* the chapter on Nimrud crested iron helmet fragments).

As can be seen the figurative frieze (officials in procession) around the base of helmets **Cat. nos. 10-11** is also one of the canonical motifs of Assyrian art (and helmets as well) since the 9th century B.C. (**Cat. nos. 3-4**, (**5**), and partly **Cat. no. 6**). The frieze of officials in procession between the two upper inlaid ribs (**Cat. no. 10**) and row of punched circles (**Cat. no. 11**) fits into the canonical 9th century B.C. Assyrian helmet tradition represented by helmets **Cat. nos. 3, 4, 6** (*Pls. 3-5*) discussed above, which tradition survived in the 8th century B.C. (**Cat. no. 10**, *Pls. 9-11*) and probably continued to develop in the 7th century B.C. Assyria as well.

The decorative system of helmet **Cat. no. 11** can be derived from the monumental arts of the Assyrian royal palaces, temples and public buildings as was proved in the discussion of the Nimrud Iron helmet **Cat. no. 10** (*Pls. 9-11*, especially *Fig. 2*). The only difference in the decoration of helmets **Cat. nos. 10** and **11** is the front motif above the four parallel lines (**Cat. no. 10**) or four parallel rows of punched circles (**Cat. no. 11**) around the base of these helmets.

The common origin of their decorative systems is proved by small details and motifs, as the "buds" from the "bud-and-garland" motif framing the "stele-field" on the front of the Nimrud iron helmet (**Cat. no. 10**)[312], which were used to represent the shining beams of the circular *melammu* ("aura", "nimbus", "*Strahlenkranz*") framing the figure of Ishtar on the front of the Munich Assyrian iron helmet (**Cat. no. 11**, *Pls. 12-16*). The main motif on the front of helmet **Cat. no. 11** differs from the front decoration of helmet **Cat. no. 10** in its phrasing and composition, but not in its concept and meaning. The figure of Ishtar has the same function on the front of helmet **Cat. no. 11** as stated above in case of helmet **Cat. no. 10**: it was based upon strict rules and observed basic principles of composition with a specific purpose (*i.e.* to indicate a unit or special service of the Assyrian army; military rank, *etc.*). The other function of the appearance of the figure of a deity on the front of an Assyrian helmet is the common apotropaic function, which might probably be the primary reason for all the front decoration on Iron Age Near Eastern helmets[313].

The figure of Ishtar with a *melammu* framing her bust is well-known from Assyrian art[314]. This representation of Ishtar shows one of the standard depictions of the deity[315]. Two of her attributes: the "nimbus" and the star of Ishtar are visible on the helmet.

Following H.-J. Kellner's explanation, the figure of the "eagle with aura(?)" carved secondarily into the front of helmet **Cat. no. 11** might be a sign, which indicated that this helmet was booty seized by an Urartian soldier from the Assyrians[316].

The technique of the helmet's construction and decoration looks more archaic than the Nimrud iron helmet **Cat. no. 10**. In spite of this fact, it cannot be undoubtedly declared that this helmet was made earlier than the Nimrud iron helmet. The technique of its construction (riveted together from two similar halves[317] or the two edges of a single sheet were joined by row of rivets) is not simply archaic, but cheaper and simpler than the construction of Nimrud iron helmet **Cat. no. 10**. In spite of this it is known that iron helmets of different construction and decorative techniques were used and manufactured together at the latest from the middle of the 8th century B.C. The emergence of iron helmets can be dated to the first half of the 8th century B.C.[318] and the first iron helmets decorated with bronze inlay should have appeared at latest in the middle of the century (Til-Barsip wall-paintings, *Chart 2, nos. 163-165*). The earliest dated example of iron helmets decorated with bronze inlay is from a destruction level of Zinçirli associated with the siege of the town by Sargon II in 720 B.C. (**Cat. no. 38**, *Pl. 57*) and the Nimrud iron helmet (**Cat. no. 10**) discussed above. These are the earliest known iron helmets in the world. In spite of their early date, the method of manufacture of both helmets is technically perfect. Not only because of the high level of bronze inlay of the Zinçirli helmet (**Cat. no. 38**, possibly the whole surface of the helmet was decorated with elaborate geometric motifs of bronze inlay, *Pl. 57*), but also because the Nimrud iron helmet (**Cat. no. 10**, *Pls. 9-11*) undoubtedly, and the

[312] And appear around the base of the Zinçirli iron helmet fragments (**Cat. no. 38**, *Pl. 57*).
[313] For the different functions of the front decoration *see* the chapter of conical crested iron helmet fragments, pp. 47-50.
[314] *See* for example on several cylinder seals; Delaporte 1910, no. 524; Jakob-Rost 1975, nos. 197 (VA 2153), 198 (VA 2529), 199 (VA Ass. 2463-12714).
[315] For the representation of Ishtar of Arba'il and for the representation of the different "kinds of *melammus*" *see* the chapter of Nimrud crested iron helmets (pp. 47-50), especially **Cat. no. 16** (*Pls. 25, 39, 41*). For the typology of the representations of Ishtar armed *see Fig. 5*.
[316] Kellner 1993, 325-331.
[317] For the perfect join both halves were probably hammered into the same negative mould or were hammered onto a wooden model
[318] On the 9th century B.C. (Ashur-nasir-pal II or Shalmaneser III) glazed brick fragment from Ashur (Ass. 10756, Andrae 1925, pl. 9e) only the scale armour attached to the rim of the helmet is iron (painted blue), the helmet itself is bronze (painted yellow).

Zinçirli iron helmet probably, was hammered out of one piece of metal. The absence of iron helmets made out of one piece of metal outside Assyria or North Syria emphasizes the high technical skill of ironsmiths there[319].

There is only one known Urartian iron helmet fragment, the skull-piece of which supposed to be hammered out of one piece of metal (**Cat. no. 108**; Karmir Blur; Erevan, Historical Museum of Armenia, *Pl. 108*)[320]. But as can be observed, the contour of the helmet is sharply broken at the base of the short point of the helmet. This short point is probably a solid iron bar, which was separately added to the skull-piece of the helmet, possibly hammered out of one piece of iron[321]. So this fragment can be excluded from the group of helmets hammered out from one piece of iron. Besides this helmet fragment there are three iron helmets and a few fragments known from Urartu:

1. **Cat. no. 104** (Munich, Archäologische Staatssammlung, PS 1971/2022, *Pl. 105*)[322]
2. **Cat. no. 105** (Adana Regional Museum)[323]
3. **Cat. no. 106** (Adana Regional Museum)[324]
4. **Cat. no. 107** (Munich, Archäologische Staatssammlung, PS 1971/1997, *Pls. 106-107*)[325]

All three helmets (and the helmet of fragments **Cat. no. 106**) were made by the same technique: they were made of two similar parts (two halves) joined by iron strips on the inside fastened together with iron rivets (*Pl. 107*). The point of one of the three helmets (**Cat. no. 107**, *Pls. 106-107*) is similar to the solid point of the Karmir Blur iron helmet fragment: both of them were separately added to the skull-piece of the helmet made of one piece (**Cat. no. 108**, *Pl. 108*) or two halves (**Cat. no. 107**, *Pl. 107*) of iron sheet.

The best example of an Assyrianizing iron helmet, decorated with bronze inlay, is a pointed helmet "turned up from Luristan" (**Cat. no. 48**, Bruxelles, Musées Royaux d'Art et d'Histoire, Inv. no.: IR 1241, from "Amlash", *Pls. 65-66*)[326]. The helmet, like those from Urartu mentioned above, was made of two similar parts joined by iron strips on the inside and riveted together. The decorative system of the helmet follows the decoration of the 9th century B.C. Assyrian, and 8th century B.C. Urartian, Northwest Iranian ("Luristan", "Kurdistan") and Hasanlu conical helmets. The only significant difference is that the bronze inlay decoration of the Assyrian iron helmets – maybe for lack of the necessary technical skill[327] – was replaced by riveting of the motifs cut out from a bronze sheet to the iron by bronze rivets having iron roundels at the back. There is a horizontal bronze strip running around the base of the helmet to which it was fastened by four rivets. There were at least two pairs of curves (one remained from each pair) riveted to the front of the helmet with a vertical bronze strip between them, from which only the rivet-holes survive. There was a row of vertically placed rectangular bronze plates running around the helmet above the horizontal bronze strip. Each of them was fastened to the iron with bronze rivets and they may imitate a frieze of figures (officials in procession?) similar to those on Assyrian helmets (*Pls.4-6, 11, 16*)

On the basis of the iron helmets catalogued here it is possible to arrange the surviving Near Eastern helmets of the Iron Age into seven groups on the basis of their production technique:

1. Helmets riveted together from two halves (**Cat. nos. 104, 105, 106**)

[319] As will be discussed below, of course not only helmets hammered out of one piece of iron were in use in Assyria. As the fragments of Nimrud crested iron helmets (**Cat. nos. 16-31**, *Pls. 25-43*) indicate such a complicated shape (as reconstructed on *Pls. 39, 41-43*) could not have been hammered out of one piece of iron. The iron fragment **Cat. no. 17** may have been either a conical or a crested helmet as well (*Pls. 26, 40*). It consisted of a wide iron band and an upper conical or crested skull-piece riveted on to the lower band. The manufacture of iron helmets made of pieces riveted together was of course a simpler and cheaper way of production. Nevertheless it should be realized that this conical Nimrud iron helmet (**Cat. no. 10**), hammered out of one piece of metal, was not a unique, or rare parade or ceremonial helmet, but a common helmet of everyday use, mass produced.

[320] Esaian 1986, 33, pl. 20:2.

[321] This shape of helmets is very similar to a special Neo-Babylonian (?) helmet tradition known from the time of the Assyrian Empire (*Chart 2, no. 226*, Cylinder seal of Ashur-nadin-shumi, Assyrian king of Babylon 700—695 B.C.) and later, worn by Neo-Babylonian rulers.

[322] Kellner 1976, 78, Cat. no. 137; Fuchs – Kellner 1978, 386, pl. 2; Kellner 1979, 54, note 8b; Curtis 1979, vol. I, 182; Kellner 1980, 210, note 17; Özgen 1982, 68, note 79; Vanden Berghe – De Meyer 1982, 133, no. 26 (fig.); Esaian 1986, 33, pl. 13:11.

[323] Kellner 1979, 156, note 6; Overlaet 1979, 54, note 9.

[324] Kellner 1979, 156, note 6; Overlaet 1979, 54, note 9.

[325] Kellner 1976, 78, Cat. no. 137a; Kellner 1979, 152, fig. B; Overlaet 1979, 54, note 8a; Curtis 1979, vol. I, 182; Kellner 1980, 210, note 17; Özgen 1982, 68, note 79; Vanden Berghe – De Meyer 1982, 133, no. 25 (fig.); Esaian 1986, 33, pl. 13:10; Wartke 1991, 330, fig. 4; Dezső – Curtis 1991, 125-126.

[326] van Pée 1965, 196-198, pl. on p. 196; Vanden Berghe 1967, 22; Overlaet 1979, 53-55, fig. 1, pls. 1-2.

[327] There are no iron objects decorated with bronze inlay known from outside Assyria and North Syria from this period.

2. Helmet riveted together from two halves and decorated with punched and engraved motifs **(Cat. no. 11)**
3. Helmet riveted together from two halves with riveted bronze decoration (**Cat. no. 48**)
4. Helmet riveted together from two halves and a solid point (**Cat. no. 107**)
5. Helmet made of a solid point fastened to a conical skull-piece hammered out of one piece of iron (**Cat. no. 108**)
6. Helmet made of a wide band on which the skull-piece hammered out of one piece of iron was riveted (**Cat. no. 17, (16)**)
7. Helmets hammered out of one piece of iron (**Cat. nos. 10**, and probably **38**)

10. Nimrud, Northwest Palace, Group no. **A.2.1** – H.: 30.8 cm, Diam.: 21.7 cm, W.: 3049 g. It has a pointed top and a shape which is almost conical. It was hammered out from one piece of iron. There is a series of holes around the base for the attachment probably of a lining, but possibly of a hauberk of armour scales. The helmet has been decorated with bronze inlay which takes the form of four parallel lines and an elaborate design (above the uppermost line) at the front: a king (on the left) holding a cup in his hand, other hand raised; the crown-prince (in the middle) in front of him with his hand raised (other hand and head are missing); and a fragmentary figure of an attendant (on the right) behind the crown-prince facing the king. One of his hand is raised. The scene was framed by a bud-and-garland border. Between the two upper inlaid lines there is a frieze around the helmet of 23 officials or attendants in procession inlaid with bronze. The helmet is restored from fragments (*Pls. 9-11*). Third quarter of 8th – first quarter of 7th centuries B.C. – London, British Museum, BM 22496. – Barnett 1953, 101-102, pls. 31-32; Calmeyer 1969, 90, note 295; Borchhardt 1972, 99, Cat. no. 22 A I; Pleiner – Bjorkman 1974, 291, fig. 5; Calmeyer 1975, 316; Ligabue – Salvatori 1977, 8; Gamber 1978, fig. 196; Overlaet 1979, 54, note 6; Curtis 1979, vol. I, 178-181, vol. II, 32, pl. 27; Curtis, *et al*. 1979, 385, fig. 34; Horn – Rüger 1979, 313, fig. 187; Özgen 1982, 20, 51, note 3; Gröschel 1986, 73, note 225; Dezső – Curtis 1991, 105-107, 122-126, fig. 21, pl. 19.

11. Provenance unknown (said to come from "Urartu, East Turkey"), Group no. **A.2.2** – H.: ca. 36 cm. Fragmentary iron helmet. It has a conical shape. It was hammered out from two similar halves, which were joined by two vertically placed, *ca.* 2 cm wide iron bands with a row of rivets on the sides of the helmet. The vertical row of rivet holes following the edge of the iron sheet still exists. There are no traces of holes around the rim of the helmet for fastening the lining. The helmet has been decorated with punched and engraved motifs. There are four parallel rows of small punched circles with a punched point in the middle of every circle around the base of the helmet. Between the two upper rows of circles there is an engraved frieze of officials in procession (from the right to the left). There were 18 figures around the helmet from which only five remained visible on the fragment. The height of the figures is 19 mm. On the front of the helmet, above the four parallel rows of circles there is a punched and engraved figurative design showing the figure of Ishtar within a circular border: This circular border motif (a sun-disk or melammu) consists of a row of punched circles with 14 buds around the circle and with a star symbol on the top of this bordering circle. Within this border there is the engraved figure of a standing female deity – probably of Ishtar – with right hand raised. Ishtar wears a long dress and the upper part her figure's head is missing. She wore probably no headdress, only her star is visible above her head, outside the circular border. There is a *melammu* (shining circle, "aura" or "radiance") around the trunk with four "rays" behind and in front of the deity. There is a secondary used motif crossing the upper punched row of circles on the front of the helmet. It consists of a bird (eagle?) head with similarly depicted "shining aura" around its head. The body of the eagle is marked similarly with lines (*Pls. 12-16*). (Third quarter of 8th – first quarter of 7th centuries B.C.) – Munich, Archäologische Staatssammlung, PS. 1980/6227-6228. – Kellner 1993, 325-331.

Group A.3: Assyrian conical crested bronze helmets (Cat. nos. 12-15)

The earliest known crested helmets of the Iron Age are those shown on the sculptures of Carchemish (*Charts 3-4, nos. 101-104*)[328]. The reliefs depicting crested helmets are from the reign of Katuwash (*ca.* 900—873 B.C.)[329], king of Carchemish, a contemporary of Ashur-nasir-pal II (883—859 B.C.). Their characteristics are the almost conical shape with a small crest at the top decorated with tassels dangling on the back. The first two helmets have a conical shape with three horizontal ribs around the base (*Charts 3-4, nos. 101-102*). There are no horizontal ribs on the third helmet, but on the front of it appears the small projection (a boss or knob – maybe a horn) characteristic of so many of the Carchemish helmets (*Charts 3-4, no. 103*). The fourth helmet (*Charts 3-4, no. 104*) shows the first signs of the later development of conical crested helmets with its point curving forward ("Phrygian shape"). The shape (almost conical) and construction of helmets (small crest at the top and tassels on the back) depicted on the sculptures follow the tradition of Late Bronze Age Hittite and North Syrian helmets known from the sculptures of the "King's Gate" at Boğazköy (*Chart 1, no. 37*)[330] and from Egyptian wall-paintings depicting Syrians, *e.g.* from the Tomb of Amenmose at Thebes (*Chart 1, nos. 20-23*)[331]. D. Rittig traced back the origin of the conical crested helmets ("*Raupenhelm*") to the same area[332]. This Late Bronze Age Hittite – North Syrian helmet tradition reached Mesopotamia through

328 Hogarth 1914, pls. B.2a, B.2b; Woolley 1921, B.26c; Woolley – Barnett 1952, pls. B.44, B.45, B.46.
329 Woolley – Barnett 1952, 245; Orthmann 1971, 29ff.; Hawkins 1976-1980, 439-441.
330 Akurgal 1961, pls. 64-65.
331 de G. Davies 1933, 29, pls. XXXIV-XXXV.
332 Rittig traced back the origin of the "*Raupenhelm*" tradition to the 17th century B.C. Hurrian-speaking communities of Alalah, Kizzuwatna and Mitanni, from where it was transmitted to the Hittite Empire. The "*Raupenhelm*"-tradition in Central-Anatolia

North Syrian Neo-Hittite states (*e.g.* Carchemish) in the second half of the 9th century B.C. The same "Carchemish-type" of crested helmet is shown on a similar orthostat slab from Kültepe (?) from the 9th century B.C. (*Charts 3-4, no. 105*)[333]. The difference between that piece and those on the Carchemish reliefs is that the Kültepe helmet has two projections (bosses) on the front, and its crest is on the back of the helmet. What may be variants of the "Carchemish-type" of crested helmet appear on the early 8th century B.C. Zinçirli orthostat reliefs (*Charts 3-4, nos. 128-129*).

This type of crested helmet would have been not only the predecessor of the Mesopotamian (Assyrian) crested helmets, but it was also widespread in late 9th – early 8th century B.C. Urartu, and from North Syria influenced the development of helmets in the East Mediterranean (apparently during the decline in the Dark Age), (*e.g.* Crete: bronze belts from Fortetsa[334]: *Chart 3, nos. M.3.1-2*; Kato-Symi: *Chart 3, no. M.3.3*).

On contemporary early 9th century B.C. palace reliefs of Ashur-nasir-pal II (883—859 B.C.) there are no representations of crested helmets at all. The first appearance of crested helmets in Neo-Assyrian art is on the Balawat gates of Shalmaneser III (858—824 B.C.) worn by soldiers identified as Urartian troops and as the troops of Sangara of Carchemish[335]. However, this type of **hemispherical crested helmet** with its round (hemispherical) cap (with or without earflaps) and with a crest fitted centrally across the top differs from the **conical type of crested helmet** of the Carchemish tradition. This hemispherical type of crested helmet originated in the 14th century B.C.[336] and it was widespread like the conical type of crested helmets in the territories north of Mesopotamia. Several representations of this type of helmet are known (first of all from Urartu or from other territories showing Urartian warriors) from the second half of the 9th century B.C.[337]. There are at least five examples of this type excavated at Hasanlu (Level IVB, Iron II Period, *ca.* 800 B.C., **Cat. nos. 52-56**, *Pls. 72-79*). These helmets or rather the Zinçirli chariot relief (dated to the first half of the 8th century B.C., *Charts 3-4, no. 127*) could have been the latest examples of the hemispherical type of crested helmets. But, as has already been mentioned, from the Balawat Gates of Shalmaneser III it is known that soldiers, said to be troops of Sangara, king of Carchemish (*ca.* 873—850 B.C.), sometimes wore this type.[338] If so, there was a change in the helmet tradition of Carchemish under Sangara. The early 9th century B.C. conical crested helmets of the sculptures of Katuwash (*ca.* 900—873 B.C.) were, possibly under Urartian influence, replaced with this hemispherical type of crested helmet, known from Hasanlu (**Cat. nos. 52-56**, *Pls. 72-79*). Finally this type of helmet was replaced in Urartu by conical/pointed helmets coming from Assyria, and by conical crested helmets spreading from North Syria in the late 9th century B.C.

The place of the development of later conical crested helmets – originating from the Late Bronze Age and early 9th century B.C. Carchemish crested helmets – is not likely to have been exclusively Carchemish. Other North Syrian states also possibly played an important role in the evolution of the conical type of crested helmets. On the basalt sculptures of Karatepe[339] there are local warriors wearing this type of crested helmet fighting with Assyrian (?) soldiers wearing pointed

---

(especially in Tabal) survived the collapse of the Hittite Empire, and during the early 1st millennium B.C. was transmitted to Assyria and Urartu. Rittig 1994, 247-254.

333 Bittel 1976b, fig. 321.

334 There are four helmets illustrated on the two "belts" from Fortetsa (Iraklion, Archaeological Museum, Inv. nos. 1568, 1569). These four helmets represent three helmet types the joint appearance of which is anachronistic or at least difficult to explain. Namely, two of the four helmets represent the **Carchemish-type of crested helmet** mentioned above (reign of Katuwash, *ca.* 900—873 B.C.; *Chart 3, nos. M.3.1-2*); the third helmet is an **"Urartian-type" of hemispherical crested helmet** known from the second half of the 9th century B.C. Urartu; Hasanlu, and disappeared in the early 8th century B.C. (Hasanlu): *Chart 3, no. M.7.1*; the fourth helmet however is a beautiful example of the developed **North Syrian and Assyrian conical crested helmet** (known from Urartu as well), (*Chart 3, no. M.4.1*). This type of helmet appeared in the last quarter of the 9th century – early 8th century B.C. (**Cat. nos. 13, 14, 113, 114**). While the first and second types appear together only in the middle of the 9th century B.C., the second and third occur together only at the end of the 9th – beginning of the 8th centuries B.C. So the four helmets on the two belts (dated to the second half of the 8th century B.C.) represent more than a hundred years (from early 9th century B.C. to the early 8th century B.C.) of the evolution of North Syrian conical crested helmets and the hemispherical crested helmets. Consequently these three helmet-types mentioned above 1.) would have been depicted as a "wandering motif" from a set of stereotyped motifs; 2.) or in 8th century B.C. Crete (and Eastern Mediterranean) there were archaizing helmets in use; 3.) or we are forced to raise the date of the Fortetsa belts from the second half of the 8th century B.C. to the first half or early 8th century B.C. (of course without any effect on the dating of the tombs in which they were found).

335 King 1915, pls. 3-4, 7, 9-12, 37-42; de Schauensee 1988, 55.

336 Ltshashen: *Chart 1, nos. 46-49*; Lori-Berd: *Chart 1, nos. 50-51*; Shirakavan: *Chart 1, no. 52*.

337 On Nimrud ivories (*Chart 3, no. 106*); Balawat gates (*Chart 3, nos. 107-110*); on a bronze statuette of a warrior from Khurvin (Iran): *Chart 1, no. 53* (Vanden Berghe 1959, 124, pls. 156b-c); on a horse's breastplate from Urartu – Northwest Iran (*Chart 3, nos. 111-112*); bronze statuettes from Toprak-Kale (*Chart 3, no. 113*); bronze statuette of a warrior from Maraş (*Chart 3, no. 114*); a fragmentary bronze belt from Urartu (*Chart 3, no. 115*); Zinçirli relief (first half of 8th century B.C., *Chart 3, no. 127*).

338 King 1915, pl. 42; de Schauensee 1988, 55.

339 The dating of the Karatepe sculptures is questionable. Th. Bossert (Bossert *et al.* 1950, 62) dated them to the reign of Tiglath-Pileser III (745—727 B.C.). On the other hand D. Ussishkin (Ussishkin 1969, 121-137) proposed a middle of 9th century B.C. date. For detailed description of the problem *see* Ussishkin 1969. Following Orthmann 1971, the present writer follows a first half of 8th century – middle of 8th century B.C. date.

helmets. These crested helmets (*Charts 3-4, nos. 177-180*) may be traced back to the Late Bronze Age helmet tradition from which the Carchemish helmets are also derived, and are late examples of the same North Syrian helmet tradition to which the Carchemish helmets (*Charts 3-4, nos. 101-104*) also belonged. Their shape is similar to the half a century earlier two Assyrian (**Cat. nos. 13-14**, *Pls. 18-23*) and to two Urartian crested helmets of exactly the same type (**Cat. nos. 109-110**, *Pls. 109-110*) inscribed with the name of Ishpuini (*ca.* 830—810 B.C.). Karatepe was probably one of the Neo-Hittite states, which transmitted the early conical crested helmet tradition to Assyria. The earliest representation of the developed version of this type of helmet is shown on an Urartian wall-painting from Erebuni, on which possibly the Urartian king, Argishti I (789—766 B.C.) himself wears the helmet (*Charts 3-4, nos. 118-119*)[340]. This helmet is very similar to the conical crested helmets (**Cat. nos. 109-110**) of Ishpuini, king of Urartu. This type of crested helmet remained in use in Urartu through the 8th – 7th centuries B.C. (*Charts 3-4, 120-124*)[341] or even in the 6th–5th centuries B.C. as well[342].

The first appearance of crested helmets in the Assyrian army is on the palace-reliefs of Tiglath-Pileser III (745—727 B.C.). Unfortunately between the reigns of Shalmaneser III (858—824 B.C.) and Tiglath-Pileser III there are no Assyrian representations available. Only the two crested helmets, the Munich (**Cat. no. 13**, *Pls. 18-19*) and Karlsruhe (**Cat. no. 14**, *Pls. 20-23*) crested helmets, said to be found in Urartu (Eastern Anatolia) but attributed by the present author to Assyria[343], show that this type of crested helmet should had appeared in Assyria at least in the last quarter of the 9th century B.C. However, there is a 50 years gap which cannot be filled in, between these helmets and the first Assyrian representations of the type on the sculptures of Tiglath-Pileser III, on which there are at least 15 developed forms of crested helmets (*Charts 3-4, nos. 145-159*). A special type of crested helmet (with pointed crest-support and crescent-shaped crest) visible on the Til-Barsip wall-paintings (*Charts 3-4, no. 172*) has its best parallels on the palace-reliefs of Tiglath-Pileser III (*Charts 3-4, nos. 154-155*) and together with an example of this type now in the British Museum (**Cat. no. 15**, *Pl. 24, Fig. 3*) can be dated to that period.

As is shown on Assyrian sculptures, crested helmets are always worn by spearmen carrying round shields and wearing breastplates (*kardio phylakes*). Not only the Assyrian light infantry, or troops said to be North Syrian allies or associated with Anatolian[344] or Phrygian[345] auxiliaries of the Assyrian army, but also their enemies, recruited when defeated as auxiliary units, wore crested helmets[346]. But, while at the end of the 9th century B.C. – first half of 8th century B.C. the crested helmet indicated the nationality of its wearer (Syrian, Anatolian or Phrygian troops), later, from the reign and army reform of Tiglath-Pileser III (745—727 B.C.) it simply indicated/demonstrated the membership of a special service, the light infantry of the Assyrian imperial army. The Assyrian horsemen, charioteers and chariot warriors, and even the heavy infantry always wore conical/pointed helmets. From that time soldiers of any nationality could be recruited to the light infantry, the auxiliary units of the Assyrian army. This is the reason why the fragments of the Nimrud crested iron helmets (**Cat. nos. 16-31**, *Pls. 25-43*) from the last quarter of the 8th century B.C. can be identified as Assyrian helmets.

The evolution of crested helmets, however, raises an important question. Assyrian written sources from the reign of Tiglath-Pileser III mention a people, the Gurraeans (lú*gur-ru*)[347]. Some scholars convincingly identified the Gurraeans with the auxiliary spearmen wearing crested helmets on the Assyrian palace reliefs[348]. But, the North Syrian and Southeast Anatolian origin of these helmets does not support entirely this theory. As has already been mentioned, there are 15 developed forms of crested helmets on the palace reliefs of Tiglath-Pileser III (11 of them Assyrian, 4 of them enemy, *Charts 3-4*, nos. 145-159). This refers to the fact, that this type of helmet can not be connected with a single people, its evolution was the result of several parallel traditions and took place in several North Syrian and Southeast Anatolian urban centers. Furthermore, it is unbelievable that a single, Aramaic or

[340] Hodjash *et al.* 1979, pls. 40, 46; Esaian 1986, pl. 9:9.
[341] Paravakar: *Chart 3, nos. 120-123*, Esaian 1976, figs. 148:1-3, 149:2; Esaian 1986, pls. 2:1-2, 2:4, 9:10-12.
[342] For example the helmets of the bronze statuettes from Ayrum (Armenia) dated to the 6th–5th centuries B.C.: Esaian 1976, fig. 146:3; Esaian 1986, pls. 2:3, 9:14.
[343] H.-J. Kellner, who published the Karlsruhe helmet (**Cat. no. 14**) first, attributed it to Urartu, and considered its decoration (regarded by the present writer as Assyrian) as "orientalisches Gemeingut auf in stark assyrierender Ausprägung" (Kellner 1980, 212). S.-G. Gröschel (Gröschel 1986, 74, notes 229-233) and M. Maass (Maass 1987, 68) considered the helmet as Assyrian.
[344] Barnett – Falkner 1962, xix, pls. XXXVI, XXXVIII, xxiii, pl. LI.
[345] Barnett – Falkner 1962, xxiii, pl. LII.
[346] Barnett – Falkner 1962, xix-xx, fig. 2:1, pl. XLI (Eastern campaign of Tiglath-Pileser III, 2nd and 9th *palû*); xx-xxiv, fig. 2:2, pl. LI (Anatolian [Phrygian?] campaign of Tiglath-Pileser III, 3rd *palû*); xxii-xxiv, fig, 2:13, pl. LXV (Urartian warrior?), (Urartian campaign of Tiglath-Pileser III, 3rd *palû*) and pl. LXI.
[347] Lanfranchi – Parpola 1990, 53, 215, 277; Fales – Postgate 1992, 5, 115.
[348] Reade 1972, 105-106; most recently Postgate 2000, 103-104; Fales 2000, 42.

non-Aramaic, nomadic or semi-nomadic people – origin of which is entirely unknown[349] – supplied the Assyrian imperial army with auxiliary spearmen, number of which can reached tens of thousands (approximately one fifth of the Assyrian infantry)[350]. The characteristic standardization of helmets (and military dress) in the Assyrian imperial army (especially from the reign of Sennacherib, *Fig. 1* and *Chart 2* for conical/pointed helmets and *Charts 3-4* for crested helmets) was probably the result of a central policy (an army reform) and not of a spontaneous process taking place in a nomadic or semi-nomadic tribe. It is possible, however, that the Gurraeans – if they were a nomadic tribe – adopted this, characteristically not nomadic type of weaponry in the Assyrian army, and became one of the most numerous groups of auxiliary spearmen[351].

After the reign of Tiglath-Pileser III (745—727 B.C.) the types and forms of crested helmets became standardized, and for the reign of Sargon (721—705 B.C.) and Sennacherib (704—681 B.C.) only variants of two major types remained in use: 1. **with crest curving forward** (*Charts 3-4, nos. 187-188, 190-192, 218-225*); 2. **with a crescent-shaped crest on the top of a pointed crest support** (*Charts 3-4, nos. 189, 214-217*). In the reign of Sennacherib the earflaps became permanent parts of the helmets. Representations of the two major types of crested helmets mentioned above are known not only from Assyria[352] and elsewhere in the Near East, but also from the Eastern Mediterranean under strong Oriental/Assyrian influence from the middle of the 8th century B.C. through the 7th century B.C. The best example of the first type (crest curving forward) is an orientalizing helmet from Praisos, Crete, now in Hamburg[353], several representations of such helmets are known from the Aegean/Eastern Mediterranean[354] (*e.g. Chart 3, no. M.3.5*). A well-known example of the second type (crescent-shaped crest) is the Argos crested helmet of Geometric Age[355]. In the reigns of Esarhaddon (680—669 B.C.) and Ashurbanipal (668—627 B.C.) the crested helmets became uniform (*Charts 3-4, nos. 229, 247-255*) and for the first time in the history of Assyrian helmet representations the continuous arcs (semicircles) framing a scene of figures appeared on the front (*Charts 3-4, no. 253-254*) and on the back (*Charts 3-4, no. 255*) of the crested helmets. Previously it was an exclusive characteristic of conical/pointed helmets. This change was the last step of the standardization of the Assyrian army, soldiers of which – at least in their uniform – finally lost the remains of their national characteristics. In spite of the only and late representation of this feature on the front of a crested helmet mentioned above, there are other examples proving its use on crested helmets. One of them is the late 9th – early 8th century B.C. Munich Asssyrian crested helmet (**Cat. no. 13**, *Pls. 18-19*) with two curves and a vertical rib on the front. Its date is so early that the decorative system of crested helmets was not yet developed. The second example is one of the Nimrud iron helmet fragments (**Cat. no. 16**, *Pls. 25, 39, 41*) showing continuous arcs on the front of possibly a crested helmet, presumably from the last quarter of the 8th century B.C., when the standardization of Assyrian army equipment was undertaken.

The original or actual crests made of perishable materials (feathers, horsehair) and fastened between the bronze or iron crest-pieces by rivets[356] or similar methods are disappeared from helmets. The original function of horsehair or feather crests and plumes of the Late Bronze Age might be to threaten the enemy with the help of becoming similar to a wild animal (bull, horse, *etc.*). An other original function of the crests on helmets might be the apothropaic function, which would be very appropriate on a helmet.

In the Assyrian imperial army of the second half of the 8th century B.C. (Tiglath-Pileser III) however, the crests of crested helmets might have a new function, which can be connected to the standardization (crested helmets = light infantry) of the "new model Assyrian army" and the standardization of its weaponry and helmets (reign of Sargon II and Sennacherib, *see* above): the different shapes of crested helmets, or at least the colour of their crests might indicate that their wearers belonged to certain units of the arm of light infantry. This might be the case with the crests of the horse trappings which are often visible on Assyrian palace reliefs, and might distinguish the different cavalry and chariot units of the Assyrian army.

---

[349] Postgate 2000, 103.

[350] Tiglath-Pileser III: 15.5 %; Sargon II: 22.4 %; Sennacherib: 22.8 %; Ashurbanipal: 28.7 %. These data are based on a statistical analysis of Assyrian palace reliefs made by the present author, and will be published in a future article.

[351] Postgate 2000, 104. For other candidates see Fales 2000, note 33.

[352] First type: **Cat. nos. 12-14**; second type **Cat. no. 15**.

[353] Hamburg, Museum für Kunst und Gewerbe, Inv. no.: 1970.26c; H.: 44.6 cm, H. of the crest: 10 cm, W. of the crest: 1.5-3.0 cm; Hoffmann 1972, 5-6, 42, pl. 13.

[354] On models and pendants: Hoffmann 1972, pl. 41; the incised representation of this kind of helmet can be seen on a "face-plate" attached secondarily to the helmet mentioned above.

[355] Argos, Tomb 45, Argos, Archaeological Museum, H.: 48 cm; second half – last third of 8th century B.C.; Borchhardt 1972, 65, Kat. 13 I, pls. 33:1-3; Pflug 1988a, 12-14, figs. 1-3.

[356] Holes for fastening the crest between crest-pieces of metal are visible around the edges of crestpieces of all the crested helmets: **Cat. nos. 12-15, 52, 55-56, 113-114, 128**.

Unfortunately there are only a few representations of crests in Assyrian art from which the colours can be reconstructed. On one of the palace reliefs of Sargon II (721—705 B.C.) from Khorsabad the red and blue paint of the feathers (or horsehair plumes) of a crested helmet remained visible at the time of the excavations[357]. The same colours appeared on three other slabs representing the red and blue crests of horsetrappings[358]. Another variant of crest decoration is known from the horses' crests on the Til-Barsip wall-paintings[359]: the characteristic black and white chequered motif which might be considered as local, North-Syrian variant as it appears in the bronze inlay decoration of the Zinçirli conical iron helmet fragments (**Cat. no. 38**, *Pl. 57*).

There are one bronze crest-piece and three crested helmets which may be identified as Assyrian products. Helmet crest **Cat. no. 12** is the only Assyrian crested helmet which may be dated exactly. It was found in the debris of the destruction of Lachish in 701 B.C. by Sennacherib (705—681 B.C.). This type of crest is well known from that time, and from the reigns of Esarhaddon (680—669 B.C.) and Ashurbanipal (668—627 B.C.) (*Charts 3-4, nos. 218-221, 229, 247-255*). This type of crested helmet was made of two separate parts: the crest was riveted on to the top of the (hemispherical) skull-piece of the helmet. Although riveting seems to be more archaic, both techniques (riveting the crest on to the helmet or hammering out from the same piece of metal) were used in parallel from late 9th century B.C. through the 8th century B.C. As an early example the crest of the Munich helmet (**Cat. no. 13**, *Pls. 18-19*) was made separately and was riveted to the skull-piece of the helmet, while that on the Karlsruhe helmet (**Cat. no. 14**, *Pls. 20-23*) of (nearly) the same date was hammered out from the same sheet of bronze as the skull-piece (only the bovine's head was added separately). The Nimrud crested iron helmets (reconstructed here from fragments, following the shape of the Karlsruhe crested helmet) were riveted together from separate parts, because it was impossible to hammer out such an elaborate shape from one piece of iron (*Pls. 41-43*).

Both the construction and decoration of helmet **Cat. no. 13** are unique. Its decorative system is the same as on the 9th century B.C. Assyrian conical bronze helmets, which under Assyrian influence spread widely in the Near East and remained in use on the periphery of the Assyrian Empire through the 8th century B.C., while in Assyria at the latest from the middle of the 8th century B.C. it changed from curves bending inwards and terminating in animal's head to continuous arcs framing a stele-like scene of figures (*Fig. 1*, second phase). This motif appears only on three representations of crested helmets (*Charts 3-4, nos. 253-255*) from the reign of Ashurbanipal (668—627 B.C.), but can be seen on the front fragment of a Nimrud crested iron helmet (**Cat. no. 16**) mentioned above and on an other Nimrud iron fragment probably from a crested helmet (**Cat. no. 17**). So there is no correspondence between the helmet representations and the actual finds.

It is very hard to decide whether the helmet is *a* modern product (with a secondary addition of the crest to the previously conical base with the characteristic design of 9th century B.C. Assyrian conical helmets) or a product of the 80 years between the reigns of Shalmaneser III (858—824 B.C.) and Tiglath-Pileser III (745—727 B.C.), from which we do not have any dated helmet representations in the Assyrian art. On the palace reliefs of Ashur-nasir-pal II (883—859 B.C.) and Shalmaneser III there are no helmet representations of this (conical) type of crested helmets at all. The appearance or evolution of crested helmets in Assyria can consequently be dated to the last quarter of the 9th century B.C., contemporary with the first known Urartian crested helmets (**Cat. nos. 109-110**) with the inscription of king Ishpuini (*ca.* 830—810 B.C.).

Presumably at the time of early evolution or adaptation of crested helmets in Assyria, the decorative systems of crested helmets were not yet fixed, and the helmets were decorated with motifs borrowed from the 9th century B.C. decorative system of Assyrian conical helmets emphasizing the importance of that special kind of decoration. But the attribution of this helmet (**Cat. no. 13**) remains uncertain. Since the Assyrian (conical) and North Syrian (conical crested) helmet types spread widely by the end of the 9th century B.C.[360], the Near East became a *koine* in respect of manufacturing conical and crested helmets. Nevertheless this helmet can be connected with an early tradition of Assyrian crested helmets: its form is known only from the palace reliefs of later Assyrian kings[361] and its front decoration is first known from Assyrian conical helmets. The structure and technique of the helmet (riveted together from separate parts) seems to be archaic, but as mentioned above, the two techniques (riveting and hammering out from one piece of bronze) were used together from the earliest known

---

[357] Botta – Flandin 1849, pl. 61.
[358] Botta – Flandin 1849, pls. 62, 63, 65.
[359] Thureau-Dangin – Dunand 1936, pl. LIII.
[360] The earliest known conical helmets outside Assyria are from Urartu (inscribed with the name of king Ishpuini, *ca.* 830—810 B.C., (**Cat. nos. 62-64**) and from Hasanlu (**Cat. nos. 49-51**), *ca.* 800 B.C. The earliest known examples of conical crested helmets are from Urartu (**Cat. nos. 109-110**, with the inscription of king Ishpuini) and from Assyria (**Cat. nos. 13-14**).
[361] *See* the Lachish crest of an Assyrian helmet (**Cat. no. 12**) made by the same technique.

examples (**Cat. nos. 13-14**) through the 8th century B.C. The proposed date of the helmet is late 9th century – early 8th century B.C.

The incised figural decoration around the base of helmet **Cat. no. 14** can be dated to the second half of the 9th century B.C. The centre scene (tree-of-life under a winged sun-disk, on either side of the tree the king with his attendant behind him) is the same as on the 9th century B.C. Assyrian conical helmet (**Cat. no. 3**, *Pls. 3-4*) and helmet fragments (**Cat. nos. 4-5**, *Pls. 5-6*) and on the probably Assyro-Babylonian conical helmet (**Cat. no. 8**) mentioned above. This decorative system seems to remain canonical on the front of helmets at least to the middle of the 8th century B.C. (reign of Tiglath-Pileser III, 745—727 B.C.). The incised scenes without *repoussé* are characteristic of the 9th century B.C. Assyrian helmets (*see* **Cat. nos. 3-7**). Parallels for the incised frieze around the helmet (the king receives tribute from Western Syria or Phoenicia[362]) are well known from two conical bronze helmets **Cat. nos. 6, 8** (*Pls. 7-8*) and from the palace reliefs of Ashur-nasir-pal II (883—859 B.C.)[363], from the Rassam Obelisk[364], from the bronze reliefs of his Balawat Gates[365] and from the bronze reliefs of the gates of the Northwest Palace, Nimrud[366]. This scene remained in frequent use under his successor, Shalmaneser III (858—824 B.C.) and is shown on the bronze reliefs of the Temple of Adad in Ashur[367] and of the Balawat Gates[368], on the Black Obelisk[369], on the throne-base of Shalmaneser III[370] and on ivory plaques from Nimrud[371]. The most exact parallels for the scene: an attendant holding a baldachin over/above the head of the king is known from the upper register of the Rassam Obelisk and from the second register of the Black Obelisk. In the reign of Tiglath-Pileser III (745—727 B.C.) the garment of Phoenician tribute bearers is modified a little, as can be seen on the palace reliefs. The only curiousity is a small detail on the helmet: the attendants of the king wear their apron-like garment at the back of their waist, whereas the attendants of the crown-prince wear it at the front. A similar phenomenon is shown on the figurative frieze of helmet **Cat. no. 6** as well. This is the only example from Assyrian art on which this kind of apron is worn on the front.

Muscarella[372] also questioned the originality of helmet **Cat. no. 14**. He wrote that the decoration „consists of oddly proportioned, irregularly, differently dressed, armed, bearded and footed men bearing strange gifts, the whole obviously inexpertly executed"[373]. His reasons – as usual – are merely stylistic. Muscarella does not questioned the authenticity of the helmet itself, perhaps, because of the striking similarity of its structural details to the bronze inlay details of the Nimrud crested iron helmet fragments (**Cat. nos. 16-31**) published some ten years later[374] (*Pls. 20-22* and *41-42*). It seems, that this similarity proves the authenticity of the helmet itself. Muscarella concludes, that the helmet is a kind of pastiche: original helmet with a modern incised decoration. If so, a metallographic analysis of the corrosion would solve the problem. It looks however, that the space between the upper and lower pairs of embossed ribs is intended to contain a kind of decoration.

Michael Maass has dated the Karlsruhe helmet to the third quarter of the 9th century B.C.[375]. This seems to be a little early in the absence of supporting evidence from contemporary Assyrian art. The figural decoration of the helmet follows the 9th century B.C. tradition, but the high technical level of its construction (rows of pseudo-rivets) imitating earlier helmets of the same form made in separate pieces, may also indicate a little later, end of 9th century B.C. date. This helmet shows that our type of crested helmet may possibly have descended from the Neo-Hittite Carchemish helmet tradition (*Charts 3-4, no. 104*) and the Balawat Gates of Shalmaneser III are a *terminus post quem* for this type of crested helmet. Consequently the suggested date for this helmet is the last quarter of 9th century – first decades of 8th century B.C.

---

[362] Wäfler 1975, 72ff., figs. 35-46: Israel; 83ff., fig. 57: Phoenicia; 121ff., fig. 64: Damascus; 163ff., fig. 93: Hattina; 200ff. figs. 120-121, 127: Bit-Adini; 225ff., fig. 143: Carchemish. For a very detailed discussion of Assyrian tribute and tribute bearer scenes *see* the recent study of J. Bär (Bär 1996).

[363] Budge 1914, pl. XXVIII; Barnett – Falkner 1962, pl. CXXIV.

[364] London, British Museum, WA 90925 + 118800 + 132013 + 136897-8; Börker-Klähn 1982, 183-184, no. 138, pls. 138a-t.

[365] London, British Museum, WA 124687-124700: 1st, 2nd and 6th registers, WA 124685-124698: 1st, 4th registers.

[366] London, British Museum, WA N 2063-2065, upper register.

[367] Andrae 1977², fig. 195.

[368] King 1915, pls. XIII, XIV, XXII, XXV-XXVIII, XXXI-XXXIV, XL, XLII-XLV.

[369] London, British Museum, WA 118885, 31st year of Shalmaneser III (827 B.C.); Börker-Klähn 1982, 190-191, no. 152, pls. 152a: 2nd and 5th registers; 152b: 2nd and 5th registers; 152c: 2nd registers; 152d: 1st, 2nd and 5th registers.

[370] Baghdad, Iraq Museum, ND 11000; Orthmann 1975, fig. 209b.

[371] Mallowan – Davies 1970, nos. 67-69, 71-78, 81-85, 96d-e, 100-102.

[372] Muscarella 2000, 150, no. 19.

[373] Muscarella 2000, 150.

[374] Dezső – Curtis 1991, 107-126, figs. 18-20.

[375] Maass 1987, 68. (*See* Gröschel 1986, 74). Kellner 1980, 211-212 dated the helmet to the late 9th – early 8th century B.C.

Following S.-G. Gröschel's division[376], the scenes on the Assyrian helmets may be divided into seven separate groups:

1. *King worshipping the "tree-of-life"* (under a winged sun-disk or with a deity [Ashur] in the winged sun-disk) with one or more attendants behind him. Following Gröschel's arrangement, there are five helmets included in this group: **Cat. nos. 3, 4, 5, 8** and **Cat. no. 14**.
2. *Tribute bearers* (**Cat. nos. 6, 8, 14**). The characteristic scene of Western Syrian or Phoenician tribute bearers is well known from the Assyrian art of the 9th–8th centuries B.C. (from the reign of Ashur-nasir-pal II (883—859 B.C.) to Tiglath-Pileser III (745—727 B.C.))[377].
3. *Officials in procession* (**Cat. nos. 3, 4, 6, 10, 11**)[378]. This scene symbolizing the Assyrian imperial bureaucracy is shown on many examples of the 9th–8th century B.C. Assyrian art[379].
4. *The king offers sacrifice* (libation) in the presence of the crown prince (**Cat. no. 10**). This group was formerly misunderstood ("libation in front of the king") following the wrong reconstruction in the first interpretation of the helmet[380]. Namely, that it is not the crown-prince who offers sacrifice (libation) in front of the king but, as is known from other representations in Assyrian art, the king himself holds the libation cup and offers a sacrifice.
5. *Deity on the front of the helmet* (**Cat. no. 11, 16**, *Pls. 16; 25, 39, 41*). This group represents the results of new examinations[381] (*see* below).

(6. *"Coronation of the king"* (**Cat. no. 6**). Unique representation of the coronation of the Assyrian king by Ishtar in front of Ashur.)

(7. *"Praying king"* (**Cat. no. 7**). Unique representation of the Assyrian king praying in front of a mountain-triangle and the winged sun-disk (probably of Shamash). U. Seidl proposed an interpretation: the Assyrian king as "the sun of all the peoples"[382].)

This division of groups shows us whether the decorative systems of Assyrian helmets followed strict rules periodically (1st group, 9th century B.C.) or represented the belonging of its wearer to a special service or unit of the Assyrian army (7th group).

The symbolism of 9th century B.C. Assyrian helmets (representing the first phase of the evolution of Assyrian state and army, the "network empire"[383]) follows the following pattern:

1. *Winged sun-disk / Ashur – tree-of-life – king* (**Cat. nos. 3, 4, 5, 8, 14**): This scene is the crucial and most important scene not only of the 9th century B.C. Assyrian helmets, but the whole Assyrian art. Consequently this scene is the central scene of the Assyrian state and royal propaganda (a kind of "royal coat-of-arms"), the symbol of the sacral and unquestionable power of the Assyrian king (the earthly regent of god Ashur) and ambitious Assyrian kingdom.
2. *Tribute bearers* (**Cat. nos. 6, 8, 14**): This scene represented the military power and successes of Assyria and the westward expansion of the kingdom (Phoenician and Syrian tribute bearers as the personification of vassals and conquered territories).
3. *Officials in procession* (**Cat. nos. 3, 4, 6, 10, 11**): Probably the personification of the Assyrian military bureaucracy.

(4. *"Coronation of the king"* (**Cat. no. 6**). Unique representation of the coronation of the Assyrian king by Ishtar in front of Ashur.)

(5. *"Praying king"* (**Cat. no. 7**). Unique representation of the Assyrian king praying in front of a mountain-triangle and the winged sun-disk (probably of Shamash).)

---

376 S.-G. Gröschel divided the figurative scenes of Assyrian helmets into three groups (Gröschel 1986, 76). The five group division presented here represents the results of new examinations (Dezső – Curtis 1991, 105-126); Kellner 1993, 325-331.

377 *See* pp. 31, 42, and notes 262-271, 362-370.

378 S.-G. Gröschel arranged groups 3 and 4 in one group: "Procession", and called our 4th group "unarmed tribute bearers in front of soldiers". Gröschel arranged two helmets (**Cat. nos. 3, 4**) into this group, but new examination of the Nimrud conical iron helmet (**Cat. no. 10**) shows, that there is a scene around the base of the helmet inlaid with bronze. The Munich conical Assyrian iron helmet **Cat. no. 11** (Kellner 1993, 325-331) can be arranged in this group as well.

379 *See* on pp. 22, 26, 27, 34-35.

380 Barnett 1953, 101-102, pls. 31-32 (helmet from "Nineveh"); Gröschel's first group (Gröschel 1986, 76).

381 Dezső – Curtis 1991, 105-126.

382 Born – Seidl 1995, 50.

383 Liverani 1988.

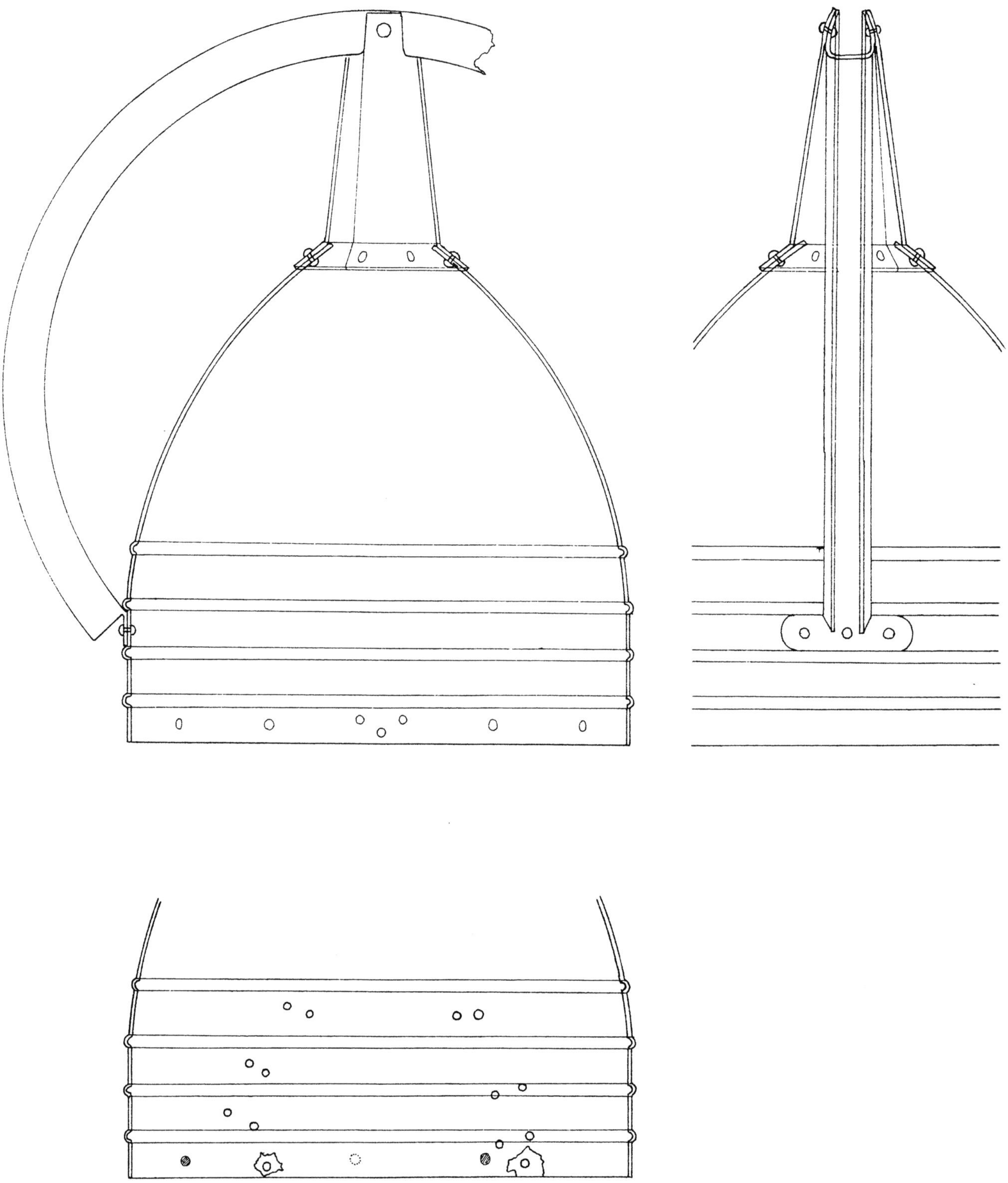

Fig. 3: Schematic drawing of helmet **Cat. no. 15** (*scale 1:2*)

The decorative system of 9th century B.C. Assyrian helmets (**Cat. nos. 3-7, 8, 14**) show us much more than the two important central aspects of the Assyrian power: the Assyrian divine kingdom and the glory of the Assyrian power and army. As have been mentioned, some scholars discussing the technique of the decoration of helmets **Cat. nos. 8, 14** (*repoussé* figures with incised details) attributed them to Assyria. Other scholars referring their "provincial execution" (and supposed Urartian provenance) considered them Urartian manufacture. But as we shall see, the decorative system of Urartian helmets is completely different. Although – similarly to other fields of Urartian art – they borrowed the iconographic and symbol set of their helmets together with the Assyrian concept of kingdom from Assyria, Urartu transformed the original Assyrian concept, and did not use the canonical Assyrian symbols in their original (ideological and propagandistic) form. Taking the Assyrian symbols out of their original context, they used them only in their notional or artistic context. They disintegrated the Assyrian propagandistic concept and from its separate elements and symbols they formed an independent Urartian symbol system. As a consequence, it is not possible to attribute helmets **Cat. nos. 8, 14** as of "assyrianizing Urartian" manufacture. Furthermore, it is not possible that Assyrian helmets with well known hostile propagandistic iconography would have been used in the Urartian army in their original form. Three aspects of the Assyrian concept "kingship" (*šarrūtu*) will be discussed in the chapter on Urartian helmets.

Helmet **Cat. no. 15** is a rare example of a special type of crested helmet used widely in the Assyrian imperial army in the middle of the 8th century B.C. (*Charts 3-4*). Several representations of this type are shown on the palace reliefs of Tiglath-Pileser III (745—727 B.C.), (*Charts 3-4, nos. 153-155*) and on the wall-paintings of Til-Barsip (*Charts 3-4, no. 172*) of the same period. The main characteristic of this helmet (which distinguishes it from other types of crested helmets) is that it consists of a pointed base (a conical cap with a tubular crest-support on it) on which the crescent shaped crest was riveted. As shown on *Fig. 3* the crest was probably flattened on the front of the helmet forming additional decoration. Only two small fragments of the lower edge of the crest remained in position. As can be judged from the position of the rivet holes (3 x 2 antithetical pairs) this front decoration might have formed either a pair of curves bending towards the centre or the 'stele-field' characteristic of Assyrian conical/pointed helmets from the second half of the 8th century B.C.

This helmet type spread widely not only in the ancient Near East, but in the Mediterranean as well. A parallel to its shape is known from an orientalizing helmet of the Geometric Period from Argos[384]. The Argos crested helmet with its earflaps together with similarly orientalizing Aegean and Eastern Mediterranean helmets derive from the 9th–8th century B.C. Assyrian and North Syrian helmet tradition (see below in the chapter of "North Syrian helmets").

12. Lachish (Destruction level III, together with remains of the siege, *e.g.* iron arrow heads), Group no. **A.3.1** – Large curved crest of a helmet or horse trappings. It has an U-shape cross-section. There is a series of holes on both sides of the crest along the edge. It was fixed to a helmet or horse trappings by two rivets (*Pl. 17*). End of the 8th century (probably 701) B.C. – Jerusalem, The Israel Museum, Inv. no.: 34.133/1-2 – Tuffnel 1953, 387, pl. 39:1-2; Wright 1958, 165, fig. 119; Ussishkin 1982, 57, fig. 49; Shanks 1984, fig. on p. 63.

13. Provenance unknown (said to come from "Eastern Turkey"), Group no. **A.3.2** – H.: 30.8 cm, Diam.: 20.4—22.3 cm. Conical helmet with a crest curving forward. The helmet was made from five pieces: the "skull" of the helmet, three parts of the crest and a bovine's head at the end of the crest. The separate parts of the helmet were fixed together with rivets. There is a series of holes around the base of the helmet probably for the attachment of a lining. There is a horizontal rib around the base of the helmet. This rib ends in a pair of curves ("crooks") on the front of the helmet. There is a vertical rib between the curves. The crest of the helmet was made from three parts and the different parts were fixed together with rivets (one bottom- and two side-pieces). The side-pieces were decorated with a pair of embossed ribs around the edge of both pieces. There is a series of holes along the upper edges of the crest-pieces to fix the horse-hair crest. Restored (*Pls. 18-19*). Late 9th century – early 8th century B.C. – Munich, Archäologische Staatssammlung, Inv. no.: PS 1979/1181. – Kellner 1980, 212-213, pls. 15-16; Pflug 1988, 24, note 75; Calmeyer 1991, no. 10 (pl.); Dezső – Curtis 1991, 114-121, pl. 17.

14. Provenance unknown (said to come from "Eastern Turkey"), Group no. **A.3.3** – H.: 28.5 cm, Diam.: 19-22 cm, H. of the frieze: 5.7 cm, W.: 790 g. Conical helmet, with a crest curving forward. The crest ends in a bovine's head, fastened to the crest by rivets. There is a series of holes on both sides of the crest-support for fixing the horse-hair crest. On the back, the crest ends at half the height of the helmet. There is a pair of parallel embossed ribs around the edge of the crest. Between the ribs there is a series of embossed pseudo-rivets. On the front of the helmet there are two vertical rows of similar pseudo-rivets, recalling the construction of the earlier helmets of the same type. There is a series of holes around the base of the helmet, probably for a lining. There are two pairs of embossed horizontal ribs around the base of the helmet. Inside the pairs of ribs there are rows of pseudo-rivets. Between the two pairs of ribs there is a frieze of incised figures. In the middle, on the front of the helmet there is a "tree-of-life" under the figure of a god (Ashur?) within winged sun-disk. On either side of the tree there are figures of an Assyrian king with his attendant behind him. The king wears the Assyrian crown, a ritual dress (*Schalgewand* Nr. 2) and raises his right hand in *ubāna tarāṣu* position. His attendant is armed with a mace, sword, bow, and quiver. There is a continuous scene around the helmet: the king with four of his attendants behind him and a procession of officials, attendants and tribute-

384 Argos, Tomb 45, Argos, Archaeological Museum; H.: 48 cm; second half – last third of 8th century B.C.; Borchhardt 1972, 65, Kat. 13 I, pls. 33:1-3; Pflug 1988a, 12-14, figs. 1-3.

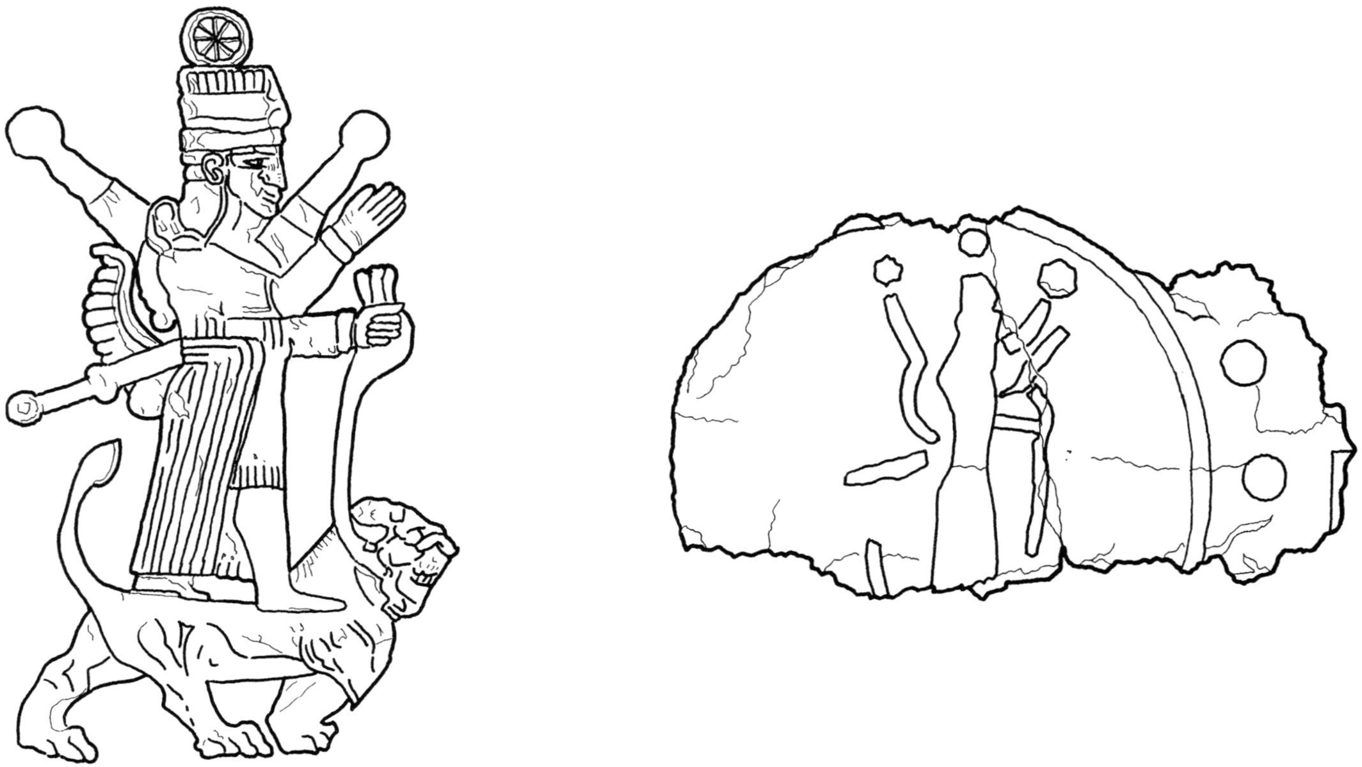

Fig. 4: *Left:* Stele dedicated to Ishtar of Arba'il, Louvre, AO 11503; *Right:* Bronze inlay decoration showing Ishtar of Arba'il on helmet fragment **Cat. no. 16**

bearers led by the crown-prince The king is armed with a sword and holds a bow. The first attendandt behind the king holds a baldachin, the second is armed with a spear, the third and fourth are armed with mace, sword, bow and quiver. The crown-prince and three of his attendants wear the same elaborate dress with long ribbons and are armed with sword. The fourth official behind the crown-prince leads the procession of western tribute-bearers. The first of them is their leader, while the others carry metal rhyton/calix, bucket, bowl, a bag (of precious metal?), a situla, two trays of unique objects, and a bunch of spears. (*Pls. 20-23*). Late 9th – early 8th century B.C. – Karlsruhe, Badisches Landesmuseum, Inv. no.: 89/12 – Kellner 1980, 211-212, pls. 13-14; Gröschel 1986, 74, notes 229-233; Maass 1987, 65-71, pls. 1:2, 2, 3; Maass 1988, 170, no. 97, fig. 97; Pflug 1988, 24, note 75; Maass 1989, 195; Dezső – Curtis 1991, 114-121, pl. 18.

15. Provenance unknown, Group no. **A.3.4** – H.: 24.6 cm, Diam.: 19.6 cm. Conical helmet with a crescent-shape crest. There are four ribs around the base of the helmet. There are 11 holes (the distance between them is 3.8-4.5 cm) along the base probably for a lining. There are three holes on either sides of the helmet for the chinstrap (or earflaps). The height of the skull of the helmet is 16.5 cm. The crest-support was made from a separate piece of bronze sheet and curved in the shape of a conical tube. The base of the crest-support was riveted to the top of the skull of the helmet by 7 rivets. There are 7 other holes (from an earlier riveting) visible inside the top of the helmet. The crescent-shape crest is broken. It has a U-shape cross-section. The crest was riveted to the crest-support by two rivets (one on either sides) and to the skull-piece (between the 2nd and 3rd ribs) by three rivets. There is a series of embossed knobs along the edges of the crest. There are 16 holes (eight pairs) on the front of the helmet for fixing the crest (now broken) to the helmet. Only two rivets survived in position, either with a small fragment of the crest (*Pl. 24*). 8th century B.C. – London, British Museum, Inv. no.: 2841 (Richard Payne Knight Bequest, 1824) – Walters 1899, 349, cat. no. 2841; British Museum 1929, 77, 79, fig. 69, no. 229; Pflug 1988a, 19-20, fig. 9, note 43.

Group A.4: Assyrian conical crested iron helmets (Cat. nos. 16-31)

In the autumn of 1846, Sir Austin Henry Layard excavating Chamber I of the Northwest Palace at Nimrud, found several crested iron helmets[385]. 16 iron fragments decorated with bronze inlay from that find are now in the British Museum, under the numbers 48-11-4, 113-115.The figure of Ishtar on the fragment is an exact parallel to the representation of Ishtar of Arba'il on a stele from Til-Barsip[386] (*Fig. 4*), dedicated to the goddess by Ashur-dur-paniya, governor (?) of Kar-Shulmanu-asharedu[387]. Ishtar is shown in the same battle dress: there are two quivers on her back (one on each shoulder, crossing each other on her back). There is a shield or *melammu* hanging on her back[388] and a sword at her waist. She wears a horned *polos* cap with a star on it. Her right hand is raised, her left hand holds the reins of the lion on whose back she stands[389].

The style of this stele is debatable. It is said to be in the North Syrian tradition[390] but, as it draws upon very ancient Mesopotamian iconographic tradition for Ishtar, there is no reason to distinguish it from the general Assyro-Babylonian imagery for this goddess.

In one of his dreams, Ashurbanipal saw Ishtar of Arba'il with the same attributes as on the Til-Barsip stele and on our helmet[391]: a quiver hanging on each of her shoulders, a sword at her waist; but in her right hand she carried a bow instead of the reins of the lion.

---

385 Layard 1850, 341-342.

386 Paris, Louvre, AO 11503; Thureau-Dangin – Dunand 1936, 156-157, pl. 14:1; Börker-Klähn 1982, 226, no. 252, pl. 252; Hrouda 1965, pl. 58, no. 5 (he dated the stele to the reign of Tiglath-Pileser III, 745—727 B.C.).

387 Thureau-Dangin – Dunand 1936, 157, pl. 14:1:

1) *ana* ^{d}*Ištar*(XV) *a-ši-bat* ^{uru}*Arba-ìl*
2) *bēlti*(GAŠAN)-*šú* ^{I}*Aš-šur-d*[*ūr*(BÀD)]-*pāni* (IGI)-*a* ^{I}[^{ú}...]
3) [...] ^{uru}*Kar*(KAR)-^{Id}*Šul-ma-*
4) *nu-ašarēdu*(MAŠ) *ana balaṭ*(TÌL) *napšāte*(ZI.ME)-*šú*
5) *iqqî*(BAL)?-*eš*?

388 The identification is debatable. The Annals of Ashurbanipal (Cyl. A, Col. IX, ll. 79-80) mentions Ishtar of Arba'il as dressed in fire and wearing *melammu* ("aura", "radiance"): Rawlinson 1909, pl. 9, Col. IX, ll. 79-80: 79) ^{d}*Ištar*(XV) *a-ši-bat* ^{uru}*Arba-ìl*, 80) ^{d}*Gíra*(GIŠ.BAR) *lit-bu-šat me-lam-me na-šá-a-ta*. Seidl 1976—1980, 88 writes: "*hinter ihr ist eine art Nimbus wiedergegeben*". But this kind of *melammu* is clearly identifiable on several representations of gods as a flaming circle around the figure of the deity. For example see the representation of Ishtar on helmet **Cat. no. 11** (*Pls. 12-16*) or on an incised metal plaquette (von Luschan *et al*. 1893, 43, fig. 12) showing the goddess on the back of her lion, in one hand holding the reins and with other hand raised. Around the goddess there is the *melammu* resembling a flaming circle. But there could be another type or conception of *melammu*-representation known from an Explanatory List of Gods (King 1909, pl. 10, col. II, ll. 2-6; pl. 15, obv. col. II, ll. 5-6): 5) *šá me-lam-me tul*$_5$*-la-tu*, 6) *ki-is-su-ra-at* ("(a goddess) girt with hanging *melammus*". The goddess is unknown. We know only that she is "*šar-rat ap-si-i*". She could be Damkina or Bēlit-ilāni or maybe Ishtar. For *šarrat apsî* and *Ištar-apsî* *see*: Tallqvist 1938, vol. I, 238, vol. II, 339). Seidl 1976—1980, 88 writes: "*Götterstalten – ist oft von einem Strahlenkranz umgeben (bei der stele aus Til-Barsib ist ein Teil davon auf dem Rücken dargestellt), der vielleicht das melammu der I. wiedergibt*". However, this is not simply a technical question of depicting the *melammu* ("*ein Teil davon auf dem Rücken dargestellt*"), but a question of the canonical representation of one of the goddess's cult statues: they wanted to represent a shield or a shield-shaped (hanging) *melammu* only on the back of the figure (**Cat. no. 16**). *See* for example *Fig.5:10*.

389 In this case, when we have not just a similar but exactly the same representation of a deity on two different objects (one of them is dedicated to the goddess) from different sites, there can be no doubt that they derive from the same source, possibly from a well known statue – a cult-statue – of the deity (Ishtar of Arba'il). Presumably this is a canonical representation of Ishtar of Arba'il derived from one of her cult statues.

390 Börker-Klähn 1982, 225-226.

391 Ashurbanipal, Cyl B, col. V, 49-53; Rawlinson 1870, pl. 32, 49-53.

The iconography and attributes of the 3rd millennium B.C. Ishtar armed[392] changed at the latest in the Old-Babylonian period. The two bunches of maces and axes crossing each other on the back of the goddess[393] were replaced by a pair of quivers similarly crossing each other on the back of Ishtar. The reason for this change might be the increasing role of the bow in the warfare of the early 2nd millennium B.C. (*Fig. 5*).

There is a series of terracotta plaques from Kish (and from Larsa), from the Old-Babylonian period showing Ishtar armed[394] On these plaques (for example: Louvre, AO 10479[395]; *Fig. 5:4*) Ishtar is shown in stepping position standing on the back of her lion. She wears the same dress and two quivers on her shoulders as on the Til-Barsip stele and on the helmet fragment. Her hands hold a mace (sceptre)[396] or bow[397], or a scimitar[398]. These terracotta plaques show the very early development of the iconography and attributes of this type of representation of Ishtar (*Fig. 5*).

Ishtar, whether of Arba'il or another city or region is not certain without an inscription, appears on many Neo-Assyrian seals, which have never been systematically studied in order to isolate traits which might identify regional variations (*e.g.* stamp seal, Oxford, Ashmolean Museum 1891.331[399] (*Fig. 5:8*); stamp seal, Berlin, Vorderasiatisches Museum VA 5887, Ass. 1119[400] (*Fig. 5:9*); cylinder seal, London, British Museum, BM 89769[401] (*Fig. 5:10*)). But, as Barrelet has argued[402]: "*L'Ishtar assyrienne armée et aptère est bien la fille de l'Ishtar guerrière hammurabienne*", which is represented on the above mentioned terracotta plaques.

The iconography and attributes of Ishtar armed on the Neo-Assyrian seals differ in small details from the Kish terracotta plaques[403]. Ishtar stands on the back of a recumbent lion (*Fig. 5:10*) or of a *lion passant* (*Fig. 5:8-9*). The quivers on her shoulders are standard. In her hands she holds the rein to the lion's nose (*Fig. 5:8*), or a bow and an arrow (*Fig. 5:10*), or a stick (*Fig. 5:8-9*). On the front of helmet **Cat. no. 6** she wears a long dress, a horned *polos* with a star on it, and there are two quivers on her back crossing each other and terminating in stars. These differences, with the variations on the Old-Babylonian plaques, might originate from different written traditions, certain episodes from mythological texts, sculptures, or cult statues of the goddess.

At the latest in the Neo-Assyrian period the pair of quivers crossing each other appear on representations of other deities, showing their warlike character (for example a stele of Adad from Arslan-Tash[404], two others from Ashur[405], and a stele of god Ashur from Ashur[406] not only in the Assyrian Empire (Mesopotamia, North Syria), but even in Urartu[407].

Ishtar of Arba'il held an important position in the Assyrian pantheon. Her warlike character would explain her appearance on the front of the helmet. In addition to its apotropaic function, in common with other ancient Near Eastern helmets decorated with representations of uncertain, unidentifiable figures and symbols of gods[408], the appearance of a specific goddess (or god) on the front of an Assyrian helmet raises other questions.

As Menzel[409], Dalley and Postgate[410] have pointed out, particular units of the Assyrian army marched under the aegis of particular deities[411]. They suggested that Ishtar was patron of the king's

---

49) ^{d}*Ištar*(XV) *a-ši-bat* ^{uru}*Arba-ìl e-ru-ba-am-ma*

50) *imittu*(XV) *šumēlu*(2,30) *tu-ul-*[*la-a-ta*] *iš-pa-a-ti*

51) [*tam-ḫa-at*] *qašti*(^{giš}PAN) *i-na i-di-šá*

52) [*šal-pat nam-ṣa*]*-ru zaq-tu* [*šá*] *e-peš ta-ḫa-zi*

53) [*ma-ḫar-šá ta*]*-zi-iz*

392 For a detailed study of the 3rd—2nd millennium B.C. iconography of Ishtar armed *see* Colbow 1991.

393 Barrelet 1955, 222ff.

394 Opificius 1961, 86ff., nos. 189ff.; Barrelet 1968, 332, nos. 575 (Larsa), 623ff. (Kish); Moorey 1975, 84-86, pls. XXc, XXIa.

395 Louvre, AO 10479, from Kish; Barrelet 1968, 332, no. 623.

396 Moorey 1975, pl. XXIa.

397 Barrelet 1968, no. 623.

398 Barrelet 1968, no. 623.

399 Oxford, Ashmolean Museum, 1891.331, Buchanan – Moorey 1988, 364, pl. XII, no. 364.

400 Berlin, VA 5887, Ass. 1119, Assur, Tomb 811, von Haller 1954, fig. 88.

401 London, BM 89769, Collon 1988, 166-167, no. 773.

402 Barrelet 1955, 258-259.

403 Moorey 1975, 85.

404 Paris, Louvre, AO 13092, Thureau-Dangin *et al.* 1931, 60, 65f., pl. 2:1; Börker-Klähn 1982, 223, no. 250, pl. 250.

405 1. Bachmann 1927, 36, fig. 23; 2. London, British Museum, BM 115694 (Ass. 8262): Andrae 1938, 50, pl. 21; Börker-Klähn 1982, 223, no. 242, pl. 242.

406 Berlin, Vorderasiatisches Museum, VA 8750 (Ass. 9015); Börker-Klähn 1982, 223, no. 243, pl. 243.

407 Tasyürek 1975, 154-155, pls. XXXVIa-b; Tasyürek 1987, vol. II, 940-955, figs. 7, 10-13.

408 For example: the winged sun-disk, the "lightning fork of god Teisheba", and the gods and genii on Urartian helmets.

409 Menzel 1981, vol. II, T. 85.

410 Dalley – Postgate 1984, 40-41.

411 We do not have detailed information on the marching and battle order of ancient Near Eastern armies, except for an extract from the Epic of Zimri-lim (Durand *et al.* 1988, 393) saying that Adad goes on the left wing of the battle order and Erra (Nergal) goes on the right; the "left and right wings" of the army or militia of Nuzi; the four divisions of the Egyptian army at the battle of

immediate entourage, maybe patron of the body-guard, or patron of the royal troops taking a place in the middle of the battle order. One passage from the inscription of Esarhaddon mentions her as "Goddess of Conflict and Battle who marches at the Side of the King"[412].

The appearance of the figure of Ishtar (even if not of Arba'il) on the front of an other Assyrian iron helmet (**Cat. no. 11**) could even more emphasize this statement and the role of the deity in the Assyrian army.

The appearance of Ishtar of Arba'il on this helmet would imply that it was made for the royal body-guard or the troops under the king's command[413]. Since this is a crested helmet (*Pls. 41-42*), it was the helmet of a foot soldier (body-guard), because the cavalry and chariotry always wear conical helmets and never crested helmets, while the infantry – especially the spearmen with round shields and breast- and back-plates (*kardio phylakes*) – often wear crested helmets.

However, it is also known that troops in the Assyrian army were organized in territorial units[414]. One of these "city units" is the "Arba'il unit" ([uru]*arba-ìl-a-a*)[415]. For example from the Nimrud Horse Lists we know that the personal names in this unit are characterized by a high proportion compounded with the name of Ishtar, the city's patron goddess[416]. Presumably Ishtar was the patron deity of this unit as well as of the individual soldiers. So there are at least two possibilities to explain the appearance of Ishtar of Arba'il on the front of the helmet: the status (*ša qurbūte*, royal body-guard) or origin of the unit (Arba'il unit of the *kiṣir šarrūti*) from which its wearer came.

The Nimrud crested iron helmets are of the same type as the Assyrian crested bronze helmets (**Cat. nos. 13, 14**) and the Urartian crested bronze helmets with the inscription of king Ishpuini (*ca.* 830—810 B.C.) (**Cat. nos. 109, 110**). Altogether we are dealing with at least four helmets. Two helmets (*helmets 1-2*) are indicated by the two front pieces (**Cat. nos. 16, 17**) and the presence of at least three helmets (*helmets 1, 3, 4*) can be reconstructed from the four crest fragments (**Cat. nos. 20-23**). Our suggested reconstruction is:

helmet 1 = **Cat. nos. 16, 19, 20, 21, 24-25-26, 27, 31**
helmet 2 = **Cat. no. 17**
helmet 3 = **Cat. no. 22**
helmet 4 = **Cat. nos. 23, 28**

The shape of the crest would be similar to the crest of the Karlsruhe helmet (**Cat. no. 14**). The width of the inlaid bands and the diameter of the dots are the same on the crest-pieces **Cat. nos. 20, 21, 22** and on the crest-base fragment **Cat. no. 19**[417]. These pieces are from the same series of helmets and fragment **Cat. no. 19** is possibly from the same helmet as fragments **Cat. nos. 20, 21** (*helmet 1*). *Pl. 42* shows that the fragment **Cat. no. 19** is located in the same position on the base of the crest on the back of the helmet as the ribs curving at right angles on the back of the crest-base of the bronze helmets **Cat. nos. 13, 14, 109, 110**, *Pls. 18-23, 109-110*). There are two holes at the lower edge of the fragment, the holes of the rivets fixing the base of the crest to the body of the helmet (*Pls. 41-43*). The Munich helmet (**Cat. no. 13**) is similar in this feature, but on our helmet the lower part of the crest overlaps the top of the body of the helmet (*Pl. 43*).

The ribs on the crested bronze helmets (**Cat. nos. 13, 14, 109, 110**, *Pls. 18-23, 109-110*) – both on the base and on the crest of the helmet – are to strengthen the sheet bronze. In the case of the iron helmet the ribs would be redundant because of the greater strength of the iron. Therefore the ribs were replaced on the iron helmet by inlaid bronze bands – as decoration replacing structural necessity. This replacement represents the strict rules and archaizing tradition of Assyrian helmet-production. The first iron crested helmets were made following the model and decoration of crested bronze helmets, namely they were made with the technique of manufacturing bronze helmets. In that early period there were no independent "iron-shapes". These independent shapes better suited to the character of sheet iron appear only on the palace reliefs of Sennacherib (704—681 B.C.): characteristic conical shapes (*Fig. 1; Chart 2, nos. 193-213*).

The vertical rows of pseudo-rivets on the front of the crest of the Karlsruhe helmet (**Cat. no. 14**) evoke the earlier construction of this type of crested helmet: the parts were riveted together. On the

---

Qadesh, and the *turtānu*s of the left and right: 1) [lú]*tur-ta-nu imitti*(XV), 2) [lú]*tar-tan-nu šumēli*(2,30); (Civil 1969, 238, col. I, 1-2; von Soden 1965—1981, 1332, *s.v. ta/urtānu*).

412 Borger 1956, 96, Monument A, Vs. 9: [d]*Ištar*(XV) *be-let qabli*($MURUB_4$) *ù taḫāzi*(MÈ) *a-l*[*i*]*-kat i-di-ia*.

413 *Ša qurbūte* (left and right), *ša šēpē* units. These troops consisted of cavalry and chariotry (Dalley – Postgate 1984, 39) and supposedly infantry as well.

414 Dalley – Postgate 1984, 36-37.

415 Presumably they consisted not only of cavalry but of infantry as well.

416 Dalley – Postgate 1984, no. 102. iii. 17'-18'; no. 108. ii. 20, 22, 23; no. 110. iii. 10', 12'.

417 This fact indicates that these helmets were mass-produced.

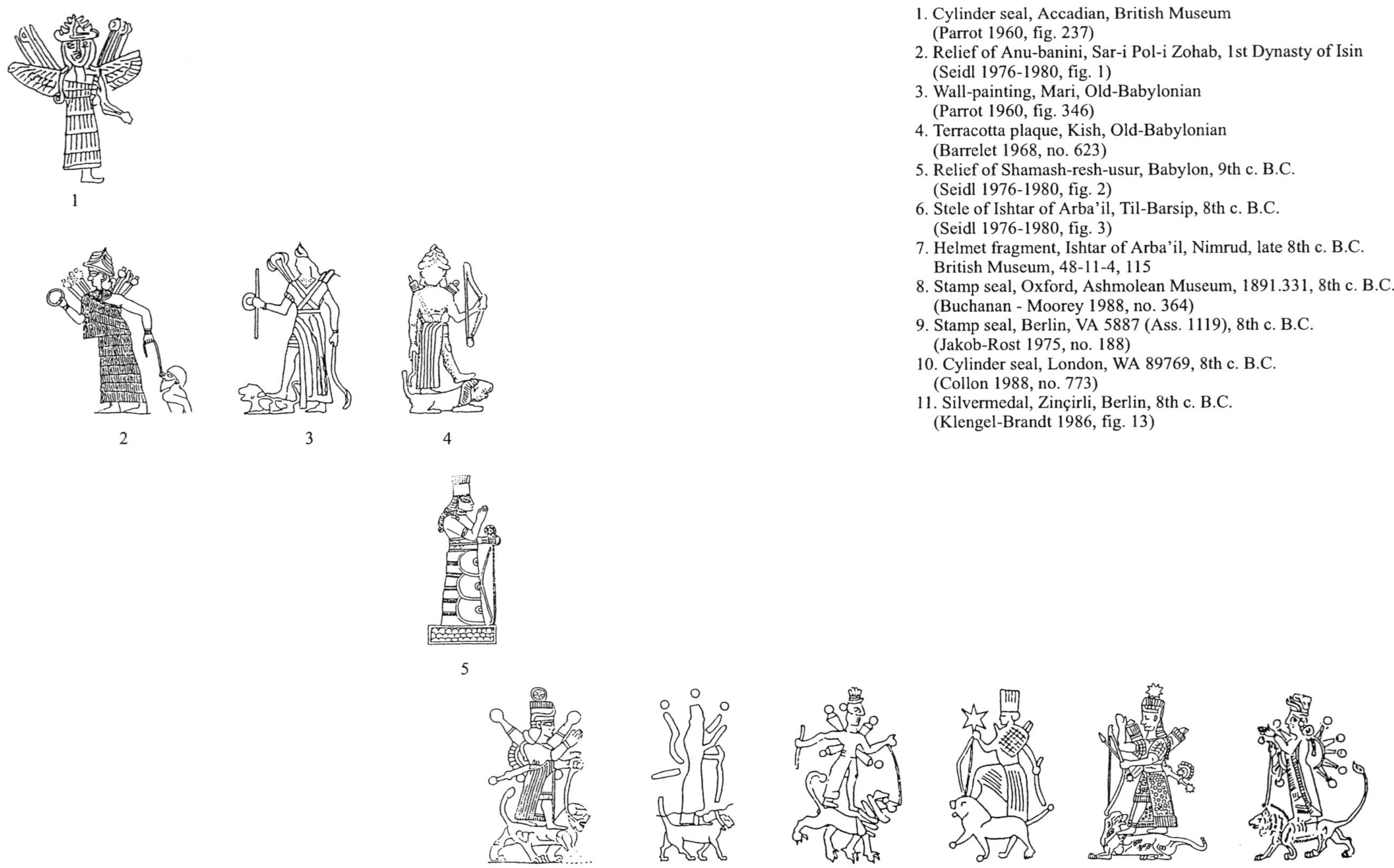

1. Cylinder seal, Accadian, British Museum (Parrot 1960, fig. 237)
2. Relief of Anu-banini, Sar-i Pol-i Zohab, 1st Dynasty of Isin (Seidl 1976-1980, fig. 1)
3. Wall-painting, Mari, Old-Babylonian (Parrot 1960, fig. 346)
4. Terracotta plaque, Kish, Old-Babylonian (Barrelet 1968, no. 623)
5. Relief of Shamash-resh-usur, Babylon, 9th c. B.C. (Seidl 1976-1980, fig. 2)
6. Stele of Ishtar of Arba'il, Til-Barsip, 8th c. B.C. (Seidl 1976-1980, fig. 3)
7. Helmet fragment, Ishtar of Arba'il, Nimrud, late 8th c. B.C. British Museum, 48-11-4, 115
8. Stamp seal, Oxford, Ashmolean Museum, 1891.331, 8th c. B.C. (Buchanan - Moorey 1988, no. 364)
9. Stamp seal, Berlin, VA 5887 (Ass. 1119), 8th c. B.C. (Jakob-Rost 1975, no. 188)
10. Cylinder seal, London, WA 89769, 8th c. B.C. (Collon 1988, no. 773)
11. Silvermedal, Zinçirli, Berlin, 8th c. B.C. (Klengel-Brandt 1986, fig. 13)

Dezsö 2000

Fig. 5: Ishtar armed

fragments **Cat. nos. 21, 22** (*Pls. 30, 31*) there is an inlaid dot under the lower inlaid band. It would be a piece from the series of dots running down along the inlaid bronze band – similar to the ribs on the front of the crest of the Mainz helmets (**Cat. nos. 109, 110**) and to the vertical rows of pseudo-rivets on the front of the crest of the Karlsruhe helmet (**Cat. no. 14**). The inlaid dots on the fragment **Cat. no. 19** (*Pl. 28*) are similar to the pseudo-rivets between two ribs on the crest of the Karlsruhe helmet. On the Munich helmet (**Cat. no. 13**) it is the same (*Pls. 18-19*), but on the front edge of the crest-pieces there is a series of rivets fastening the two crest-pieces to the U-shaped bronze sheet holding the crest-pieces together (the same method was used in the case of **Cat. no. 23** (*Pl. 32*) = *helmet 4*). The crest of our helmet (**Cat. nos. 20-22**, *Pls. 29-31, 42*) had the same structure, but the crest-pieces were welded not riveted to the U-shaped iron sheet. There is no trace of rivets, only on fragment **Cat. no. 23**.

On the Karlsruhe helmet (**Cat. no. 14**) there are two pairs of horizontal ribs around the helmet with a row of pseudo-rivets between them (*Pls. 20-22*). Our fragment **Cat. no. 31** closely resembles these bands. On the Karlsruhe helmet these ribs and pseudo-rivets recall the earlier construction of this type of helmet – the upper band was the juncture of the crest and the body of the helmet (**Cat. no. 17** fragment is from a helmet of similar structure, *Pls. 26, 40*). On our fragment **Cat. no. 31** there are two inlaid bronze dots (their function is only decorative similar to the pseudo-rivets) and two small holes for the rivets fixing the band to the lower edge of the crest and to the body of the helmet (*Pls. 42, 43*).

Fragments **Cat. nos. 24-25-26, 27** (*Pls. 33-34, 39, 42*) are probably from the side of the same helmet. The inlaid bands replaced the ribs of the bronze helmets, but on the lower edge of fragment **Cat. no. 24-25-26** there is an iron band hammered to the rim around the helmet to strengthen it (or the trace of an ancient repair).

Fragments **Cat. nos. 16, 17** are from the fronts of two different helmets. These continuous arcs replacing the pairs of curves on the front of Assyrian helmets are characteristics of the 2nd and 3rd phase of evolution of Assyrian conical/pointed helmets (*Fig. 1*). There are two holes surviving along the finished upper edge of fragment (**Cat. no. 17** from a series of holes for rivets fixing the lower part of the helmet to the upper (crested or pointed) "cap" of the helmet (*Pl. 40*).

The best parallels for the reconstructed shape of the Nimrud crested iron helmets – with or without earflaps – appear on the palace reliefs of Tiglath-Pileser III (745—727 B.C.)[418], Sargon II (721—705 B.C.)[419] and Sennacherib (704—681 B.C.)[420]. The shape and construction of crested helmets have been changed and standardized by the reigns of Esarhaddon (680—669 B.C.)[421] and Ashurbanipal (668—627 B.C.)[422]. On the sculptures of Ashurbanipal there are only hemispherical helmets with earflaps and small crest presumably made of a separate piece of metal fixed to the "skull-piece" of the helmet by rivets. The first examples of front decoration (continuous arcs, semicircles without vertical rib) on crested helmets appears on these palace reliefs (*Charts 3-4, nos. 253-255*).

Since the Nimrud crested helmets have an older shape than the shape of helmets depicted on the palace reliefs of Ashurbanipal, and the figural decoration of Ishtar of Arba'il on **Cat. no. 16** fragment is the same as on the stele from Til-Barsip, we can date them to the second half of the 8th century B.C. Many of the objects in the Northwest Palace seem to have been stored in the late 8th century B.C. and simply left there throughout the 7th century B.C. to be excavated from the debris of the sack of Nimrud in 614/612 B.C.

The technique of inlaying iron with bronze is unique. The Assyrian iron helmets (**Cat. nos. 10, 16-31**) together with the Zinçirli iron helmet fragments (**Cat. no. 38**) are the earliest and only examples of this technique from the ancient Near East. The earliest representational evidence for this sophisticated decorative technique is shown on the Til-Barsip wall-paintings (*Chart 2, nos. 163-165*), where the helmets were painted light blue (iron) and the decoration was painted yellow (bronze).

Bronze inlay as a decorative method is reasonable only on iron objects with a relatively big surface. The weapons, arms and armour are certainly the best objects for this reason. Undoubtedly the iron weapons, armour and helmets – as the most important and mass-produced objects – furthermore the Assyrian weapon industry with its continuous and enormous orders of iron weapons, armours and tools was the base, on which the Assyrian (and North Syrian) iron industry reached the highest standard in the ancient world. The best example of this is **Cat. no. 10** (*Pls. 9-11*), the high pointed Assyrian iron helmet hammered out from one piece of iron and decorated with bronze inlay. Both hammering out a (pointed) helmet from one piece of iron and inlaying the iron with bronze are unknown from other territories of the ancient Near East and emphasize the overall technical skill of the Assyrian blacksmiths and the highest standard of the Assyrian iron (and consequently the Assyrian weapon)

---

[418] *Charts 3-4, no. 150.*
[419] *Charts 3-4, no. 192.*
[420] *Charts 3-4, nos. 222-225.*
[421] *Charts 3-4, no. 229.*
[422] *Charts 3-4, nos. 247-255.*

industry. Consequently it can be assumed that the background of the production of iron helmets is not necessarily a local artistic tradition, but much more the new tendency of mass production of iron implements (mainly arms and armour) and the needs of the customer, which customer was most probably a huge army[423]. Its production needed highly developed ironworking tradition, technical skill and background, which existed probably only in Assyria and North Syria. Experts are divided concerning the technical level of Assyrian ironworking industry and the achievement of Assyrian ironsmith[424]. It has to be mentioned, however, that after the collapse of the Assyrian and Babylonian empires the production of iron helmets ceased for a long period, although the Persians might inherite part of the tradition. Iron helmets were virtually unknown for hundreds of years (till the Hellenistic period) and were very rare in later periods of the antiquity as well. One of the reasons of the lack of iron helmets was probably the lack of the necessary technical skill.

The bronze inlay has a purely decorative function[425]. On Assyrian iron helmets which were archaizing in decoration it was only decoration replacing structural necessity[426]. The technique of inlaying bronze (or other nonferric metals) into the iron is known in the literature as "Damascene inlay"[427]. The groove of the design in the iron was probably undercut to keep the bronze in position[428] (*Fig. 6*). The surface of the groove might be "toothed" by a chisel[429], or as can be seen on the drawings of x-ray photographs of helmet fragments **Cat. nos. 20-22, 24-25-26** (*Pls. 29-31, 33, 39, 42*) small pins projecting from the groove in the iron were left or most probably small holes were drilled or punched into the surface of the groove to provide better conditions (surface) to keep the bronze inlay in position (*Fig. 6*). This additional method (pin projecting from the groove or hole in the surface of the cavity) was used only for inlaying long narrow strips of bronze into the iron. It can be assumed, that the linear designs were inlaid with the "Damascene inlay" technique: a bronze wire or rod was hammered into the undercut groove, while the undercut grooves of the sophisticated figurative designs (figures of the king, crown-prince and attendant on helmet **Cat. no. 10**; the figure of the goddess Ishtar of Arba'il on helmet fragment **Cat. no. 16**; and the figure of the lion on helmet fragment **Cat. no. 18** were probably filled with moulded bronze[430] and after cooling the surface was polished.

16. Nimrud, Northwest Palace, Chamber I, Group no. **A.4.1** – H.: 5.4 cm, W.: 9.1 cm. Fragment from the front of an iron helmet (probably crested helmet) (*helmet no. 1*) inlaid in bronze with the figure of Ishtar of Arba'il standing on the back of a lion. Ishtar is in full battle dress: there is a quiver and a shield (or *melammu*, "aura", "radiance") hanging on her back, her sword is clearly visible at her waist. Another quiver projects from her back in front of her face. Ishtar wears a polos (flat top cap). There are three dots inlaid above the figure: one dot – the star of Ishtar? – is above or on her cap, and another one above the quivers of the goddess perhaps representing the quiver-caps. Her right hand is raised, her left hand holds the reins of her lion, from which only the tail is extant. The scene was framed by a double line of arcs wirh a row of dots between them. A longer and a shorter section of the inner arc with two dots and a shorter section of the outer arc are extant on the right side of the fragment (*Pls. 25, 39, 41*). Late 8th – early 7th century B.C. – London, British Museum, Inv. no.: 48-11-4, 115 – Layard 1850, vol. I, 341-342; Curtis 1979, vol. I, 179, vol. II, 31; Dezső – Curtis 1991, 105-122, no. 1, figs. 1, 16, 18.

17. Nimrud, Northwest Palace, Chamber I, Group no. **A.4.2** – H.: 5.2 cm, W.: 6.7 cm. Fragment from the front of an iron helmet (*helmet no. 2*). There are a horizontal inlaid bronze band and small pieces of a curving inlaid band (meeting each other beyond this fragment) on the fragment. The arc was probably the outer line of a double-line arc framing a figural scene on the front of the helmet. There are three dots – remaining from a row of dots between the two lines of the arc – under the curve. The width of the inlaid bands distinguishes this fragment from the others. There are two small holes for rivets along the upper finished edge of the fragment. This fragment is part of a wide iron band forming the lower part (edge) of the helmet onto which the upper (crested or conical) "skull" of the helmet was riveted (*Pls. 26, 40*). Late 8th – early 7th century B.C. – London, British Museum, Inv. no.: 48-11-4, 115 – Layard 1850, vol. I, 341-342; Curtis 1979, I, p. 179, vol. II, 31; Dezső – Curtis 1991, 105-122, no. 2, figs. 3, 17.

18. Nimrud, Northwest Palace, Chamber I, Group no. **A.4.3** – H.: 5.0 cm, W.: 7.5 cm. Fragment from the crest (?) of an iron helmet with inlaid bronze decoration. There are remains of a "*lion passant*" on the fragment. The head and one of the back legs are missing. The animal was probably surrounded by an inlaid frame of which only a vertical line is extant on the left side of the fragment (*Pl. 27*). Late 8th – early 7th century B.C. – London, British Museum, Inv. no.: 48-11-4, 115 – Layard 1850, vol. I, 341-342, Curtis 1979, vol. I, 179, vol. II, 31; Dezső – Curtis 1991, 105-122, no. 3, fig. 4.

---

[423] The Assyrian army needed probably thousands of new weapons and helmets (as well as other equipment) in every year, and a huge repairing capacity to repair the equipment which was damaged on the annual campaigns. Figures of the Assyrian imperial army (second half of 8th century – 7th century B.C.) can be estimated around 150.000 soldiers. The supply of the army needed an enormous weapon (and of course other) industry, which was one of the reasons of the deportations of skilled craftsmen from several parts of the empire.

[424] Pleiner – Bjorkman 1974, 283-313; Curtis *et al.* 1979, 369-390; Piaskowski – Wartke 1989, 89-114; Born – Seidl 1995; Moorey 1999, 278-292.

[425] Polished yellow metal on shining iron surface.

[426] Replacing the *repoussé* ribs of bronze helmets which provided a greater strength to the bronze sheet. The ribs would have been redundant on iron helmets becouse of the greater strength of the iron.

[427] Untracht 1975, 153-155, esp. p. 155.

[428] Untracht 1975, fig. on p. 155; Rostoker – Bronson 1990, 175-176.

[429] Untracht 1975, 155.

[430] The iron helmet might be heated for the better join.

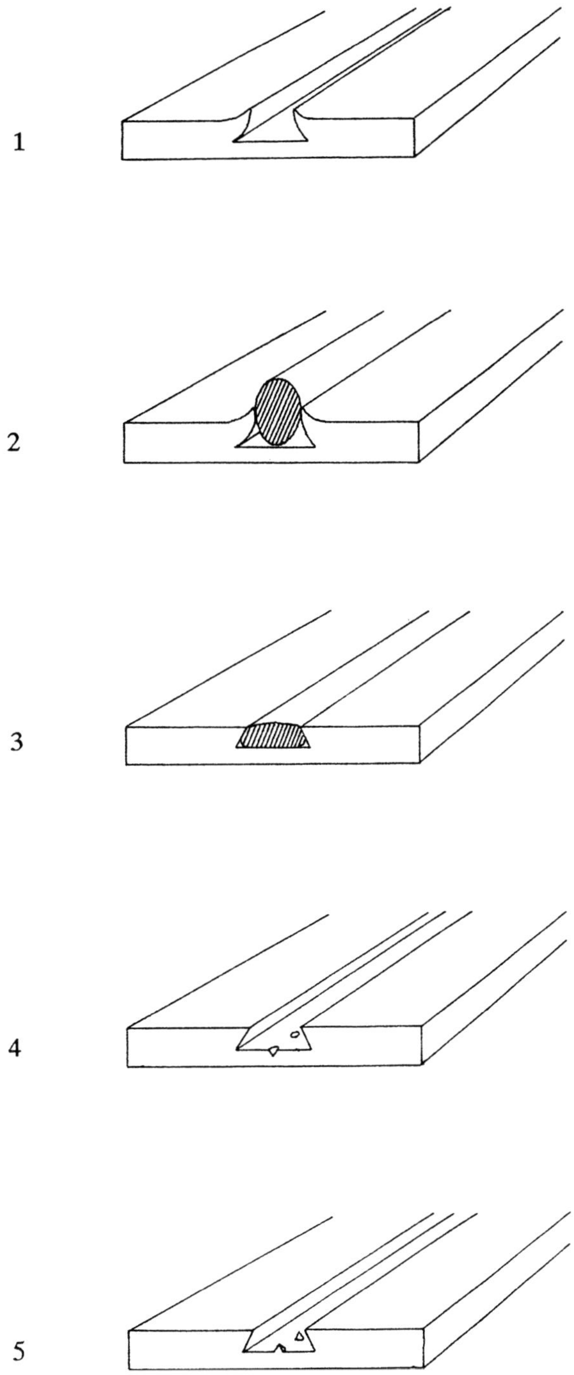

Fig. 6: Drawing of bronze inlay technics: 1. The undercut groove; 2. Hammering the wire in the groove; 3. The finished inlay; 4-5. Additional Assyrian methods: holes and/or pins in the groove.

19. Nimrud, Northwest Palace, Chamber I, Group no. **A.4.4** – H.: 4.7 cm, W.: 5.5 cm. Iron fragment from the base of the crest of a helmet (probably *helmet no. 1*) with inlaid bronze decoration. The lower edge of the cfragment is curved at a right angle, but is broken at the curve. This right angle curve was the point of contact between the crest and the "skull" of the helmet. The outward curving edge of the crest overlapped the edge of the body of the helmet. There are two holes in the curve for rivets fixing the crest and the body of the helmet together. The decoration of the fragment consists of two horizontal inlaid bronze bands, the lower is curved upwards at a right angle, but does not reach the upper band because of the vertical part is missing. There are two inlaid dots on the fragment: one between the horizontal bands on the left side of the vertical band and the other on the right side of the vertical band (*Pls. 28, 42*). Late 8th – early 7th century B.C. – London, British Museum, Inv. no.: 48-11-4, 115 – Layard 1850, vol. I, 341-342, Curtis 1979, vol. I, 179, vol. II, p. 31; Dezső – Curtis 1991, 105-122, no. 4, figs. 5, 20.

20. Nimrud, Northwest Palace, Chamber I, Group no. **A.4.5** – H.: 6.5 cm, W.: 9.0 cm. Iron fragment of the crest of a helmet (*helmet no. 1*) decorated with bronze inlay. The fragment is from a crest curving forward. The peak of the crest is missing. There are 8 holes along the edge of the crest-fragment for fixing the plume between the two crest-pieces of the helmet. The crest shows traces of an ancient repair: the broken second hole from the left was replaced by a smaller hole out of line. There are two inlaid bronze bands following the edges of the crest and a third branching from the lower band parallel with the upper band. Several small circles are visible on the x-ray photograph in the traces of the bronze bands, probably small iron pins or projections from the iron, or small holes in the iron for the better fixing of the inlay into the iron. There is a dot inlaid with bronze between the bands (*Pls. 29, 42*). Late 8th – early 7th century B.C. – London, British Museum, Inv. no.: 48-11-4, 115 – Layard 1850, vol. I, 341-342; Curtis 1979, vol. I, 179, vol. II, 31; Dezső – Curtis 1991, 105-122, no. 5, figs. 6, 20.

21. Nimrud, Northwest Palace, Chamber I, Group no. **A.4.6** – H.: 3.6 cm, W.: 6.8 cm. Iron fragment of the crest of a helmet (*helmet no. 1*) decorated with bronze inlay. Similar to **Cat. no. 20** probably the pair of it. The peak is extant with two holes and an inlaid dot. There are two inlaid bronze bands following the edges of the crest. There are five small iron pins (or holes in the iron: only small circles are visible on the x-ray photograph) in the traces of the upper band for the better fixing of the bronze inlay into the iron (*Pl. 30*). Late 8th – early 7th century B.C. – London, British Museum, Inv. no.: 48-11-4, 115 – Layard 1850, vol. I, 341-342; Curtis 1979, vol. I, 179, vol. II, 31; Dezső – Curtis 1991, 105-122, no. 6, fig. 7.

22. Nimrud, Northwest Palace, Chamber I, Group no. **A.4.7** – H.: 5.8 cm, W.: 9.7 cm. Iron fragment of the crest of a helmet (*helmet no. 3*) decorated with bronze inlay. Similar to **Cat. nos. 20-21**, but this piece is from a third helmet (*helmet no. 3*). The peak is extant. There are 7 holes along the edge of the crest for fixing the plume. There are two inlaid bronze bands following the edges of the crest reaching each other at the peak. There is a dot between the bands and another under them near the peak. There are 7 small circles visible in the traces of the bronze bands on the x-ray photograph (small iron pins or iron holes in the iron) for fixing the inlay into the iron. On the inner surface of the fragment there is an iron band following the edges of the crest. The broken edge of the iron band at the lower edge of the crest shows that this band was part of an U-shaped iron sheet with its vertical sides holding the two crest-pieces. This iron sheet was the front piece of the crest. The crest-pieces and this support-piece were welded together without rivets (*Pl. 31*). Late 8th – early 7th century B.C. – London, British Museum, Inv. no.: 48-11-4, 115 – Layard 1850, vol. I, 341-342; Curtis 1979, vol. I, 179, vol. II, 31; Dezső – Curtis 1991, 105-122, no. 7, fig. 8.

23. Nimrud, Northwest Palace, Chamber I, Group no. **A.4.8** – H.: 5.8 cm, W.: 9.7 cm. Iron fragment of the crest of a helmet (*helmet no. 4*) decorated with bronze inlay. Similar to **Cat. nos. 20-22**, probably from a fourth helmet. The peak of the crest is extant. There are 7 holes along the upper edge of the fragment to fix the plume. There are two other holes along the lower edge of the fragment probably for the rivets holding the two crest-pieces of the helmet to an U-shaped iron sheet holding them together. There are three inlaid bronze dots on the fragment (*Pl. 32*). Late 8th – early 7th century B.C. – London, British Museum, Inv. no.: 48-11-4, 115 – Layard 1850, vol. I, 341-342; Curtis 1979, vol. I, 179, vol. II, 31; Dezső – Curtis 1991, 105-122, no. 8, fig. 9.

24-25-26. Nimrud, Northwest Palace, Chamber I, Group no. **A.4.9-10-11** – H.: 9.2 cm, W.: 10.8 cm. Three fragments from the side of an iron helmet with inlaid bronze decoration. There are three horizontal lines on the fragments inlaid in bronze (the distance between them is 2.3 cm) with a row of dots between them. There is one extant dot in the upper row and there are two in the lower row. There is a 1.2 cm wide horizontal iron band hammered on the bottom of the outer surface of the fragment, probably to strengthen the edge. There are two small circles in the traces of the lower and two in the traces of the middle inlaid bands visible on the x-ray photograph. These are the traces of the same iron pins projecting from the cavities carved in to the iron, or small holes in the iron, both for fixing the bronze into the iron. There are no traces of rivets used for this purpose on the inner surface of the fragments. On the inner surface of the fragments there are traces of an oblique iron band (W.: 1.2 cm). It was probably for ancient repairs on the broken edge of the helmet (*Pls. 33, 39, 42*). Late 8th – early 7th century B.C. – London, British Museum, Inv. no.: 48-11-4, 115 – Layard 1850, vol. I, 341-342; Curtis 1979, vol. I, 179, vol. II, 31; Dezső – Curtis 1991, 105-122, nos. 9-10-11, figs. 10, 16, 20.

27. Nimrud, Northwest Palace, Chamber I, Group no. **A.4.10** – H.: 5.4 cm, W.: 9.2 cm. Fragment from the side of an iron helmet (*helmet no. 1*) decorated with bronze inlay. There is a horizontal line inlaid with bronze on the fragment. There are 4 dots – one under the line, one above the line and two smaller dots above the line as part of a row of dots. The decoration of this fragment closely connects it to fragments **Cat. nos. 24-25-26**, probably from the same helmet. The distance between the line and the lower dot, the diameter of the bigger dots, and the width of the inlaid band are the same on each fragments (*Pls. 34, 39, 42*). Late 8th – early 7th century B.C. – London, British Museum, Inv. no.: 48-11-4, 115 – Layard 1850, vol. I, 341-342; Curtis 1979, vol. I, 179, vol. II, 31; Dezső – Curtis 1991, 105-122, no. 12, figs. 11, 16.

28. Nimrud, Northwest Palace, Chamber I, Group no. **A.4.13** – H.: 4.7 cm, W.: 8.6 cm. Fragment from the side of an iron helmet. There are three inlaid bronze dots placed at the vertices of an equilateral triangle on the fragment. The diameter of the dots is the same as on the fragments **Cat. nos. 24-25-26, 27**. Similar decoration is shown on the crest-piece fragment **Cat. no. 23** (*Pl. 35*). Late 8th – early 7th century B.C. – London, British Museum, Inv. no.: 48-11-4, 115 – Layard 1850, vol. I, 341-342; Curtis 1979, vol. I, 179, vol. II, 31; Dezső – Curtis 1991, 105-122, no. 13, fig. 12.

29. Nimrud, Northwest Palace, Chamber I, Group no. **A.4.14** – H.: 4.6 cm, W.: 6.5 cm. Fragment from the side of an iron helmet. There is an inlaid bronze band on the fragment from the linear decoration of the helmet (*Pl. 36*). Late 8th – early 7th century B.C. – London, British Museum, Inv. no.: 48-11-4, 115 – Layard 1850, vol. I, 341-342; Curtis 1979, vol. I, 179, vol. II, 31; Dezső – Curtis 1991, 105-122, no. 14, fig. 13.

30. Nimrud, Northwest Palace, Chamber I, Group no. **A.4.15** – H.: 3.6 cm, W.: 5.5 cm. Fragment from the side of an iron helmet. There is an inlaid bronze band on the fragment probably from the linear decoration of the base of the helmet (*Pl. 37*). Late 8th – early 7th century B.C. – London, British Museum, Inv. no.: 48-11-4, 115 – Layard 1850, vol. I, 341-342; Curtis 1979, vol. I, 179, vol. II, 31; Dezső – Curtis 1991, 105-122, no. 15, fig. 14.

31. Nimrud, Northwest Palace, Chamber I, Group no. **A.4.16** – H.: 2.3 cm, W.: 4.7 cm. Iron band from an iron helmet with inlaid bronze decoration. There is an iron band (W.: 1.7 cm) on the fragment with a small piece from the side of the helmet. There are two dots inlaid with bronze on the band and two small holes for rivets fixing the lower edge of the crest and this covering band to the body of the helmet (*Pls. 38, 42*). Late 8th – early 7th century B.C. – London, British Museum, Inv. no.: 48-11-4, 115 – Layard 1850, vol. I, 341-342; Curtis 1979, vol. I, 179, vol. II, 31; Dezső – Curtis 1991, 105-122, no. 16, figs. 15, 20.

NORTH SYRIAN HELMETS

Group NS.1: North Syrian conical bronze helmets (Cat. nos. 32-37)

As has already been mentioned North Syria together with Assyria (as well as Urartu and Northwest Iran) was one of the most important metalworking centres of the ancient Near East.

The place of the mid 2nd millennium B.C. evolution of pointed helmets is uncertain. The earliest known representation is Assyrian (*Chart 1, no. 41*). But as the evolution of conical crested helmets derived from the Late Bronze Age Hittite and North Syrian helmet tradition, it can be clearly connected with North Syria (*Chart 4*); but only through representational evidence as there is not a single crested helmet known from North Syria. Nor is there a single helmet known from North Syria from the first phase of evolution of conical/pointed helmets (*Fig. 1*).

On basis of their provenance 5[431], by reason of their form and stylistic features 2[432] conical/pointed helmets can be placed in this group. But all of the seven helmets can be dated to the second half of the 8th century B.C., when North Syria was already under Assyrian control. Consequently these helmets cannot unreservedly be considered as North Syrian helmets. Much rather they should be regarded as the products of Neo-Assyrian imperial art and metalworking, which gradually assimilated the culture and crafts of the nations of the empire creating a *koine* at least in case of warfare, arms and armour. The shape of the helmet itself is Assyrian since there is no evidence for conical/pointed helmets in North Syria before the Neo-Assyrian Period. There are two bronze statuettes wearing the 9th century B.C. type of Assyrian conical helmet (*Fig. 1*, Phase I) from South Syria or rather Phoenicia: "Beqaa Valley" (*Chart 2, no. 90*[433]); "Tartus" (*Chart 2, no. 91*[434]). Since their provenance and dating is uncertain[435] they should not be considered as evidence for the evolution of conical/pointed helmets.

Three of the five helmets (**Cat. nos. 32, 33, 38**) from controlled excavations were found at Zinçirli. These three helmets are from the destruction level which can be connected with the siege of the city by Sargon II in 720 B.C. There is no evidence to indicate whether these helmets were worn by Assyrians or local soldiers. On the palace reliefs the enemies of Assyria never wear conical/pointed helmets[436], only crested ones. On the other hand only one of the three Zinçirli helmets is paralelled on contemporary palace reliefs of Assyria (Tiglath-Pileser III), namely the **Cat. no. 32** (*Pl. 44*). However the decorative system of the other bronze helmet (**Cat. no. 33**) is the evidence on which the identification of two other helmets of the group was based (**Cat. no. 34**, without provenance; **Cat. no. 35**, from Cyprus): winged sun-disk[437] between pairs of curves bending inward on the front of the helmet (*Fig. 7*). On 9th century B.C. Assyrian helmets the winged sun-disk is only a complementary motif above the tree-of-life. There is not a single Assyrian helmet known on which the winged sun-disk was an independent motif. Unfortunately there is no evidence for winged sun-disks on helmets shown on Assyrian sculptures, because contemporary sculptors of the Assyrian reliefs did not sculpt incised details to represent the incised decoration of helmets. However, they may have used paint to represent the details. This would explain why helmets on the palace reliefs of Tiglath-Pileser III (745—727 B.C.) and of Sargon II (721—705 B.C.) now appear to be undecorated.

Similarity to the helmets mentioned above, its shape and iconographic features, allows a further helmet (**Cat. no. 34**, said to be come from "Luristan", now in the British Museum) to be ranged among the helmets of this group. The origin of this helmet, one of the most beautiful, is doubtful. It has been connected with Western Iran[438], but Calmeyer, who was among the first to publish the helmet has already emphasized the Zinçirli parallels of the winged sun-disk[439]. However the height (37.5 cm) and

---

[431] **Cat. nos. 31-34, 36, 37**.
[432] **Cat. nos. 35, 38**.
[433] Beirut, National Museum; Seeden 1980, 109, no. 1723, pl. 102.
[434] Paris, Louvre, AO 22265; Seeden 1980, 109-110, no. 1725, pls. N, 103.
[435] Seeden 1980, 109-110 dated the statuettes around the late 2nd – early 1st millennium B.C. This date seems a little early, since the statuettes wear the Assyrian type of conical helmet with the characteristic 9th century – early 8th century B.C. motifs on the front of the their helmets. Amiet 1980, 434, no. 727 dated the statuettes to the 6th–5th centuries B.C. It seems to be too late on the same basis as mentioned above. J. Crouwel and M.A. Littauer (Collon – Crouwel – Littauer 1976, 79) dated the Louvre statuette to the Achaemenid Period on the basis of the Persian Age parallels of its chariot. On the other hand D. Collon (Collon – Crouwel – Littauer 1976, 79-81) insists on the 14th–13th century B.C. dating of the statuette. In the present writer's view a late 9th, or rather first half of 8th century B.C. date can be suggested for the statuettes.
[436] With a single exception on the sculpture of Sennacherib showing some defenders of Lachish wearing conical helmets (WA 124906).
[437] Calmeyer 1969, 91; Curtis 1979, vol. I, 182; Gröschel 1986, 74, note 236.
[438] Barnett 1967, 38, pl. 12a.
[439] Calmeyer 1969, 89, 91, Cat. no. 44C, fig. 91.

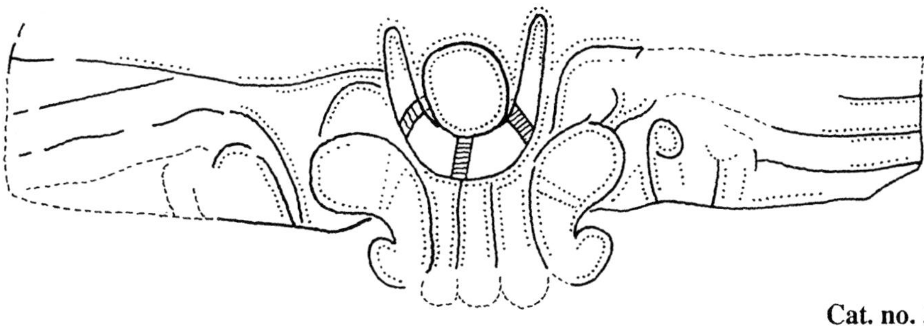

Cat. no. 33

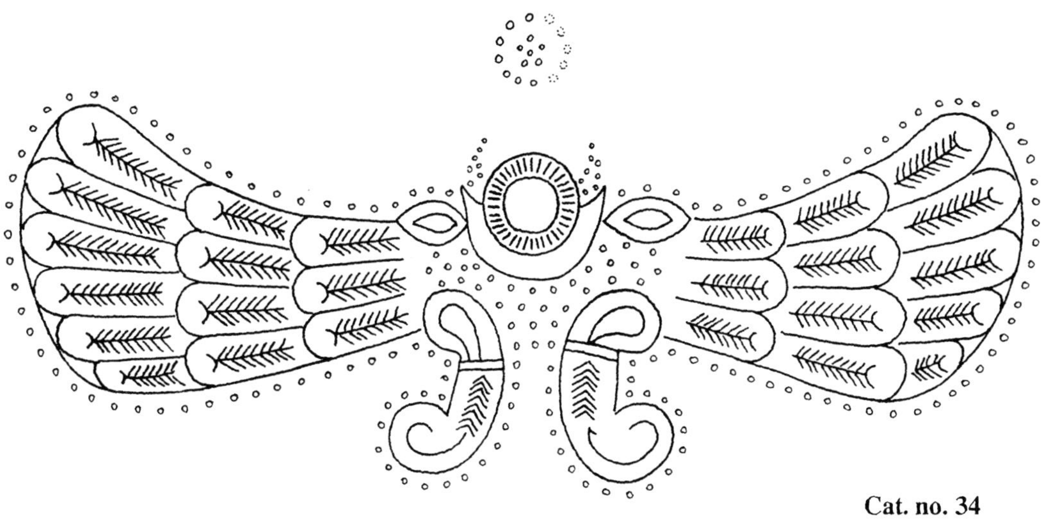

Cat. no. 34

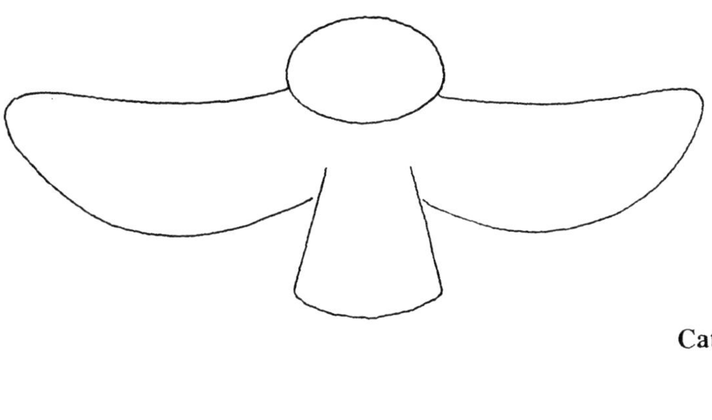

Cat. no. 35

Fig. 7: Winged sun-disks on the front of North Syrian conical helmets

the very long tubular point of the helmet is unfamiliar amongst the Western Iranian helmets. This is rather a characteristic feature of Assyrian (*Chart 2, nos. 139-143*) and North Syrian helmets (*Chart 2, nos. 163-171*: Til-Barsip; *no. 173*: Arslan-Tash; *nos. 175-176*: Karatepe). The other feature which may connect the helmet to North Syria is the winged sun-disk between the curves on the front of the helmet.

Although the winged sun-disk was particularly important in Syria it is not definitive indication that this helmet was made there, as the symbol was widespread in the Near East ("*altorientalisches Gemeingut*"). It might have been made at a number of places in the Near East in the 8th century B.C. and travelled with soldiers on campaign.

This seems to be the case with two helmets of North Syrian origin (or one of them of North Syrian inspiration) which were found in Cyprus (**Cat. nos. 35, 36**): the North Syrian (Assyrian) origin or at least inspiration is unquestionnable. However, helmet **Cat. no. 35** was probably originally of North Syrian manufacture (*see* the winged sun-disk and the composition of the curves not misinterpreted yet) but helmet **Cat. no. 36** is probably a local product and was made by copying a North Syrian or Assyrian helmet directly in the late 8th – early 7th century B.C.[440] These helmets are good examples of the direct influence of North Syria (under Assyrian rule) on the Eastern Mediterranean. There is a hiatus in the development of Aegean/Mediterranean helmets in the 10th–9th centuries B.C.[441]. The Bronze Age Aegean helmet tradition ended in the 11th century B.C. and the new Mediterranean helmet tradition from the late 9th century B.C. developed under the influence of the 9th–8th centuries B.C. Syro-Assyrian helmet tradition.

The different levels of the 8th–7th century B.C. Syro-Assyrian influence in the Aegean can be reconstructed from the finds of orientalizing helmets[442]:

Level 1: Direct import[443]

Cat. no. 35 (Cyprus, Palaepaphos, Kouklia, Mavrommatis Tomb (Tomb 7-8)) probably an original North Syrian helmet imported from the Near East and found in Cyprus.

This helmet is probably an original North Syrian helmet imported from the Near East and found in Cyprus. P. Schauer suggested[444] that this helmet might have arrived in Cyprus with some troops of Sargon II (*see* the Kition stele of Sargon II above). The winged sun-disk is one trait identifying this helmet as Syro-Assyrian, manufactured probably in North Syria in the second half of the 8th century B.C. The identification of two other helmets as Syro-Assyrian was based on this motif: there are three helmets in the group decorated with this motif at the front. One of them is the Palaepaphos helmet, the second is a bronze helmet from Zinçirli (**Cat. no. 33**), and third is from unknown provenance but said to come from "Luristan" now in London (**Cat no. 34**). Von Luschan and Andrae recognized[445] on the fragments of the Zinçirli iron helmet (**Cat. no. 38**) the elements of a winged sun-disk, which is a characteristically North Syrian motif[446]. On 9th century B.C. Assyrian helmets the winged sun-disk is only a complementary motif above the tree-of-life. There is not a single Assyrian helmet known on which the winged sun-disk was an independent motif. So as has been mentioned we can assume that these helmets (which were made in the last third of the 8th century B.C., when almost the whole territory of North Syria was already under Assyrian rule) were made probably in North Syrian workshops for the Assyrian army, or for let us say North Syrian contingents of the Assyrian imperial army (which might contain Assyrian and North Syrian troops as well). The winged sun-disk as an independent motif for some unknown reason was connected with Urartian helmets[447]. After this explanation the Palaepaphos helmet (for the same unknown reason) became an "Urartian helmet"[448]. But from the middle of the 8th century B.C. there was no Urartian influence exerted on North Syria any more. Furthermore, I must say there is not a single Urartian helmet decorated with winged sun-disk, and there is not a single unprovenanced helmet decorated with winged sun-disk which has been or would be attributed to Urartu. These helmets of course might not have reached different parts of the Mediterranean exclusively via military campaigns. They could reach territories outside their area of origin as booty, or with actual objects of trade. Schauer sited a bronze fragment from

440 Both of the helmets were found in tombs which can be dated to the Cypro-Archaic I Period (750—650 B.C.).
441 Schauer 1988, fig. 9.
442 For detailed study *see* Dezső 1998.
443 Dezső 1998, 11-12, ill. 6.
444 Schauer 1983, 186-187.
445 von Luschan – Andrae 1943, 78.
446 I. Winter phrased the winged sun-disk of the bronze Zinçirli helmet as a "good 'North-Syrian style' winged sundisk"; Winter 1988, 194.
447 Borchhardt 1972, 100, 103-105; Pflug 1988b, 34.
448 Maier – Karageorghis 1984, 175, pl. 162.

the point of a helmet from the Ría de Huelva weapon deposit as a fragment of an oriental helmet, which might have reached Spain with Phoenician traders.[449] Hencken however showed that the piece is from an Europaean type of crested helmet which was used from Spain to Germany and Italy[450].

Level 2: Copying and formal reinterpretation of oriental patterns[451]

Cat. no. 36: (Cyprus, Alassa, "Pano-Mandilaris", Tomb 16) a direct copy of a Syro-Assyrian helmet from Cyprus[452] with a local, Mediterranean formal reinterpretation: eye/face cutout.

The Syro-Assyrian inspiration of the structure of this helmet is unquestionnable (*see* for example the squat conical shape, the pair of ribs around the base, the curve and the vertical rib at the front). This helmet is a local, Cypriote product, and was made by copying the structure of a Syro-Assyrian helmet in the late 8th – early 7th century B.C. Both of the helmets (Palaepaphos (**Cat. no. 35**) and Alassa (**Cat. no. 36**)) were found in tombs which can be dated to the Cypro-Archaic I Period (750—650 B.C.). But the widening out *repoussé* curve and the eye or face cutout at the front of the helmet are clear signs of local reinterpretation, especially in the shape. As will be discussed later, there was another, iconographical type of reinterpretation of Oriental patterns known from an Argos helmet.

The eye or face cutout is an entirely new feature appearing for the first time on this helmet[453]. The origin of this eye cutout is clearly the reinterpretation (and/or misinterpretation) as eyebrows of the oriental *repoussé* curves on the front of the helmet in a foreign cultural background. This helmet is probably the forerunner of the later Anatolian and Cypriote tradition (late 7th – early 6th century B.C.) of composite iron and bronze helmets of radial type furnished with similar eye cutouts[454]. This eye cutout, or the reinterpretation of the oriental curves appear on contemporary Scythian helmets of Kuban-type[455], which probably had a connection with the Anatolian radial type (**Cat. nos. 142, 143**) of composite iron and bronze helmets. This helmet type is part of the Mediterranean "covered face" helmet tradition, which is one of the best known differences between the Near Eastern "open face"[456] and the Mediterranean "covered face" traditions. The Alassa helmet might be a transitional form between the two traditions and the two regions as well. It shows the first sign of the deepening of the sides of the helmet for better protection, which culminated in the Tamassos and "Geryon"-types[457] of helmets in Cyprus, and in the Corinthian and its attached types of helmets in the Greek mainland[458]. Furthermore it shows us that the Mediterranean covered face helmet tradition may be traced back to the Oriental helmet tradition.

This helmet therefore – together with the Tamassos-type helmets – is one of the best pieces of evidence for the innovative role of Cyprus in the new Eastern Mediterranean helmet tradition of the 8th and 7th centuries B.C., and the intermediatory role of the island in the connection of the Near East and the Mediterranean.

Level 3: Copying and iconographical reinterpretation of oriental patterns[459]

– Argos, Odos Perseous 41, warrior grave of Geometric Age[460] (third quarter of 8th century B.C.). Greek version of an oriental conical helmet. The oriental features are shown very well on the helmet: (squat) conical shape, the vertical rib on the front, a pair of curves bending inward; but the original oriental curves were not understood and were interpreted as eyebrows so two apotropaic eyes were incised under them.

---

449 Schauer 1983, 185-186, fig. 5b.
450 Hencken 1971, 72-74, fig. 48a-b.
451 Dezső 1998, 12-13, ills. 7, 8.
452 *See* the conical helmets of the warriors of Cypriote terracotta chariot models, Dezső 1998, 16-21, ill. 4.
453 Appears earlier only on the 12th–11th century B.C. Iranian helmets decorated with figures of deities on the front (**Cat. nos. 122-125**).
454 From Sardis (**Cat. no. 142**), and probably from Old-Smyrna (**Cat. no. 143**), Idalion (*see* p. 109; Dezső 1998, 17) and Lindos (*see* p. 109; Dezső 1998, 17, ill. 3) as well.
455 *See* pp. 105-106, Group I.5 (Irano-Scythian hemisperical iron helmet, **Cat. no. 138**).
456 As have been mentioned the 9th century B.C. (and occasionally in the 8th century B.C.) Assyrian helmets were furnished with a scale armour faceguard, which was attached to the rim of the helmet. This, scale armour type of face protection was virtually unknown in the Mediterranean.
457 For the detailed description of the Tamassos and "Geryon"-types of helmets *see* Dezső 1998, 17-21, esp. 18-19, ill. 2.
458 For the early evolution of the Corinthian and its attached types of Greek helmets *see* Dezső 1998, 47-49, ill. 3.
459 Dezső 1998, 13-15, ill. 9.
460 Protonotariou-Deilaki 1973, 97-99, pl. 95e; Gröschel 1986, 71-72, fig. 4; Dezső 1998, 13-15, ill. 9.

Level 4: Cypriote and Greek helmets inspired by oriental (Syro-Assyrian) helmet types

1. Cyprus[461]:

– Tamassos, Royal Tomb XI (Berlin, Antikensammlung, Inv. Misc. 8142,620)[462]

– Tamassos, Dromos of Royal Tomb IV (Nicosia, Cyprus Museum, Inv. no. 98/1971)[463]

– "Cyprus" (Istanbul, Archaeological Museum, Inv. no.: 16)[464]

One of the characteristics of the three "Tamassos-type" helmets mentioned above is their tubular point bulging like a knob at the top. There are two horizontal ribs around the base to strengthen the rim of the helmet. Originally all of the three helmets were furnished with detachable earflaps and face-masks. There are fragments of the neck-guard belonging to the Berlin helmet, but the Istanbul helmet has lost its earflaps and face-mask. Second half of 7th century B.C.

There are several representations of this "Tamassos-type" of pointed helmet known from Cypriote art. They are shown on two fragmentary terracotta heads from the Samian Heraion[465] and a third terracotta head from Idalion, Cyprus, now in the Antikenmuseum, Berlin[466].

This type of helmet is a characteristic example of the oriental inspiration or the stimulus diffusion of the oriental conical helmet types. This type of helmet, in spite of its later date may be traced back to the late 8th century B.C. conical helmets of the Near East, namely Assyria. Its shape with the long tubular point appears in Assyrian art in the middle of the 8th century B.C. On the Til-Barsip wall-paintings (*Chart 2, nos. 163, 164*)[467] there are Assyrian helmets with high, tubular points and, another new feature, a neckguard. The points of these helmets were sealed by a rectangular element, which might be added separately to these (iron) helmets. The tubular point bulging like a knob on the Tamassos helmets may be a late ancestor of this form. Earlier representations of conical helmets with a kind of bulging knob are visible on a Karatepe sculpture depicting an Assyrian horseman[468] probably from the first half of the 8th century B.C. (*Chart 2, no. 175*). But this kind of tubular point bulging like a knob is clearly a local development on the Tamassos helmets. Only the shape is oriental, but further details, like the detachable earflap combined with a face-mask are local. The face-mask, or earflaps covering parts of the face on the early Corinthian helmets and their Cretan versions[469], became a characteristic Greek feature. This kind of face mask, as part of the defensive armour was never known in the Near East. The Assyrians in the 9th century B.C. used scale armour attached to the rim of their helmets to guard parts of their faces. These scale armour "face-masks" are characteristically 9th century B.C. Assyrian features and appear in the palace reliefs[470] of Ashurnasirpal II (883—859 B.C.), on the Balawat gates[471] of Shalmaneser III (858—824 B.C.), and on a wall-painting fragment discussed above[472]. The oriental "open faced helmet tradition" can be contrasted with the "covered face Aegean helmet tradition" which was introduced probably with the early Corinthian helmets at the end of the 8th century B.C. The face mask might have originated in the development of earflaps to cover larger parts not only of the ears but also face. The Lindos earflaps[473], one of which was considered by Pflug as parallel for an Olympian earflap[474], is from a helmet made of five parts. The Lindos-type of earflaps (from an unknown type of probably Eastern Greek helmet[475]) with the eye and mouth cut-outs might be inspired by the mainland Greek "covered

461 Dezső 1998, 16-22, ills. 14, 15.

462 H.: 42.3 cm; Buchholz 1978, 197-200, figs. 45a-e; Pflug 1988b, 27-41; Pflug 1989, Cat. no. 1, 17-18, 49, fig. 11, pl. on p. 49; Dezső 1998, 16-28, ills. 1, 2.

463 H.: 42.3 cm; Diam.: max: 27.4 cm, min.: 14 cm; Buchholz 1972, 185, pl. 32:2; Buchholz 1973, 334, 388, figs. 36a-d; Karageorghis 1973, 665, fig. 101; Borchhardt 1977, pls. E V b-c; Pflug 1988b, fig. 2; Dezső 1998, 16-21, ills. 1, 2, 14, 15.

464 Pflug 1988b, 30, figs. 3-4; Dezső 1998, 16-21, ills. 1, 2.

465 1. Samos, Heraion, terracotta head (Samos Museum, Inv. no. T.2637, H.: 9.1 cm); Schmidt 1968, pl. 44; Borchhardt 1977, pl. E Va; Pflug 1988b, figs. 5-6; Dezső 1998, 16-17, ills. 2, 4, no. M.2.19; 2. Samos, Heraion, terracotta head (Samos Museum, Inv. no. T.2741, H.: 7.5 cm); Schmidt 1968, pls. 44, 46; Dezső 1998, 16-17, ills. 2, 4, no. M.2.20.

466 Idalion, Cyprus, terracotta head (Berlin, Antikenmuseum, Inv. no. 499x); Schmidt 1968, pl. 44; Pflug 1988b, figs. 7-8; Dezső 1998, 18, ills. 2, 4, no. M.2.18.

467 Wall-paintings, Til-Barsip, Room XXIV, reign of Tiglath-Pileser III (745—727 B.C.); Thureau-Dangin – Dunand 1936, pls. XLIX, upper register, LI

468 Karatepe, "Obere Grabung", Bossert *et al.* 1950, pl. XIII, no. 64; Orthmann 1971, 494 (Sph III), pl. 18f.

469 Hoffmann 1972, *passim.*

470 Budge 1914, *passim.*

471 King 1915, *passim.*

472 Ass.10756; Andrae 1925, pl. 9e.

473 Blinkenberg 1901, 189-190, nos. 571-576, pls. 22-23.

474 Pflug 1988a, 21.

475 Pflug 1988a, 21. However, the general form of this type is indicated by a miniature votive bronze crested helmet found on the Lindos Acropolis (Blinkenberg 1931, 391, no. 1564, pl. 63:1564). This votive miniature helmet was made of two parts –

face tradition" appears in its developed form with the early Corinthian helmets, or might be a parallel development. However, the earliest known example or the archetype of the earflap/cheek-piece with eye and mouth cut-out is an Urartian earflap inscribed with the name of king Ishpuini (*ca.* 830—810 B.C.)[476]. These Lindos earflaps have a clear similarity to the earflaps of the terracotta representations of the Tamassos-type of helmet (*see* above). One of the helmets of these terracotta heads is furnished with a detachable earflap – face-guard[477], while the other has a solid earflap – face-guard[478]. There is a row of rivet holes at the upper edge of one of the Lindos earflaps[479] which means that the construction of the helmet to which this earflap belonged to was similar to the Greek Geometric helmets[480] made of five separate parts (*see* below). So it is clear that in case of the Tamassos-type of helmet there were a few (local?) variants in use at the same time:

1. With solid earflap (cheek-piece) – face-mask: terracotta head (Berlin)[481].
2. With detachable cheek-piece – face-mask: Nicosia helmet and a terracotta head (Samos)[482].
3. With detachable cheek-piece and detachable face-mask: Berlin helmet.
4. The helmet type of the Lindos earflaps[483] (only the earflap construction is similar to the 2nd group).

The relatively late date[484] of the Tamassos-type of helmets makes it possible that the face-guard as idea came from the Corinthian helmets of earlier date, but one can argue for a parallel, East Greek development (*see* the Lindos earflaps) as well[485].

The history of (8th–) 7th–6th century B.C. genuine Cypriote helmets can be partly reconstructed from several representations (terracotta sculptures, bronze statuettes, *etc.*) of different types of headdress or helmet. But the identification of the headdresses is difficult. In several cases it is not possible to judge whether the headdress represented on a terracotta sculpture is a metal helmet or a leather helmet or a linen cap[486]. At this point we have to exclude all the headdresses and "helmets" made of soft, perishable materials (leather, linen) and concentrate only on the metal helmets. The identification of metal helmets needs some additional features (*e.g.* a crest, or definite earflaps). "*It should be noted, that simple conical headdresses, without such additional identifying features, appear on many terracottas of otherwise unarmed horseback riders. Here it must remain uncertain whether helmets or caps were intended.*"[487] There are two major groups of terracotta and bronze statuettes representing helmets or different types of headdress which were interpreted in earlier studies as helmets:

1. "Civilians" or figures without additional indication of the warlike character of the representation.
2. Warriors with additional indication of their warlike character: warriors armed with a shield and/or a weapon; horsemen armed with shield and/or a weapon; chariot models with armed warriors.
3. A third group contains all the representations of known types of helmet (*e.g.* the terracotta representations of the Tamassos-type of pointed helmet; representations of crested helmets; and the "Geryon"-type[488] of Cypriote helmets. And there are the

---

similarly to the Cretan miniature votive helmets: Gortyn (Hoffmann 1972, 2, pl. 41:4), Praisos (Bosanquet 1901—1902, 258, pl. 10; Benton 1939—1940, pl. 31:17) and Palaikastro (Benton 1939—1940, pl. 28:31). The two-piece construction of the actual helmets of this type was proved by the Hamburg Cretan crested helmet from Praisos (*see* below). The characteristic form of the Lindos earflaps is clearly visible on the Lindos miniature helmet.

476 Belli 1993, 65; Rittig 1994, Abb. 6; *see* pp. 95-96.

477 Terracotta head (Samos, Heraion, T.2637), H.: 9.1 cm: Schmidt 1968, pl. 44; Pflug 1988b, 32-33, figs. 5-6; Dezső 1998, 16-17, ills. 2, 4, no. M.2.19.

478 Terracotta head (Berlin 499x), H.: 8.4 cm: Schmidt 1968, pl. 44; Pflug 1988b, 32-33, figs. 7-8; Dezső 1998, 18, ills. 2, 4, no. M.2.18.

479 Blinkenberg 1901, 189-190, no. 571, pl. 22.

480 Pflug 1988a, 21; Pflug 1988b, 32.

481 Terracotta head (Berlin 499x), H.: 8.4 cm: Schmidt 1968, pl. 44; Pflug 1988b, 32-33, figs. 7-8; Dezső 1998, 18, ills. 2, 4, no. M.2.18.

482 Terracotta head (Samos, Heraion, T.2637), H.: 9.1 cm: Schmidt 1968, pl. 44; Pflug 1988b, 32-33, figs. 5-6; Dezső 1998, 16-17, ills. 2, 4, no. M.2.19.

483 Dezső 1998, 18; 1. Lindos, Acropolis, (right side earflap); H.: 14.5 cm; W.: 11.6 cm; Blinkenberg 1931, 189-190, no. 571, pl. 22; 2. Lindos, Acropolis (left side earflap); H.: 13.4 cm; W.: 10.9 cm; Blinkenberg 1931, 190, no. 572, pl. 22.4; 3. Lindos, Acropolis; H.: 13.3 cm; Blinkenberg 1931, 190, no. 573.

484 Pflug 1988b, 40-41.

485 Pflug 1988a, 21.

486 For detailed study *see* Törnkvist 1974, 21-32; Tatton-Brown 1979, 281-284; moreover, Pflug 1988b, 36.

487 Crouwel – Tatton-Brown 1988, 81.

488 For the definition of the "Geryon"-type of helmet *see* Dezső 1998, 18-21; for the evolution of Cypriote helmet types, the Tamassos and "Geryon"-types, *see* Dezső 1998, 16-21, and esp. ill. 2.

representations of metal helmet forms with definite additional features, like profiled earflaps, which would be unnecessary or redundant on leather caps.

There is no definite coincidence between the representations of headresses on the first two groups: there are, for example, no crested helmets on the heads of the first group, and there are no profiled earflaps on the headdresses of this group. However, there are "earflaps" and plumes on the caps of this group, but certainly not made of metal. The shape of their headdresses, however, sometimes similar or the same as that of the metal helmets, but without any strict indication of their material they must be excluded from further study here. Furthermore it is well known that there were leather helmets in use in Cyprus at this period. There are several representations of leather helmets in the second group, the definite warlike character of which would otherwise exclude the use of leather or linen caps. However, there is a coincidence between these leather helmets and some of the representations of the first, "not warlike" group[489]. On the chariot groups for example there is an observable tendency for the chariot warrior (or at least one member of the crew) to wear a helmet of metal shape, while others of the crew wear helmets of "leather shape"[490]. In a few cases all the two or three members of the crew wear headdresses which can probably be identified as leather helmets[491].

So the remaining helmet representations ("metal forms") provide a well known picture of the early oriental or orientalizing types of pointed helmets (*Fig 8*):

1. pointed helmet without earflaps[492] (*Chart 2, no. M.2.1*)
2. pointed helmet with solid earflaps (without indication of hinges)[493] (*Chart 2, no. M.2.2*)
3. pointed helmet with detachable earflaps[494] (*Chart 2, no. M.2.3*)
4. pointed helmet with a tubular point bulging like a knob and with earflaps (probably an early example or predecessor of the Tamassos-type of pointed helmets)[495] (*Chart 2, no. M.2.4*)

On the second level there are three interesting representations of warriors helmets which represent the intermediate phase in the evolution of Cypriote helmets between the early oriental or orientalizing and the developed Cypriote helmet types (*Fig. 8*):

1. Tubular pointed helmet with earflaps with cut-outs for the eyes and mouth[496]. This helmet with its characteristic earflaps might be an early version of the later Tamassos-type of pointed helmets (*Chart 2, no. M.2.7*).
2. Tubular pointed helmet with a point bulging like a knob and furnished with a nose-guard and long, curving earflaps[497]. This helmet with its nose guard and curving earflaps might be the predecessor of the later "Geryon-type" of Cypriote helmets (*Chart 2, no. M.2.5*).
3. Tubular pointed helmet with a point bulging like a knob and furnished with long earflaps curving in a right angle to form a cheek-piece[498]. Similarly to the second representation, this helmet might be an intermediate stage between the original oriental form and its modified Cypriote version, the "Geryon-type" (*Chart 2, no. M.2.6*).

---

[489] Compare for example the leather helmets of the crew of chariot no. 1781+798 (Gjerstad *et al.* 1935, pl. CCXXXV:3) with the leather cap (with long projection hanging at the back of the cap and "soft earflaps") of a terracotta sculpture of the first group (Gjerstad *et al.* 1935, pl. CCVII:1-2, no. 1509).

[490] Gjerstad *et al.* 1935, pls. CCXXXIV:4 (no. 1715); CCXXXIV:5 (no. 2000); CCXXXIV:6 (no. 1046); CCXXXV:1 (no. 1125); CCXXXV:2 (no. 1124); CCXXXV:3 (no. 1781+798); CCXXXV:4 (no. 1170); CCXXXV:5 (no. 1998). Second half of Cypro-Archaic I, mid-7th century B.C.

[491] Gjerstad *et al.* 1935, pl. CCXXXIV:2 (no. 1166); Crouwel 1987, pls. XXXVIII:1a-b, XXXVIII:3, XXXIX:2. Second half of Cypro-Archaic I, mid-7th century B.C.

[492] Ayia Irini, terracotta chariot group, Nicosia, Cyprus Museum; Gjerstad *et al.* 1935, pl. CCXXXIV:5 (no. 2000). Second half of Cypro-Archaic I, mid-7th century B.C.

[493] 1. Ayia Irini, terracotta chariot group, Gjerstad *et al.* 1935, pl. CCXXXIV:6 (no. 1046); 2. Ayia Irini, terracotta chariot group, Gjerstad *et al.* 1935, pl. CCXXXV:5 (no. 1998); 3. Ayia Irini, terracotta sculpture of a warrior with shield, Gjerstad *et al.* 1935, pl. CCXXXII:7 (no. 1032); 4. Ayia Irini, terracotta head, Gjerstad *et al.* 1935, pl. CCXXXII:15 (no. 2384); 5. "Cyprus", terracotta horseman, Edinburgh, Royal Scottish Museum 1921.354, Crouwel – Tatton-Brown 1988, pl. XXVI:4. Second half of Cypro-Archaic I, mid-7th century B.C.

[494] 1. Ayia Irini, terracotta chariot group, Gjerstad *et al.* 1935, pl. CCXXXIV:6 (no. 1046)?; 2. "Cyprus", terracotta horseman, London, British Museum, 1876.4.9.92, Crouwel – Tatton-Brown 1988, pl. XXVI:2. Second half of Cypro-Archaic I, mid-7th century B.C.

[495] 1. Ayia Irini, terracotta chariot group, Nicosia, Cyprus Museum, Gjerstad *et al.* 1935, CCXXXV:3 (no. 1781+798); 2. Ayia Irini, terracotta chariot group, Nicosia, Cyprus Museum, Gjerstad *et al.* 1935, pl. CCXXXV:4 (no. 1170). Second half of Cypro-Archaic I, mid-7th century B.C.

[496] Ayia Irini, terracotta sculpture of a warrior armed with a spiked shield, Gjerstad *et al.* 1935, 721-722, pl. CXCIV, no. 1385+1530.

[497] "Cyprus", terracotta statuette of a warrior, Heuzey 1880, fig. F.

[498] "Cyprus", terracotta statuette of a warrior, Heuzey 1880, fig. G.

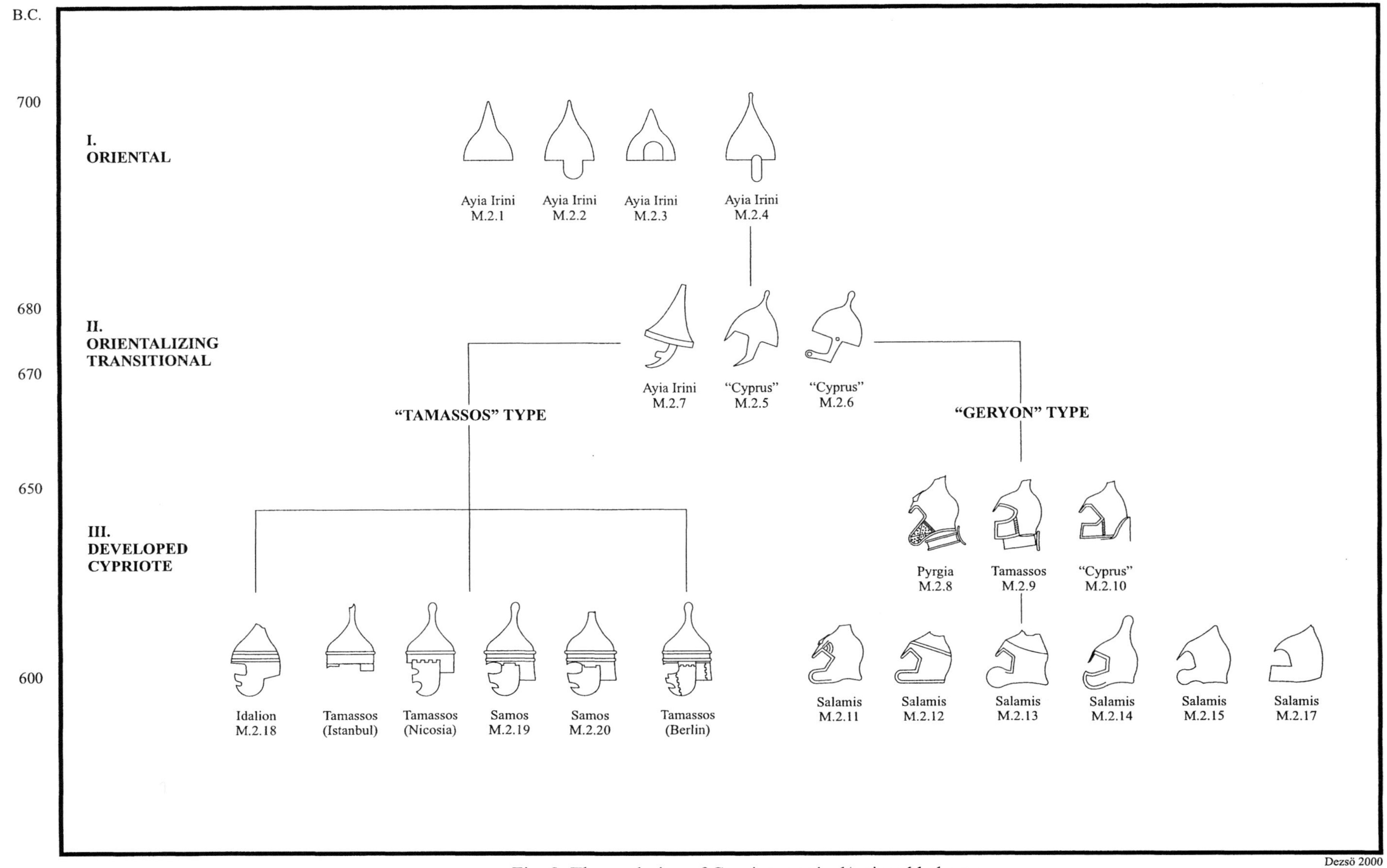

Fig. 8: The evolution of Cypriote conical/pointed helmets

On the third level there are the later (late 7th – early 6th century B.C.) Cypriote types, helmets which represent only the oriental inspiration combined with characteristic local or Aegean features (curving cheek-pieces and face-mask, *Fig 8*):

1. Representations of the Tamassos-type of pointed helmets discussed above in detail.
2. Representations of the Cypriote "Geryon"-type of helmets[499].

2. Crete[500]:

– Praisos (Hamburg, Museum für Kunst und Gewerbe, Inv. no.: 1970.26c)[501]

High crested helmet inspired by oriental crested helmets. Its Assyrian archetypes can be dated to the middle, or third quarter of the 8th century B.C.[502], but this type is well known from the late 8th – early 7th century B.C. Aegean art as well[503]. Late 8th – early 7th century B.C.

The Praisos helmet represents a later phase of this evolution, the second phase of orientalization. Its best representational parallels are known mainly from Crete (open face helmets with high crestholder and neckguard): miniature votive bronze helmets from Gortyn (5 pieces) (*Chart 3, no. M.5.2*); from Praisos (6 pieces) (*Chart 3, nos. M.5.3*) and from Palaiokastro (*Chart 3, no. M.5.4*). These votive helmets represent the same technique as the Praisos helmet, they were joined together from two similar halves. Further representational evidence is known from other media of Cretan art: from the incised decoration of a bronze corslet (*Chart 3, no. M.4.7*), from a Late Orientalizing alabastron from Fortetsa (Sphinx, *Chart 3, no. M.4.5*), from a Melian hydria from Rheneia (Sphinx, *Chart 3, no. M.4.6*) and from an early orientalizing pithos from Knossos (*Chart 3, no. M.5.18*). The recently discovered warrior stelai from Prinias (*Chart 3, nos. M.4.8-13*) and a terracotta plaque of a warrior from Gortyn (*Chart 3, no. M.4.14*) and a later bronze plaquette from the Geometric Period temple at Dreros, representing Athena (*Chart 3, no. M.4.22*) show the same helmet type. All the representational evidence corresponds with the mid 7th century B.C. (second half of 7th century B.C.) date of the helmet. Snodgrass dated the use of this type, open faced crested helmet to the 7th century B.C., and emphasizes its primary Cretan distribution[504]. This or at least a very similar helmet type was known in the Greek Mainland and on the Greek Islands as well[505]. Its representations are known from a votive terracotta shield from Tiryns (Amazonomacheia, *Chart 3, nos. M.4.17-19*) and from a votive terracotta shield from Samos (Amazon, *Chart 3, no. M.4.16*). A parallel helmet type without neckguard has the same distribution in the Greek Mainland and on the Greek Islands but this type is unknown on Crete. Its representations are known from a terracotta head from Amyklae, Sparta (*Chart 3, no. M.4.20*), from a Lakonian Sub-Geometric sherd (Sphinx, *Chart 3, no. M.4.15*), from a Theran rock carving (*Chart 3, no. M.5.5*) and from the famous relief pithos from Mykonos (*Chart 3, nos. M.4.3-4*). A similar helmet type, if not the same, is represented extensively in the bronzework of the Greek Mainland in the late 8th century and 7th century B.C. onwards. It can be seen with neckguard on a bronze statuette ("*Pferdeführer*") from Olympia (*Chart 3, no. M.5.12*) and on a bronze statuette now in New York (*Chart 3, no. M.5.20*); without neckguard on a bronze statuette from Amorgos (*Chart 3, no. M.5.15*), series of bronze statuettes from Olympia (*Chart 3, nos. M.5.6-8, 13-14, 17*), from a bronze tripod leg from Olympia (*Chart 3, no. M.5.21*), from a Samian bronze group (*Chart 3, no. M.5.19*), from a Delphi bronze warrior (*Chart 3, no. M.4.21*), from a bronze statuette of a warrior from the Acropolis, Athens (*Chart*

[499] For the first studies of this type *see* Tatton-Brown 1979; Pflug 1988b, 37-39; Dezső 1998, 20-21, ill. 2.
[500] Dezső 1998, 22-30.
[501] H.: 44.6 cm; H. of the crest: 10 cm; Hoffmann 1972, 5-6, 42, pl. 13; Dezső 1998, 22-30, ill. 1.
[502] *Charts 3-4, nos. 147-148, 150* (Tiglath-Pileser III [745—727 B.C.]); *Charts 3-4, nos. 191-192* (Sargon II [721—705 B.C.]); Dezső 1998, ill. 5.
[503] Hoffmann 1972, pl. 41; Borchhardt 1977, pls. E VIa-c; Dezső 1998, 28, ill. 5, M.3.5; for detailed discussion *see* Dezső 1998, 22-30; *Chart 3, no. M.3.5* (Kavousi).
[504] Snodgrass 1964, 17: "Its life seems to extend sporadically throughout the seventh century, but effectively it must be superseded by c. 650. From beginning to end Crete and the islands predominate in the distribution of the type, and it was retained there later than on the mainland".
[505] This helmet type was in some way (perhaps because of its definite Oriental character) connected with the representations of sphinxes and Amazons, which themselves had some Oriental origins in the Greek world and mythology. Snodgrass 1964, 18 writes: "It is worth noticing the frequent association of this helmet with the Sphinx... This confirms the Oriental connections of the crest form, and also suggests, that the helmet had acquired semimythical associations, and might therefore be represented long after it was in actual use." As a unique helmet, one variant became the standard helmet of the early representations of Athena (probably because of the "open face" tradition of representations of Greek deities – in most cases the face of an armoured Greek deity, *i.e.* Athena, has to be uncovered). As Gröschel 1986, 71-78 has already pointed out, the idea of the figurative decoration of Athena's helmet in the *Iliad* (V, 743-744) can be found in the Near Eastern decorated helmet tradition. No doubt, the Near East, and especially Assyria was far enough (several months of cruise) to become the home of certain semimythical elements in the Greek mythology and iconography.

3, no. M.5.22). One bronze handle attachment from Olympia shows this type of helmet without neckguard but furnished with elaborate earflaps covering parts of the face as well (*Chart 3, no. M.5.24*). Another representation of a similar helmet on a bronze sheet – said to be Oriental in origin and in some details reworked in Greece – decorated in *repoussé* and incised figurative scenes, shows very similar helmets to the helmets of the 7th century B.C. Assyrian army (with small rounded earflaps, *Chart 3, no. M.5.23*)[506]. The bronze sheet however is not an oriental work or at least not from the Assyrian Empire, since the helmets depicted on this sheet are worn by horsemen wearing scale armour as well. In the Assyrian army, however, horsemen never wore crested helmets only pointed ones, and the light infantry wearing crested helmets never wore scale armour, only small round breast plates (*kardio phylakes*). So the details of this bronze sheet are against the Near Eastern or at least the Assyrian tradition, and were probably made outside the Assyrian Empire, where such conventional details were not known. D. Rittig considered this bronze sheet as an Anatolian work[507]. As can be seen later, the well known Greek Geometric helmet type with its characteristic earflaps and five piece construction[508] had two variants with a crest curving forward: the Ordona variant (*see* below, IIIA.)with a hemispherical cap and a high crestholder curving forward and riveted onto the top of the helmet, and the conical variant with a small crest curving forward mounted onto the point of the helmet known from a miniature votive bronze helmet from Delos (*see* below, IIIB.). However, three bronze statuettes from Olympia (*Chart 3, nos. M.5.9-11*) show a probable third variant (*see* below, IIIC.) of this type of Geometric Greek helmet with the characteristic earflaps, but with a high point probably curving forward.

3. Greek mainland[509]:

I. Greek Geometric helmets with pointed cap[510]

– Budapest, Museum of Fine Arts, Inv. no. 8442[511] (*Pls. 55-56*)

– Argos, Odos Diomidous (Argos, Archaeological Museum)[512]

The two helmets mentioned above are the best examples of pointed helmets of the Greek mainland. Their characteristic is the conical skull-piece hammered out from one piece of bronze, to which the other parts of the helmet (front plate, neck-guard and two pointed earflaps) were riveted. This basic helmet type has three variants: pointed (mentioned above), pointed with a crescent-shaped crest (below) and hemispherical with a high crest curving forward (below). Last quarter of 8th century B.C.

II. Greek Geometric helmets with crescent-shaped crest[513]

– Argos, Tomb 45 (Argos, Archaeological Museum)[514]

– Olympia (Inv. Br. 10533)[515]

– Olympia (without number)[516]

– Olympia (Inv. B 51)[517]

– Olympia (Inv. B 4717)[518]

– Athen, Kerameikos, Grave V[519]

Variant of the conical type discussed above: on the top of the elongated point of the helmet there is a crescent-shaped crest curving back almost to the side of the skull-piece.

---

506 Olympia, "Brunnen 17", Olympia Museum, Inv. B 5085a; Kunze 1961—1962, 115-116, pl. 30b.

507 Rittig (1994, 253-254, Abb. 6-7) emphasized the similarity of the small earflap of the helmet (Rittig 1994, Abb. 7) to the Urartian earflaps with eye and mouth cutouts known from Urartian sites (Belli 1993, 65) and inscribed with the name of Ishpuini, king of Urartu (*ca.* 830—810 B.C.). However, the original Urartian and Mediterranean (Dezső 1998, ill. 3: B 357, B 4228, Lindos 571, M.8.4, and M.8.10) earflaps with eye and mouth cutout (for detailed discussion *see* pp. 95-96) must have covered the face of the soldier. So this small "earflap" indicates the ear of the horseman or represents an earflap out of its original context, which context – similarly to the original context of the helmet itself – was obviously unknown for the coppersmith.

508 And originated probably in Argos.

509 Dezső 1998, 30-40, ills. 16-23.

510 Dezső 1998, Type I, 32-35.

511 Kunze 1958, 118-125; Kunze 1967, 112-113, fig. 38; Pflug 1988a, 14; Dezső 1998, 32-35, ills. 16, 17.

512 Protonotariou-Deilaki 1971, 81-82, fig. 13, pl. ζ; Protonotariou-Deilaki 1982, 33-48; Pflug 1988a, 14, fig. 4; Dezső 1998, 32-35, ill. 18.

513 Dezső 1998, Type II, 35-37

514 Courbin 1957, 322ff., figs. 19, 39-45, pls. 1, 4; Borchhardt 1977, pl. E IV b; Pflug 1988a, 12, figs. 1-3; Dezső 1998, 35-37, ill. 19.

515 H.: 28.7 cm; Kunze 1958, 119, no. 1, pls. 33-34; Pflug 1988a, 112-114; Dezső 1998, 35-37, ill. 20.

516 H.: 20.8 cm; Kunze 1958, 119, no. 2, pl. 35; Pflug 1988a, 112-114; Dezső 1998, 35-37, ill. 21.

517 H.: 18 cm; Kunze 1958, 119, no. 3; Pflug 1988a, 112-114; Dezső 1998, 35-37, ill. 22.

518 Kunze 1967, 111, no. 3a, pl. 50:1; Pflug 1988a, 12, note 12; Dezső 1998, 35-37, ill. 23.

519 Brückner – Prentice 1893, 108; Borchhardt 1972, 65, Kat. 13 III 1; Pflug 1988a, 20; Dezső 1998, 35-37.

This kind of crested helmet was widespread in Assyria[520] and North Syria[521] in the third quarter of the 8th century B.C. There is only a single known Assyrian or North Syrian example of this type of crested helmet: **Cat. no. 15** (*Fig. 3, Pl. 24*).

III. Greek Geometric helmets with crest curving forward

IIIA. With hemispherical cap and a separately added high crest curving forward[522]

- Ordona (South Italy), Collection of A. Ceccanti, Florence, Inv. no. CC 436[523]
- Ordona (South Italy)[524]
- Venice, Private Collection[525]

These helmets are the only examples of the third variant: a high crest was riveted on to the top of a hemispherical skull-piece. Like the Cretan crested helmet discussed above, its oriental prototypes are known from the the middle of the 8th century B.C. on Assyrian palace reliefs.

IIIB. With a separately added small crest curving forward[526]

The only surviving example of this type is a 6 cm high miniature bronze helmet[527]. It shows the distinctive traits of the basic type: it has a conical cap, two characteristic earflaps, a front plate and a back plate (neck guard). The characteristic feature of this type is, however, the separately added small crest curving forward. Parallels to this small crest are known from the Assyrian palace reliefs from the reign of Tiglath-Pileser III (745—727 B.C.): *Charts 3-4, nos. 145-147*, but mainly from the reign of Sennacherib (704—681 B.C.): *Charts 3-4, nos. 218-219*, Esarhaddon (680—669 B.C.): *Charts 3-4, no. 230* and Ashurbanipal (668—627 B.C.): *Chart 3, nos. 247-255*, when (during the 7th century B.C.) the standardization of the crested helmets made progress as well. Representations of this type in Greek art cannot be separated from the representations of helmet type IIIA. because these types are differently manufactured versions of the same basic type, the helmet with crest curving forward.

IIIC. With a high cap curving forward[528]

This group represents the helmet type of the so called Steiner bronzes found in Olympia:

- Bronze statuette of a warrior, Olympia, Inv. no. B.2000, H.: 21 cm, Late or Sub-Geometric (first quarter of 7th century B.C.)[529]. (*Chart 3, no. M.5.9*)
- Bronze statuette of a warrior, Olympia, Inv. no. B.1999, H.: 17.3 cm, Late or Sub-Geometric (first quarter of 7th century B.C.)[530]. (*Chart 3, no. M.5.10*)
- Bronze statuette of a warrior, Olympia, Inv. no. B.1701, H.: 16.9 cm, Late or Sub-Geometric (first quarter of 7th century B.C.)[531]. (*Chart 3, no. M.5.11*)

The basic structure of their helmets is the same as the helmets of groups discussed above: the characteristic earflaps, and neck-guard are clearly visible. Unfortunately the points of their helmets are missing. Therefore we can only suppose that their helmets were furnished with a high point or an unstilted crest curving forward[532]. There is a further damaged example of the type known from Olympia (*Chart 3, no. M.5.12*). There are, however, several other representations of the complete form of this type: from Olympia (*Chart 3, nos. M.5.6-8, 13-14, 21*), from the Greek Mainland (*Chart 3, nos. M.5.16-17, 20, 22*), from Crete (*Chart 3, no. M.5.2-4, 18*) and from the Greek Islands (*Chart 3, nos. M.5.5, 15, 19*). Kunze proposed a possible Argive manufacture for the "Steiner bronzes"[533]

As *Charts 2-4* show, there were at least two chronological phases in the influence exerted by the Near Eastern helmet tradition on the Mediterranean helmet tradition:

---

520 *Charts 3-4, nos. 153-155* (Tiglath-Pileser III [745—727 B.C.]).
521 *Charts 3-4, no. 173* (Tiglath-Pileser III [745—727 B.C.]).
522 Dezső 1998, Type IIIA, 37-39
523 Pflug 1988a, 14-15, figs. 5-6; Dezső 1998, 37-39.
524 Egg 1986, 15, pl. 17; Pflug 1988a, 15; Dezső 1998, 37-39.
525 Sotheby's Catalogue 19.6.1961, 28, no. 117; Kunze 1967, 115f., fig. 39; Lepore 1984, 330, note 7; Pflug 1988a, 14; Dezső 1998, 37-39.
526 Dezső 1998, Type IIIB, 39.
527 Deonna 1938, 208, pl. 69:554; Kunze 1967a, 114-115; Gröschel 1986, 64, note 166.
528 Dezső 1998, Type IIIC, 39-40.
529 Kunze 1944, 120, pls. 38-39; Dezső 1998, ill. 5, no. M.5.9.
530 Kunze 1944, 122, pls. 43-44; Dezső 1998, ill. 5, no. M.5.10.
531 Kunze 1944, 122, pls. 45-46; Dezső 1998, ill. 5, no. M.5.11.
532 Snodgrass 1964, 15: "... their crests (now lost) were apparently the unstilted, forward curving type in contrast with the Argos crest."
533 Kunze 1944, 120-122.

1. The influence of the independent North Syrian ("Carchemish-type" and connecting) conical crested helmets[534]. Second half of 9th century – early 8th century B.C.

From: North Syria (Carchemish: *Charts 3-4, nos. 101-104*; Kültepe(?): *Charts 3-4, no. 105*; Zinçirli: *Charts 3-4, nos. 128-129*; and probably Karatepe: *Charts 3-4, nos. 177-180*)

To: (almost exclusively to) Crete (Fortetsa: *Chart 3, nos. M.3.1-2, M.4.1*; Kato-Symi: *Chart 3, no. M.3.3*; Khaniale Tekke: *Chart 3, no. M.3.4*; Kavousi: *Chart 3, no. M.3.5*)[535]

2a. The influence of the Syro-Assyrian or Assyrian conical crested helmets. Second half of 8th century – 7th century B.C.

From: the Assyrian Empire (already including North Syria). Assyrian palace reliefs: Nimrud, reign of Tiglath-Pileser III (745—727 B.C.): *Charts 3-4, nos. 145-155*; Khorsabad, reign of Sargon II (721—705 B.C.): *Charts 3-4, nos. 187-192*; Nineveh, reign of Sennacherib (704—681 B.C.): *Charts 3-4, nos. 214-225*; reign of Esarhaddon (680—669 B.C.): *Charts 3-4, no. 230*; reign of Assurbanipal (668—627 B.C.): *Charts 3-4, nos. 247-255*; Til-Barsip wall-paintings (Tiglath-Pileser III): *Charts 3-4, no. 172*

To: Cyprus, Greece (islands and mainland), Western Anatolia: *Chart 3, nos. M.4.1-23* (conical/pointed helmets with crest curving forward); *Chart 3, nos. M.5.1-24* (conical/pointed helmets with unstilted crest curving forward); *Chart 3, nos. M.6.1-24* (conical/pointed helmets with crescent-shaped crest).

2b. The influence of Assyrian conical/pointed helmets. 9th–7th century B.C. (especially second half of 8th century – 7th century B.C.) (parallely with the Near Eastern spread of the helmet type)

From: Assyria (later from the Assyrian Empire)[536].

To: Eastern Mediterranean (*Chart 2, nos. M.2.1-32*).

As can be seen, the first "wave of orientalization" reached Crete probably in the early 8th century B.C. The only evidence for the spread of Near Eastern helmet types in Crete – the pictorial representations of helmets on sheet metalwork – raises the question, whether these representations refer to the actual spread or use of Near Eastern helmet types only in this part of the Eastern Mediterranean and nowhere else, or just simply refer to a kind of artistic connection, the existence of a kind of Near Eastern, let us say North Syrian, artistic tradition in Crete in the first half or middle of the 8th century B.C. If these Near Eastern helmets were used on Crete in the 8th century B.C. – well out of the date of their Near Eastern use – it was a kind of archaizing helmet tradition[537]. However, therefore it is much more probable that the connection between the Near East and Crete in the first half of 8th century existed only in the arts. But the level of artistic influence (these pieces, exclusively (sheet) metalwork 1. represent the actual presence of Near Eastern (North Syrian) craftsmen on Crete ("migration"); 2. represent the use of Near Eastern motifs (would have been depicted as "wandering motifs" from a set of stereotyped motifs, from "motif books" or wooden moulds used for sheet metalwork) on Crete; 3. or

---

534 The origin of Mediterranean hemispherical crested helmets (hemispherical cap with a crest fitted centrally across the cap), (*Chart 3, nos. M.7.1-8*) is questionnable. As have been mentioned this type of helmets had a short life in North Syria (*see* above). The single early representation on the Fortetsa "belt" might be the result of this influence (*Chart 3, no. M.7.1*; Brock 1957, pl. 115; Borchhardt 1972, 70, Kat. 15 II 2;.). But the fitting of a crest to a hemispherical cap might be a local feature as well. There is no reason to see for example Near Eastern (let us say Urartian) influence in the crest structure of the corinthian helmets furnished with this kind of crest. It might be for example a variant of the helmets with crescent-shaped crest without crest holder. The contemporary diversity of the early Mediterranean (Greek) helmet types is shown for example on the Mykonos relief pithos: helmets with high, forward curving crest (*Chart 3, nos. M.4.3-4*); their crest variants, the hemispherical crested helmets (*Chart 3, nos. M.7.2-3*); and probably an early variant of the helmet type, which survived almost exclusively in Corinthian helmets ("covered face tradition" originated from the oriental "open face tradition", Dezső 1998, ill. 3, M.8.1-2). For the early variants of the Greek "covered face tradition" *see* Dezső 1998, ills. 2-3.

535 It must be mentioned that – at least in its form – an almost similar type of helmet was used in the territories around the head of the Adriatic Sea from the late 8th century to the first half of the 7th century B.C. The exact origin of this Adriatic type is not known, but Hencken (1972, 163-168, figs. 132-133) and Egg (1988, 218-221, figs. 9-11) assumed, that this type was partly the result of Oriental influence, which spread indirectly via the Aegean. The geographically very limited use of it makes the origin of this type very curious.

536 For the nearly hundred (mainly Assyrian) representations of the variants of this helmet type *see Chart 2*; Dezső 1998, ill. 4. For detailed study *see* later.

537 For the chronological problems of the three different helmet types on the Fortetsa "belts" and the Kavousi bronze openwork *see* Dezső 1998, 23-30. Snodgrass 1964, 12 wrote on the Kavousi bronze sheet that "it can perhaps be dated to the late eighth century, less by its provenance (a tholos of Late Geometric or Early Orientalizing date) than by its decorations". Furthermore he wrote on the Fortetsa girdle (Snodgrass 1964, 12): "which illustrates, along with helmets of Oriental type, a series of archers in what appear to be early Corinthian helmets, with cheek-pieces and ridge crest". The Cretan origin of the Corinthian helmets would be eliminated since the helmets mentioned above almost certainly represent the hemispherical crested helmets known from the Near East, second half of 9th – early 8th century B.C. date of which fits into the chronological framework of the Near Eastern helmets.

simply refers to Near Eastern trade contacts (in this case the metalwork was made in North Syria). The artistic connection may owe something to all three possibilities. In the present writers view this early phase of orientalization existed only in the "archaizing representations" of these helmets and not in the "archaizing – well out of the date of their Near Eastern – use" of them. There is no other type of evidence available for the existence of early (9th century B.C.) oriental helmet tradition in the Mediterranean, or especially on Crete. There is no sign that development of further helmet types originated in this early 8th century B.C. (and probably pictorial) Cretan helmet tradition. The two later helmets, the Fortetsa (*Chart 3, no. M.4.1*) and Kavousi (*Chart 3, no. M.4.2*) helmets, with a crest curving forward refer to the 8th century B.C. development of this helmet type (originated in the Carchemish and North Syrian conical crested helmet tradition, *see* above). These helmets would lower the date of the metalwork on which they were represented, but we know, that the evolution of this helmet type started in the late 9th – early 8th centuries B.C.[538]. It is possible that the evolution of the Eastern Mediterranean crested helmets started with this first stage of orientalization, but more probably that the helmet types of the 8th and 7th century B.C. Eastern Mediterranean originated in the influence of the complete set of Assyrian helmet variants, and consequently in the second (second half of 8th century B.C.) phase of orientalization.

One of the most interesting helmets in this group is a pointed helmet (**Cat. no. 37**, *Pl. 50*) which appeared in an auction at Sotheby's in 1976[539]. The helmet is unique. However, the similarity of its shape to the helmets shown on the Til-Barsip wall-paintings associates it with the group of North Syrian helmets. The long tubular point of the helmet is characteristic of the second half of the 8th century B.C. (*see* on the other helmets of the group). Its best parallels are shown on the Til-Barsip wall-paintings (*Chart 2, nos. 163-171*). The top of their point is flat, moreover there are flat-top knobs on some of their points (*Chart 2, nos. 163-164, 169*). Similarity of the same kind is observable between the broad neckguard of the helmet and the neckguards of the helmets on the Til-Barsip wall-paintings (*Chart 2, nos. 163-165, 167-169*). The decorative system of the helmet is unique. The vertical rib running from the top to the front of the helmet is known from Near Eastern helmets, but there is no parallel for the schematic human faces. In spite of this curiosity the helmet can be associated with the group of North Syrian helmets.

32. Zinçirli, North Palace, Room K1. Group no. **NS.1.1** – H.: 37 cm, Diam.: *ca.* 21.2 cm (von Luschan – Andrae 1943, 163, fig. 83) or H.: 33 cm, Diam.: 26 cm (*ibid.* pl. 40?). Bronze helmet. It has a conical shape with a slight curve and pointed top. The only decoration is three horizontal embossed ribs around the base of the helmet. Restored from pieces (*Pl. 44*). *Ca.* 720 B.C. – Berlin, Staatliche Museen zu Berlin, Preussischer Kulturbesitz, Vorderasiatisches Museum, Inv. (Excav.) no.: S.3976 – von Luschan – Andrae 1943, 75-76, 163, fig. 83, pl. 40b; Barnett 1953, 102; Borchhardt 1972, 100, Cat no. 22 B I 1, pl. 34:5; Salvatori 1975, 260, note 14; Curtis 1979, vol. I, 182-183; Özgen 1982, 20, 52 note 5; Dezső 1998, 11.

33. Zinçirli, North Palace, Room K3. Group no. **NS.1.2** – H.: 33 cm, Diam.: 26 cm. Bronze helmet. It has a conical shape with a slight curve and in a reed-form cut top. There are two (or four?) embossed ribs around the base of the helmet. The upper ends on the front in three pairs of embossed arcs ("crooks"). Only the lower parts of the arcs exist. Between the arcs, on the front of the helmet there is an engraved winged sun-disk (*Fig. 7*). Fragmentary (*Pl. 45*). *Ca.* 720 B.C. – Berlin, Staatliche Museen zu Berlin, Preussischer Kulturbesitz, Vorderasiatisches Museum, Inv. (Excav.) no.: S.3695 – von Luschan – Andrae 1943, 75-76, 162-163, figs. 84-85, pl. 40a; Barnett 1953, 102; Calmeyer 1969, 91, fig. 93; Borchhardt 1972, 100, Cat. no. 22 B I 2, pl. 34:6; Calmeyer 1975, 316; Salvatori 1975, 260, note 14; Curtis 1979, vol. I, 182-183; Özgen 1982, 20, 52 note 5; Maass 1987, 72-73, note 34; Dezső 1998, 11.

34. Provenance unknown (said to come from Luristan). Group no. **NS.1.3** – H.: 37.5 cm, W.: 700 g. Pointed bronze helmet with slightly curved contour decorated in *repoussé* with linear design and with punched motifs. There are four horizontal ribs running around the base of the helmet, the upper of which is interrupted by three pairs of curves on the front of the helmet. The curves end in rams' heads. At the height of 7 cms there is a break in the curve of the contour of the helmet which ends at the front in pairs of incised lines on the curves. There is a punched design imitating the wool of the rams (?) above the curves. Between the upper (4th) rib and the break in the contour of the helmet there is a punched decoration made by a special U-shaped tool: on a basis of six rows of U-shaped scales there are six rows (of similar scales) of high "hills" with punched rosettes between them. Above the break in the contour of the side of the helmet there are four further rows of punched scales. In the field between the three pairs of curves bending inward there is a winged sun-disk in *repoussé* with incised and punched details (*Fig. 7*). Above the winged sun-disk there is a punched rosette (consists of double line of punches) and an incised "T"-shaped motif. There are three holes on either side of the helmet to fasten the chin-strap. It has a separately added cast top (the break in the contour of the side of the helmet) (*Pl. 46*). 8th century B.C. – London, British Museum, Inv. no.: WA 134611 – Barnett 1967, 38, pl. 12a; Lloyd 1967, 116, fig. 123; Calmeyer 1969, 89, no. 44C, fig. 91; Borchhardt 1972, 108, Cat. no. 24 VIII 2, fig. 12; Barnett – Curtis 1973, 133, 136, pl. 64b; Salvatori 1975, 260, note 15; Muscarella 1977, 80; Overlaet 1979, 57; Curtis 1979, vol. I, 181; Özgen 1982, 20, 52 note 6; Maass 1987, 73, note 35; Born – Seidl 1995, Abb. 27; Dezső 1998, 11.

35. Cyprus, Palaepaphos, Kouklia, Mavrommatis Tomb (Tomb 7-8). Group no. **NS.1.4** – H.: 24.6 cm, Diam.: 21.4 cm. It has a conical shape with pointed top. There is a series of holes around the rim for attachment of a lining. The helmet has been decorated with *repoussé* technique. There are two horizontal ribs around the base of the helmet which end on the front in one pair of arcs ("crooks") with rounded extremities (animal heads). There is a winged sun-disk between the arcs (*Fig. 7*). Complete (*Pl.*

[538] For example *see* the Munich (**Cat. no. 13**), Karlsruhe (**Cat. no. 14**) and Mainz (**Cat. nos. 109, 110**) crested helmets.
[539] The present writer is much indebted to P.R.S. Moorey for the photograph of this helmet.

47). Cypro-Archaic Period I (750—650 B.C.) – Nicosia, Cyprus Museum, Inv. no. 1965/XI-29/62 – Karageorghis 1966, 321-322, fig. 55; Karageorghis 1967, 234-235, figs. 20, 24; Borchhardt 1972, 99, Cat. no. 22 A II; Overlaet 1979, 55, no. 1; Tatton-Brown 1979, 105, Cat. no. 327, fig. 327; Özgen 1982, 20, 53 note 8; Pflug 1988b, 34-36, fig. 9; Dezső 1998, 11-12, ill. 6.

36. Cyprus, Alassa, "Pano-Mandilaris", Tomb 16. Group no. **NS.1.5** – H.: 18.5 cm, W.: 21.3 cm. Fragment from the front of a conical helmet. There are two horizontal embossed ribs around the base of the helmet. The lower rib ends on the front, the upper rib continues in a widening out curve (arc). There is a series of holes (only five are extant on the fragment) around the base of the helmet in regular intervals of 65 mm. At the front of the helmet there was an eye or face cutout. The finished edge of the cutout is still visible on the fragment. There is a fifth hole for fastening the lining along the edge of the cutout. This cutout followed the contour of the curving arc on the front of the helmet and probably ended in a nose projection between the two curves. On the top part (at the point) of this fragmentary helmet a small section of a vertical rib running from the top to between the two curves still exists. There are six further fragments belonging to this helmet: **1**. undecorated fragment from the side of the helmet (15.5 cm x 8.0 cm); **2**. undecorated fragment from the side of the helmet (4.6 cm x 4.3 cm); **3**. undecorated fragment from the side of the helmet (5.0 cm x 3.7 cm); **4**. fragment from the rim of the helmet with th elower rib and four holes (8.7 cm x 3.0 cm); **5**. fragment from the base of the helmet, the base of the antithetic curve of the front is visible on this fragment (6.2 cm x 6.3 cm); **6**. fragment from the front of the helmet, perhaps the right point of the eye/face cutout with an existing hole for the lining (2.9 cm x 2.5 cm). (*Pls. 48-49*). Cypro-Archaic Period I (750-650 B.C.). – Nicosia, Cyprus Museum, Inv. no. T16/1 – Karageorghis 1988, 798, no. 1, fig. 6; Dezső 1998, 12-13, ills. 7-8.

37. Provenance unknown. Group no. **NS.1.6** – H.: 35.4 cm. Conical bronze helmet with a tube-shape point ending in a flat top. It has a wide neck guard at the back hammered out from the same piece of metal as the helmet. (The helmet was hammered out from one piece of metal.) There is a rib along the edge of the helmet following its contours. There is a series of holes along the front edge of the helmet probably for a lining. There is a vertical rib on the front of the helmet running from the top. There are two large embossed knobs on either sides of the vertical rib. The knobs were decorated with schematic embossed human heads (eyes, nose, mouth)? (*Pl. 50*). Second half of 8th century B.C. – private collection – Sotheby's 1976, 33, no. 188, pl. 17.

Group NS.2: North Syrian conical iron helmet (Cat. no. 38)

The third Zinçirli helmet, of iron (**Cat. no. 38**), has already been discussed. Its inlaid bronze decoration, if it has been reconstructed correctly, does not have parallels in Assyrian art. Consequently only the tendency of its elaborate decoration connects it to the 3rd phase of the evolution of conical/pointed helmets (*Fig. 1*). Von Luschan and Andrae recognized on the fragments the elements of a winged sun-disk[540], which is a characteristically North Syrian motif (**Cat. nos. 33, 34, 35**). The chequered motif of the lower three registers on the fragment is known from quivers and the crests of the horseharness on the Til-Barsip wall paintings[541]. However, the row of buds shown along the rim of the fragment is a typically Assyrian motif. This motif frames the stele-like front decoration of the Nimrud conical iron helmet (**Cat. no. 10**, *Pls. 9-11*) and also reperesents the shining beams of the "aura" (*melammu*) on the front of helmet **Cat. no. 11** (*Pls. 12-16*).

38. Zinçirli, North Palace. Group no. **NS.2.1** – 46 fragments of an iron helmet. The whole helmet has been elaborately decorated with bronze inlay. There is a row of (typically Assyrian) buds around the base of the helmet. There are at least ten (but probably more) rows of inlaid bronze decoration. In the first three rows there are horizontally placed quadrate plates alternately inlaid with bronze forming a chequered motif.[542] In the following five rows there are also horizontally placed rectangular plates: two smaller quadrates are placed in the middle of two bigger quadrates; in the following quadrate there are six small rectangles (*Pl. 57*). Between the rows mentioned above there are two undecorated lines. The decoration of the two upper (9th and 10th) rows consists of horizontally placed rectangles separated by vertical lines. (*Pl. 57*). *Ca.* 720 B.C. – Berlin, Staatliche Museen zu Berlin, Preussischer Kulturbesitz, Vorderasiatisches Museum, Inv. (Excav.) no.: S.3964 – von Luschan – Andrae 1943, 76-79, 163, fig. 88, pl. 41; Barnett 1953, 102; Borchhardt 1972, 100, Cat. no. 22 B I 3; Overlaet 1979, 54 note 7, 55; Curtis 1979, vol. I, 182-183; Özgen 1982, 20, 52 note 5; Dezső – Curtis 1991, 124-125, fig. 22, Dezső 1998, 11.

540 von Luschan – Andrae 1943, 78.

541 Thureau-Dangin – Dunand 1936, pls. XLIX, LIII.

542 This checquered motif is similar to the checquered decoration of the horses' crests on the Til-Barsip wall-paintings (Thureau-Dangin – Dunand 1936, pl. LIII) and can be considered as a North Syrian feature.

NORTHWEST IRANIAN HELMETS

Group NWI.1: Northwest Iranian conical bronze helmets (Cat. nos. 39-47)

In spite of several examples, this is the most mysterious group of Iron Age Near Eastern helmets. Unfortunately there is not a single helmet in this group, whose provenance is known[543]. Most of the helmets of this group appeared on the art market with fictitious or too general provenance such as "Luristan", "Kurdistan", "near Amlash" or "Kirmanshah – Saqqiz", *etc*. The local people told the 1977 expedition of the Belgian Archaeological Mission to Iran that a year before in one of the Djub-i Gauhar looted tombs, a similar conical helmet was found[544]. If it is true, most of the helmets of the group might have been found in tombs. The tombs of the Djub-i Gauhar cemetery can be dated to the Iron Age III Period (800/750—600 B.C.). This interval is so long that it cannot be compared with the Assyrian (or Urartian) "microchronology". Without provenance, exact chronology and information based on other sources (representations, texts), the helmets of the group are examined here only on the basis of technical, formal and typological characteristics.

As mentioned earlier, Calmeyer derived the pairs of curves (terminating in animal heads) decorating the fronts of helmets from the Zagros area. He assumed that the Urartians borrowed the motif from that region in the early 8th century B.C.[545] But the earliest dated examples of this type of helmet from the Zagros-area are the Hasanlu helmets (**Cat. nos. 49-51**), which can be dated only about 800 B.C.[546] But as Salvatori[547] and Overlaet[548] have argued, the earliest examples of this type of helmets decorated with pairs of curves at the front are shown on the palace reliefs of Ashur-nasir-pal II (883—859 B.C.) and the Hasanlu helmets reflect the Assyrian influence. However, Overlaet assumed that – since the curves on the sculptures of Ashur-nasir-pal II have rounded extremities – the curves terminating in animal heads appeared only after his reign[549]. But in the present author's view several small details on the reliefs (such as the incised decoration of the helmets) were reproduced in paint and have now disappeared. So it can be assumed that the curves bending to the centre on the front of Assyrian helmets ended in animal heads (**Cat. nos. 3-5**), otherwise the curves would have lost their original function. It is hard to believe that the first curves ("snakes") on the front of the helmets appeared without animal heads.

The main characteristic of Northwest Iranian helmets is the 2 or 3 pairs of curves bending to the centre and terminating in animal heads with a vertical rib between them ending in an animal head as well. This decorative system remained standard and did not change as on the Assyrian helmets (*Fig. 1, Phase II*). The changes of the shape and decorative system of these helmets can not be observed to such an extent as on the Assyrian and Urartian helmets.

The first five helmets are decorated with 4 (**Cat. no. 39**) or 3 (**Cat. nos. 40-43**) horizontal ribs around the base (to strengthen the rim) and with 3 pairs of curves with a vertical (strengthening) rib between them. The seventh (**Cat. no. 45**) and eighth (**Cat. no. 46**) helmets differs from the others. Helmet **Cat. no. 45** is similar to the 9th century B.C. Assyrian helmets (conical shape, two horizontal ribs around the base ending in a pair of curves at the front). This helmet would be the earliest helmet of the group. Its shape and decoration (similar to its exact parallel from Hasanlu, **Cat. no. 49**) could be derived directly from Assyrian helmets. The eighth helmet (**Cat. no. 46**) is a modern *pastiche*, so it can be placed into this group only with reservations. However the shape of the ninth helmet (**Cat. no. 47**) is a parallel of the shape of helmet **Cat. no. 46**, and forms probably a separate, second half of 8th century formal variant. The comparatively homogeneous material of the group can be dated to the 8th century B.C. only with the help of formal analysis.

The origins of helmet **Cat. no. 44** is questionable. Its decorative system connects it to the group of Northwest Iranian helmets. The characteristic design: pairs of curves at the front do not appear on Urartian helmets as an independent motif. On the front of Urartian helmets (conical bronze helmets with figural decoration, our Group no. U.1.5, **Cat. nos. 91-103**) this motif always accompanied with

[543] The Northwest Iranian Hasanlu helmets – the only helmets which were found on controlled excavations – are a separate group.
[544] "When the Belgian Archaeological Mission in Iran was excavating in 1977 at Djub-i Gauhar in Luristan, they were told that a pointed helmet with *repoussé* decoration of "snakes" (crooks) with ram heads had been found in one of the looted graves. As all the excavated graves at Djub-i Gauhar are dated in the Iron Age III (800/750—650 B.C.), the helmet probably belongs to this period also."; Overlaet 1979, 60, note 25.
[545] Calmeyer 1969, 90-91.
[546] Hasanlu IVB, Iron II Period; *see* Dyson – Muscarella 1989, 19-20.
[547] Salvatori 1975, 260-261.
[548] Overlaet 1979, 58-61.
[549] Overlaet 1979, 60.

additional figurative decoration. The Urartian cuneiform inscription on the helmet makes its origin questionable. There are two possibilities to explain the origin of the inscription: 1. The helmet is Urartian and was inscribed for particular reasons (dedicatory inscription [votive gift] or property sign ["This helmet belongs to the arsenal of RN"]; 2. the helmet is Northwest Iranian and as booty was inscribed with one of the two types of Urartian helmet inscriptions mentioned above. In this case an Urartian soldier might brought the helmet from Northwest Iran, or a Northwest Iranian ally or mercenary serving in the Urartian army might use it.

The characteristic double-line border of the curves at the front of this helmet unknown from Urartian helmets connects it to Northwest Iran (**Cat. no. 41**) and Hasanlu (**Cat. no. 51**) as well.

39. Provenance unknown (reported as from Luristan). Group no. **NWI.1.1** – H.: 29.7 cm. Pointed helmet decorated in *repoussé* with details in linear design. The long tubular point of the helmet has a flat top. There are two embossed ribs around the base on which stand three pairs of curves bending to the centre. Two more horizontal ribs are interrupted by the curves. All the curves and a vertical rib running from the top of the helmet to between the *repoussé* curves end in animal (ram's) heads. Small circles behind each head indicate the skin or fleece of the animal. Complete (*Pl. 58*). 8th–7th century B.C. – Mariemont, Musée Royal de Mariemont, Inv. no.: Ac. 72/27 – Anonymus 1973, 59-60, no. 4 (fig.); Evrard-Derriks 1977-78, 38-39, no. 22 (fig.); Overlaet 1979, 58.

40. Provenance unknown (reported as from Luristan). Group no. **NWI.1.2** – H.: 33.8 cm.; Diam.: 14.5 – 25 cm. Pointed helmet decorated in *repoussé* with a vertical rib on the front and three pairs of curves which stand on three ribs around the base of the helmet. The curves (arcs) and the vertical rib end on the same level in animal (lion) heads. Complete (*Pl. 59*). 8th–7th century B.C. – Paris, Musée du Louvre, AO 22123 – Vingt ans d'acquisitions, 1967, no. 27; Amiet 1971, 148; Amiet 1978, 67; Overlaet 1979, 57-58, pl. 3; Vanden Berghe – De Meyer 1982, 129, no. 19 (fig.); Esaian 1986, 23-24, pl. 13:12.

41. Provenance unknown (reported as from Kurdistan, Kirmanshah – Saqqiz). Group no. **NWI.1.3** – H.: 32.6 cm; Diam.: 21.2 cm. Pointed helmet decorated in *repoussé* with linear design. There are three horizontal ribs around the base on which stand three pairs of curves (arcs). The innermost pair of curve ends in rounded extremities, the others with a vertical rib between them end in animal (sheep) heads. Just behind the heads the fleece is depicted. Between the curves there are remains of other decoration. Complete (*Pls. 60-61*). 8th century B.C. – Altenessen, Museum Altenessen, Inv. no. 72/246A – Calmeyer 1969, 88, no. 44A, fig. 89; Borchhardt 1972, 107, Cat. no. 24 VIII 1, fig. 9; Calmeyer 1975, 316; Salvatori 1975, 260, note 18; Kellner 1976, 78, Cat. no. 141, pl. 11; Özgen 1982, 20, 52 note 6; Vanden Berghe – De Meyer 1982, 130, no. 20 (fig.); Esaian 1986, 26, pl. 13:8; Schumacher 1986, 62, colour pl. on p. 61; Völcker-Janssen 1992, 51-52, no. 93, pl. 32.

42. Provenance unknown (reported as from Kurdistan, Kirmanshah – Saqqiz). Group no. **NWI.1.4** – H.: 28 cm. Pointed helmet decorated in *repoussé* with linear design and punched circles. There are three horizontal ribs around the base on which stand three pairs of curves (arcs). All the curves and a vertical rib end in sheep heads. Just behind the heads small punched circles depict the fleece. Complete (*Pl. 62*). 8th–7th century B.C. – Copenhagen, National Museum, Inv. no. 14495 – Buhl 1962, 30-31, colour pl. on p. 31; Anonymus 1968, pl. on p. 51; Calmeyer 1969, 90, no. 44D, fig. 92; Borchhardt 1972, 108, Cat. no. 24 VII 2, fig. 10; Overlaet 1979, 57; Özgen 1982, 20, 53 note 6.

43. Provenance unknown (reported as from Luristan). Group no. **NWI.1.5** –H.: 30 cm; Diam.: 23 cm. Pointed helmet with a *repoussé* decoration of three ribs around the base on which stand three pairs of curves (arcs). The curves and a vertical rib at the front of the helmet all ends in one horizontal line. The rib is fragmentary. 8th–7th century B.C. – Teheran, Foroughi Collection – Ghirshman 1961, no. 482; Ghirshman 1963, 336, fig. 418; Ghirshman 1964, fig. 418; Huff 1968, fig. 4; Vanden Berghe 1968, fig. 2:4; Belloni – Dall'Asèn 1969, fig. on p. 25; Calmeyer 1969, 88, no. 44B, fig. 90; Borchhardt 1972, 108, Cat. no. 24 VIII 1, fig. 11; Özgen 1982, 20, 52 note 6.

44. Provenance unknown (reported as from "Karmir Blur", "Urartu"). Group no. **NWI.1.6** – "*Rare casque en bronze repoussé ... timbre conique, orn, de cannelures concentriques ... la base du timbre et d'une cannelure transversale verticale vers laquelle s'incurvent en relief quatre rainures issues de la base. Une inscription gravée en caracters cunéiformes parcourt le bord frontale du timbre. Accidents. Ourartou, Karmir Blour, 8e siècle av. J.-C.*" (Paris – Hotel Drouot 1991, 13) Pointed helmet with a *repoussé* decoration of three horizontal ribs around the base which end at the front in two pairs of *repoussé* curves. The curves at the front have a double contour in *repoussé* and terminate in curved extremities. There is an embossed vertical rib running from the broken top of the helmet to between the curves. There is an Urartian inscription in a panel at the front of the base of the helmet. Broken at the top and between the curves. 8th century B.C. – Private Collection – Paris – Hotel Drouot 1991, 13, no. 57, pl. on p. 13.

45. Provenance unknown (reported as from "Iran"). Group no. **NWI.1.7** – H.: 27 cm; Diam.: 20.3-19.1 cm. Pointed helmet with *repoussé* decoration. There are two horizontal ribs around the base of the helmet. The ribs end at the front in one pair of curves (arcs) with rounded extremities. Between the curves, there is a vertical rib on the front of the helmet. There is a series of holes around the base of the helmet for the attachment probably of a lining. The front of the helmet is restored (*Pl. 63*). 8th–7th century B.C. – Mainz, Römisch-Germanisches Zentralmuseum, O 39591 – Kellner 1976, 78, Cat. no. 142; Vanden Berghe – De Meyer 1982, 129, no. 18 (fig.); Esaian 1986, 28, pl. 13:6.

46. Provenance unknown (reported as from "Hasanlu", "Iran"). Group no. **NWI.1.8** – H.: 35 cm. "*The helmet consists of a hammered globular cap surmounted by a cast point with a flat top. Holes are pierced through along the edge at intervals of ca. 20 mm. The helmet is consisted of a number of soldered fragments. Some of these proved to be modern. One of these new fragments is the upper part of the globular cap of the helmet and this may have changed the general shape of the helmet slightly. Probably the connection between the point and the globular cap was more fluent and as a result the helmet may have been slightly higher. A bronze plaque, which was attached to the helmet, proved to be a recently added element and therefore had to be removed. It is a rectangular plaque of bronze with a thickness of ca. 1.1 mm, flaring out at the lower end into a curved bottom, which is pierced through at regular intervals. It is decorated with a vertical rib. The top of the plaque has been cut off, obviously because it was too long to fit on the helmet. It was a horse's forehead plaque.*" (Overlaet 1979, 61-63). *Pasliche.* 8th century B.C. – Bruxelles, Musées Royaux d'Art et d'Histoire, Inv. no.: IR 1361 – Vanden Berghe 1967, 22; Overlaet 1979, 61-63, fig. 3, pl. 4; Dewanckel 1979, 64-65.

47. Provenance unknown (reported as from "Northern Iran"). Group no. **NWI.1.9** – H.: 34 cm; Diam.: 19.2—21.6 cm. High, undecorated pointed helmet. There is a series of holes around the rim of the helmet (at 15 mm intervals) probably for a lining. The contour of the helmet has a characteristic S-curve, which separates the rounded (almost hemispherical) cap-piece and the tapering point of the helmet (*Pl. 64*). 8th century B.C. – London, Houshang Mahboubian Family Collection – Mahboubian 1997, 287, no. 363.

Group NWI.2: Northwest Iranian conical iron helmet (Cat. no. 48)

As we have seen earlier, this helmet (**Cat. no. 48**) can be placed into the 3rd group according to the technique of manufacturing ancient Near Eastern iron helmets (*see* pp. 35-37). Characteristic is its manufacture of two similar parts (halves) riveted together. This technique is known mainly from the periphery of the Assyrian Empire[550]. The elaborate decoration of the Assyrian and North Syrian helmets inlaid with bronze was replaced by the simpler technique of riveting, although the archaizing decorative system of the helmet follows the 9th – early 8th century B.C. decoration of Assyrian helmets. However the elongated, pointed shape dates this helmet to the middle of the 8th century B.C. It can be assumed that outside Assyria (and North Syria under Assyrian rule), on the periphery of the Assyrian Empire the techniques for hammering out a helmet from one piece of iron (**Cat. no. 10**) or inlaying the iron with bronze (**Cat. nos. 10, 16-31, 38**) were not employed in the 8th–7th centuries B.C.[551] This illustrates the high standard of Assyrian and North Syrian metalworking of the 8th–7th centuries B.C.

However, accepting the suggested reconstruction of the figurative scenes, this helmet is exactly comparable to the 9th – early 8th century B.C. Assyrian helmets, and shows a strong Assyrian influence on the Northwest Iranian helmet industry, if its origin has been correctly reported. However, it must be added that – as has been mentioned earlier – the technique of riveting an iron helmet from more parts or two halves together was – as a well defined stage of the evolution of the technique of manufacturing iron helmets – undoubtedly used in Assyria as well. This technique together with the riveted bronze decoration – because it was much cheaper than hammering out an iron helmet from a single piece of iron and inlay it with bronze – was used later times probably not only on the peripheries, but in Assyria as well. So the possibility that this helmet indeed a middle of 8th century B.C. Assyrian manufacture and was taken to Northwest Iran as booty, present, *etc.*, can not be entirely excluded.

48. Provenance unknown (reported as from "Amlash", "Luristan"). Group no. **NWI.2.1** – H.: 32.5 cm; Diam.: 22 cm. "The helmet is made of two similar parts joined by iron strips on the inside, fastened with iron rivets. The bronze decorations were fastened on the helmet with bronze rivets having iron roundels at the back. They have now partly vanished, but those rivets and roundels which are preserved, indicate where originally bronze strips were attached. Parallel to the lower edge runs a bronze strip, fastened with four rivets, above which was a band of vertically placed bronze rectangular plates, each attached with one rivet. Some of these plates have now disappeared but the remaining rivets leave no doubt that this band ran all round the helmet. Two pairs of curved bronze strips or crooks, from which only one of each pair remains, decorated the front of the helmet. The space between these crooks was probably ornamented with a second band of vertical rectangular plates but with a doubled number in comparison to the lower band. Only the rivets and one of these plates are preserved. The remaining rivets prove that more decorations were present but these have all disappeared." (Overlaet 1979, 53-55). That there was originally a vertical strip (like the other helmets of the group) can be reconstructed from remaining rivets on the front of the helmet. The row of vertically placed rectangular bronze plates above the horizontal bronze strip around the base of the helmet would have been a figurative scene (similar to the Assyrian helmets **Cat. nos. 3, 4, 10, 11**) of officials in procession. The row of vertically placed rectangular bronze plates (from which only one extant and six rivets remained) between the two pairs of curves on the front of the helmet are the remains of a central scene known from the 9th century B.C. Assyrian helmets (**Cat. nos. 3-5**): the tree-of-life (in the middle) and a king with two of his attendants behind him (on the sides of the tree). Fragmentary (*Pls. 65-66*). Middle – second half of 8th century B.C. – Bruxelles, Musées Royaux d'Art et d'Histoire, Inv. no.: IR 1241 – van Pée 1965, 196-198, pl. on p. 196; Vanden Berghe 1967, 22; Overlaet 1979, 53-55, fig. 1, pls. 1-2; Dezső – Curtis 1991, 125-126.

---

[550] From Urartu *see* **Cat. nos. 104-107**.

[551] As we have seen, the Assyrians made iron helmets riveted together from different parts as well (*see* for example **Cat. nos. 11, 17**). The earliest iron helmets of the Near East and Assyria (probably in the late 9th – early 8th century B.C.; *see* above) were supposingly riveted together from pieces of iron sheet. But, while in Assyria the technique of hammering out a helmet from one piece of iron appeared early, at least in the middle of the 8th century B.C., on the other territories of the contemporary Near East the technique of riveting was exclusively used.

HASANLU HELMETS

The archaeological site of Hasanlu is situated on the southern shores of Lake Urmia, Northwest Iran. When the settlement was destroyed at the end of the 9th century B.C.[552] by enemy action more than 7000 metal objects were left in the debris. This has made Hasanlu (IVB) one of the most significant sites for the study of Iron Age metalwork in the Near East. There is evidence that some of this metalwork was made locally[553], perhaps most of it. However, as Hasanlu was in contact with the Urartians (who may have been the destroyers of Hasanlu IVB) with the Assyrians[554] and the Syrians[555] and with the native peoples of Western Iran[556], it is often difficult to be certain whether specific objects were imported or manufactured locally under foreign influence. The helmets illustrate this problem. They were found both in the destruction debris and in tombs; two were found close to or on skulls (**Cat. no. 50**, *Pls. 68-69*; **Cat. no. 52**[557]). Some may have belonged to the invaders not to the defenders of Hasanlu. There were two types of helmets found in Hasanlu: conical helmets and crested helmets.

Group H.1: Hasanlu conical bronze helmets (Cat. nos. 49-51)

These helmets are unambiguous evidence of 9th century B.C. Assyrian influence at Hasanlu. The first two helmets without figural or incised decoration recall the helmet representations (*Chart 2, no. 83*) of the palace reliefs of Ashurnasirpal II (883—859 B.C.). Hasanlu is one of the first places outside Assyria, where this type of helmet was worn in the 9th century B.C. The conical type of helmet had already been in use in Urartu since at least the reign of Ishpuini (*ca.* 830—810 B.C.)[558], but there is no Urartian helmet decorated simply with pairs of curves at the front. The Urartian conical helmets decorated with pairs of curves at the front always have complementary figural decoration (**Cat. nos. 57, 91-93, 96, 99-101**). So this type of helmet decoration (pairs of curves at the front without additional figural decoration) derived from Assyria, had survived in Northwest Iran into the 8th century B.C., when in Assyria it was replaced by the continuous "stele-like" arcs on the front of the helmets.

The second helmet (**Cat. no. 51**) is an exact parallel for an unprovenienced Northwest Iranian helmet (**Cat. no. 46**). There are two horizontal strengthening ribs around the base ending in one pair of *repoussé* curves on the front of both helmets. The four holes on the front of helmet **Cat. no. 45** are for some kind of decoration (*e.g.* riveting a decorated metal plate onto the front of the helmet). Both of the helmets is a "squat" conical helmet. I.J. Winter assumed[559] that this squat ("Persian") type of conical helmet can be seen on a *repoussé* warrior (or "master of animals") on a horse's breastplate from Hasanlu[560] (*Chart 2, no. 125*). She found parallels in the helmet representations of an Assyrian palace relief from the reign of Ashurnasirpal II (883—859 B.C.) showing fleeing Iranian horsemen in front of an Assyrian chariot[561]. But owing to the bad condition of the relief slab only one helmet of the Iranian fugitives can be seen, namely of a wounded man lying on the field, whose "helmet" shows a "banded structure" similar to leather caps.

Since the conical shape of these two helmets (**Cat. nos. 46, 50**) is a characteristically 9th century B.C. type, the possibility of their being manufactured as independent variants (a "subgroup") either in Assyria or Northwest Iran (Hasanlu) can not be precluded. The lower height and squat form of

---

552 The date of the destruction has been disputed. R.H. Dyson and O.W. Muscarella proposed a late 9th century date (Dyson – Muscarella 1989, 1-27). For a later dating (714 B.C.) *see* Medvedskaya 1988, 1-15 and Medvedskaya 1991, 149-162 [she defends her case strongly here]. This catalogue follows the date, about 800 B.C., argued by R.H. Dyson and O.W. Muscarella, since there is no evidence for using the hemispherical crested helmets later than the first half of the 8th century B.C. The latest known representation of this type of helmet is the Zinçirli chariot relief (*Chart 3, no. 127*) which hardly can be dated to the end of the 8th century B.C. For detailed history of hemispherical crested helmets *see* pp. 37-38, 74-77, *Fig. 9*.

553 *E.g.* horse harness (de Schauensee 1988, 47.)

554 The basis of the identification/comparison are the palace reliefs of Ashurnasirpal II (883—859 B.C.). The best examples amongst the considerable assyrianizing group of objects (architectural metalworking, different groups of metal vessels) are the four *rhyta* ending in animal heads found in Hasanlu (Dyson 1961, figs. 10-11; Winter 1980, fig. 69; de Schauensee 1988, figs. 22a-c). There are several objects of clearly Assyrian origin found at Hasanlu: ivory carvings, glazed wall tiles, cylinder seals, half of a macehead inscribed in cuneiform: "Palace of Ashur-uballit" (king of Assyria 1365—1330 B.C.); Winter 1980, 22-23, note 109; Pigott 1982, 202ff.

555 Ivory carvings and lion bowls (Winter 1980, 22, note 108).

556 *E.g.* bronze vessels riveted together from several separate parts (de Schauensee 1988, 50, fig. 21).

557 Muscarella 1988, 50: "These fragments, all found together by the head of a skeleton".

558 Helmets with the "lightning fork symbol of Teisheba" on their front: **Cat. nos. 62-64**.

559 Winter 1980, 24.

560 HAS 74-241; Winter 1980, figs. 1, 17-31.

561 BM 124559; Winter 1980, fig. 11.

these helmets may be other characteristic features of this subgroup, which flourished in Northwest Iran (with Hasanlu) in the late 9th – early 8th centuries B.C.

The third helmet (**Cat. no. 51**) may be much rather considered as an early example of Northwest Iranian helmets. The two pairs of curves distinguish it from the other conical Hasanlu helmets mentioned above. The double contour line of the outer curve at the front is similar to the contour lines of the inner curve on the front of a Northwest Iranian helmet (**Cat. no. 42**) and to the curves of another one (**Cat. no. 45**). This is the earliest example of such a double contour line which, since it is without any parallels from other territories of the ancient Near East, might be a Northwest Iranian characteristic. On the other hand, the embossed rosettes decorating the helmet clearly show Assyrian influence. There are rosettes on a 9th century B.C. Assyrian conical helmet (**Cat. no. 3**, *Pls. 3-4*) as well.

These three helmets are the earliest known examples of the emergence of Northwest Iranian helmets in the late 9th century B.C., when Assyrian influence was so strong (their parallels can be seen on the Assyrian palace reliefs[562]), that it can not be decided exactly whether they were made locally or came from Assyria.

49. Hasanlu, Burned Building II, probably Room 5, northwest quadrant. Group no. **H.1.1** – H.: 27 cm; Diam.: 22 cm. Conical helmet with pointed top. There are three horizontal ribs around the base of the helmet. The uppermost ends in a pair of curves (arcs) with rounded extremities. Between the curves there is a vertical rib runing from the top. Fragmentary (*Pl. 67*). *Ca.* 800 B.C. (Hasanlu Period IV [Iron II]) – Tehran, Iran Bastan Muzeh, Inv. no.: HAS 60-883 – Dyson 1961, 536, fig. 14; Azarpay 1968, pl. 9; Calmeyer 1969, 90, note 298; Borchhardt 1972, 107, Cat. no.: 24 IV 2; Salvatori 1975, 259, note 8; Overlaet 1979, 51, note 1, p. 55; Curtis 1979, vol. I, 183; Winter 1980, 24, note 124, fig. 60; Özgen 1982, 53, note 6; Muscarella 1988, 50.

50. Hasanlu, Burned Building II, Room 5, north central quadrant. Group no. **H.1.2** – H.: 26.2 cm, Diam. (flattened): 30 cm. Squat conical helmet with pointed top. There are two horizontal ribs around the base of the helmet. At the front of the helmet the ribs end in one pair of curves (arcs) with rounded extremities. Between the curves there is a vertical rib coming from the top. There is a series of holes around the base for the attachment probably of a lining. Fragmentary (*Pls. 68-69*). *Ca.* 800 B.C. (Hasanlu Period IV [Iron II]) – Philadelphia, University of Pennsylvania Museum of Archaeology and Anthropology, Inv. no.: UM 61-5-352 (HAS 60-1030) – Dyson 1961, 536, fig. 14; Overlaet 1979, 56, note 15; Winter 1980, 24, fig. 61; Muscarella 1988, 50.

51. Hasanlu, Period IV house, above Pd V house, cleanup. Group no. **H.1.3** – 22 fragments of the front and the top of a conical helmet. There were two horizontal ribs around the base of the helmet and a row of holes for the attachment probably of a lining. There were two pairs of curving bands. No median ridge is now apparent, but it would have been structurally necessary. There was a row of rosettes around the base of the helmet (two exist) and there were rosettes between the curving bands (two exist on each side). There is a big rosette in the middle of the front between the curving bands (*Pls. 70-71*). *Ca.* 800 B.C. (Hasanlu Period IV [Iron II]) – Tehran, Iran Bastan Muzeh, Inv. no.: HAS 72-69 – Winter 1980, 24, note 124; de Schauensee 1988, 55, pl. 54; Muscarella 1988, 50.

Group H.2: Hasanlu hemispherical crested helmets (Cat. nos. 52-56)

The first known appearance of the archetype of Near Eastern crested helmets is on the so called "Mari tropaion" (*Chart 1, no. 11*)[563] of Old-Babylonian period (18th century B.C.). This type of crested helmet might be the archetype of both conical and hemispherical crested helmets.

The earliest known representations of hemispherical crested helmets (*Fig. 9*) are known from 14th–11th centuries B.C. Transcaucasian bronze statuettes (chariot models found in tumuli): Ltshashen (14th–13th centuries B.C., *Chart 1, nos. 46-49*), Lori Berd (14th–13th centuries B.C., *Chart 1, nos. 50-51*) and Shirakavan (13th–11th century B.C., *Chart 1, no. 52*). The main characteristic of these helmets is that the crest was fastened centrally across the top of the hemispherical skull-piece of the helmet. That helmet type emerged in the Late Bronze Age is shown on a 14th century B.C. Anatolian cylinder seal (*Chart 1, no. 39*)[564] on which a warrior wearing an early conical helmet (*Chart 1, no. 38*) defeats a warrior wearing this type of helmet.

The second and final flourishing of this type of helmets was in the 9th century B.C. (*Fig. 9*). At that time mainly Urartian soldiers wore it, but it has to be mentioned that this type was widespread in all the territories north of Assyria, and was replaced by Assyrian conical and North Syrian conical crested helmets only by the first decades of the 8th century B.C. The hemispherical type of crested helmet originated in Transcaucasia in the Late Bronze Age, and from that region spread to Urartu and Northwest Iran (Hasanlu, Khurvin: *Chart 1, no. 53*) as well (*Fig. 9*).

On the bronze reliefs of the Balawat gates of Shalmaneser III (858—824 B.C.) there are Urartian soldiers wearing this type of crested helmet (*Charts 3-4, nos. 107-110*) and defeated by or

[562] **Cat. no. 49**: *Chart 2, no. 83*; **Cat. no. 50**: *Chart 2, nos. 81-82*.
[563] A shell inlay "tropaion", Damascus, Archaeological Museum, Inv. no. 1915; Borchhardt 1972, 96, fig. 8, Kat. 21 I 4.
[564] Boston, Tyszkiewicz Collection, Terrace 1962, fig. 12.

fighting with Assyrian troops. On the same bronze reliefs the troops of Sangara, king of Carchemish also wear this type of crested helmet[565]. If so, for a short period in the reign of Sangara, in Carchemish the earlier Carchemish type of conical crested helmet (*Charts 3-4, nos. 101-104*) was replaced by the hemispherical type of crested helmet. The same helmet type is shown on a 9th century B.C. ivory carving from Nimrud, on which a defeated Urartian soldier holds the reins of his Assyrian enemy's horse (*Chart 3, no. 106*). It is clear from this representation that the crest of the helmet might be made not only from horse hair but from feathers, or feather-like metal sheets as well. There are several representations of Urartian soldiers wearing this type of crested helmet not only from Assyria but Urartu as well.

On an unprovenienced "assyrianizing" horse's bronze breastplate[566] there is a chariot (with warriors wearing conical helmets[567]) pursuing a horseman wearing this hemispherical type of crested helmet (*Chart 3, nos. 111-112*). The two parties are unfortunately unidentified. Seidl without any certain evidence for identification suggested[568] that an undefined enemy attacks Urartians. Dyson and Muscarella identify the scene as Urartians attacking an unknown enemy[569]. A similar horseman is shown on an unprovenienced Urartian bronze belt fragment (*Chart 3, no. 115*) as well[570].

This breastplate can be dated to the last quarter of the 9th century B.C. on the basis of the helmets depicted: conical helmets with vertical fin (**Cat. nos. 59-61**) and 'Hasanlu' earflaps an exact parallel of which is a probably Urartian (or Northwest Iranian) earflap **Cat. no. 112**; hemispherical crested helmets, a 9th century B.C. type in Assyrian (*Chart 3, nos. 106-110*) and Urartian art (*Chart 3, nos. 113-115*), but with wide distribution in Iran, including Hasanlu IV (**Cat. nos. 52-56**), Khurvin (*Chart 1, no. 53*).

Since the conical helmet with vertical fin (on the chariot warriors, *Chart 2, nos. 116-117*) is known only from Urartu (**Cat. nos. 59-61**), and the long earflap type was known both in Northwest Iran (Hasanlu, **Cat. no. 53**) and Urartu (**Cat. no. 112**) it cannot be argued that Assyrians or other nations attack Urartians in the scenes on this breastplate. It is more probable that Urartians attack an enemy of Transcaucasian or Northwest Iranian origin, where the hemispherical type of crested helmet was also widespread[571].

The same type of crested helmet is shown on Urartian bronze statuettes from Toprak Kale (*Chart 3, no. 113*) and Maraş (*Chart 3, no. 114*) and on the Urartian bronze plate (belt?) fragment mentioned above (*Chart 3, no. 115*). As has been mentioned this type of helmet appeared in the 9th century B.C. in North Syria as well. One of its latest (first half of 8th century B.C.) representations is on a chariot scene on a relief slab from Zinçirli (*Chart 3, no. 127*). It is interesting that this type of helmet appears on a second half of 8th century B.C. bronze belt from Fortetsa, Crete (*Chart 3, no. M.7.1*) as an outdated helmet together with two early 8th century B.C. Carchemish type of conical crested helmets (*Chart 3, nos. M.3.1-2*) and with an advanced conical crested helmet (*Chart 3, no. M.4.1*) which developed from the Carchemish type of crested helmet at the end of the 9th century B.C.[572]. So the hemispherical crested helmet on the Fortetsa belt[573] might be an outdated motif borrowed from North Syria.

There are two groups of Hasanlu hemispherical crested helmets:

1. Bronze helmets (Helmets with bronze skull-piece, to which the bronze crest-piece and the bronze/iron earflaps were fastened): **Cat. nos. 52-54** (*Pls. 72-77*)
2. Leather helmets (Helmets with leather skull-piece furnished with bronze crest-supports): **Cat. nos. 55-56** (*Pls. 78-79*)

In view of their crest-structure these five helmets can be divided into further two groups:

1. Helmets with horsehair or feather crests fastened between crescentic bronze crest-supports, whose rim was perforated to fix the horsehair or feathers: **Cat. nos. 52, 55-56** (*Pls. 72-73, 78-79*)

[565] King 1915, pl. 42; de Schauensee 1988, 55.
[566] Seidl 1986, *passim*.
[567] The main characteristics of these conical helmets are the vertical fin, which is known only from Urartian conical helmets (**Cat. nos. 59-61**), and the long earflaps, which are known only from Hasanlu crested helmets (**Cat. nos. 52-54**). Seidl 1986, fig. 3, *Chart 2, nos. 116-117.*
[568] Seidl 1986, 229ff., figs. 1-2.
[569] Dyson – Muscarella 1989, 19-20.
[570] Kellner 1991, no. 447, pl. 88.
[571] Dyson – Muscarella 1989, 20; There is a bronze statuette of a warrior wearing this type of crested helmet from Khurvin (*Chart 1, no. 53*), Vanden Berghe 1959, 124, pls. 156b-c.
[572] *See* p. 38, note 334.
[573] Fortetsa, Tomb P, no. 1568, Iraklion, Archaeological Museum, Inv. no. 1568; Brock 1957, pl. 115; Borchhardt 1972, 70, Kat. 15 II 2.

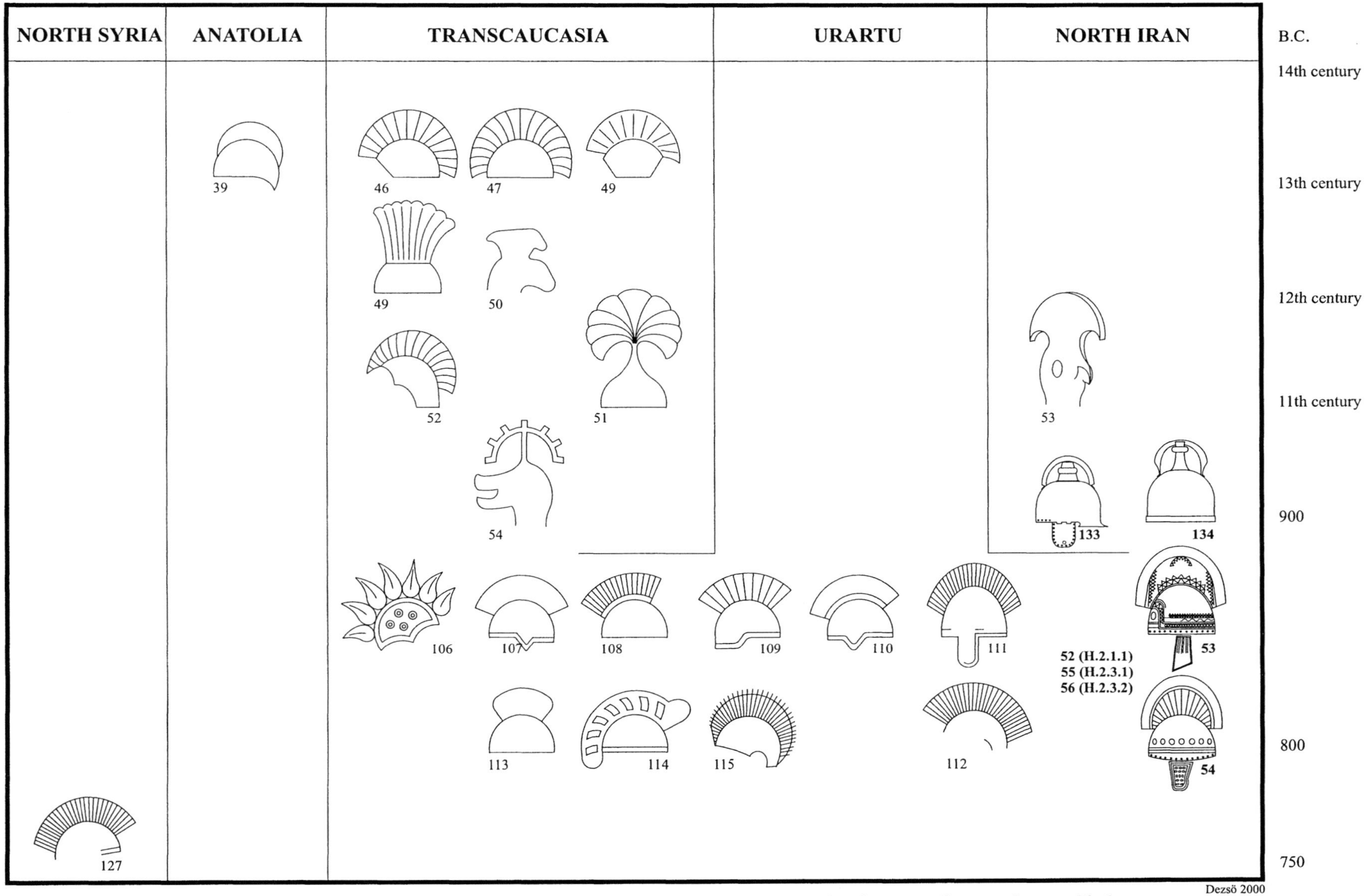

Fig. 9: The evolution of Near Eastern hemispherical crested helmets and North Iranian (Group I.3) type of crested helmets

2. Helmets with bronze crest (probably without horsehair or feather crest): **Cat. nos. 53-54** (*Pls. 74-77*)

Unfortunately the helmets furnished with horsehair or feather crest were with the exception of **Cat. no. 52** mainly helmets with leather skull-pieces. Consequently there is no possibility of reconstructing their decoration. This probably cheaper helmet type might be the more widely distributed.

Examining the decorative system of helmet **Cat. no. 53**, it is clear that there is not a single motif characteristic of Urartian helmets to be found on this helmet. The earflaps of the helmet are unique. The only existing parallel of this type is an unprovenienced Urartian ("Transcaucasian"[574]) or Northwest Iranian earflap (**Cat. no. 112**) and the same earflaps are shown on the representation of conical helmets (*Chart 2, nos. 115-116*) on an unprovenienced horse's breastplate discussed above[575]. The "assyrianizing" scene of the breatsplate depicts chariot riders wearing conical helmets and pursuing horsemen wearing crested helmets of this type. The earflaps of these conical helmets in their form are exact parallels of the earflaps of the helmet (**Cat. no. 53**, *Pls. 74-75*). The men wearing broad belts on the representations of the horse's breastplate might be Urartians, but the pursuing warriors and the pursued horsemen have the same type of belt[576], and their double zig-zag motifs parallel similar, incised motifs on helmet **Cat. no. 53**. The geometric designs of helmet **Cat. no. 53** are well known not from Urartian art, but from Caucasian[577] and Transcaucasian[578] bronze belts (late 2nd – early 1st millennium B.C.[579]) and from scabbards from Ltshashen[580] (13th–12th century B.C.[581]). The incised snake motifs appear on the same objects as well[582]. This decorative system of incised geometric and figurative motifs spread presumably to Northern Iran (Iranian Azerbaijan) as well in the late 2nd – early 1st millennium. So the distribution zone of this type of hemispherical crested helmet and its decorative system was between Transcaucasia and Northern Iran (Hasanlu).

These helmets are the only existing examples of hemispherical crested helmets. This type emerged probably in Transcaucasia[583] (a little north from the later Urartian homeland) and spread in the late 2nd – early 1st millennium through ethnic and cultural boundaries to the West (Urartu, and from Urartu to North Syria) and to South-East (Northern Iran, Hasanlu), as can be seen from their representations (*see Map*). Consequently there is no reason to associate it with a single ethnic group (as might happen in case of the Balawat Gates), and there is no possibility for building up ethnic theories on different types of helmets in Hasanlu. The hemispherical type of crested helmets was used in the 9th century B.C. both in Urartu and in Hasanlu.

This type of crested helmet was replaced only by conical helmets and conical crested helmets more regularly in the late 9th century B.C. The Hasanlu helmets are the best examples of the coexistence of conical and hemispherical crested helmets in the late 9th century B.C.

Group H.2.1: Hemispherical crested helmet with bronze skull-piece, horsehair or feather crest and bronze earflaps (Cat. no. 52)

52. Hasanlu, Burned Building II, Room 5, north central quadrant. Group no. **H.2.1.1** – Diam. of crest at base: 28 cm; Width of fragments: 4.7 – 5.0 cm; Length of earflaps: 14.3 cm; Width of earflaps: 7.5 cm. "Recognizable among the fragments are a complete earflap (or cheek guard), almost straight on one side, curved on the other, with a raised border and flat rim pierced throughout with small holes (No. 60a); an almost complete fragment of a second earflap (fragments No. 60b and c join); fragments of two crescent-shaped crest supports (No. 60d and e, which join and f, g, h, j); one other fragment (No. 601) could be a part of the crest supports, as it tapers and has holes along the border; amorphous fragments that might be parts of the crushed helmet (No. 60i, k, m). These fragments, all found together by the head of a skeleton, belong to a helmet furnished with detachable earflaps and two units that once held the crests, which were probably composed of such perishable materials as feathers or animal hair. The crest supports rested directly on the cap of the helmet, following its contour." (Muscarella 1988, 50-52.). (*Pls. 72-73*). Ca. 800 B.C. (Hasanlu Period IV [Iron II]) – New York, Metropolitan Museum of Art, Inv. no. MMA 61.100.39a-m [HAS 60-620] – Winter 1980, 24; Muscarella 1988, 50-52, no. 60, figs. 60a-m (on p. 510).

574 Kellner 1976, 78, Cat. no. 144, fig. 46.
575 Seidl 1986, figs. 1-3.
576 Seidl 1986, fig. 3.
577 Tehov 1977, figs. 97-100.
578 Odzun, Esaian 1984, fig. 40; Hatshbulag, Esaian 1984, fig. 47.
579 Esaian 1984, 110-119.
580 Esaian 1966, 62, fig. 1.
581 Esaian 1966, 62.
582 Belts: Tehov 1977, figs. 97-98; scabbard: Esaian 1966, fig. 1.
583 *See* on p. 74 (Ltshashen, Lori-Berd and Shirakavan) and *Fig. 9*.

Group H.2.2: Hemispherical crested helmets with bronze skull-piece, bronze crest and bronze or iron earflaps (Cat. nos. 53-54)

53. Hasanlu, Burned Building II, north central quadrant. Group no. **H.2.2.1** – H.: 27 cm (with crest); Diam.: 18.5 cm; Diam. of crest at base: 23 cm; L. of earflaps: 23 cm; W. of earflaps: 7 cm. Crested bronze helmet with (detachable) earflaps. The helmet was decorated with elaborate incised geometric design. The crescent-shaped crest supports rested directly on the cap of the helmet, following its contour. There is a series of holes around the base of the helmet for attachment probably of a lining. There are four incised bands – each of them made of a double row of punched points – around the base of the helmet they end at the front of the helmet in four curves (arcs). Between the lower bands there is an incised double saw-tooth motif with punched points between them. Between the two upper bands there is an incised snake-motif (waving). On the upper band there is an incised, punched double saw-tooth motif with punched points. On the front of the helmet under the arcs there is a fragmentary, incised decoration of a circle (eye-motif) with a bordering zig-zag band. There are remains of other incised decorations under the circle on the front of the helmet. The base of the crescent-shaped crest supports (made of flat bronze sheets) is decorated by a triple line of saw-tooth motif. The sides and the middle of the crest supports are decorated with similar (double) saw-tooth motifs. The earflaps have an elongated shape with a straight front edge and curved back edge. There is an embossed rib and a series of holes for attachment probably of a lining around the edge of the earflap. There are three vertical incised bands decorated with saw-tooth motif on the earflap. Fragmentary. (*Pls. 74-75*). *Ca.* 800 B.C. (Hasanlu Period IV [Iron II]) – Tehran, Iran Bastan Muzeh, Inv. no.: HAS 60-528 – Barnett – Falkner 1962, xx, xxii, fig. 2:15; van Loon 1966, 119, note 195; Borchhardt 1972, 107, Cat. no. 24 IV 1, pl. 40:2; Muscarella 1980, 29, note 55; Muscarella 1988, 50; de Schauensee 1988, 55, pl. 55; Born – Seidl 1995, Abb. 57b.

54. Hasanlu, Burned Building II, Room 5, southwest quadrant. Group no. **H.2.2.2** – H.: 27 cm (with crest); Diam.: 23 cm; Diam. of crest at base: 26 cm; L. of earflaps: 16.6 cm; W. of earflaps: 12.6 cm. Crested bronze helmet with (detachable) earflaps. The crescent-shaped crest supports rested directly on the cap of the helmet, following its contour. There is a series of holes around the base of the helmet for attachment probably of a lining. There are three incised (or embossed) bands around the base of the helmet and a series of incised (or embossed) circles above them (14 ?). The crest supports were made of plain bronze sheet and are decorated with 30 incised lines in the shape of a fan. The **iron earflaps** have an elongated shape with straight front edge and curved back edge. There are two embossed ribs following the contours of the edge and a series of holes (probably for a lining) are on the earflaps. There are nine rows of horizontally and vertically placed embossed ovals in the middle field of the earflaps. The helmet was discarded (?), the earflaps are in fragmentary condition. (*Pls. 76-77*). *Ca.* 800 B.C. (Hasanlu Period IV [Iron II]) – Inv. no.: HAS 60-906, discarded in field – Winter 1980, 24; M. de Schauensee, personal communication.

Group H.2.3: Hemispherical crested helmets with leather skull-piece and horsehair or feather crest (Cat. nos. 55-56)

55. Hasanlu, Burned Building II, north central quadrant. Group no. **H.2.3.1** – H. of crest: 18.4 cm; Diam. of crest at base: 30 cm; W. of crest: 7.4 cm. Two crest supports of a crest. The crescent shaped crest supports rested directly on the cap of the helmet, following its contour. There is an embossed rib and a series of holes along the upper edge of the crest supports. (*Pl. 78*). – *Ca.* 800 B.C. (Hasanlu Period IV [Iron II]) – Philadelphia, University of Pennsylvania Museum of Archaeology and Anthropology, Inv. no.: UM 63-5-247 [HAS 62-581] – Winter 1980, 24.

56. Hasanlu, Burned Building II, probably Room 13 or 14. Group no. **H.2.3.2** – H. of crest: 12 cm; Diam. of crest at base: 23 cm; W. of crest: 6 cm. Two crest supports of a crest. The crescent shaped crest supports rested directly on the cap of the helmet, following its contours. There is an embossed rib and a series of holes along the upper edge of the crest supports. (*Pl. 79*). *Ca.* 800 B.C. (Hasanlu Period IV [Iron II]) – Tehran, Iran Bastan Muzeh, Inv. no. HAS 60-1078 – Winter 1980, 24.

URARTIAN HELMETS

More than half the amount of ancient Near Eastern helmets known are Urartian. Of course this does not mean that the Urartians used more helmets than other nations of the Near East. Urartian helmets have been found in large quantities in two types of findspot: votive stores of Urartian temples (*e.g.* Altintepe, south portico of the temple[584]) and arsenals of royal palaces (Karmir Blur[585]). An unprovenienced hoard of bronze and iron objects[586] which probably contained nearly 30 Urartian bronze helmets[587] might have come from one or other of these two types of findspot. Another bigger find of helmets was excavated at Çavuştepe, Uçkale[588], where the destruction level in which these helmets were found showed that heavy fighting had probably been going on, and these helmets might be the helmets of Urartian soldiers fallen in the battle. There were 5 Urartian helmets excavated in the Urartian fortress at Burmageçit[589]. These helmets probably formed a part of an arsenal or were the helmets of fallen soldiers. Undoubtedly in Assyria there were helmets stored in arsenals (*ekal mašarti*) and in storerooms of temples (as votive offerings), but while Assyria was totally sacked by the Medes and Babylonians in 614/612 B.C.[590], Urartu was never so totally plundered, and there helmets and shields remained in the debris of the buildings destroyed in the last wars of Urartu with the Scythians. Only two finds of Assyrian helmets are known from Assyrian capitals. One of them was excavated by Layard in 1846 at Nimrud, in the Northwest Palace, Chamber I (only **Cat. nos. 10, 16-31** remained from the several helmets which were found there)[591]. The other find which contained several bronze helmets of different shapes, is mentioned by Place at Khorsabad (Dur-Sharrukin), Palace of Sargon II (721—705 B.C.), Rooms 16, 17[592]. However, immediately after the Gulf War iraqi archaeologists discovered "a whole network of underground galleries and chambers" under the citadel of Nimrud. These chambers were used as "magazines and were packed with dozens of untouched bronze objects, including shields and helmets." Their number is unknown.[593]

Unfortunately there is not a single existing example of the earliest known Urartian helmet type, the hemispherical crested helmet. As has been mentioned, this type of crested helmet possibly originated in the 14th century B.C. in Transcaucasia, and was widespread in the 9th century B.C. Northwest Iran (Hasanlu), Urartu and under Urartian influence in North Syria until the first half of 8th century B.C. But at that time other helmet types have been used for more than half a century. The first of them was the conical helmet, borrowed from Assyria. The oldest surviving examples are **Cat. nos. 62-64** inscribed with the inscription of king Ishpuini, king of Urartu (*ca.* 830—810 B.C.). The other new type was the conical crested helmet, which originated in North Syria (*Charts 3-4*) and was widespread from the last quarter of the 9th century B.C. in Urartu as well as in Assyria. The earliest dated examples of this type are **Cat. nos. 109, 110**, similarly inscribed with the inscription of Ishpuini. The hemispherical crested helmets were replaced by the conical helmets at the latest in the early 8th century B.C.

Group U.1: Urartian conical bronze helmets (Cat. nos. 57-103)

Group U.1.1: Plain conical bronze helmets (Cat. nos. 57-58)

There are only two undecorated conical bronze helmets known from Urartu. It seems that the undecorated conical bronze helmets were in use in smaller numbers than the decorated ones[594]. There are no mass finds of undecorated conical bronze helmets from Urartian arsenals, magazines or temples, and moreover one of the two existing helmets of this type (**Cat. no. 57**) is considered by E. Özgen as

584 Özgüç 1966, 41: "Other votive offerings were found near the door in the south portico: bronze helmets of Assyrian shape, shields ...".
585 Piotrovsky 1967, 45. 20 helmets were found in Karmir Blur: 11 pieces with the "lightning fork symbol of Teisheba" and 9 pieces with figural decoration.
586 Kellner 1979, fig. A: few hundred bronze and iron objects from a hoard; Born – Seidl 1995, Abb. 43.
587 Kellner 1980, pl. 1: nearly 10-15 helmets smashed by the earth; Born – Seidl 1995, Abb. 43.
588 Erzen 1973, 58: "*Ausserdem sind in der Uçkale die jetzt zwei durch eine Lehmziegelmauer getrennte grosse Wohnräume zeigt, sehr schöne Freskofragmente, bronzene reliefplatten, Helme und Inschriftsteine gefunden worden.*"
589 Three of which were decorated with "lightning fork symbol of god Teisheba". Belli 1993, 62-63.
590 *See* also Calmeyer 1991a, 123.
591 Layard 1850, vol. I, 341-342.
592 Place 1867—1870, vol. I, 64-65.
593 Personal communication with excavator Muzahim Mahmoud, London, 1992.
594 Probably the cheapest undecorated conical iron helmets (**Cat. nos. 104-108**), most of them were riveted together from two similar halves, replaced them.

booty taken from the Assyrians[595]. He has probably right, because the undecorated conical helmets with three parallel ribs around the base are known only from the Assyrian art: from the palace reliefs of Shalmaneser III (858—824 B.C., *Chart 2, no. 92*) and Tiglath-Pileser III (745—727 B.C., *Chart 2, no. 134*).

The best parallel for the shape of helmet **Cat. no. 58** is in the University Museum, Philadelphia (**Cat. no. 102**, *Pls. 103-104*). Examining the incised lion decoration of this helmet E. Özgen dated it to the 7th century B.C. The pointed shape, curved contour and unique technique[596] of these two helmets, unknown from other Urartian helmets, suggest also a late, 7th century B.C. date.

57. Karmir Blur, Citadel Palace, Room 18, Group no. **U.1.1.1** – Plain conical helmet with pointed top. There are three horizontal ribs around the base of the helmet. On each side there is a loop for the chin-strap. Fragmentary. (*Pl. 80*) – reign of Menua (*ca.* 810—790 B.C.) or Rusa II (732—714 B.C.) – Erevan, Historical Museum of Armenia – Piotrovsky 1950, 56, fig. 33; Barnett – Watson 1952, 143; Piotrovsky 1955, 25; Piotrovsky 1959, 165, fig. 27; Piotrovsky 1966a, 242, fig. 27; van Loon 1966, 119; Piotrovsky 1967, 45; Borchhardt 1972, 107, Cat. no. 24 V 1; Curtis 1979, vol. I, 182; Özgen 1982, 28, 67-68, note 79; Esaian 1986, 29, pls. 9:19, 20:1

58. Provenance unknown (said to come from "Urartu (Eastern Turkey)"), Group no. **U.1.1.2** – H.: 28.2 cm. Bronze helmet of hemispherical body and conical peak (both hammered out separately and joined together?). There are two ribs and a row of holes (probably for the attachment of a lining) around the base of the helmet. Complete. (*Pl. 81*) – Late 8th – early 7th century B.C. – Berlin, Staatliche Museen zu Berlin, Preussischer Kulturbesitz, Museum für Vor- und Frühgeschichte, Inv. no. XIc 4826 – Calmeyer 1991a, no. 20 (pl.).

Group U.1.2: Conical bronze helmet with vertical fin and figural decoration (Cat. no. 59)

This helmet represents one of the most interesting types of ancient Near Eastern helmet. The vertical fin following the axis of the helmet is unique. The only parallel for this crest-like fin is known on two other Urartian helmets decorated with the "lightning fork symbol of Teisheba" in *repoussé* (**Cat. nos. 60, 61**). The only representation of this type of helmet appears on the unprovenienced horse's breastplate[597] already mentioned. The helmets of the charioteers (*Chart 2, nos. 116-117*) represent the same type of helmet with vertical fin. Since the only examples of this type of helmet can be described as Urartian and the hemispherical crested helmets of the fleeing enemy on the breastplate were earlier known as Urartian helmets, the identification of the two parties is very difficult. However, it can be assumed that probably Urartian chariot warriors pursued non-Urartian (North Iranian) horsemen. On the other hand it has to be declared that both types of helmets were also used in Urartu, Transcaucasia and Northern Iran (Hasanlu) in the late 9th century B.C. Consequently the scenes of the breastplate can be explained as stereotyped motifs without any historical background or ethnic implications. The essential rules of ancient Near Eastern representation systems did not allow depiction of enemies in the same type of helmets[598].

If the parallel between the Hasanlu hemispherical crested helmets and those shown on the horse's breastplate (*Chart 3, nos. 111-112*)[599] may be taken to indicate a date for it, then its representation of conical helmets with vertical fins suggests **Cat. no. 59** is late 9th or early 8th century B.C. in date. This dating can be proved by two Urartian conical helmets furnished with similar vertical fin and decorated with the "lightning fork symbol of Teisheba" in *repoussé* (**Cat. nos. 60, 61**). One of

595 Özgen 1982, 28: "This unique find may be booty taken from the Assyrians."
596 Hemispherical body and conical peak made separately and probably forged together.
597 Berlin, Sammlung Axel Guttmann, AG 230; Seidl 1986, 229-233, figs. 1-4; Born – Seidl 1995, Abbs. 45-53.
598 In Assyrian and North Syrian art the enemies of Assyria never wear the same type of helmet as the Assyrian soldiers. The only exceptions might be the siege reliefs of Assyrian art, where the besiegers and defenders were separated by the citywalls, and there was no need to distinguish them with the different types of headgear. There are examples of conical helmets *contra* hemispherical crested helmets (Balawat Gates); conical helmets *contra* conical crested helmets (Zinçirli reliefs); conical helmets *contra* men without helmets; conical crested helmets *contra* men without helmets. There is not a single example to show Assyrian, North Syrian, Urartian or Northwest Iranian warriors wearing conical/pointed helmet (which was used all over the Near East in the 8th century B.C.) and fighting with each other. So another possibility is to see "wandering motifs" without historical implications in the two types of helmet on the breastplate, *i.e.* the two parties had to be distinguished by the most characteristic part of their weaponry, the helmets. On the Assyrian palace reliefs the distinction is always clear. In the 9th century B.C. the Assyrian soldiers always wear conical helmets while their enemies wear no helmets or wear hemispherical crested helmets. In the 8th century B.C. the (probably Assyrian) heavy infantry (in scale armour), charioteers and horsemen always wear conical/pointed helmets. The light infantry (probably of foreign origin) with round shield and spear always wear crested helmets. The enemies of Assyria wear crested helmets as well. So there is not a single representation in Assyrian art, in which a horseman wears a crested helmet (as on the breastplate) or a conical helmet with vertical fin. So this breastplate is very unlikely to have any close association with Assyria.
599 The long earflaps of the chariot warriors' helmets with vertical fin are known only from a Hasanlu crested helmet (**Cat. no. 53**, *Pls. 74-75*).

these two helmets, **Cat. no. 60** is inscribed with the inscription of king Argishti I (789—766 B.C.). On the other hand the decoration ("lightning fork") appeared in the last third of the 9th century B.C.[600].

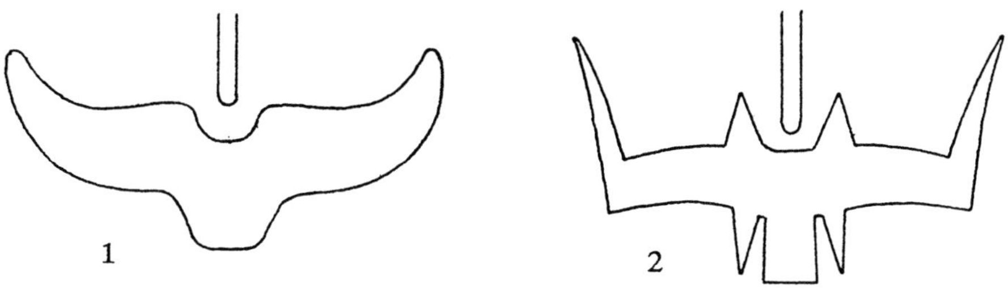

Fig. 10: 1. The horn emboss on the front of helmet **Cat. no. 59**;
2. The "lightning fork symbol of Teisheba"

The decorative system of the helmet is unique as well. The *repoussé* horn with the fin replacing the vertical rib at the front of the helmet might be an independent, but parallel motif with the antler-like "lightning fork symbol of Teisheba"[601] and not simply the *paraphrasis* or archetype of it (*Fig. 10*). This seems to be supported by a very interesting ivory fragment from Nimrud[602] showing a male head (probably a deity) wearing a conical helmet (*Chart 2, no. 133*). The helmet is furnished with a solid earflap and a curious neckguard made of different material (probably of leather). The helmet is decorated with the same *repoussé* horns and a snake (winding along the base of the helmet) as shown on helmet **Cat. no. 59**. Mallowan, who published this ivory fragment mentioned a rock-cut relief of god Tarhuntas[603] as a possible parallel for the helmet of the ivory fragment. The pair of horns (as sign of the divinity) and the snake may refer to a special deity[604]. The shape of the horns and their position, however, differ from the Mesopotamian tradition (horned crown of deities) and would indeed suggest Anatolian, Neo-Hittite origin.

The rib running around the base of the helmet and terminating at the front in snake heads has the same function as the ribs strengthening the rim of other helmets. This is the only Urartian example of curves or ribs ending in snake heads. The *repoussé* curves on the front of other Urartian helmets terminate in lion heads (**Cat. nos. 91-97, 98, 99**) bovid heads (**Cat. no. 100**) or unidentified animal heads (probably ram heads **Cat. no. 105**) but never in snake heads. The incised decoration of the *repoussé* horn at the front of the helmet shows a hero ("master of animals") holding two bulls. The scene of a hero holding two animals is well known from the art of the Near East. I.J. Winter examining a horse's breastplate from Hasanlu[605], which was decorated with a similar scene in very high *repoussé*[606], has examined the role of this motif in the art of North Syria[607], Assyria[608] and Iran[609] in the 9th century B.C. One of the best parallels for this scene is shown on the incised garment decoration of Ashurnasirpal II (883—859 B.C.)[610]. The hero in the middle holds the reins or fleece of a bull on either side. This scene similar to the incised and *repoussé* decoration of winged griffins, winged bulls and lions on the helmet was inspired clearly by Assyrian art. This scene was widespread mainly outside Urartu, and shows a foreign (probably Assyrian) influence in the decoration of this helmet. On the other hand the shape (conical helmet with vertical fin) and decorative system (*repoussé* rib terminating in snake heads; *repoussé* horns at the front) indicate the Urartian manufacture of the helmet.

---

600 The earliest examples of helmets decorated with this motif, **Cat. nos. 62-64** bear the inscription of king Ishpuini (*ca.* 830—810 B.C.).

601 Helmets decorated with this motif are groupped in group U.1.3 (**Cat. nos. 60, 61**) and U.1.4 (**Cat. nos. 62-90**). *See* below.

602 Nimrud, Burnt Palace, ND 1605; Mallowan 1966, 216, fig. 165.

603 Ivriz, Turkey. The relief was made by King Urpalla (Warpawalas) of Tabal, a vassal of Tiglath-Pileser III around 740 B.C.; Akurgal 1961, pls. 140, XXIV; Orthmann 1971, pl. 14e.

604 *See* p. 83 and notes 627-629.

605 HAS 74-241, Winter 1980.

606 Winter 1980, figs. 17-30, foldout drawing.

607 Winter 1980, 11-13.

608 Winter 1980, 13-16.

609 Winter 1980, 16-21.

610 Relief slab, Metropolitan Museum of Art, 32.143.4; Layard 1853, pl. 48:1; Winter 1980, fig. 40. For further representations of the "master of animals" on incised garment decorations on the palace reliefs of Ashurnasirpal II *see* Layard 1853, pls. 6, 8, 9, 44:1, 44:3, 47:2.

H.J. Kellner, following J. Borchhardt's theory, derives this type of helmet from the Near Eastern and Aegean conical crested helmets ("*Kammhelm*") which emerged in the middle of the 2nd millennium B.C.[611]. However P. Schauer regarded this type as the archetype of the European (Etruscan) conical crested helmets[612]. Since this type of helmet (together with helmets **Cat. nos. 60, 61**) clearly descended from the Assyrian and Urartian conical helmets of the 9th century B.C. (**Cat. nos. 60, 61**) with the geometric motif, the "lightning fork symbol of Teisheba" at the front is only a variant of the conical helmets of the same decoration (but without vertical fin). As it does not have any direct links to the middle of the 2nd millennium B.C. crested helmets, it can be assumed that this helmet is only a variant of the 9th century B.C. Near Eastern conical helmets. On the other hand there is no evidence of links between this type of helmet and European crested helmets of the Iron Age.

59. Provenance unknown (said to come from "Eastern Turkey"), Group no. **U.1.2.1** – H.: 28 cm; Diam.: 20.0-23.5 cm; H. of the fin: 3.4 cm. Conical helmet with a vertical fin (crest). The fin begins on the front of the helmet 6 cm above the rim and ends on the back of the helmet 5.5 cm above the rim. There is a horizontal rib (a body of a snake) around the base of the helmet, 1.6 cm above the rim. The rib ends on the front in two snake heads. The gap between them is 16 cm. Above the rib there is an engraved scene (frieze) of passing animals (2 x 3): lions, winged bulls and winged griffin with lion's head. On the front of the helmet there is an embossed wing or hornshaped device, and incised figures, in which a male deity holds two winged bulls. This scene is in the centre of the row of animals. Complete (*Pls. 82-83*). – Late 9th – early 8th century B.C. – Tokio, Ancient Orient Museum, Galvano-copy: Munich, Archäologische Staatssammlung, Inv. no. PS. 1975/3176 – Schauer 1979—1980, 533, figs. 16-17; Kellner 1980, 210-211, pls. 10-12; Hori – Miyashita – Ishida 1982, 69, 73, no. 7, pl. 7; Esaian 1986, 30, pl. 15; Born – Seidl 1995, 94, Abbs. 84-85.

Group U.1.3 Conical bronze helmets with vertical fin and symbol of Teisheba (Cat. nos. 60-61)

These two helmets the shape of which connects them to the conical helmet with vertical fin mentioned above (**Cat. no. 59**) and the decoration of which is the same as the decoration of the most numerous group of known Urartian helmets (*see* below), are unique as well. This might be the best example of the difference between the function of the shape and decoration. Both **Cat. nos. 59** and **60-61** may have been ceremonial helmets, indicated by the fin[613]; in the case of helmets **Cat. nos. 60, 61** their decoration ("lightning fork symbol of Teisheba", *see* below) suggests more clearly that they may have belonged to a special group of votive helmets. If they were functional, they may have been worn by soldiers in a special service section[614] of the Urartian army.

60. Karmir Blur, Group no. **U.1.3.1** – Conical helmet with vertical fin and a geometric design in *repoussé*, the "lightning fork of Teisheba" at the front. There are three horizontal ribs around the base of the helmet. There are three holes on either side of the helmet for fastening the chin-strap. Instead of the vertical vertical rib at the front of the similar helmets **Cat. nos. 62-90** there is a vertical rib extending on the back of the helmet as well. Complete. There is an *Urartian inscription* of king *Argishti I* (789—766 B.C.) around the base of the helmet. – Reign of king Argishti I. (789—766 B.C.) – Erevan, Historical Museum of Armenia – Lloyd 1967, fig. 122 (back right).

61. Provenance unknown (said to come from "Urartu (Eastern Turkey)"), Group no. **U.1.3.2** – Conical helmet with vertical fin and a geometric design in *repoussé*, the "lightning fork of Teisheba" at the front. There are three horizontal ribs around the base of the helmet, the upper two of which end at the front (in a loop) at the base of the geometric design. There are seven small holes on the edges of the geometric design, which are probably rivet holes for fastening a precious metal plaque, a "lightning fork" to the front of the helmet. There are three holes on either side of the helmet for fastening the chin-strap. Complete. – First half of 8th century B.C. – Berlin, Axel Guttmann Collection, inv. no. AG.165 – Born – Seidl 1995, 92-100, Abbs. 81-82.

Group U.1.4: Conical bronze helmets with symbol of Teisheba (Cat. nos. 62-90)

This group of helmets can be separated from other Urartian helmets by its decoration. There is a geometric, antler-like motif in *repoussé* at the front of the helmets, joined with a vertical rib, which is an integral part of the motif, running from the point of the helmets at the front. There are three horizontal ribs running around the base of the helmet as on other helmets.

This is the earliest known type of Urartian conical helmet. The earliest known examples are **Cat. nos. 62-64** with the Urartian inscription of king Ishpuini (*ca.* 830—810 B.C.). The inscriptions are standard in form declaring the proprietary rights of the king: "the arsenal of RN". P. Calmeyer assumed that this motif ("lightning symbol") at the front of the helmets with or without inscription was the royal

---

[611] Kellner 1980, 210.

[612] Schauer 1979—1980, 533, figs. 16-17.

[613] Helmet **Cat. no. 59** with similar vertical fin and a very elaborate decoration may be a ceremonial helmet, however on the front of helmet **Cat. no. 60** there are the holes of a metal plaque (probably of precious metal) which was riveted onto the front of the helmet in shape of the "lightning fork symbol of Teisheba".

[614] *E.g.* royal bodyguard.

property mark. Since this group of helmets is the most numerous group known from Urartu[615], and most of the helmets of this group are inscribed with the name of an Urartian king, it can be assumed that this type of helmet was the helmet of the royal troops or royal bodyguard, and was not simply a votive item. It was used not only during the last third of the 9th century B.C. but also throughout the 8th century B.C. as well. Inscribed pieces of this type are known only from the reigns of Ishpuini (*ca.* 830—810 B.C.), Menua (*ca.* 810—790 B.C.) and Argishti I. (789—766 B.C.). O. Belli concluded[616] that there was a change in the Urartian decorative tradition during the reign of Argishti I. and this geometric motif was replaced by the figurative design of genii, gods, horsemen, chariots, etc. (our helmets of Group U.1.5). This trend is undoubtedly observable, but there is no evidence which would refer to the ceasing or stopping of its manufacture. Helmets found in recent Turkish excavations[617] will probably solve this problem.

There are several theories about the origin of the antler- or lightning-like motif on the front of these helmets[618]. Piotrovsky argued that this is the "lightning fork of Teisheba"[619]; on the other hand Riemschneider proposed that this is the fork/harpoon of Haldi[620]. But: "*Nun ist es durchaus möglich, dass das seltsame Ornament von der syrischen Blitzlanze des Teschup und der Hieroglyphe für den hethitischen Tarchund beeinflußt ist. Aber die hohcragende Spitze zeigt deutlich, daß man sich an das Wettergottzeichen, das mehr einem 'W' ähnelt, nicht erinnert fühlen soll.*". Nagel[621], van Loon[622] and Borchhardt[623] assumed that this motif shows the antler of a deer or horns. The identification with a schematic tree-of-life[624] seems wrong. Seidl raised the question that the motive is a geometric or stereometric version of a winged sun-disk[625]. P. Calmeyer was probably right, when he wrote that "the abstraction has gone so far as to make it impossible for us to recognize the original picture"[626].

However, as has been mentioned, the *repoussé* horns on the front of helmet **Cat. no. 59** (*Pls. 82-83*) may be a parallel symbol with the "lightning fork of Teisheba" or at least may derived from a common archetype. The interesting ivory fragment (*Chart 3, no. 133*) mentioned above – an almost exact parallel of the iconographic system of helmet **Cat. no. 59** – seems to support this idea. If Mallowan[627] was right, and the horns of the ivory fragment (and consequently the *repoussé* horns of helmet **Cat. no. 59**) can be related to the headdress of god Tarhuntas[628] (Neo-Hittite Luwian name of the Hurrian/Hittite storm god Teshup, who was known in Urartu as Teisheba), this type of front decoration would really be connected with divine symbols, or a certain deity (with horn and snake attributes[629]). However, the symbols (the "lightning fork" or the horn and snake) must be abstractions of the divine headdress, and would refer only to the fact, that the wearer of the helmet (or his unit) can be connected to a special service of the army – as in the case of Assyrian helmets decorated with the figure of Ishtar.

A distinctive characteristic of helmet **Cat. no. 62** is an incised bull's head under the left wing of the *repoussé* antler or "lightning fork". Similar incised bull's heads can be seen on two Urartian umbos[630]. R.D. Barnett assumed that Urartian hieroglyphic signs did not occur earlier then the reign of Menua (*ca.* 810—790 B.C.)[631]. This evidence disproves that theory and proves that the hieroglyphic writing had been in use at least as early as the reign of Ishpuini. The meaning of the sign is unknown. There are two possibilities to interpret its function: 1. maker's mark; 2. property label (the sign of its wearer, or the king, or the deity, to whom the helmet might be offered).

---

615 Apart from the 22 pieces mentioned above Piotrovsky 1967, 45 mentiones that 9 helmets from the 20 helmets found in Karmir Blur were of this type; and Kellner 1980, 206, pl. 1 mentiones that most of the *ca.* 30 helmets of a looted hoard belonged to this type as well. Furthermore Belli 1993, 63 mentiones that three of the five Urartian helmets found in the Urartian fortress at Burmageçit belonged to this group, and there are several fragments of helmets of this type in the store rooms of the Van Museum (Belli 1993, 65).

616 Belli 1993, 63.

617 Altintepe, Çavustepe, Burmageçit, *see* above.

618 Özgen 1982, 28.

619 Piotrovsky 1967, 45.

620 Riemschneider 1965, 78-79.

621 Nagel 1959—1960, 146ff., note 5.

622 van Loon 1966, 119.

623 Borchhardt 1972, 106, note 472.

624 Hančar 1966, 102.

625 Born – Seidl 1995, 92-96.

626 Calmeyer 1991b, 316.

627 Mallowan 1966, 216, fig. 165, ND 1605.

628 Rock relief at Ivriz, Turkey; Akurgal 1961, pls. 140, XXIV; Orthmann 1971, pl. 14e.

629 The Mesopotamian version of the Hurrian/Hittite Teshup was probably Tishpak, who replaced Ninazu, the deity of Eshnunna as early as in the late 3rd millennium B.C. and inherited the Sumerian deity's divine animal, the snake dragon as well.

630 König 1955, 22, nos. 112B, 112D, pl. 106/107.

631 Barnett 1974, 51-52; p. 50, fig. 4: 39.

62. Provenance unknown (said to come from "Urartu (Northwest Iran)"), Group no. **U.1.4.1** – H.: 29.5 cm, W.: 862 g. Conical helmet with pointed top. There is a series of holes around the base for the attachment probably of a lining. There are three horizontal embossed ribs around the base. On the front of the helmet there are is a vertical embossed rib above the "lightning fork symbol of god Teisheba" in *repoussé*. There is an *Urartian inscription* of king *Ishpuini* between the two lower ribs around the base of the helmet (*[I]Iš-pu-ú-i-ni-ni-e-i ú-ri-iš-ḫu-si-ni-e-*[*i*]). Complete (*Pl. 84*) – reign of Ishpuini (*ca.* 830—810 B.C.) – London, British Museum, BM 135061 – Barnett – Curtis 1973, 133, 136; Frankel 1979, 20, fig. 12; Curtis 1979, vol. I, 181; Özgen 1982, 28, 68, note 81; Vanden Berghe – De Meyer 1982, 128, no. 17 (fig.); Esaian 1986, 29-30, pl. 13:3; Maass 1987, 73, note 36.

63. Provenance unknown (said to come from "Urartu (Eastern Turkey)"), Group no. **U.1.4.2** – Similar to helmet **Cat. no. 62**. There is an *Urartian inscription* of king *Ishpuini* around the base of the helmet. Complete. – reign of Ishpuini (*ca.* 830—810 B.C.) – Mainz, Römisch-Germanisches Zentralmuseum, Inv. no. O.39778/154 – Kellner 1980, 205, pl. 1; Egg 1984, 618, 621, 649, fig. 39.

64. Provenance unknown (said to come from "Urartu (Eastern Turkey)"), Group no. **U.1.4.3** – Similar to helmet **Cat. no. 62**. There is an *Urartian inscription* of king *Ishpuini* around the base of the helmet. Complete. – reign of Ishpuini (*ca.* 830—810 B.C.) – Mainz, Römisch-Germanisches Zentralmuseum, Inv. no. O.39778/155 – Kellner 1980, 205, pl. 1; Egg 1984, 618, 621, 649, fig. 39.

65. Provenance unknown (said to come from "Urartu (Eastern Turkey)"), Group no. **U.1.4.4** – Conical helmet with pointed top. There are three horizontal ribs around the base of the helmet. On the front of the helmet there are a vertical rib and the "lightning fork symbol of god Teisheba" in *repoussé*. Flattened and crushed. There is an *Urartian inscription* of king *Menua* around the base of the helmet (*be-li-šú [I]Me-nu-a*). – reign of Menua (*ca.* 810—790 B.C.) – Mainz, Römisch-Germanisches Zentralmuseum, Inv. no. O.39778/156 – Kellner 1980, 205, pl. 1; Egg 1984, 618, 621, 649, fig. 39.

66. Provenance unknown (said to come from "Urartu (Eastern Turkey)"), Group no. **U.1.4.5** – Conical helmet with pointed top. There are three horizontal ribs around the base of the helmet. On the front of the helmet there are a vertical rib and the "lightning fork symbol of god Tiesheba" in *repoussé*. Flattened and crushed. There is an *Urartian inscription* of king *Menua* around the base of the helmet (*be-li-šú [I]Me-nu-a*). – reign of Menua (*ca.* 810—790 B.C.) – Mainz, Römisch-Germanisches Zentralmuseum, Inv. no. O.39778/157 – Kellner 1980, 205, pl. 1; Egg 1984, 618, 621, 649, fig. 39.

67. Provenance unknown (said to come from "Urartu (Eastern Turkey)"), Group no. **U.1.4.6** – Conical helmet with pointed top. There are three horizontal ribs around the base of the helmet. On the front of the helmet there are a vertical rib and the "lightning fork symbol of god Tiesheba" in *repoussé*. Flattened and crushed. There is an *Urartian inscription* of king *Menua* around the base of the helmet (*be-li-šú [I]Me-nu-a*). – reign of Menua (*ca.* 810—789 B.C.) – Mainz, Römisch-Germanisches Zentralmuseum, Inv. no. O.39778/158 – Kellner 1980, 205, pl. 1; Egg 1984, 618, 621, 649, fig. 39.

68. Provenance unknown (said to come from "Urartu (Eastern Turkey)"), Group no. **U.1.4.7** – H.: 33.3 cm, Diam.: 19-20.5 cm. Conical helmet with pointed top. There are three horizontal ribs and a series of holes (probably for the attachement of a lining) around the base of the helmet. On the front of the helmet there are a vertical rib and the "lightning fork symbol of god Teisheba" in *repoussé*. Complete. There is an unintelligible *Urartian inscription* of king *Menua* around the base of the helmet (*be-li-šá [I]Me-nu-a*). (*Pl. 85*) – reign of Menua (*ca.* 810—790 B.C.) – Munich, Archäologische Staatssammlung, Inv. no.: PS. 1985/521 – Kellner 1987, 21, pl. 22; Calmeyer 1991a, no. 16 (pl.); Kellner 1991, 13, fig. 7.

69. Karmir Blur, Citadel Palace, Room 24, Group no. **U.1.4.8** – Conical helmet with pointed top. There are two horizontal ribs around the base of the helmet. On the front of the helmet there are a vertical rib and the "lightning fork symbol of god Teisheba" in *repoussé*. There is a loop on each side for the chin-strap. Complete. (*Pl. 86*) – reign of Menua (*ca.* 810—790 B.C.) or Rusa II (732—714 B.C.) – Erevan, Historical Museum of Armenia – Piotrovsky 1950, 59, fig. 38; Barnett – Watson 1952, 143, fig. 21; Piotrovsky 1955, 25; Piotrovsky 1959, 165, fig. 26; Nagel 1959-1960, fig. 4; Piotrovsky 1966a, 242, fig. 26; van Loon 1966, 119; Piotrovsky 1967, 45; Borchhardt 1972, 107, Cat. no. 24 V 2; Curtis 1979, vol. I, 182; Özgen 1982, 28, 68 note 82; Esaian 1986, 29, pls. 9:20, 20:3.

70. Burmageçit, Urartian Fortress, Group no. **U.1.4.9** – H.: 31.5 cm, W.: 0.1 cm. Fragment of a pointed helmet. There are three horizontal ribs around the base of the fragment. On the front of the helmet there are a vertical rib and the "lightning fork symbol of god Teisheba" in *repoussé*. There are two holes at the side of the helmet, around the rib for fastening of chin-strap loops or earflaps. There is a fragmentary *Urartian inscription* of king *Menua* (*ca.* 810—790 B.C.) around the base of the helmet: *[I]Me-nu-a-i-e* [*ú-ri-iš-ḫu-si-ni-e*] ("From the magazine of Menua."). – reign of Menua (*ca.* 810—790 B.C.) – Elazig Museum, Inv. no. 85.TMKC. 1-72 – Belli 1993, 61-67, fig. 3a, pl. 13.

71. Burmageçit, Urartian Fortress, Group no. **U.1.4.10** – Fragment of a pointed helmet. There are three horizontal ribs around the base of the fragment. On the front of the helmet there are a vertical rib and the "lightning fork symbol of god Teisheba" in *repoussé*. There are two holes at the side of the helmet, around the rib for fastening of chin-strap loops or earflaps. – reign of Menua? (*ca.* 810—790 B.C.) – Elazig Museum – Belli 1993, 61-67

72. Burmageçit, Urartian Fortress, Group no. **U.1.4.11** – Fragment of a pointed helmet. There are three horizontal ribs around the base of the fragment. On the front of the helmet there are a vertical rib and the "lightning fork symbol of god Teisheba" in *repoussé*. There are two holes at the side of the helmet, around the rib for fastening of chin-strap loops or earflaps. – reign of Menua? (*ca.* 810—790 B.C.) – Elazig Museum – Belli 1993, 61-67

73. Provenance unknown (said to come from "Urartu (Eastern Turkey)"), Group no. **U.1.4.12** – H.: 30.5 cm, Diam.: 19.5-21.5 cm. Conical helmet with pointed top. There are three horizontal ribs around the base of the helmet. There are three bigger (D.: 5 mm) and one smaller holes on each side of the helmet under the lower rib, for the chin-strap or earflaps. On the front of the helmet there are a vertical rib and the "lightning fork symbol of god Teisheba" in *repoussé*. There is an *Urartian inscription* of king *Menua* on the helmet. Complete. (*Pl. 87*) – reign of Menua (*ca.* 810—790 B.C.) – Karlsruhe, Badisches Landesmuseum, Inv. no.: 89/13 – Maass 1987, 71, 72-74, pl. 1:2 bottom right

74. Provenance unknown (said to come from "Urartu (Eastern Turkey)"), Group no. **U.1.4.13** – H.: 30.2 cm, Diam.: 19-20 cm. Conical helmet with pointed top. There are three horizontal ribs around the base of the helmet. There are three holes (D.: 3.5 mm) on the right side of the helmet. On the left side there is only one existing hole from the three. The holes are for a chin-strap or earflaps. On the front of the helmet there are a vertical rib and the "lightning fork symbol of god Teisheba" in *repoussé*. There is an *Urartian inscription* of king *Menua* on the helmet. Complete. (*Pl. 88*) – reign of Menua (*ca.* 810—790 B.C.) – Karlsruhe, Badisches Landesmuseum, Inv. no.: 89/14 – Maass 1987, 71, 72-74, pl. 1:2 bottom left

75. Provenance unknown (said to come from "Urartu (Eastern Turkey)"), Group no. **U.1.4.14** – H.: 30.3 cm, Diam.: 21-21.5 cm, W.: 811 g. Conical helmet with pointed top. There are three horizontal ribs around the base of the helmet. On the front of the helmet there are a vertical rib and the "lightning fork symbol of god Teisheba" in *repoussé*. (*Pl. 89*). First half of 8th century B.C. – Karlsruhe, Badisches Landesmuseum, Inv. no.: 84/69 – Born – Seidl 1995, Tabelle III.

76. Karmir Blur, Citadel Palace, Room 37, Group no. **U.1.4.15** – H.: 30 cm. Conical helmet with pointed top. There are three horizontal ribs around the base of the helmet. On the front of the helmet there are a vertical rib and the "lightning fork symbol of god Teisheba" in *repoussé*. Fragmentary. There is an *Urartian inscription* of king *Argishti I* around the base of the helmet: ^{d}*Ḫal-di-e e-ú-ri-e i-ni ku-bu-še-*[*e*] ^{I}*Ar-*[*gi-iš-ti*]*-še* ^{I}*Me-nu-a-ḫi-ni-*[*še uš*]*-tú-ni* ("Argishti, son of Menua offered this helmet to Haldi, the lord.") (*Pl. 90*) – reign of Argishti I (789—766 B.C.) – Erevan, Historical Museum of Armenia – Piotrovsky 1955, 25-26, fig. 16; Barnett 1959, 16; Melikishvili 1960, 270, no. 148a; Piotrovsky 1966a, 242; van Loon 1966, 119; Piotrovsky 1967, 45; Lloyd 1967, fig. 122; Azarpay 1968, 17; Piotrovsky 1970, pl. 43; Rubinstein 1975, fig. 11; Curtis 1979, vol. I, 182; Özgen 1982, 28-29, note 82; Esaian 1986, 29, pls. 9:18, 19; Maass 1987, 73, note 37

77. Provenance unknown (said to come from "Urartu (Eastern Turkey)"), Group no. **U.1.4.16** – H.: 30 cm, Diam.: 21.2-23.4 cm, W.: 807 g. Conical helmet with pointed top. There are three horizontal embossed ribs around the base of the helmet. There is a series of holes around the base of the helmet for the fastening of a lining and probably for the chin-strap or earflaps. On the front of the helmet there are a vertical rib and the "lightning fork symbol of god Teisheba" in *repoussé*. At the base, between the lower and middle ribs there is an *Urartian inscription* of king *Menua*: ^{I}*Me-nu-a-i ú-ri-iš-hé* ("Property of Menua"). However, under the lower rib there is a later *Urartian inscription* of king *Argishti I* (789—766 B.C.) around the right side of the base of the helmet: $^{[I]}$*Ar-gi-iš-ti-i ú-ri-iš-ḫé* ("Property of Argishti"). The inscription (the signs *iš* and *ú*) was partly damaged by the later insertion of the holes. Complete. – reign of Menua (*ca.* 810—790 B.C.) – Berlin, Axel Guttmann Collection, AG 339 – Born – Seidl 1995, 90-92, 107-124, Abbs. 94, 95, Taf. I, XII, Tabelle II.

78. Rutchi-Tig (Verhnyaya Rutkha, Georgia), Group no. **U.1.4.17** – H.: 29.5 cm, Diam.: 23 cm. Conical helmet with pointed top (broken). There are three horizontal ribs around the base of the helmet. On the front of the helmet there are a vertical rib and the "lightning fork symbol of god Teisheba" in *repoussé*. The left side of the helmet is fragmentary. There is an *Urartian inscription* of king *Argishti I* along the base of the helmet: *Ar-gi-iš-ti-i ú-ri-iš-ḫé* ("Property of Argishti"). (*Pl. 91*) – reign of Argishti I (789—766 B.C.) – Berlin, Staatliche Museen zu Berlin, Preussischer Kulturbesitz, Museum für Vor- und Frühgeschichte, Inv. no. III d 4882 – Nagel 1959-1960, 144-147, figs. 1-3; Kunst im Handwerk früher Völker 1961, no. 183; Nagel 1963, 13-14, pl. 4, no. 9; Riemschneider 1965, 78-79, fig. 12; Sulimirski 1970, pl. 50; Borchhardt 1972, 107, Cat. no. 24 VI, pl. 35:1-2; Salvatori 1975, 259, note 13; Kellner 1980, 78, Cat. no. 138; Curtis 1979, vol. I, 183; Özgen 1982, 29, note 83; Vanden Berghe – De Meyer 1982, 128, no. 16; Kohlmeyer – Saherwala 1984, 38, fig. 26; Esaian 1986, 31-32, pl. 13:4; Maass 1987, 73, note 38

79. Provenance unknown (said to come from "Urartu (Eastern Turkey)"), Group no. **U.1.4.18** – Conical helmet with pointed top. There are three horizontal ribs around the base of the helmet. On the front of the helmet there are a vertical rib and the "lightning fork symbol of god Teisheba" in *repoussé*. Flattened and crushed. – 8th century B.C. – Mainz, Römisch-Germanisches Zentralmuseum, Inv. no. O.39778/159 – Kellner 1980, 205, pl. 1; Egg 1984, 618, 621, 649, fig. 39

80. Provenance unknown (said to come from "Urartu (Eastern Turkey)"), Group no. **U.1.4.19** – Conical helmet with pointed top. There are three horizontal ribs around the base of the helmet. On the front of the helmet there are a vertical rib and the "lightning fork symbol of god Teisheba" in *repoussé*. Flattened and crushed. – 8th century B.C. – Mainz, Römisch-Germanisches Zentralmuseum, Inv. no. O.39778/160 – Kellner 1980, 205, pl. 1; Egg 1984, 618, 621, 649, fig. 39

81. Provenance unknown (said to come from "Urartu (Eastern Turkey)"), Group no. **U.1.4.20** – Conical helmet with pointed top. There are three horizontal ribs around the base of the helmet. On the front of the helmet there are a vertical rib and the "lightning fork symbol of god Teisheba" in *repoussé*. Flattened and crushed. – 8th century B.C. – Mainz, Römisch-Germanisches Zentralmuseum, Inv. no. O.39778/161 – Kellner 1980, 205, pl. 1; Egg 1984, 618, 621, 649, fig. 39

82. Provenance unknown (said to come from "Urartu (Eastern Turkey)"), Group no. **U.1.4.21** – H.: 28.5 cm, Diam.: 19.2-22.8 cm. Conical helmet with pointed top restored from 20 pieces. There are three horizontal ribs around the base of the helmet. On the front of the helmet there are a vertical rib and the "lightning fork symbol of god Teisheba" in *repoussé*. Only the lower end of the vertical rib exists, the upper end is damaged. Restored. (*Pl. 92*) – 8th century B.C. – Munich, Archäologische Staatssammlung, Inv. no. PS 1971/1767 – Anonymus 1975, 221, fig. 5; Kellner 1976, 78, Cat. no. 140, pl. 14; Overlaet 1979, 51, note 1; Özgen 1982, 29, 69 note 84; Vanden Berghe – De Meyer 1982, 127, no. 15 (fig.); Esaian 1986, 29, pl. 13:2

83. Provenance unknown (said to come from "Urartu (Eastern Turkey)"), Group no. **U.1.4.22** – H.: 27 cm, Diam.: 20.5-21.6 cm. Conical helmet with pointed top. There are three horizontal ribs around the base of the helmet. There is a series of holes around the base of the helmet for attachment probably of a lining. On the front of the helmet there are a vertical rib and the "lightning fork symbol of god Teisheba" in *repoussé*. Complete. – 8th century B.C. – Les Arcs, Ebnöther Collection – Vanden Berghe – De Meyer 1982, 127, no. 14 (fig.); Esaian 1986, 29, pl. 13:1.

84. Provenance unknown (said to come from "Urartu (Eastern Turkey)"), Group no. **U.1.4.23** – H.: 28.5 cm, Diam.: 20.5 cm. Conical helmet with pointed top. There are three horizontal ribs around the base of the helmet. On each side there are two holes for the chin-strap. On the front of the helmet there are a vertical rib and the "lightning fork symbol of god Teisheba" in *repoussé*. Complete. – 8th century B.C. – Japan, Private Collection – Hori – Miyashita – Ishida 1982, 69, no. 8, pl. 7; Esaian 1986, 30, pl. 20:5.

85. Provenance unknown (said to come from "Urartu (Eastern Turkey)"), Group no. **U.1.4.24** – H.: 31 cm, Diam.: 21.7 cm (max.). Conical helmet with pointed top. There are three horizontal ribs around the base. On the sides of the helmet there are holes (3-3) for the chin-strap loops. On the front of the helmet there are a vertical rib and the "lightning fork symbol of god Teisheba" in *repoussé*. Complete. – 8th century B.C. – New York, Ariadne Galleries, Inc. – Calmeyer 1991a, no. 13, pl. on p. 129 (sic! mistakenly mentioned as no. 12).

86. Provenance unknown (said to come from "Urartu (Eastern Turkey)"), Group no. **U.1.4.25** – H.: 29.5 cm, Diam.: 19.5 cm, W.: 0.767 kg. Conical helmet with pointed top. There are three horizontal ribs around the base. On the front of the helmet there is the "lightning fork symbol of god Teisheba" in *repoussé*. The vertical rib ends in a ram's head. There are two pairs of holes on each sides of the helmet for the chin-strap loops. One of the wire loops exists on the helmet. Complete. (*Pl. 93*) – 8th century B.C. – Boston, Museum of Fine Arts, Inv. no.: 1981.25 – Kendall 1981, 213, note 45; Anonymus 1987, no. 103; Calmeyer 1991a, no. 15 (pl.).

87. Provenance unknown (said to come from "Urartu (Eastern Turkey)"), Group no. **U.1.4.26** – H.: 30.5 cm, Diam.: 21.3-23 cm, W.: 801 g. Conical helmet with pointed top. There are three horizontal ribs around the base. On the front of the helmet there are a vertical rib and the "lightning fork symbol of god Teisheba" in *repoussé*. There are two pairs of three holes on each sides of the helmet for chin-strap loops or earflaps. Complete. – 8th century B.C. – Berlin, Axel Guttmann Collection, Inv. no. AG 70 – Born – Seidl 1995, 90-92, 107-124, Taf. I, Tabelle II.

88. Provenance unknown (said to come from "Urartu (Eastern Turkey)"), Group no. **U.1.4.27** – H.: 31 cm, Diam.: 21-23 cm, W.: 820 g. Conical helmet with pointed top. There are three horizontal ribs around the base. On the front of the helmet there are a vertical rib and the "lightning fork symbol of god Teisheba" in *repoussé*. There are two pairs of three holes on each sides of the helmet for chin-strap loops or earflaps. Complete. – 8th century B.C. – Berlin, Axel Guttmann Collection, Inv. no. AG 71 – Born – Seidl 1995, 90-92, 107-124, Taf. I, Tabelle II.

89. Provenance unknown (said to come from "Urartu (Eastern Turkey)"), Group no. **U.1.4.28** – H.: 29 cm, Diam.: 18.8-21.5 cm, W.: 987 g. Conical helmet with pointed top. There are three horizontal ribs around the base. On the front of the helmet there are a vertical rib and the "lightning fork symbol of god Teisheba" in *repoussé*. There are two pairs of holes on each sides of the helmet for the existing chin-strap loops made of bronze wire. Complete. – 8th century B.C. – Berlin, Axel Guttmann Collection, Inv. no. AG 168 – Berlin, Axel Guttmann Collection, Inv. no. AG 168 – Born – Seidl 1995, 90-92, 107-124, Taf. I, X, XI, Tabelle II.

90. Provenance unknown (said to come from "Urartu (Eastern Turkey)"), Group no. **U.1.4.29** – H.: 31.5 cm, Diam.: 17.5-23.5 cm, W.: 725 g. Conical helmet with pointed top. There are three horizontal ribs around the base. On the front of the helmet there are a vertical rib and the "lightning fork symbol of god Teisheba" in *repoussé*. There is a series of holes around the base of the helmet for the fastening of a lining and probably for chin-strap loops or earflaps. Complete. – 8th century B.C. – Berlin, Axel Guttmann Collection, Inv. no. AG 385 – Born – Seidl 1995, 90-92, 107-124, Taf. I, IX, Tabelle II.

Group U.1.5: Conical bronze helmets with figural decoration (Cat. nos. 91-103)

The number of conical bronze helmets with figural decoration is relatively large. This type of helmet might have appeared at the same time as the helmets decorated with the "lightning fork symbol of Teisheba", in the late 9th century B.C. The first dated examples of this group (**Cat. nos. 91-92**) are inscribed with the name of the Urartian king Argishti I (789—766 B.C.). G. Ligabue and S. Salvatori have pointed out that both the shape and (figurative) decoration of Urartian conical helmets had been borrowed from Assyria[632]. Examining the decorative system of Urartian conical helmets the original motifs borrowed from Assyria can be separated from the later genuine Urartian motifs[633]. The 9th – early 8th century B.C. Assyrian motif, pairs of curves on the front of the conical helmets, remained in use in Urartu through the 8th century, when in Assyria the continuous arcs had already replaced them.

This motif appears in Urartu exclusively on the conical helmets decorated with incised figurative scenes, and disappears only in the 7th century B.C. (**Cat. no. 102**, *Pls. 103-104*). S.-G. Gröschel identified six motifs in the decoration of these helmets:

1. Freestanding deity or genius
2. Adoration: tree-of-life with genii
3. Two genii on either side of a rosette
4. Lions
5. Procession of horsemen and chariots
6. Hunting scene with horsemen and chariots

However, this does not provide the basis for a classification of the helmets, since two of these motifs may appear on one helmet. The following typology of the different designs on Urartian conical helmets with figurative decoration can be proposed:

---

632 Ligabue – Salvatori 1977, 7.

633 Only the "lightning fork of Teisheba" can be considered as authentic Urartian motif.

U.1.5.1. *Front* : adoration: trees-of-life and genii (11 trees with 11 pairs of genii) within 4 pairs of curves. *Sides and back*: Procession of horsemen and chariots. **Cat. nos. 91, 93-95**; questionable: **Cat. nos. 92, 96-97**. All of them are from Karmir Blur.

U.1.5.2. *Front*: Freestanding deity or genius + trees-of-life with adoring genii within 4 pairs of curves. *Sides and back*: Procession of horsemen and chariots. **Cat. no. 98**.

U.1.5.3. *Front*: Adoration: a tree-of-life with two genii + procession of horsemen and chariots within 4 pairs of curves. *Sides and back*: Procession of horsemen and chariots. **Cat. no. 99** + Karmir Blur[634].

U.1.5.4. *Front*: 2 genii facing each other + hunting scene within four pairs of curves. *Sides and back*: Hunting scene (horsemen hunting ibexes). **Cat. nos. 100-101**.

U.1.5.5. *Front*: 2 genii on either side of a rosette. *Sides and back*: lions. **Cat. no. 102**.

U.1.5.6. *Front*: 3 pairs of curves + vertical rib. The upper part of the helmet is decorated entirely with vertical ribs (cannelures). **Cat. no. 103**.

Group U.1.5.1:

This group comprises only helmets found at Karmir Blur. B.B. Piotrovsky mentioned that 9 of the 20 excavated helmets bear figurative decoration: 4 complete and 5 fragmentary[635]. There are published photographs of only 4 of them (**Cat. nos. 91, 93-95**). Two inscriptions have been published (**Cat. nos. 92, 96**) and one (**Cat. no. 97**) just mentioned. The decoration of only the four (almost) complete helmets is known. There is no exact information whether the helmets of which only the inscriptions were published were decorated or not. However, one of the 9 helmets with figurative decoration can be placed into the 3rd group[636] of the present writer. The decoration of the four complete helmets is the same: conical bronze helmets with four pairs of curves terminating in lion heads at the front. There are 11 sacred trees in three registers (1 + 5 + 5) with pairs of genii at their sides within the curves. Two types of pairs of adoring genii at the sides of the sacred trees alternate with each other: pairs of bearded winged genii alternate with pairs of beardless, wingless genii (*Pl. 94*). On the sides of the helmet in two registers there are two scenes of alternating chariots and horsemen[637] in procession. The upper register contains: horseman – chariot – horseman – chariot (facing from the middle of the back of the helmet to the front). The lower register contains: horseman – chariot – horseman – chariot – horseman (facing similarly from the back of the helmet to the front). The two friezes are framed with the characteristic border motif of Urartian art, the double saw-tooth motif.

The sacred tree and the adoring genii are clearly Assyrian motifs in origin. Furthermore P. Calmeyer regarded the connecting scenes of processing chariots and horsemen as the embodiment of the Assyro-Babylonian conception of *šarrūtu* ("royalty"). He supposed that the king himself stands in the chariot[638]. It is very interesting that these helmets were all found in Karmir Blur. Although no far-reaching conclusions can be drawn, it is very likely that these helmets are examples of a Karmir Blur workshop existing in the reigns of Argishti I and Sarduri II and examples of an iconographical (and probably ideological) tradition originating in this royal workshop (Karmir Blur) of Argishti I (789—766 B.C.). As these helmets were found in the magazines of the Karmir Blur palace and were not *ex votos* in Urartian temples, they are likely to have been worn by soldiers of the royal bodyguard or the king's immediate entourage.

91. Karmir Blur, Citadel Palace, Room 28, Group no. **U.1.5.1.1** – H.: 29.5 cm, Diam.: 26.5 cm (max.). It has a conical shape and pointed top. The top is restored. There are four pairs of embossed curves (ribs) ending in lions' heads on the front of the helmet. The curves (arcs) stand on an incised saw-tooth band running around the base of the helmet. Behind the lions' heads there is an incised design representing the lion's mane. Between the curves, in three registers, there are 11 alternating pairs of winged, bearded genii adoring (pollinating) a tree-of-life and wingless, beardless deities adoring (pollinating) a tree-of-life. Between two bands of saw-tooth motif, in two registers, there are ten pairs (five-five) of genii and deities, however the eleventh is above them, on the upper saw-tooth motif, between the lions' heads. This is a pair of wingless, beardless deities. Around the helmet in two registers, between pairs of saw-tooth bands, there is an alternating scene of chariots (with a warrior and driver, both of them wear conical helmets) and horsemen (carrying spears, round shields and wearing conical helmets) in procession. They move from the back of the helmet in opposite directions, in the directions of the curves on the front of the helmet. The upper register contains four chariots and four horsemen (in horseman – chariot – horseman – chariot order on either side). The lower register contains four chariots and six horsemen (in horseman – chariot – horseman – chariot – horseman order on either side). There is an *Urartian inscription* of king *Argishti I* around the base of the helmet: *Ḫal-di-e e-ú-ri-e i-ni ku-bu-še-e* [I]*Ar-gi-iš-*

---

634 Piotrovsky 1967, 46: "Sometimes, instead of a number of sacred trees, there is a single large one, and sometimes there is a chariot and a galloping horseman, as can be seen on some of the fragments."

635 Piotrovsky 1967, 45-46.

636 *See* notes 634-635.

637 This motif is one of the most characteristic scenes of the Urartian art, know from not only helmets, but other objects (*e.g.* belts) as well.

638 Calmeyer 1986, 84-85.

ti-še ^{I}*Me-nu-a-<ḫi>-ni-še* [*uš-tú*]*-ú-ni* ("Argishti, son of Menua offered this helmet to Haldi, the lord."). (*Pls. 94, 95*) – reign of Argishti I (789—766 B.C.) – Erevan, Historical Museum of Armenia – Piotrovsky 1952, 38-39, fold-out drawing between pp. 40-41, 50 (inscription), pls. 11-15; Barnett – Watson 1952, pls. 3:2 (sic!), 33:2 (sic!, mistakenly attributed to Sarduri II); Barnett 1953, 102 (sic!, mistakenly attributed to Sarduri III); Piotrovsky 1955, 26; Pallottino 1955, 112-113, pl. 44; König 1955, 18, 114, pl. 106/107, no. 100A; Piotrovsky 1959, 167, pls. 36, 38:2, 41:2 (sic!, mistakenly attributed as detail of Argishti's quiver); Barnett 1959, 8; Mongait 1959, pl. opposite p. 226; Melikishvili 1960, 270, no. 148; Piotrovsky 1962, 70-71, pls. 20-21; Riemschneider 1965, 78, pl. 33; Piotrovsky 1966a, 242, pls. 36, 38:2, 41:2 (sic!, mistakenly attributed as detail of Argishti's quiver); van Loon 1966, 119-121, pl. 29 (sic!, mistakenly attributed to Sarduri II); Herrmann 1966, 106-107, fig. 25; Piotrovsky 1967, 45-47, fig. 30 (sic!), pl. 20a-b; Lloyd 1967, fig. 122 (front right); Piotrovsky 1968, figs. 111 (sic!), 115 (sic!, mistakenly said to be in Leningrad); Azarpay 1968, 17, 27-28, pls. 10-13; Khondakarian – Amirian – Hovhannissian 1968, 10; Piotrovsky 1969, 179, pls. 94-95; Piotrovsky 1970, pls. 44, 46, 47; Borchhardt 1972, pl. 34:4 (sic!, mistakenly attributed to Sarduri II); Hovhannissian 1973, 53, figs. 37, 40; Piotrovsky 1974/1977, pl. 33:9; Barnett 1974, 48; Martirosyan 1974, fig. 5; Rubinstein 1975, figs. 23-24; van Loon 1975, 463, pl. 390b; Salvatori 1975, 259, note 12; Ligabue – Salvatori 1975, 7; Overlaet 1979, 58, no. 10; Curtis 1979, vol. I, 182; Özgen 1982, 29, 69-71, note 86; Vanden Berghe – De Meyer, 1982, 78, fig. 33; Özgen 1983, 111, note 5; Gröschel 1986, 75, notes 238-239; Calmeyer 1986, 79, note 2, fig. 1; Magen 1986, pl. 25:5; Esaian 1986, 25-26, pl. 16.

92. Karmir Blur, Group no. **U.1.5.1.2** – Fragmentary helmet found in 1958. There is an *Urartian inscription* of king *Argishti I* around the base of the helmet: [^{d}*Ḫal-di-e e-ú-*]*-ri-e i-ni k*[*u*]*-bu-še-e* ^{I}*Ar-gi-iš-ti-*[*še* ^{I}*Me-nu-a-ḫi-ni-še uš-tú-ú-ni*] ("Argishti, son of Menua offered this helmet to Haldi, the Lord"). (*Pl. 94*) – reign of Argishti I (789—766 B.C.) – Erevan, Historical Museum of Armenia (?) – Melikishvili 1960, p. 459, no. 77.

93. Karmir Blur, Citadel Palace, Room 10, Group no. **U.15.1.3** – Fragmentary helmet. It had a conical shape and pointed top. There are four pairs of curves (arcs) ending in lions' heads on the front of the helmet. The curves stand on an incised saw-tooth band running around the base of the helmet. Behind the lions' heads there is an incised design representing the lion's mane. Between the curves, in three registers, there are 11 alternating pairs of winged, bearded genii adoring (pollinating) on the sides of a tree-of-life and wingless, beardless deities adoring (pollinating) on the sides of a tree-of-life. Between two bands of saw-tooth motif in two registers there are ten pairs (five-five) of genii and deities, however the eleventh is above them, on the upper saw-tooth motif, between the lions' heads. This is a pair of wingless, beardless deities. Around the helmet in two registers, between pairs of saw-tooth bands there is an alternating scene of chariots (with a warrior and driver, both of them wearing conical helmets) and horsemen (carrying spears, round shields and wearing conical helmets). They move from the back of the helmet in opposite directions, in the direction of the curves on the front of the helmet. The upper register contains four chariots and four horsemen (in horseman – chariot – horseman – chariot order in procession on either side). The lower register contains four chariots and six horsemen (in horseman – chariot – horseman –chariot – horseman order in procession on either side). There is an *Urartian inscription* of king *Sarduri II* around the base of the helmet: ^{d}*Ḫal-di-e e-ú-ri-e i-*[*ni ku-bu-še* ^{I}]^{d}*Sar*$_{5}$*-du-ri-še* ^{I}*Ar-giš-ti-ḫi-ni-še uš-tú-ni ul-gu-si-ia-ni e-di-ni* ("Sarduri, son of Argishti offered this helmet to Haldi, the lord, for his life."). (*Pls. 94, 96*) – reign of Sarduri II (765—733 B.C.) – St. Petersburg, Hermitage Museum – Piotrovsky 1948, 150 with photo, H. 4, figs. 8-10 (inscription); Piotrovsky 1949, 89; Piotrovsky 1950, 63-67, figs. 40, 40a, 41B, pl. 12; Piotrovsky 1952, 50 (inscription); Barnett – Watson 1952, 139, fig. 15; Barnett 1953, 102 (sic!, mistakenly attributed to Sarduri III); Piotrovsky 1955, 26; König 1955, 22, 135, pl. 86, no. 112a; Piotrovsky 1959, 167, pl. 38:1 (sic!, mistakenly attributed to Argishti I); Melikishvili 1960, 319, no. 174; Piotrovsky 1966a, 242, pl. 38:1 (sic!, mistakenly attributed to Argishti I); van Loon 1966, 119, pl. 26:2; Piotrovsky 1967, 45-47; Piotrovsky 1968, fig. 108; Azarpay 1968, 30, 36, pls. 16-17; Piotrovsky 1969, 48, pl. 93; Piotrovsky 1970, no. 121 (colour plate); Borchhardt 1972, 107, Cat. no. 24 V 3; Barnett 1974, 48; Ligabue – Salvatori 1977, 7; Overlaet 1979, 58, no. 10; Curtis 1979, vol I, 182; Özgen 1982, 29, no. C 1, 71-72, note 86; Eichler 1984, 19 (sic!, mistakenly mentioned in the Historical Museum of Armenia), 39, 53-54, 65, 68; Gröschel 1986, 75, notes 238, 241; Calmeyer 1986, 79, note 3, fig. 2; Esaian 1986, 25-26

94. Karmir Blur, Citadel Palace, Group no. **U.1.5.1.4** – It has a conical shape with pointed top. Complete. There are four pairs of curves (arcs) ending in lions' heads on the front of the helmet. The curves stand on an incised saw-tooth band running around the base of the helmet. Behind the lions' heads there is an incised design representing the lion's mane. Between the curves in three registers there are 11 alternating pairs of winged bearded genii adoring (pollinating) on the sides of a tree-of-life, and wingless, beardless deities adoring (pollinating) on the sides of a tree-of-life. Between two bands of saw-tooth motif, in two registers, there are the 10 pairs (five-five) of genii and deities, however the eleventh is above them, on the upper saw-tooth motif, between the lions' heads. This is a pair of wingless, beardless deities. Around the helmet, in two registers, between pairs of saw-tooth bands there is an alternating scene of chariots (with a warrior and driver, both of them wear conical helmets) and horsemen (carrying spears, round shields and wearing conical helmets). They move from the back of the helmet in opposite directions, in the direction of the curves on the front of the helmet. The upper register contains four chariots and four horsemen (in horseman – chariot – horseman – chariot order in procession on either side of the helmet).The lower register contains four chariots and six horsemen (in horseman – chariot – horseman – chariot – horseman order in procession on either side of the helmet). There is an *Urartian inscription* of king *Sarduri II* around the base of the helmet: ^{d}*Ḫal-di-e* EN(*bēli*)*-šu i-ni ku-bu-še* ^{I}*Sar*$_{5}$*-du-ri-še* ^{I}*Ar-giš-ti-ḫi-ni-še* NÍG.BA ("Sarduri, son of Argishti offered this helmet to Haldi, the lord, for his life."). (*Pls. 94, 97*) – reign of Sarduri II (765—733 B.C.) – Erevan, Historical Museum of Armenia – Piotrovsky 1959, 167; Melikishvili 1960, 459, no. 78 (inscription); Piotrovsky 1962, 70-71, figs. 41, 79, pls. 16-19; Piotrovsky 1965, 412, pl. 96:2; Riemschneider 1965, 78, pls. 31-32; Piotrovsky 1966a, 242; van Loon, 1966, 119, pls. 27-28; Piotrovsky 1966b, 1062, fig. 1193 (sic!, mistakenly mentioned in Leningrad); Piotrovsky 1967, 45-47, pls. 16-19; Lloyd 1967, fig. 122 front left; Piotrovsky 1968, fig. 118 (sic!, mistakenly attributed to Argishti II); Khondakarian – Amirian – Hovhannissian 1968, 10; Piotrovsky 1969, 48, pl. 96 (sic!, mistakenly attributed to Argishti II); Piotrovsky 1970, pls. 45 (sic!, mistakenly attributed to Argishti I), 48; Salvini 1973, 884, fig. 870 (sic!, mistakenly mentioned in Leningrad); Piotrovsky 1974/1977, pl. 30:3; Overlaet 1979, 58, no. 10; Curtis 1979, vol. I, 182; Özgen 1982, 29, no. C 1, 71-72, note 86; Özgen 1983, 111, note 5; Eichler 1984, 19 (sic!), 39, 53-54, 65, 68; Gröschel 1986, 75, notes 238, 240; Calmeyer 1986, 79, note 3; Esaian 1986, 25-26, pl. 18 (?)

95. Karmir Blur, Citadel Palace, Group no. **U.1.5.1.5** – It has a conical shape with pointed top. The top and the left side of the helmet were broken at the rim. Restored. There are four pairs of curves (arcs) ending in lions' heads on the front of the helmet. The curves stand on an incised saw-tooth band running around the base of the helmet. Behind the lions' heads there is an incised design reprtesenting the lion's mane. Between the curves, in three registers, there are 11 alternating pairs of winged, bearded genii adoring (pollinating) a tree-of-life, and wingless, beardless deities adoring (pollinating) on the sides of a tree-of-life.

Between two bands of saw-tooth motif there are ten pairs (five-five) of genii and deities, however the eleventh is above them, on the upper saw-tooth motif, between the lions' heads. This is a pair of wingless, beardless deities. Around the helmet in two registers, between pairs of saw-tooth bands there is an alternating scene of chariots (with a warrior and driver, both of them wear conical helmets) and horsemen (carrying spears, round shields and wearing conical helmets).They move from the back of the helmet in opposite directions, in the direction of the curves on the front of the helmet. The upper register contains four chariots and four horsemen (in horseman – chariot – horseman – chariot order in procession on either side of the helmet). The lower register contains four chariots and six horsemen (in horseman – chariot – horseman – chariot – horseman order in procession on either side of the helmet). There is an *Urartian inscription* of king *Sarduri II* around the base of the helmet. (*Pl. 94*) – reign of Sarduri II (765—733 B.C.) – Erevan, Historical Museum of Armenia – Piotrovsky 1959, 167; Piotrovsky 1966a, 242; van Loon 1966, 119; Piotrovsky 1967, 45-47; Lloyd 1967, fig. 122 (behind right); Khondkarian – Amirian – Hovhannissian 1968, 10; van Loon 1975, fig. 391; Overlaet 1979, 58, no. 10; Curtis 1979, vol. I, 182; Özgen 1982, 29, no. C 1, 71-72, note 86; Gröschel 1986, 75, notes 238, 240; Calmeyer 1986, 79, note 3; Esaian 1986, 25-26.

96. Karmir Blur (?), Group no. **U.1.5.1.6** – It was found in 1958. There is an *Urartian inscription* of king *Sarduri II* around the base of the helmet: ^{d}*Ḫal-di-e* EN(*bēli*)-*šu i-ni ku-bu-še* ^{I}*Sar$_5$-du-ri-še* ^{I}*Ar-giš-ti-ḫi-ni-še* NÍG.BA ("Sarduri, son of Argishti offered this helmet to Haldi, the lord, for his life."). (*Pl. 94*) – reign of Sarduri II (765—733 B.C.) – Erevan, Historical Museum of Armenia (?) – Melikishvili 1960, 459, no. 78.

97. Karmir Blur, Citadel Palace, Room 36, Group no. **U.1.5.1.7** – (*Pl. 94*) – late 8th century B.C. – Erevan, Historical Museum of Armenia (?) – Piotrovsky 1955, 26; Barnett 1959, 15-16.

Group U.1.5.2:

As can be seen from the division into groups, the decoration of Urartian helmets, built up from standard motifs, followed the decorative principles of Assyrian helmets. The various motifs, though different from those on Assyrian helmets, were used in a meaningful not a random way, as in Assyria, to indicate whether a specific helmet was made as a votive object or whether to be worn by soldiers belonging to particular sections of the Urartian army, whether royal bodyguard (*see* the chapter of helmets decorated with the "lightning fork symbol of Teisheba").

The only example of Group U.1.5.2 is a variant of the helmets of Group U.1.5.1. But, as distinguished from the 11 sacred trees and 11 pairs of genii of the Karmir Blur helmets of the first group, there are only 6 sacred trees in a lower register bordered by a double saw-tooth motif. On either side of the sacred trees there are alternating pairs of bearded, winged and beardless, wingless genii. Only the right-wing genius of the extreme right sacred tree is missing. Above this scene, within the standard four pairs of curves terminating in lions' heads, there is a 7 cms high standing figure of a bearded, winged genius. There are two canonical friezes of chariots alternating with horsemen in procession in two registers bordered by the saw-tooth motif around the base of the helmet. In the upper register there are five horsemen and two chariots, however in the lower register there are four horsemen with four chariots. One of the interesting features of the decoration of the helmet is the representation of two Assyrian type spiked shields hanging on the back of two of the chariots. This feature is known from Assyrian art in the reigns of Ashurnasirpal II (883—859 B.C.), Shalmaneser III (858—824 B.C.) and Tiglath-Pileser III (745—727 B.C.). According to E. Özgen, the chariots belong to the 8th century B.C. type (8-spoked wheels) of Urartian chariots[639]. The similarity of this helmet to the helmets of the first group (**Cat. nos. 93, 95-97**) also dates it to the 8th century B.C. (reigns of Argishti I and/or Sarduri II).

Muscarella also questioned[640] the originality of helmet **Cat. no. 98**. He criticized again the poor execution of the incised figurative scene, the inconsistencies in spacing of the figures (*e.g.* the pair of the righ hand side heraldic figure is missing, *Pl. 100*), proportions of figures, the misunderstanding of some motifs (the bow in the hand of the central figure). However he has no opinion about the age of the chariot scene, so he proposes a laboratory examination to establish its corrosion history. He concludes, that this piece is a *pastiche*: a plundered (consequently original) „plain" helmet was decorated by a forger with the incised figurative scenes[641]. However – as far as the present writer knows – there is not a single „plain" Urartian helmet decorated with pairs of curving bands ending in animal heads known from excavation (and from the art market as well). It looks, that such a type simply did not exist in Urartu[642]. It seems, that only those helmets were decorated with such *repoussé* curves ending in animal heads, which had additional embossed/incised figurative decoration. And indeed, all of the similar helmets (Group no. U.1.5, **Cat. nos. 91-103**) are decorated with embossed/incised figurative scenes. Two Urartian helmets which were termed by the present writer as

639 Özgen 1983, 118.

640 Muscarella 2000, 149-150, no. 17.

641 Muscarella 2000, 149.

642 Helmets decorated with pairs of curving bands ending in animal heads without additional figurative decoration are known only from Iran: Hasanlu: (Group H.1) and unprovenanced pieces said to come from „Northwest Iran" (Group NWI.1). It is possible however, that the 9th century B.C. Assyrian helmets represented on the palace reliefs of Ashurnasirpal II (*Chart 2, nos. 76-86*) were similarly decorated and there were not additional figurative scenes on them.

„plain Urartian helmets" (Group no. U.1.1, **Cat. nos. 57-58**) are decorated simply with horizontal *repoussé* ribs around the helmet base, and not with embossed curving bands („crooks") ending in animal (in Urartian art: lion) heads. In this case Muscarella's opinion needs to be revised (*pro* or *contra*).

98. Provenance unknown (said to come from "Urartu (Eastern Turkey)"), Group no. **U.1.5.2.1** – H.: 27.3 cm, H. of the main figure: 7 cm, Diam.: 21.5-22 cm. Conical helmet with pointed top. There are two holes and a loop (or ring) on either side of the helmet for the chinstrap. There is an incised "double zigzag band" around the base of the helmet. There are four pairs of curves (arcs) in *repoussé* ending in lions' heads on the front of the helmet. Behind the heads there is an incised decoration of the lion's mane. Between the curves there is the main figure of the decoration of the helmet: four-winged, bearded god, wearing a horned crown, carrying a bow in his left hand, standing (taking a step to the right) on a "double zigzag band". His height is 7 cm. Under this figure, between two "double zigzag bands" there is an alternating scene of pairs of bearded, winged genii standing (and adoring) on eihter side of a tree-of-life and beardless, wingless gods adoring on eihter side of a tree-of-life. There are two and a half pairs of bearded genii and three pairs of beardless gods. The height of the frieze is 2.5 cm. There is an incised double frieze between two pairs of "double zigzag bands" on the sides and on the back of the helmet. The upper frieze contains horsemen and chariots moving to the left: horseman – horseman – chariot – horseman; and moving to the right: horseman – chariot – horseman – horseman. Each horseman has a conicalk helmet, round shield and two spears. In each chariot (with 8-spoked wheels) there are two men wearing conical helmet: a bearded warrior (?) and a beardless driver. There is a ("Assyrian", thorny – protected by metal bristles) shield hanging on the back of each chariot (only one is missing). The height of each frieze is 3.5-3.6 cm. Complete. (*Pls. 98-100*) – 8th century B.C. – Karlsruhe, Badisches Landesmuseum, Inv. no. 89/1 (Galvano copy: Munich, Archäologische Staatssammlung, Inv. no. PS 1975/3067 – Kellner 1980, 207-208, pls. 2-6; Özgen 1982, 30, 73 note 89; Özgen 1983, 114, note 31, figs. 11-12; Eichler 1984, 19: H.1, 37 (fig.), 39 (figs.), 53-54, 65-66, 68, pls. 8, 9:1; Gröschel 1986, 75-76, note 243, pls. 15-20; Calmeyer 1986, 79-86, fig. 4, pls. 15-20; Esaian 1986, 27, pl. 14; Maass 1987, 71, 72-74, pl. 1:2 middle (helmet no. 2); Calmeyer 1991a, no. 9, fig. 8, pl. 9; Born – Seidl 1995, Abbs. 25-26.

Group U.1.5.3:

This group is a further variant of Groups U.1.5.1-2. Its characteristic is that on the front of the helmet within the four pairs of curves terminating in lion's heads there is a sacred tree in the upper register with two beardless, wingless genii on either side. In the lower register instead of the sacred tree there is a symmetrical scene of a chariot which is followed by two horsemen. Around the base of the helmet there is the well known scene of horsemen and chariots in procession. The fragmentary state of the helmet unfortunately makes the reconstruction of the exact number of horsemen and chariots impossible. The different registers are separated by the double saw-tooth motif.

In addition to this fragment there is another unpublished helmet from Karmir Blur, mentioned by the excavator, B.B. Piotrovsky[643], which belongs probably to this group. After examination of the first three groups it may be concluded that their decorative systems illustrate one aspect of the Assyro-Babylonian concept of royalty (*šarrūtu*). Unfortunately there is no evidence to indicate how the helmets of these three groups relate to the different (probably royal) workshops and arsenals of Urartian towns (*e.g.* to Karmir Blur where helmets of groups 1, 3 were found). It may be assumed that helmets of Groups U.1.5.1-3 are variants of the same subject made in the workshops of different Urartian towns.

99. Provenance unknown (said to come from "Urartu (Eastern Turkey)", "near Lake Van"), Group no. **U.1.5.3.1** – Fragment from the front of a conical helmet. "In the centre of the front of the helmet a life-tree is inscribed in cartouche. There are two female figures on each side of the life-tree facing it. On both the right and left sides of the helmet beside the female figures are four mythological creatures which have the heads of lions and the bodies of snakes. The snake bodies are curved so that the lion heads are facing downward. Also on both sides of the helmet and next to the mythological creatures Urartian warriors and cavalrymen are depicted. They are wearing helmets and are pictured as going to battle with their chariots or on horseback. Each chariot is pulled by a pair of ornamented horses. The cavalrymen are side by side in pairs. All figures are shown moving toward the center." (Tasyürek 1974, 177-178). The scenes are bordered by "double zigzag bands" (saw-tooth motif). (*Pl. 101*) – 8th century B.C. – Gaziantep, Gaziantep Museum, Inv. no.: 64.3.1973 – Tasyürek 1974, 177-181, fig. 1, pl. 1; Özgen 1982, 29, 72 note 87; Gröschel 1986, 75, note 242; Calmeyer 1986, 79.

Group U.1.5.4:

Helmets of this group differ from helmets of Groups U.1.5.1-3 not in their decorative system but in the theme represented. On the front of the helmets within the four pairs of curves terminating in lions' heads, there are two genii holding situlas and facing each other in the upper register. In the lower register there is a symmetrical scene of ibex-hunting from a chariot. The same scene appears in two registers around the base of the helmet: horsemen hunting ibexes (six horsemen pursue six ibexes). The registers are bordered by bud-and-garland motifs. P. Calmeyer has emphasized the strongly Assyrianizing character of the hunting scene[644]. The first examples of similar hunting scenes are known

[643] Piotrovsky 1967, 46.
[644] Calmeyer 1986, 85.

from Assyria, from the reign of Ashurnasirpal II (883—859 B.C.), when the hunting scenes (bull-hunting, lion-hunting) were very important elements of the Assyrian royal iconography (propaganda and cult). The best parallels can be seen on the hunting scene ornaments of Assyrian garments on the palace reliefs of Ashurnasirpal II[645].

Helmets decorated with hunting scenes – similar to the royal hunting scenes of the Assyrian palace reliefs – also fit into the iconographical system of the royal ideology. The Urartian iconography of the *šarrūtu* ("kingship") is borrowed from Assyria and has the same three (one religious and two "secular") aspects as can be seen on the Assyrian palace reliefs:

1. The religious aspect of *šarrūtu*
   Assyria: Winged sun-disk, tree-of-life, adoring king, libation, *etc.*
   Urartu: Groups U.1.5.1-2: sacred trees with genii
2. The military aspect of *šarrūtu*
   Assyria: Battle scenes of Assyrian palace reliefs
   Urartu: Group U.1.5.3: chariots and horsemen
3. The "heroic" aspect of *šarrūtu*
   Assyria: Lion and bull hunting scenes of the Assyrian palace reliefs
   Urartu: Group U.1.5.4: ibex hunting

As mentioned above, the Assyrian *šarrūtu* has the same aspects. On the Assyrian palace reliefs however, appear representations of *šarrūtu,* which are missing from Urartian helmets: king adoring the tree-of-life with attendants and genii behind him and the crown prince leading the procession of tribute bearers. Referring to a different ideological background or tradition, the Urartian king for example never appears in a prominent form on Urartian helmets. On the other hand the Urartian helmets often were inscribed with inscriptions or simply with the names of Urartian kings. The reason might be religious (dedicatory inscriptions) or secular (symbolic ownership of the helmet or the soldier who wore it).

The fact that the iconographic systems of these Urartian helmets can be defined confirms the assumption that the helmets decorated with figurative designs were made by royal order, or followed unwritten rules and were worn by the royal attendants, military officials or body guard. These helmets were made surely not for votive purposes.

100. Provenance unknown (said to come from "Urartu (Eastern Turkey)"), Group no. **U.1.5.4.1** – H.: 27.5 cm, Diam.: 18.3-22.7 cm. Conical helmet with pointed top and *repoussé* decoration. "Here the usual horseman – chariot – horseman scheme is replaced with hunting scenes. Each frieze contains six horsemen alternating with six ibexes. In the center field, flanked by four semicircular arcs terminating in animals' heads, the upper register shows two standing genii. It is interesting to note that these situla-carrying figures face each other with the usual gesture of benediction despite the absence of the sacred tree on the middle. In the lower register are shown two chariots following two ibexes from opposite directions. Other different features are the use of bud garland motif as border decoration of the friezes instead of the familiar opposing saw-tooth design, and the presence of two holes, ca. 2 cm apart, on each side intended for bronze attachment loops of the chinstrap." (Özgen 1982, 29). Traces of gold leaf are visible. (*Pls. 102*) – 8th century B.C. – Private Collection, Galvano-copy: Munich, Archäologische, Inv. no. PS 1971/1782a – Kellner 1976, 78, Cat. no. 139; Kellner 1980, 208-210, pls. 7-9; Özgen 1982, 29, 67, note 78, 73, note 88; Vanden Berghe – De Meyer 1982, 131, no. 21 (fig.); Gröschel 1986, 76, note 244; Calmeyer 1986, 85, note 37; Esaian 1986, 26, pl. 14; Roaf 1990, colour picture on p. 171; Calmeyer 1991a, no. 14 (pls.).

101. Çavuştepe, Uçkale, Turkey, Group no. **U.1.5.4.2** – Urartian conical bronze helmets. "Dr. H.-J. Kellner kindly informed me that one of these yet unpublished helmets was inscribed with the name of Sarduri II and its decoration ("*Krummstab*", two situla-carrying genii facing each other in the center field, scenes of horsemen hunting ibexes) resembles closely to an uninscribed helmet in a private collection, PS Inv. 1971. 1782a." (**Cat. no. 100**). (Özgen 1982, 28). There is an *Urartian inscription* of king *Sarduri II* around the base of one of the helmet. – reign of Sarduri II (765—733 B.C.) – Erzen 1973, 58; Kellner 1980, 205, note 2; Özgen 1982, 28, 67, note 78.

Group U.1.5.5:

This group represents a relatively late development of the Urartian helmets decorated with figurative motifs. The characteristic of the only helmet belonging to this group (**Cat. no. 94**) is its construction. This helmet was made up of an approximately hemispherical body and a conical peak, both hammered out and then joined together, like the plain conical helmet **Cat. no. 58** mentioned above. The decorative system of this helmet is unique. The curves terminating in animals' heads have already disappeared from the front of the helmet. The decoration consists of friezes of *lion passants*, which are known from Urartian shields. In the upper register there are 2 x 3 lions separated by a 14-petalled rosette at the back and by a 12-petalled rosette between two kneeling winged genii at the front of the helmet. It is very

---

[645] Layard 1853, pls. 48:6, 49:4.

difficult to reconstruct the decoration of the fragmentary lower register. The decoration of the existing parts suggests the same decoration of 2 x 3 lions, which were separated from one another by a freestanding palmette under the rosette at the front of the helmet. The two registers are bordered by double saw-tooth motifs. The central scene at the front of the helmet consists of the torso of a deity facing to the right in a winged sun-disk.

E. Özgen, who published this helmet first[646], dated it to the late 7th century B.C. on the basis of the similarity of the incised lions to the lions on the shields of Rusa III (629—615 B.C.)[647]. The decorative system of Urartian helmets probably changed in the 7th century B.C. Only a few of the 8th century B.C. characteristics remained in use (frieze of two registers bordered with saw-tooth motif, central field at the front). In contrast with the 8th century B.C. examples, the decoration of this helmet is hardly explicable in terms of the concept of royalty (*šarrūtu*). It seems that the strict rules of 8th century B.C. decorative systems (the iconography of royal propaganda on helmets) became more free/informal for the 7th century B.C., so far as can be reconstructed from a single helmet.

102. Provenance unknown (said to come from "Urartu (Eastern Turkey)"), Group no. **U.1.5.5.1** – H.: 27.6 cm (restored), Diam.: 22 cm. "Conical bronze helmet, made up of an approximately hemispherical body and a conical peak, both hammered out and then joined together. Badly broken around the rim, only a short stretch remaining; several cracks extend upward from the broken edge. The decoration is in repouss, with all details incised/chased from the front with a sharp pointed tool. Since only a stretch of ca. 5 cm of the actual rim with a singular repouss, ridge is preserved it is impossible to determine whether it was originally provided with a dedicatory royal inscription as well as with holes and/or loop designed for the chinstrap. The most notable feature is the absence of the familiar semicircular arcs ending in animals' heads. The decoration is in two friezes running horizontally around the lower part of the helmet. Each frieze is bordered by an opposing saw-tooth band. The upper frieze is largely preserved. Its left and right halves are occupied by two groups of three passant lions each, striding towards the center. Preceeding these lions in each half there is a 12-petaled rosette flanked by two kneeling winged figures. A 14-petaled rosette separates these two groups of lions at the back of the helmet. The center field is dominated by a large deity figure facing right. His torso is surrounded by a winged disk. He extends both above and below the limits of the upper frieze and is shown performing a gesture of benediction. The lower frieze can be partially reconstructed on the basis of preserved decoration elements despite its grately damaged condition. It is apparent that the groups of three passant lions are repeated here as well. But unfortunately the figural decoration below the kneeling genii as well as below the figure of the deity in the center cannot be determined precisely. The same kneeling figure – rosette – kneeling figure scheme may have been repeated here in conjunction with the lions or the lions with additional identical figures may have continued their procession towards the center. It is also very likely that the figure of the deity with the winged sun-disk once stood on an animal. The two groups of passant lions in the lower frieze are separated from one another by means of a free-standing palmette just below the 14-petaled rosette of the upper frieze." (Özgen). (*Pls. 103-104*) – 7th century B.C. – Philadelphia, University Museum, UM 67-39-21 – Özgen 1982, 5, 13-15, 30-37, figs. 1-3; Gröschel 1986, 76, note 245.

Group U.1.5.6:

This group contains only a single helmet. This unprovenanced helmet is unique. Though it has been identified by several scholars as an Urartian helmet, the identification is not so clear. The vertical rib terminating in an animal's head is very rare on Urartian helmets (the only parallel is **Cat. no. 86**, *Pl. 93*, decorated with the "lightning fork symbol of Teisheba", which helmet itself is unique for this feature). It can be assumed that a vertical rib appears on the front of helmets decorated with *repoussé* lightning fork (**Cat. nos. 62-90**) as part of the motif. The vertical rib is entirely missing from Urartian helmets decorated with figurative scenes. The only exception is **Cat. no. 103**, which can not be undoubtedly identified as an Urartian helmet. This feature was originally Assyrian and widespread in the 9th century B.C. (*see* for example **Cat. no. 3**, *Pls. 3-4*). In the 8th century B.C. it is known only from Northwest Iranian helmets. The second feature which makes the identification of this helmet debatable is the decoration of the entire upper half with cannelure-like *repoussé* ribs. This technically very difficult feature has no parallels from the ancient Near East. Without parallels or representations it can only be assumed that this helmet might be of individual manufacture, probably made in the late 9th – early 8th centuries B.C. in Assyria, or in Urartian territories contiguous with Northwest Iran, where in the 8th century B.C. vertical ribs terminating in animal's (especially in ram's) heads remained in use.

103. Provenance unknown (said to come from "Urartu (Eastern Turkey)"), Group no. **U.1.5.6.1** – H.: 31.8 cm, Diam.: 19.8-20.6 cm. Conical helmet with pointed top. The lower half of the helmet is decorated with five horizontal ribs of which the upper four end in rams' heads. On the front of the helmet there are three pairs of curves (arcs, "*Krummstab*") in *repoussé*, and a vertical rib coming from the top of the helmet. All of them end in rams' heads. Between the curves there are two incised figures and a winged genius (?). There are two loops on the rim of the helmet for the chin-strap. The upper part of the helmet is decorated entirely with vertical ribs running from the top and ending in a wavy horizontal rib above the arcs, at half the height of the helmet. Complete. – 8th – 7th century B.C. – Les Arcs, Ebnöther Collection – Vanden Berghe – De Meyer 1982, 132, no. 23 (fig.); Calmeyer 1991a, no. 12 (pl.).

---

[646] Özgen 1982, 5, 13-15, 30-37, figs. 1-3.
[647] Shields from Toprak Kale: Berlin, V.A. 805; British Museum, BM 22481, BM 22482.

There is a further decorated bronze fragment of a helmet in the University Museum, Philadeplhia[648]. E. Özgen wrote: "*It is very likely to be a "non-Urartian" product on the basis of its artistic execution despite the presence of some Urartian motifs*"[649] Further Urartian conical bronze helmets were found at Altintepe, at the South Portico of the temple[650], which might also have been decorated with figurative motifs.

Group U.2: Urartian conical iron helmets (Cat. nos. 104-108)

The earliest appearance of iron helmets in Urartu can be dated to the first half of the 8th century B.C. As mentioned above (pp. 36-37) regarding their technique of manufacture, the Urartian iron helmets are classed among the 1st, 4th, and 5th groups of ancient Near Eastern iron helmets.

It is characteristic of helmets of the 1st group that they were riveted together from two, separately fashioned halves. The two halves were joined by two vertical iron bands (one on each side) inside the helmet secured with iron rivets (*Pl. 105*). There are three helmets in this group (**Cat. nos. 104-106**). In the 8th century B.C. in Urartu, not only iron helmets, but also iron vessels (*e.g.* jugs[651] and askoi[652]) were made with the same technique. Wartke assumed that the shape of the peaked helmets and the vessels made it impossible for them to be formed in a single piece[653]. It is right in the case of iron vessels, however, as have been discussed above (pp. 35-37) in Assyria the blacksmiths solved the problem of hammering a pointed iron helmet in a single piece (and decorated them with bronze inlay: **Cat. no. 10** (*Pls. 9-11*). The technique – may be because of the lack of necessery technical skill – is unknown from outside Assyria. The technique of shaping separately and riveting together might have been much cheaper than hammering in one piece. The more difficult shapes, such as conical crested helmets were riveted together from separate parts in Assyria as well (*see* pp. 50-52, **Cat. nos. 16-31**, *Pls. 25-43*)[654].

It is characteristic feature of the 4th group (**Cat. no. 107**) that the helmets were fashioned in two similar halves, riveted together as the helmets of the 1st group, but at the top a solid iron point (not cast; it must be made in some way by hammering wrought iron) held the two parts together. The single helmet of the 5th group (**Cat. no. 108**) was possibly hammered out from a single piece of iron (unfortunately only a small fragment of the upper part and the point exist), onto which the point was fastened. As the contour of the helmet breaks at the base of the point, it has to be assumed that the point was separately hammered and added to the body of the helmet. This helmet would be the most developed iron helmet known outside Assyria. Its shape with a separately added point is similar to the shape of early 7th century B.C. Assyro-Babylonian helmets (*Chart 2, no. 226, 257-262*). So it can be assumed that this technique and the characteristic shape was an early 7th century Assyro-Babylonian innovation, which determined the history of the Babylonian helmets of the 7th century B.C.

104. Provenance unknown (said to come from "Urartu (Eastern Turkey)"), Group no. **U.2.1** – H.: 24.8 cm, Diam.: 18.6-21.4 cm. Pointed helmet made of two similar parts joined by two vertical iron bands (one on each side) inside the helmet and with iron rivets. There is a series of holes around the base of the helmet for the attachment, probably of a lining. Restored from fragments. (*Pl. 105*) – 7th century B.C. – Munich, Archäologische Staatssammlung, Inv. no.: 1971/2022 – Kellner 1976, 78, Cat. no. 137a; Kellner 1979, 152, fig. B; Overlaet 1979, 54, note 8a; Curtis 1979, vol. I, 182; Kellner 1980, 210, note 17; Özgen 1982, 68, note 79; Vanden Berghe – De Meyer 1982, 133, no. 25 (fig.); Esaian 1986, 33, pl. 13:10; Wartke 1991, 330, fig. 4; Dezső – Curtis 1991, 125-126.

105. Provenance unknown (said to come from "Urartu (Eastern Turkey)"), Group no. **U.2.2** – Pointed helmet made of two similar parts joined together by two vertical iron bands (one on each side) inside the helmet and with iron rivets. – 7th century B.C. – Adana, Regional Museum – Kellner 1979, 156, note 6; Overlaet 1979, 54, note 9; Dezső – Curtis 1991, 125-126.

106. Provenance unknown (said to come from "Urartu (Eastern Turkey)"), Group no. **U.2.3** – Several fragments of one (?) iron helmet. – 7th century B.C. – Adana, Regional Museum – Kellner 1979, 156, note 6; Overlaet 1979, 54, note 9; Dezső – Curtis 1991, 125-126.

107. Provenance unknown (said to come from "Urartu (Eastern Turkey)"), Group no. **U.2.4** – H.: 25.5 cm, Diam.: 17-24 cm. Pointed helmet made of two similar parts joined together by two vertical iron bands (one on each side) inside the helmet and with iron rivets. At the top a massive (solid) point (not cast; it must be made in some way by hammering wrough iron) held together the two parts. There is a series of holes around the base of the helmet for the attachment, probably of a lining. Restored from fragments. (*Pls. 106-107*) – 7th century B.C. – Munich, Archäologische Staatssammlung, Inv. no.: 1971/1997 – Kellner 1976, 78,

[648] Philadelphia, University Museum, UM 67-39-22; Özgen 1982, 5.
[649] Özgen 1982, 5.
[650] Özgüç 1966, 41: "Bronze helmets of Assyrian shape"; Özgen 1982, 28, 67, note 77.
[651] Kellner 1976, figs. 2-3; Wartke 1991, 328, fig. 5.
[652] Kellner 1976, figs. 4-5; Wartke 1991, 328, fig. 6.
[653] Wartke 1991, 328.
[654] Dezső – Curtis 1991, 105-126.

Cat. no. 137; Fuchs – Kellner 1978, 386, pl. 2; Kellner 1979, 152, fig. 1; Overlaet 1979, 54, note 8b; Curtis 1979, vol. I, 182; Kellner 1980, 210, note 17; Özgen 1982, 68, note 79; Vanden Berghe – De Meyer 1982, 133, no. 26 (fig.); Esaian 1986, 33, pl. 13:11; Calmeyer 1991a, no. 19 (pl.); Dezső – Curtis 1991, 125-126.

108. Karmir Blur, Group no. **U.2.5** – Fragment of the upper part and point of a conical iron helmet. The helmet was probably hammered out from a single piece of iron, but the small (possibly solid) point of the helmet brakes the continuous curve of the outline of the side of the helmet. So the point might have been added separately to the helmet. There is a vertical rib embossed on the front of the helmet. There are no remains of additional embossed decoration would have been on the missing lower part of the helmet. (*Pl. 108*) – 7th century B.C. – Erevan, Historical Museum of Armenia – Esaian 1986, 33, pl. 20:2; Dezső – Curtis 1991, 126.

Group U.3: Urartian conical crested bronze helmets (Cat. nos. 109-110)

As has been mentioned earlier[655] this type of crested helmet originated in the Late Bronze Age helmet tradition from which the Carchemish type of crested helmet, the archetype of conical crested helmets, emerged in the late 10th – early 9th century B.C. (*Charts 3-4, nos. 101-105*). The developed form of this type of crested helmet appeared in Urartu (as also in Assyria) in the last quarter of the 9th century B.C. The earliest known Urartian examples are helmet **Cat. nos. 109, 110** with the inscription of king Ishpuini (*ca.* 830—810 B.C.). This type of crested helmet is known from two fresco-fragments from Erebuni (*Charts 3-4, nos. 118-119*) which can be dated to the first half of the 8th century B.C. On one of them possibly the Urartian king, Argishti I (789—766 B.C.) himself wears the helmet (*Charts 3-4, no. 118*). Although the Assyrian type of conical/pointed helmet overshadowed it, it remained in use through the 8th and 7th centuries B.C. (Paravakar[656]: *Chart 3, nos. 120-123*; Hatshbulag[657]: *Chart 3, no. 124*), and it can be seen even on 6th–5th centuries B.C. bronze statuettes of warriors found at Ayrum[658].

The place of the structure of the Mainz Urartian crested bronze helmets (**Cat. nos. 109, 110**) in the system of 9th–8th century B.C. crested helmets has already been discussed in the chapters on the Assyrian crested bronze helmets[659] and Assyrian crested iron helmets[660]. The two horizontal ribs running around the base of the helmets (*Pls. 109-110*) are characteristics of the conical/pointed bronze helmets as well. On the other hand the vertical ribs strengthening the crest (*Pl. 110*) seem to remain canonical features in the structure of conical crested helmets. There are traces of the same structural features on the half a century later Nimrud crested iron helmets (**Cat. nos. 16-31**) as well. The only difference is that the strengthening ribs of the crested bronze helmets were replaced with bronze inlay bands on the iron helmets. Such bands are an unnecessary reinforcement on the stronger iron helmets, but may have been retained as a traditional, perhaps canonical decorative feature. The ribs running vertically from the crest and curving in a right angle on the front of the Mainz crested bronze helmets (**Cat. nos. 109, 110**, *Pls. 109-110*) are replaced with bronze inlay on the crest of the reconstructed Nimrud crested iron helmet(s) (*Pl. 42*). The evolution is the same in the case of the strengthening ribs running parallel with the upper edge (and the perforation) and curving in a right angle at the back of the crest (*Pl. 110*). These ribs were replaced with bronze inlay and loose their structural function to become archaizing decorative elements on the crests of the Nimrud iron crested helmets (*Pl. 42*). The similarity of these features is not accidental. The structure of the Mainz crested bronze helmets (**Cat. nos. 109, 110**), on which the former parts are separated with ribs similar to the structure of the Karlsruhe crested bronze helmet (**Cat. no. 14**, *Pls. 20-23*) recalls the earlier construction of crested bronze helmets, in which the separate parts were riveted together (**Cat. no. 13**, *Pls. 18-19*).

109. Provenance unknown (said to come from "Eastern Turkey"), Group no. **U.3.1** – H.: 29 cm; Diam.: 27.4-16.3 cm. It has a conical shape ending at the top in a crest curving forward. There is a series of holes on both sides of the crest support for fixing the horse-hair crest (plume). On the back of the helmet the crest ends at half the height of the helmet. There is a hole for a loop or a ring at the front point of the crest (*Pls. 109-110*). There are two horizontal ribs around the base of the helmet and there is a third at one third the height of the helmet and a fourth rib around the helmet at the lower end of the crest. There are three holes on either side of the base of the helmet for the earflaps. The helmet was made of one piece. There is an *Urartian inscription* of king *Ishpuini* around the base of the helmet. Complete (*Pls. 109-110*). – Ishpuini (*ca.* 830—810 B.C.) – Mainz, Römisch-Germanisches Zentralmuseum, Inv. no. O 39702 – Vanden Berghe – De Meyer 1982, 132, no. 24 (fig.); Esaian 1986, 23, pl. 13:9; Maass 1987, 69, note 22; Pflug 1988a, 23-24, note 74, fig. 11; Egg – Waurick 1990, pl. 7; Calmeyer 1991a, 123, fig. 7; Dezső – Curtis 1991, 114-116, pl. 16.

---

655 *See* pp. 37-40.
656 Esaian 1976, 238, figs. 148:1-3, 149:2; Esaian 1986, 23-24, pls. 2:1-2, 2:4, 9:10-12.
657 Kesamanli 1966, fig. 4; Esaian 1984, fig. 14, no. 47.
658 Esaian 1986, 23-24, pls. 2:3, 2:6.
659 *See* pp. 37-40.
660 *See* pp. 50-52.

110. Provenance unknown (said to come from "Eastern Turkey"), Group no. **U.3.2** – H.: 29 cm; The helmet is an exact parallel of helmet **Cat. no. 109**. There is an *Urartian inscription* of king *Ishpuini* around the base of the helmet. Crushed. – Ishpuini (*ca.* 830—810 B.C.) – Mainz, Römisch-Germanisches Zentralmuseum, Inv. no. O 39703 – Unpublished.

Group U.4: Urartian bronze neck-guard (Cat. no. 111)

There is a bronze sheet found at Karmir Blur, which seems to be decorated as a lion's mane. Barnett came to the conclusion that it must be from the back of a helmet[661]. From Urartu (and the Near East), however there is not a single known detachable neck-guard. Such neck-guards appear only on the Til-Barsip wall-paintings (*Chart 2, nos. 163-170*) and on a probably North-Syrian helmet (**Cat. no. 37**, *Pl. 50*), but all of them seem to be rigid and hammered out from the same sheet as the helmet itself.

111. Karmir Blur, Citadel Palace, Room 38, Group no. **U.4.1** – "Fragments of bronze plates bearing a decoration representing a lion's mane, perhaps from the neck of a helmet" (Barnett 1959, 18). – 8th century B.C. – Erevan, Historical Museum of Armenia – Barnett 1959, 18; Özgen 1982, 30, 74, note 90.

Group U.5: Urartian bronze earflaps/cheek-pieces (Cat. nos. 112-114)

There are two types of earflap known from Urartu, however the same problem exists as in the case of the neck-guard: there is no a single authentic Urartian representation of such earflaps. **Cat. no. 112** (*Pl. 111*) is an unprovenienced Urartian ("Transcaucasian") or Northwest Iranian earflap which is an exact parallel of the "Hasanlu" type of long earflap known from helmet **Cat. no. 53** (*Pls.74-75*). This type of earflap is shown on conical helmets with vertical fin in the scenes of an "Assyrianizing" but probably Urartian horse's breastplate (*Chart 2, nos. 116-117*). This type of earflap might be widespread in Transcaucasia, Urartu and Northwest Iran in the late 9th and early 8th centuries B.C. As can be judged from the representational and archaeological evidence, both conical and hemispherical crested helmets might be furnished with this type of earflaps.

The second type of earflap has a rounded lower edge (**Cat. nos. 113-114**). This type of earflap appears only on the palace reliefs of Sennacherib (704—681 B.C.) (*Chart 2, nos. 197-198, 202-204, , 206, 208, 210-212; Charts 3-4, nos. 214, 219*). On other Assyrian helmet representations of the 7th century B.C., with a few exceptions which show hinged earflaps, it is hard to say whether they show detachable or hard-set earflaps.

The earliest known earflaps are the earflaps of an early 1st millennium North-Iranian crested helmet (**Cat. no. 133**, *Pl. 119*) and the earflaps of the Hasanlu hemispherical crested helmets (**Cat. nos. 52-54**, *Pls. 72-77*) from the end of the 9th century B.C. The earliest representations of earflaps on Iron Age helmets are shown on the Balawat Gates (on Urartian hemispherical crested helmets, *Chart 3, nos. 107, 109-110*) and on a horse's breastplate (Hasanlu type long earflaps on hemispherical crested and conical helmets, *Chart 2, nos. 111, 116-117* similar to **Cat. no. 112**) of probably the same age. These earflaps (**Cat. nos. 112-114**) are hinged earflaps, which technique of application is unknown from Urartian helmets. On Urartian helmets there are always three holes for the chinstrap or earflaps hanging on bronze rings. Unfortunately there is no known Urartian helmet with traces of hinges on the sides.

O. Belli mentions[662] that cheek-pieces with eye and mouth cutouts ("*E förmige Wangenklappen*") inscribed with the name of king Ishpuini (*ca.* 830—810 B.C.) were found in Urartian sites[663]. This type of cheek-piece is unknown from Urartian art. Its parallels are known from Aegean helmets of later periods (7th century B.C.)[664]: Olympia[665], Sparta[666], Samos[667] and Lindos[668]. Its latest form is known from the Tamassos-type of Cypriote helmets (*e.g.* the Nicosia helmet, *see* pp. 60-62) and its representations from Samos (*Fig. 8*; *Chart 2, nos. M.2.19-20*) and Idalion (*Fig. 8*; *Chart 2, no. M.2.18*).

---

661 Barnett 1959, 18; Özgen 1982, 30, 74, note 90.

662 Belli 1993, 65.

663 Rittig 1994, Abb. 6.

664 For detailed discussion of the problem *see* Dezső 1998, *passim*.

665 Olympia Museum, B 357, B 4228. Furtwängler 1890, no. 41, pl. 7:41; Dezső 1998, ill. 3, no. M.8.5.

666 Dawkins 1929, pls. XVI, CVIII; Dezső 1998, ill. 3, nos. M.8.3, M.8.6, M.8.8.

667 Brize 1985, pls. 15-17, 20:1, 22:2, 24:2; Pflug 1988b, fig. 12; Dezső 1998, ill. 3, no. M.8.10.

668 1. Lindos, Acropolis, (right side earflap); H.: 14.5 cm; W.: 11.6 cm; Blinkenberg 1931, 189-190, no. 571, pl. 22; 2. Lindos, Acropolis (left side earflap); H.: 13.4 cm; W.: 10.9 cm; Blinkenberg 1931, 190, no. 572, pl. 22.4; 3. Lindos, Acropolis; H.: 13.3 cm; Blinkenberg 1931, 190, no. 573; 4. Blinkenberg 1931, 391, no. 1564, pl. 63:1564; Dezső 1998, 18, ill. 3: Lindos 571, no. M.8.4.

112. Provenance unknown (said to come from "Urartu (Transcaucasia)"), Group no. **U.5.1** – L.: 15.9 cm; W.: 6 cm. Long, "Hasanlu" type of bronze earflap with rectangular upper and oblique lower edge. There are two parallel *repoussé* ribs following the contour of the earflap. There are two holes in the upper corners to fasten the aerflap to the helmet, and a third hole at the lower edge to fasten the two earflaps to each other by a chinstrap. There is an incised motif in the upper right corner of the earflap ("*Wangenschutz aus Bronze, wohl von einem Helm, in oberer rechter Ecke Darstellungen eines Pfluges eingraviert*", Kellner 1976, 78). (*Pl. 111*) – late 9th – early 8th centuries B.C. – Munich, Archäologische Staatssammlung, Inv. no. PS 1971/1829 – Kellner 1976, 78, Cat. no. 144, fig. 46.

113. Altintepe, Group no. **U.5.2** – L.: 11 cm, W.: 10.5 cm. Bronze earflap. Approximately half oval in shape. Originally a lining of some material, perhaps leather, was fixed to it with a row of rivets. Across the diameter of the oval are the remains of a hinge. On the original outer side, the heads of the rivets have been smoothed down and are invisible. – 8th century B.C. – Ankara, Museum of Anatolian Civilizations – Barnett – Gökçe 1953, 125, no. 18, pl. 18:1a; Borchhardt 1972, 109, Cat. no. 24 IX, pl. 29:1-2.

114. Altintepe, Group no. **U.5.3** – L.: 11 cm, W.: 10.5 cm. Bronze earflap. Approximately half oval in shape. Originally a lining of some material, perhaps leather, was fixed to it with a row of rivets. Across the diameter of the oval are the remains of a hinge. On the original outer side, the heads of the rivets have been smoothed down and are invisible. The right edge of the earflap is fragmentary. – 8th century B.C. – Ankara, Museum of Anatolian Civilizations – Barnett – Gökçe 1953, 125, no. 19, pl. 18:1b; Borchhardt 1972, 109, Cat.no. 24 IX, pl. 29:3-4.

CAUCASIAN HELMETS

This small and poorly attested group of helmets emerged on the Northern periphery of Urartu. During the 8th century B.C. flourishing of Urartu, several examples of Urartian helmets might have reached the territories of Abhazia and the Caucasus. The best example of these helmets is an Urartian helmet decorated with the "lightning fork symbol of Teisheba" in *repoussé* (**Cat. no. 78**) inscribed with the inscription of king Argishti I (789—766 B.C.), which was found at Rutchi-Tig (Verhnyaya Rutkha), Georgia. Without detailed information however, there is no possibility of separating the authentic Urartian helmets from the local products.

Examining helmets **Cat. nos. 115**, **118-121** it may be supposed that both helmets was made similarly to their Urartian archetypes (if they are not Urartian) of a single piece of bronze. It is known from a short reference[669] that at Hurjin Hoger (Northern Armenia) a bulldozer destroyed a tumulus, in which a conical bronze helmet and a bronze shield were smashed(?). The origin of these pieces is unknown. Rabinovich considered[670] the Faskau helmet (**Cat. no. 115**) as local product and evidence for the cultural connections between the Northern Caucasus region and the Near East, while Esaian emphasized the significant similarity of the helmet to the Karmir Blur helmets[671].

Helmets **Cat. nos. 116-117** however are unique. **Cat. no. 116** was found at Klin-Yar Mogila, in a grave (no. 186) containing a male skeleton and iron weapons. The most interesting thing is the construction of the helmet. The technique is unique. The helmet with the earflaps was "cut out" from a single sheet of bronze two ends of which was fixed together at the front by four rivets. This technique is undoubtedly a cheaper and easier method of manufacture, and refers to the lack of technical skill necessary for hammering out a helmet from a single piece of bronze and iron, and usually is found on the peripheries of major metalworking areas. As Urartu was (in case of the iron helmets) the periphery of Assyria[672], Transcaucasia and the Caucasus region for the 8th century B.C. became the periphery of Urartu and thus Assyria as well.

The shape of the helmet with rounded earflaps refers clearly to Assyrian helmets of the reigns of Tiglath-Pileser III (745—727 B.C.) (*Chart 2, nos. 138-140, 142*), Sennacherib (704—681 B.C.) (*Chart 2, nos. 193-213*)[673] and Ashurbanipal (*Chart 2, nos. 231-255*). Taking its conical shape into consideration the helmet can most probably be dated to the first half or middle of the 7th century B.C.

Helmet **Cat. no. 117** was found at Trialeti and its construction ("sewn from bronze leaf") is a rare example of a cheaper technique on the periphery of Urartu, which probably replaced the more developed technique of hammering out a helmet from a single piece of bronze used in Urartian (and other Near Eastern) metalworking centres. If the present reconstruction is correct, the helmet was made of triangular bronze plates sewn together and radiating from the point. The helmet was furnished with two elongated earflaps. These long earflaps may be the parallels of the well known long Hasanlu earflaps (**Cat. nos. 52-54**) and the earflaps shown on the unprovenienced horse's breastplate (*Chart 2, nos. 116-117*). This would confirm the existence of a Transcaucasian – Northwest Iranian helmet *koine*, one of the most important characteristic of which was the elongated earflap.

This type of 8th–7th centuries B.C. helmet, or its construction may be the archetype of two early 6th century B.C. composite (iron – bronze) helmets from Sardis: **Cat. no. 142** and Old Smyrna (Bayrakli) **Cat. no. 143** (*see* below).

115. Faskau, Georgia, Galiatskiy Mogilnik, Group no. **C.1** – H.: 31 cm. "Another (helmet), also of bronze, and apparently plain, was found in a burial at Faskau in the Caucasus. This piece is not likely to be of authentic Assyrian manufacture but is probably Urartian." (Barnett 1953, 102, note 4). – 8th–7th century B.C. – Uvarova 1900, 276, fig. 211; Piotrovsky 1944, 308; Barnett 1953, 102, note 4; Semionov 1955, 61; Rabinovich, 104

116. Klin-Yar Mogila, Grave no. 186, Group no. **C.2** – H.: 31 cm (41 cm with earflaps), H. of earflaps: 10.3 cm (right), 10.0 cm (left); W. of earflaps: 6.0-6.6 cm. Pointed helmet made of a single sheet of bronze riveted together at the front of the helmet with four rivets. There are two long, rounded earflaps made of the same bronze sheet as the two halves of the helmet. – first half of 7th century B.C. – Kislovodsk Museum – Belinskiy 1990, 190-195, figs. 1-3.

117. Beshtasheni – Safar Haraba, Trialeti, Mogila 13, Group no. **C.3** – Conical helmet made of sheet bronze plates sewn (stitched) together. It has two long earflaps with rounded extremities. ("... a cone-shaped bronze helmet sewn from bronze leaf, with long earflaps at the sides ...", Kuftin 1941, 224). – 8th century B.C. – Excav. no.: C4 163 – Kuftin 1941, 68, 159, 224, note 265, pls. 11, 40; Kuftin 1946, 345; Piotrovsky 1955, 26, note 1; Belinskiy 1990, 194.

---

[669] Esaian 1976, 219, note 18.
[670] Rabinovich 1941, 104.
[671] Esaian 1986, 31.
[672] The technique of riveting helmets together from separate parts – as one of the cheapest method in mass production – was of course known in Assyria as well.
[673] Both detachable and stiff.

118. Kolhid, Abhazia, Group no. **C.4** – Fragmentary conical bronze helmet. – 8th–7th century B.C. – Kuftin 1936; Kuftin 1949, 140, pl. 10; Piotrovsky 1955, 26, note 2

119. Artchadzorskoy Kurgan, Group no. **C.5** – Conical bronze helmets – 7th century B.C. – Otcheti 1896, 68; Esaian 1986, 32.

120. Primorskoe, Abhazia, Group no. **C.6** – H.: 31 cm. Fragment of a conical bronze helmet. – 7th century B.C. – Kuftin 1941, 140; Esaian 1986, 32; Belinskiy 1990, 194.

121. Hurjin Hoger, Group no. **C.7** – Conical bronze helmet from a kurgan destroyed by a bulldozer – 8th–7th century B.C. – Esaian 1976, 219, note 18.

IRANIAN HELMETS

In Northern Iran there was a helmet tradition which emerged from an unknown tradition of 3rd–2nd millennium B.C. Mesopotamian – Elamite helmets and is partly independent from Assyrian – North Syrian helmet tradition of the Iron Age. There are five major variants showing the influence of different cultures:

Group I.1: Hemispherical bronze helmet with eyebrow cut outs and figurative design riveted onto the skull-piece of the helmet (12th–11th centuries B.C.): **Cat. nos. 122-127**
Group I.2: Hemispherical bronze helmet with a crest riveted onto the skull-piece of the helmet (10th–9th centuries B.C.):
Group I.2.1: Ashmolean-variant (**Cat. no. 128**, *Fig. 11*)
Group I.2.2: Mainz-variant (**Cat. nos. 129-132**)
Group I.3: Hemispherical bronze helmet with a crescent shaped crest riveted onto the skull-piece of the helmet ("Meslek-i Shahr"-variant, 9th century B.C.): **Cat. nos. 133-134** (*Fig. 12*)
Group I.4: Marlik "conical" bronze helmets (9th–8th centuries B.C.): **Cat. nos. 135-137**
Group I.5: Irano-Scythian hemispherical iron helmet (7th century B.C.): **Cat. no. 138**

Group I.1: Iranian hemispherical bronze helmets (Cat. nos. 122-127)

One of the most important characteristics of this group is the hemispherical shape fitting close to the head and the eyebrow cut outs at the front of the helmets (**Cat. nos. 122-127**). Four of the six helmets in this group were decorated with figurative designs added separately to the helmets (**Cat. nos. 122-125**), while the fifth helmet was undecorated (**Cat. no. 126**) and the sixth helmet (**Cat. no. 127**) was decorated with riveted knobs. On three of the four decorated helmets the decoration still exists (**Cat. nos. 122-123, 125**), while on the fourth helmet (**Cat. no. 124**) there are only remains of the sometime decoration: three sheet bronze silhouette plates on which the bitumen figures were fastened at the front of the helmet and the base of a similarly sculpted bird, perhaps a raptorial, on the top of the helmet. The decoration of the helmets is unique. There is a row of decorative bronze rivets with silver plated heads around the base of the helmets. There are three figures on the front of the helmets. Each of the figures was sculpted from a bitumen core overlaid with silver and gold and then fastened to a human shaped bronze plate riveted to the helmet. In the central position is a bearded male deity identified as a mountain-water god (probably the Elamite Napirisha[674] who was brought in connection with the Mesopotamian Ea) by the scales on a conical background and the water flowing from the vessel he holds. He is flanked by two identical goddesses. The goddesses hold their hands open in reverence before their breasts. They, too, are placed against backgrounds with scales, which suggests that they may be mountain deities associated with the male god, who is probably dominant since his crown has multiple (four) pairs of hornes while theirs have only single pairs. On the top of the helmets there was a similarly sculpted and overlaid bird, perhaps a raptorial, with wings outstretched and head rendered in the round, centered dramatically over the water god. Behind the bird there was a tapering, tail-like bronze tube, in which the feather or horsehair plume of the helmet was fastened. One of the four decorated helmets (**Cat. no. 125**) differs from the other three (**Cat. nos. 122-124**) only in details: there is a deep rectangular "face-cut" on the front of the helmet and there is only the figure of a single female deity, which is similar to the goddesses on the front of the other three helmets mentioned above and which was sculpted similarly from bitumen core overlaid with gold and silver and then fastened to a bronze plate riveted to the helmet.

The similarity of the decoration of these helmets, and the fact that there are no other helmets known decorated with the same technique, prove the canonical character of the motifs, the scene and the decorative system of these helmets. Although all the four helmets are unprovenienced, the technique of their decoration (silver and gold foil over a bitumen core)[675] – which is known from several territories of the Near East[676], but characteristic of several objects[677], mainly bitumen roundels

[674] Muscarella 1988, 225-226 leaves the possibility of identification both with Napirisha and with Inshushinak open.
[675] For the technique and relating chronological problems *see* Moorey 1988, esp. 240-242.
[676] From Lebanon, Kāmid el-Lōz: a silver plaque overlaid with gold foil showing a frieze of gilded *repoussé* goddesses (Miron 1990, 52-56, nos. 63-79, fig. on p. 55, fig. 37, pls. 8:1-5); Assyria: a gold foil figure of an Assyrian warrior carrying a piked shield on his back from Nimrud (9th century B.C.), now in the Ashmolean Museum, Oxford, Inv. no. 1951.35 (Moorey 1988, pl. IVb); fragment of a similar figure from Nimrud, British Museum, 48-11-4, 304 (Moorey 1988, pl. IVc).

excavated at Susa[678] and found in the "Safid River Region"[679], Northern Iran – connects them to Iran, probably to the Middle Elamite Age Elam[680]. The representation of these three deities (garment, attributes) does not differ from the general representational tradition of the deities depicted in the 3rd–2nd millennium B.C. Mesopotamian art, however Muscarella correctly called attention to the Elamite features of their representation[681].

This type of helmet itself is extraordinary. A kind of cultic-apotropaic function for it may be confirmed by the single possible parallel known from a cuneiform document from the age of the Dynasty of Agade (24th century B.C.), excavated at Susa. This document (ration list) mentiones 4 helmets[682] decorated with silver or silver-plated eagles (*sagšu* [*ti*$_8$]^{mušen})[683]:

vs. II: 14. 3 sagšu [ti$_8$] ^{mušen}
15. [kù]-babbar-*śu-nu* $^1/_2$ [ma-na]
16. *in* 10 gín-[ta]
("three 'eagle helmets/headdresses', silver of which altogether $^1/_2$ *mana*, 10 *gín* each")

rs. I: 2. 1 sagšu [ti$_8$ ^{mušen}]
3. kù-babbar-*śu* 10 g[ín]
("1 'eagle helmet/headdress', (weight of) its silver is 10 *gín*")

Unfortunately there are no known representations of helmets or other types of headdresses decorated with eagles or other kind of birds[684]. As Wilkinson[685] and Muscarella[686] have pointed out, a similar hemispherical Elamite helmet with a long feather hanging down at the back is shown on a relief of Ashurbanipal (668—627 B.C.). It does not emerge from a tube (as on helmet **Cat. no. 122**), but this, mid 7th century B.C. Elamite helmet is a late descendant of the late 2nd millennium Middle Elamite helmets discussed above.

122. Provenance unknown (said to come from "Southwestern Iran"), Group no. **I.1.1** – H.: 16.5 cm, Diam.: 22.5 cm. Hemispherical bronze helmet. There are two cutaways on the front of the helmet with a small projection between them going over the nose. There is a series of silver of silver studs, with bronze pins bent back inside the helmet, around the base of the helmet. There are three figures on the front of the helmet. "Each of the figures was sculpted from bitumen core overlaid with silver and gold and then fastened to the bronze plate riveted to the helmet. In the central position is a bearded male deity, identified as a mountain-water god by the scales on a conical background and the water flowing from the vessel he holds. He is flanked by two identical goddesses. The goddesses hold their hands open in reverence before their breasts. They, too, are placed against backgrounds with scales, which suggests that they may be mountain deities associated with the male god, who is probably dominant since his crown has multiple (four) pairs of horns while theirs have only single pairs." (Muscarella 1988, 223). On the top of the helmet (similarly to the helmets **Cat. nos. 123, 124**) there was a similarly sculpted and overlaid bird, perhaps a raptorial, with wings outstretched and head rendering in the round, centered dramatically over the water god. On the back there is a tapering metal tube for the insertion of a feather or a horsehair plume. (*Pl. 112*) – late 2nd millennium B.C. – New York, Metropolitan Museum of Art, Inv. no. 63.74 (Fletcher Fund, 1963) – Wilkinson 1965, 107-108, fig. 9; Moorey 1971, 257; Calmeyer 1975, 314; Orthmann 1975, colour pl. 36; Schauer 1979-1980, 531, fig. 10; Muscarella 1984, 21, pl. 20; Muscarella 1988, 223-228, figs. on pp. 224-226; Moorey 1988, 241, pl. 5; Egg – Waurick 1990, pl. 4:2.

123. Provenance unknown (said to come from "Northern Iran"), Group no. **I.1.2** – H.: 14.6 cm, H. of the central figure: 12.5 cm. Hemispherical bronze helmet. There are two curved cutaways on the front of the helmet with a small projection between them going over the nose. There is a series of silver (?) studs, with bronze pins bent back inside the helmet (?), around the base of the helmet. There are three figures on the front of the helmet. Each of the figures was sculpted from bitumen core overlaid

---

677 For the parallel use of silver and gold in the Elamite art *see* Muscarella 1988, 223; There are two similar statuettes (Paris, Louvre), one of which was made of silver (Parrot 1983, pl. 301), while the other was made of gold (Parrot 1983, pl. 302). So the possibility of the cultic importance of the parallel or simultaneous usage of gold and silver in the Elamite art is open.

678 Amiet 1977.

679 New York, Metropolitan Museum of Art, MMA 62.115 (Wilkinson 1965, fig. 10; Muscarella 1988, fig. 16); MMA 66.31.1, MMA 64.63 (Muscarella 1988, figs. 17, 19); Norbert Schimmel Collection (Wilkinson 1965, fig. 11; Muscarella 1988, fig. 18).

680 Moorey 1988, 241: "Here, as with the roundels, a "Middle Elamite" date and a place of manufacture in Susiana may be proposed on stylistic grounds, since the treatment of the goddesses on the helmet overlay is paralleled in general and in detail on various objects of this period excavated in Susa".

681 Muscarella 1988, 225.

682 The scribe used the same word (*sagšu*) in the document for leather helmets (*sagšu kuš-gu*$_4$, altogether 61 pieces), bronze helmets (*sagšu zabar*, altogether 18 pieces) and helmets decorated with silver/silver-plated eagle (*sagšu* [*ti*$_8$]*-mušen*) designated as "*Adler-Kopfbedeckung*" by C. Wilcke (Wilcke 1975, 312). Examining the Elamite Iranian helmets (**Cat. nos. 122-125**) in the present writers view there is no reason to distinguish the "helmets decorated with silver/silver-plated eagles" from the leather and bronze helmets. The weight of the silver given (10 gín (*šiqlu*) = 83 g) per helmet is not enough even for a tiara. Consequently it may be supposed, that there was probably a leader cap or helmet, on to which the silver or silver-plated eagle was fastened. For leader helmets of that age *see Chart 1, no. 9.*

683 Scheil 1913, no. 86

684 On the other side, the decoration of helmets with precious metals is better attested in the ancient Near East. For written sources (Dynasty of Agade, Ur III Period) *see* Wilcke 1975, 312.

685 Wilkinson 1965, 108.

686 Muscarella 1988, 225.

with silver and gold and then fastened to a bronze plate riveted to the helmet. In the central position is a bearded male deity identified as a mountain-water god by the scales on a conical background and the water flowing from the vessel he holds. He is flanked by two identical goddesses. The goddesses hold their hands open in reverence before their breasts. They, too, are placed against backgrounds with scales, which suggests that they may be mountain deities associated with the male god, who is probably dominant since his crown has multiple (four) pairs of hornes while theirs have only single pairs. On the top of the helmet (like on the similar helmets in New York, **Cat. no. 122** and Mainz, **Cat. no. 124**) there was a similarly sculpted and overlaid bird, perhaps a raptorial, with wings outstretched and head rendered in the round, centered dramatically over the water god. The bird now is missing. Only studs remain where the bird should exist. – late 2nd millennium B.C. – Private collection – Trésors de l'ancien Iran 1966, Cat. no. 536, pl. 32; Moorey 1971, 257; Calmeyer 1975, 314, fig. 1; Muscarella 1988, 223-225; Moorey 1988, 241.

124. Provenance unknown (said to come from "Northern Iran"), Group no. **I.1.3** – H.: 15.6 cm, Diam.: 19.6 cm. Hemispherical bronze helmet. There are two small curved cutaways on the front of the helmet with a small projection between them going over the nose. There are holes around the base of the helmet in which gilded silver studs, with bronze bronze pins bent back inside the helmet, were fixed. There is only one existing silver (?) stud on the front of the helmet. There are three other holes visible from the series on the front of the helmet. There are three human-shaped vertical bronze plates fastened onto the front of the helmet and an elongated fourth plate is fixed on the top of the helmet. Onto these plates were fastened four figures sculpted in bitumen and overlaid by silver and/or gold. The figures are missing. In the middle there was a figure of a male deity wioth a vessel in his hand. On both sides of the male deity there was a female deity (*see* **Cat. nos. 122, 123**). Above the central male deity, on the top of the helmet, there was a similarly sculpted and overlaid bird, perhaps a raptorial, with wings outstretched and head rendered in the round, centered dramatically over the water god. Complete without figures and studs. (*Pl. 113*) – late 2nd millennium B.C. – Mainz, Römisch-Germanisches Zentralmuseum, Inv. no. O.39704 – Schauer 1979—1980, 531, fig. 11; Muscarella 1988, 223-225; Egg – Waurick 1990, pl. 4:1.

125. Provenance unknown (said to come from "Haft Tepe (Elam, South-East Iran)"), Group no. **I.1.4** – Hemispherical bronze helmet with deep rectangular "face-cut". The helmet bears similar decoration as helmets **Cat. nos. 122-124**. The only difference is that on the front this helmet there is only a figure of a single female deity, which was sculpted from bitumen core overlaid with gold and silver and then fastened to a bronze plate riveted to the helmet. The only difference is that on the front this helmet there is only a figure of a single female deity, which was sculpted from bitumen core overlaid with gold and silver and then fastened to a bronze plate riveted to the helmet. There is a row of bronze studs around the base of the helmet. – *ca.* 1300 B.C. – Private collection – Fales 1989, 189, colour plate on p. 188.

126. Provenance unknown (said to come from "Northern Iran"), Group no. **I.1.5** – H.: 18 cm, Diam.: 21 cm. Hemispherical bronze helmet with deep "face-cut" similar to **Cat. no. 125**. The helmet is decorated with 8 pieces of shield-like knobs which riveted to the skull piece of the helmet. There are three shield-like knobs in a horizontal row on the front of the helmet. These knobs have a flat rim. A fourth shield-like knob sits on the top of the front section of the helmet. This knob has a similar flat rim, but is decorated with an additional small "umbo". There are two smaller knobs on the back section of the top, and further two at the back of the helmet. These four knobs have no flat rims. The helmet is restored from fragments. – late 2nd millennium B.C. – Wilfrid Israel Museum of Oriental Art and Culture, Kibbutz Hazorea, Israel, Inv. no. P-259 – unpublished.

127. Provenance unknown (said to come from "Northern Iran ("Caspian")"), Group no. **I.1.6** – H.: 15 cm. Hemispherical bronze helmet. There are two deep, semicircular cutaways on the front of the helmet for the eyes. Between the cutaways there is a projection going over the nose. There is a rectangular hole on the top of the helmet (?). Undecorated. – late 2nd millennium B.C. – Private collection – Bronzes et terres cuites Louristan – Amlash 1981, Cat. no. 131 (fig.).

Group I.2.1: Iranian hemispherical crested bronze helmet (Ashmolean variant) (Cat. no. 128)

The shape of the Iranian helmet of the Ashmolean Museum (**Cat. no. 128**) can be reconstructed only with difficulty (*Fig. 11*). The overall shape of the skull-piece was hemispherical, similarly to the helmets mentioned above. The thickness of the bronze sheet however was so great (nearly 3 mm) that the rim of the helmet might crack[687] when it was hammered (*Fig. 11*). This crack was cut out in form of a sector which reached the centre of the helmet. After that the helmet would easily be shaped to form a hemispherical skull-piece. The long cut out was covered with a bronze crest which was probably conical with a point curving forward, and rivet holes of which can be seen along the edge of the cut out in shape of a rectangular panel reaching the rim of the helmet. There are pairs of rivet holes around the base of the helmet for fastening the lining.

128. Provenance unknown (said to come from "Northwest Iran"), Group no. **I.2.1.1** – H.: ca. 24 cm. "Piece of sheet metal curved to a shape closely resembling that of the human head. The main edge, pirced with pairs of holes set at regular intervals is cut back into a slight curve on either side at the bottom; lines of holes divide the main area of the surface into a rectangular panel. (The rectangular panel running up the centre may have been to accommodate the tube for a plumed crest.) Undecorated; badly damaged and distorted; dents could have been the result of blows received in combat." (Moorey 1971, 256-257). (*Pls. 116-117*) – early 1st millennium B.C. – Oxford, Ashmolean Museum, Inv. no. 1965.859 (Bomford no. 160) – Moorey 1971, 256-257, Cat no. 492, figs.

687 It was possibly a common fault in manufacture in the ancient bronzeworking, when the bronze sheet became very thin or rigid and "tired", and instead of stretching, the rim or edge of the sheet might easily crack.

UNIV. DUNELM

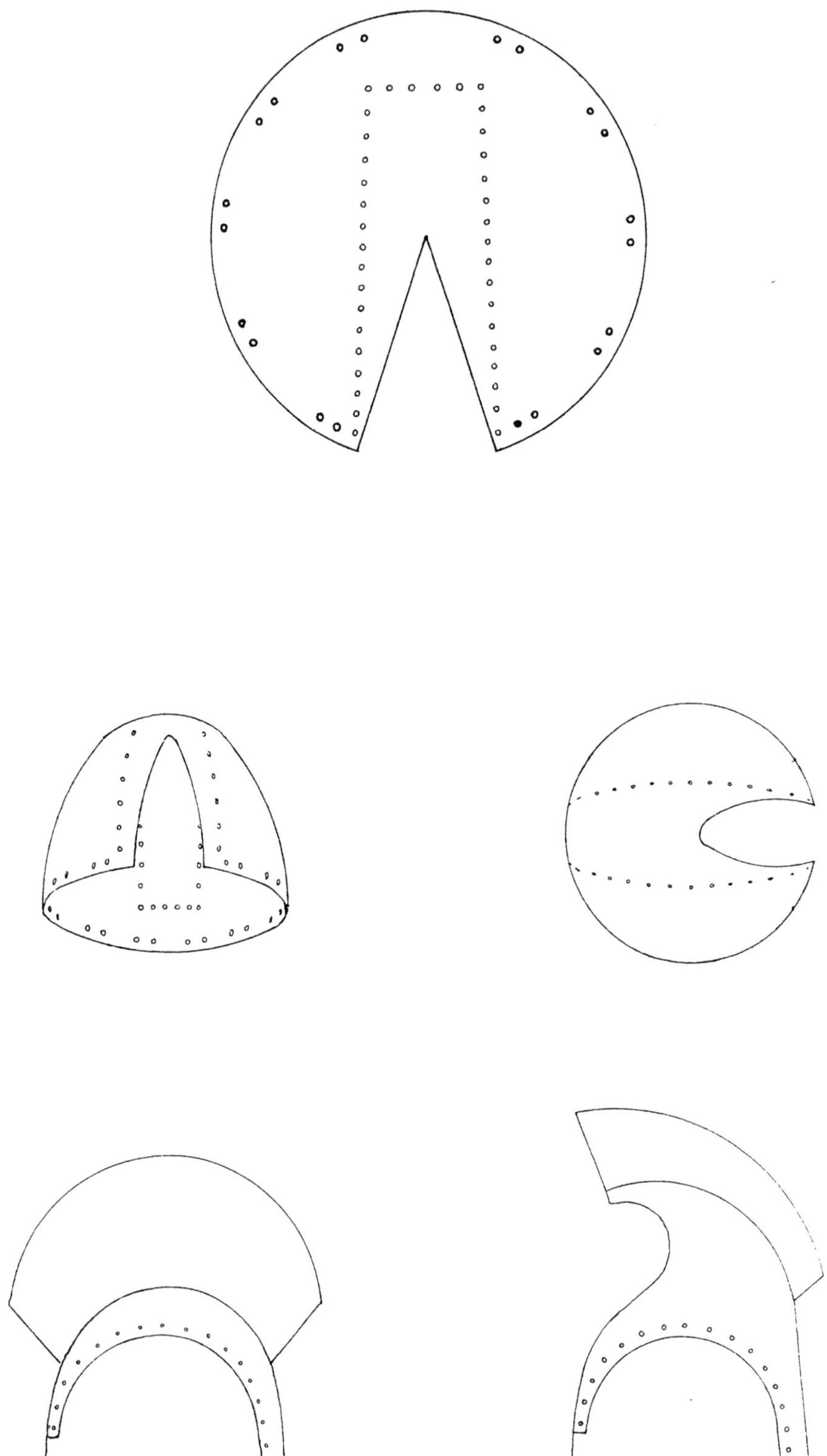

Fig. 11: Schematic drawing of reconstruction of helmet **Cat. no. 128**

Group I.2.2: Iranian hemispherical crested bronze helmets (Mainz variant) (Cat. nos. 129-132)

The skull-piece of the Mainz helmets (**Cat. nos. 129-132**) is similarly hemispherical as the Ashmolean helmet (**Cat. no. 128**) mentioned above. There are 4 (two pairs) or 6 (three pairs) or 8 holes (4 pairs) in an elliptoid-rectangular panel on the top of these helmet, which panel is similar to that on the Ashmolean helmet (**Cat. no. 128**). The holes are probably for the rivets of a tube, supporting a (crescentic or forward curving) crest or a plume.

129. Provenance unknown (said to come from "Northwest Iran"), Group no. **I.2.2.1** – H.: 13.3 cm, Diam.: 17.8-22.5 cm. Hemispherical bronze helmet. There are 8 holes (4 pairs) in an elliptic order on the top of the helmet. The holes are probably for the rivets of a tube, supporting a crest or plume. Undecorated. Complete. (*Pl. 118*) – early 1st millennium B.C. – Mainz, Römisch-Germanisches Zentralmuseum, Inv. no. O.39671 – Schauer 1979-1980, 530, figs. 9a-b.

130. Provenance unknown (said to come from "Northwest Iran"), Group no. **I.2.2.2** – H.: 15.7 cm, Diam.: 21.1-21.4 cm. Hemispherical bronze helmet with a slight rib across the top. There are four holes (two pairs on the sides of the rib) probably for the rivets of a tube, supporting a crest or plume. The crest was secured by a rivet, from which only a rivet hole remained at the back of the helmet. Undecorated. – early 1st millennium B.C. – Mainz, Römisch-Germanisches Zentralmuseum, Inv. no. 39670 – unpublished.

131. Provenance unknown (said to come from "Northwest Iran"), Group no. **I.2.2.3** – H.: 15.1 cm, Diam.: 19.1-22 cm. Hemispherical bronze helmet with a slight rib across the top. There are six holes (three pairs on the sides of the rib) probably for the rivets of a tube, supporting a crest or plume. The crest was secured by a rivet, from which only a rivet hole remained at the back of the helmet. Undecorated. – early 1st millennium B.C. – Mainz, Römisch-Germanisches Zentralmuseum, Inv. no. O.39672 – unpublished.

132. Provenance unknown (said to come from "Northwest Iran"), Group no. **I.2.2.4** – 62 bronze fragments of probably more than one helmet of the type of **Cat. nos. 129-131**. – early 1st millennium B.C. – Mainz, Römisch-Germanisches Zentralmuseum, Inv. no. O.39673/1-62 – unpublished.

Group I.3: Iranian hemispherical crested bronze helmets (Meslek-i Shahr variant) (Cat. nos. 133-134)

The construction of helmets **Cat. nos. 133-134** reported as from Meslek-i Shahr, Iranian Azerbaijan and Northern Iran, differ from Iranian helmets discussed above (*Fig. 12*). These helmets are probably *pastiches*. In spite of the *pastiche* character of these helmets' crest (crest support), the crest structure is known from the decoration of a Transcaucasian bronze belt[688] found at Asthi-Blur (near Lake Sevan, Armenia) and dated to the early 1st millennium, 9th century B.C.[689] (*Chart 1, no. 54, Fig. 9*). On this belt there are five incised and punched figures of warriors/hunters whose helmets have the same crest structure: a conical crest-support on the top of the skull-piece holds a crest curving back at the front and the back sides of the of the helmet. On the belt the horsehair or feather crest is represented by a row of incised triangles. The Meslek-i Shahr helmet (**Cat. no. 133**) and its Transcaucasian representation are further evidence for the existence of a Transcaucasian – Northwest Iranian helmet *koine* in the early 1st millennium B.C. (*Fig. 9*). The origin of this helmet can be proved by the second helmet (**Cat. no. 134**) of this type from a pre-second World War commercial excavation in Iran[690]. This helmet follows the same general principles as helmet **Cat. no. 133**. The only difference between them is their different construction. The cap-piece and the crest-support of helmet **Cat. no. 134** was hammered out from a single piece of bronze and only the crescentic crest was added separately and fastened with rivets to the crest support and the cap-piece of the helmet. So the *pastiche* character of helmet **Cat. no. 133** became questionnable, and this helmet can be regarded as technical variant of a local, North Iranian and Transcaucasian helmet type which flourished probably in the early 1st millennium B.C.

133. "Meslek-i-Shahr, Azerbaijan" (reported as from), Group no. **I.3.1** – H.: 36.4 cm, H. of the earflaps: 11.6 cm, Diam.: 22.2 cm. Hemispherical bronze helmet with conical crest-support, semicircular crest and detachable earflaps. On the front of the helmet there is a series of holes probably for the attachment of a lining. On the back of the helmet there is a narrow brim to protect the nape and the neck. There is a row of holes on the brim for the lining. The top of the helmet is flat and there is a conical crest-support riveted on it. The top of the crest-support is flat, and the whole looks like a secondary used, upside down turned cup. The top of the crest-support (the bottom of the cup) is twice as wide as the crest riveted to it. The side of the cup is embossed at half the height, which is superfluous, unfunctional in the case of a crest-support. The crest is semicircular and it has an U-shaped cross-section. There is a row of holes along both edges of the crest to fix the horse-hair plume. Both ends of the crest are riveted to the "skull-piece" of the helmet. On either sides of the helmet there is a detachable earflap fastened to the helmet with hinges. The lower corners of the earflaps are rounded. There is a series of holes along the edges of the earflaps probably for a lining. Complete. *Pastiche*. (*Pl. 119*) – early 1st millennium B.C. – London, British Museum, Inv. no. WA 134705

---

688 Esaian 1976, 244, fig. 517; Esaian 1984, 114, fig. 13, no. 42.
689 Esaian 1984, 110-119.
690 Carried out by the father of H. Mahboubian.

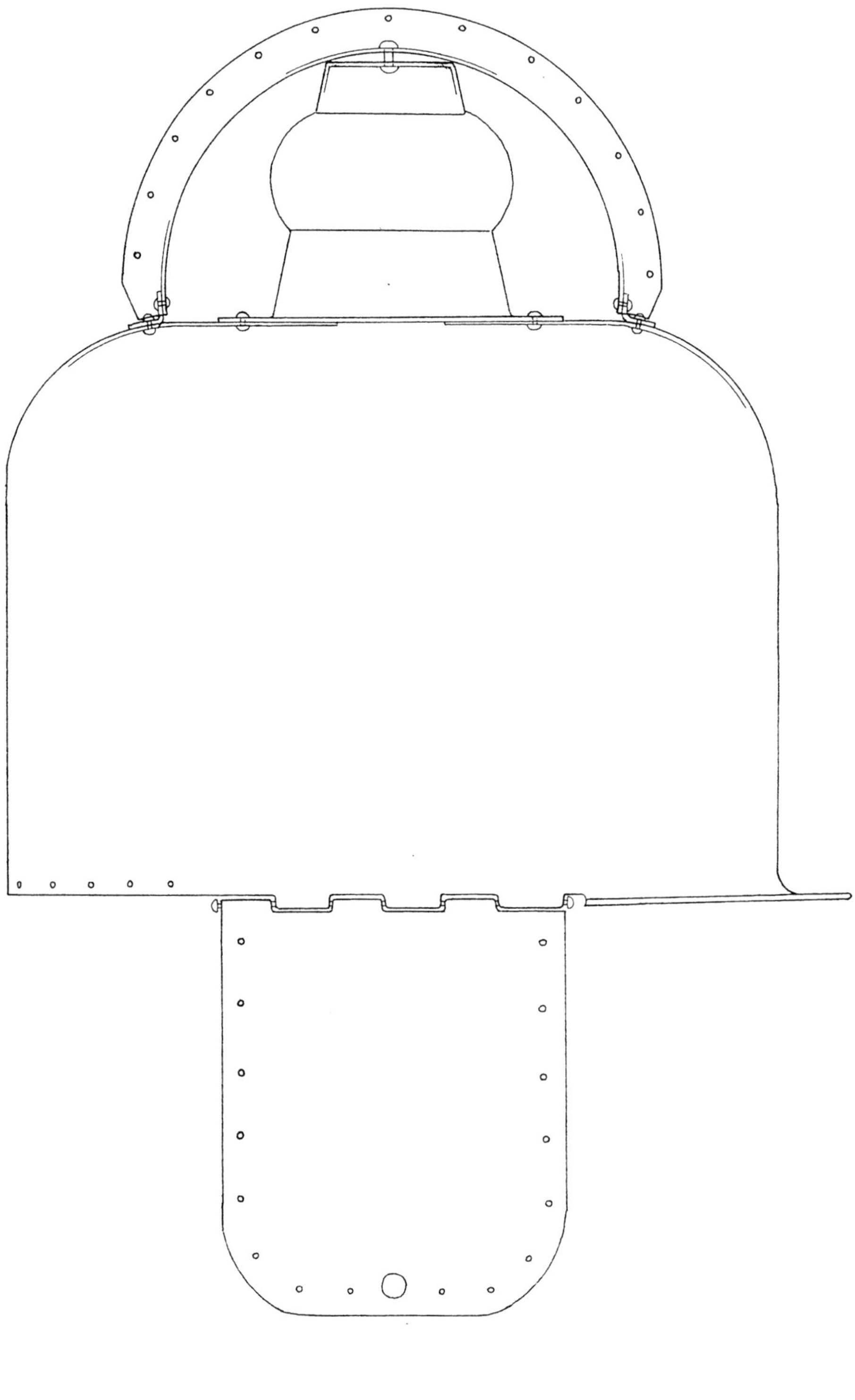

Fig. 12: Schematic drawing of helmet **Cat. no. 133** (*scale 1:2*)

– Barnett – Curtis 1973, 123-124; Gamber 1978, 195, fig. 197; Frankel 1979, 20, fig. 11.

134. Provenance unknown (said to come from "Northern Iran"), Group no. **I.3.2** – H.: 31.5 cm, Diam.: 21 cm. Hemispherical helmet with conical crest-support, and semicircular (crescentic) crest. There is a rib around the base of the helmet to strengthen its construction. There is a series of holes (in 15 mm intervals) around the rim of the helmet probably for the attachment of a lining. There are three bigger holes at the sides, around the rim of the helmet for the attachment of two earflaps. The overall shape of the helmet is similar to helmet **Cat. no. 133**. Only its construction is different: the cap and the crest-support of the helmet was hammered in one piece. The shape of the crest support (a tapering tube with a flaring out boss) is the same as the crest-support of helmet **Cat. no. 133**. The crescentic crest was riveted to the crest support and the cap of the helmet. "The crescent shaped plate was worn to follow the line of the brow as indicated by the remains of a nosepiece on the same axis, directly below" (Mahboubian 1997, 285). (*Pl. 120*) – early 1st millennium B.C. – London, Houshang Mahboubian Family Collection – Mahboubian 1997, 284-285, no. 361.

Group I.4: Marlik "conical" bronze helmets (Cat. nos. 135-137)

There were three helmets found in tombs during the excavations of Marlik-tepe (**Cat. nos. 135-137**). One of these helmets (**Cat. no. 137**) which was found in Tomb 52, Trench XXXIII G, was considered by E.O. Negahban as a possible cymbal[691] (total diameter of which is about 24 cm) with a hemispherical boss surrounded by a narrow brim. The other two helmets (**Cat. nos. 135-136**) are rather conical. An interesting feature of both helmets is the vertical rib running from the front to the back in the axis of the helmet[692]. This rib may remind the readers of the vertical fins of Urartian helmets **Cat. nos. 59-61** (*Pls. 82-83*) shown on the unprovenienced horse's breastplate (*Chart 2, nos. 116-117*)[693], and recalls a northern tradition of helmets, which by this feature (unknown in Assyria, North Syria and Northwest Iran) modified the original Assyrian conical helmets. Negahban attributed conical helmets of Marlik type similar in particular to the helmet worn by the warrior depicted on silver beaker 202 M[694], to the 9th century B.C.[695]. The Marlik conical helmets are probably the most distant examples of the Assyrian conical helmet tradition.

135. Marlik, Iran, Trench XVII B, Tomb 26, Group no. **I.4.1** – Diam.: 20 cm, W.: 300 g. "Conical helmet showing dents received from a macehead." (Negahban 1964, 45). "This helmet, the lower part including the rim base broken and missing, approximately 20.0 cm. in diameter, found in Tomb 26, Trench XVII B, is made from sheet bronze, now covered by coat of rust. Its shape is conical with a rather angular line running across the top from front to back, gradually vanishing near mid-height. One side of the helmet has a cavity, partly broken, probably caused by a blow from a macehead or other heavy object." (Negahban 1995, 92). – early 1st millennium B.C. – Tehran, Archaeological Museum, Excav. no. 92 M, Museum no. 25200/8200 – Negahban 1964, 20, 45, no. 53, fig. 53; Moorey 1971, 257; Muscarella 1988, 50; Negahban 1995, 92, fig. 81, pl. XIII:163.

136. Marlik, Iran, Trench XVII B, Tomb 26, Group no. **I.4.2** – Diam.: 20 cm. "This helmet, broken with the upper part mostly missing, approximately 20.0 cm. in diameter, found in Tomb 26, Trench XVII B, made from sheet bronze now covered by a thick coat of rust, is similar in shape to 92 M (**Cat. no. 135**). Although the upper part is missing, more of the lower half of the helmet is preserved in this example. A depression probably caused by a blow can be seen on one side of the helmet." (Negahban 1995, 92). – early 1st millennium B.C. – Tehran, Archaeological Museum, Excav. no. 145 M – Negahban 1995, 92, fig. 81, pl. XIII:164.

137. Marlik, Iran, Trench XXXIII G, Tomb 52, Group no. **I.4.3** – Diam.: 24 cm. "Spherical helmet with a narrow brim." (Negahban 1964, 20). "This helmet or cymbal, about 24.0 cm. in total diameter, found in Tomb 52, Trench XXXIII G, is made from sheet bronze which is now heavily rusted. It has a hemispherical boss surrounded by a narrow brim with traces of textile on the surface. Many broad-brimmed cymbals were found at Marlik, but only 826 M and 525 M, with their rather narrow brims, possibly could have been used as helmets, although the fact that they were found in pairs supports the likelihood of their being cymbals." (Negahban 1995, 92). – early 1st millennium B.C. – Tehran, Archaeological Museum, Excav. no. 826 M – Negahban 1964, 20; Moorey 1971, 257; Muscarella 1988, 50; Negahban 1995, 92, pl. XIV:165.

Group I.5: Irano-Scythian hemispherical iron helmet (Cat. no. 138)

This iron helmet is unique. This is the earliest known iron helmet outside the area of the traditional Near Eastern cultures. Its shape derives from the late 7th – early 6th century B.C. Scythian "Kuban" type of cast bronze helmets[696], which was inspired by the well known Near Eastern (Assyrian, Urartian and Northwest Iranian) conical helmets decorated with pairs of curves and a vertical rib at the front. The structure and decoration of the "Kuban" type of helmets derived probably from the Near Eastern

691 Negahban 1964, 20; Negahban 1995, 92.
692 Negahban 1964, fig. 53.
693 Seidl 1986, fig. 3.
694 Moorey 1971, 257.
695 Negahban 1995, 92.
696 For detailed studies of this type *see* Rabinovich 1941; Galanina 1985, 169-183, fig. 1; Galanina 1987, 7-18; Galanina 1997.

(Assyrian) conical helmet tradition: the eyebrow cutouts[697] bordered by a rib follow the line of the pair(s) of curves on the front of the earlier Near Eastern conical helmets. The vertical rib on the front of the "Kuban" type of helmets may be similarly traced back to the vertical rib between the pairs of curves on the front of Near Eastern conical helmets (Assyria, North Syria, Urartu, North-West Iran). The vertical rib of the Near Eastern helmets is similarly changed on the Marlik helmets (**Cat. nos. 135-137**) as well. The eyebrow cutout – which follows the line of the pairs of curves on the front of Near Eastern helmets – may be a reinterpretation of these curves ("crooks") as eyebrows, or follows the late 2nd millennium B.C. Northern (or Middle Elamite Western) Iranian tradition of these cutouts (*see* **Cat. nos. 122-127**). Most of these Scythian helmets bear a knob of conical shape or in the form of a ring.

So these Scythian helmets can be derived from various Near Eastern helmet traditions (Assyrian – Urartian tradition and North-West Iranian – Transcaucasian tradition) and played an important role through direct connections in the evolution of the Sardis type of composite helmets (*see* the chapter of Persian helmets).

Helmet **Cat. no. 138** and the necessary ironworking technology (hammering out a hemispherical skull-piece from a single piece of iron), the technical skill was the result of probably the cultural influence of 7th century B.C. Assyrian Empire and Urartu as well. The Scythian adoption of Near Eastern (Assyrian and Urartian) scale armour was probably the result of the same interaction. It is known that the Scythians conquered and settled Northwest Iran and Transcaucasia in the early 7th century B.C. or even somewhat earlier and left it at the end of the same century. This cultural influence resulted in three types of helmets emerging in the northern periphery (Anatolia – Transcaucasia – Northern Iran) of the traditional Near Eastern cultures (Assyria and Urartu):

1. "Kuban" type of Scythian cast bronze helmets[698]. There are altogether 17 known examples of this type of helmet[699]:
   1. Kelermes, curgan no. 1, year 1903 (H.: 18.5 cm; Diam.: 18 cm x 28 cm); (*Pl. 121*)
   2. Kelermes, curgan no. 2, year 1904 (H.: 16 cm; Diam.: 17 cm x 28 cm); (*Pl. 122*)
   3. Kelermes, curgan no. 2, year 1904 (H.: 17.3 cm; Diam.: 18 cm x 23 cm); (*Pl. 123*)
   4. Krimskaya, Vorontsovskiy curgan, year 1886 (H.: 18 cm; Diam.: 17.2 cm) (now Warsaw, Museum Wojska Polskiego)
   5. Krimskaya, curgan no. 1, year 1895
   6. Hut. Krasnoe Znamya, curgan no. 6, year 1979
   7. Gvardeyskoe, "from curgan" (H.: 16.9 cm; Diam.: 19 cm x 21 cm)
   8. Nartan, curgan no. 20, year 1978—1979 (17.1 cm; Diam.: 20 cm x 21 cm)
   9. "Stariy Petsheur" (H.: 18 cm, Diam.: 18.5 cm x 22.5 cm)
   10. "River Irtish" (H.: 20.5 cm)
   11. "Afrasiab" / Samarkand Museum (H.:24.5 cm; Diam.: 18 cm)
   12. "Kismichi" / Dzhambul Museum (H.: 20 cm, Diam.: 20 cm x 24 cm)
   13. without provenance, Moscow, State Historical Museum (H.: 18 cm, Diam.: 24 cm)
   14. without provenance, Moscow, State Historical Museum (H.: 17 cm, Diam.: 17 cm x 27 cm)
   15. without provenance, Warsaw, Museum Wojska Polskiego
2. Group I.5: Irano-Scythian hemispherical iron helmet (**Cat. no. 138**)
3. Group P.2: Composit iron and bronze helmets (**Cat. nos. 142-143**)

138. Irani Azerbaijan, next to Lake Urmia, warrior grave, Group no. **I.5.1** – H.: 16.1 cm; Diam.: 19.8-22.2 cm. Hemispherical iron helmet. There are two rounded cutouts (one on the front and the other on the back of the helmet): face-cutout and neck-cutout. There is a rib crossing centrally the top of the helmet. There is a series of 12 holes on the front and 12 holes on the back around the rim of the helmet probably for a lining. The helmet has a long, triangular earflap attached to the skull-piece by hinges. The other earflap is missing. The point of the existing earflap is broken. There are 22 holes around the rim of the earflap for a lining. – late 7th century B.C. – Mainz, Römisch-Germanisches Zentralmuseum, Inv. no. O.39854 – unpublished.

---

697 For detailed study *see* the chapter of Persian helmets.

698 Five helmets from curgans near the railway stations of Kelermes (Tshernenko 1968, 76-81; Ilinskaya – Terenozhkin 1983, 70, no. 13; Tshernenko 1984, 36-37, figs. 19:1-2; Galanina 1987, 15-16, pl. 10) and Krimskaya (Tshernenko 1984, 36, note 104). Gvardeyskiy curgan (Vinogradov 1973, 275-276; Ilinskaya – Terenozhkin 1983, 46, fig. 2; Tshernenko 1984, 36, note 105) and Nartan (Nagoev 1979, 119). Galanina in her articles on the "Kuban" type of helmet listed altogether 14 pieces of this type (Galanina 1985, 178-180), and reported (Galanina 1997) further 3 helmets of the "Kuban" type from near River Volga, from Northern Caucasus and without provenance.

699 Tshernenko 1968, 76-77, pls. 41, 42, 43:1-2, 44:1-2 and chart 3; Galanina 1985, chart on pp. 178-181.

PERSIAN HELMETS

The information available on Persian helmets is very sparse. With the exception of the helmet from the Median booty found at Olympia (**Cat. no. 141**) there is not a single helmet which can confidently be described as a Persian one.

In some of the Neo-Babylonian and Late-Babylonian documents of the Achaemenid Period concerned with military matters appears the word *karballatu* probably Old-Iranian in origin[700]. The meaning of it is obscure, "a piece of linen headgear for soldiers"[701] or "*(Spitz)Mütze*"[702]. The word has no military implications in a few Neo-Assyrian texts in which it occured[703]. However it can be translated similarly to the Neo-Assyrian texts as a kind of linen garment/headgear (*karballat kitî*) in Neo-Babylonian texts as well[704], but in some cases it certainly has the meaning of a kind of headgear for soldiers. In some of these documents concerning military service (loans to acquire military equipment for Babylonians serving in the Achaemenid army) the word *karballatu* occurs together with the word *sariam* (*siriam, šir'annu* = (scale)armour). One of these documents[705] drafted in 513 B.C. under the reign of Darius I contains a list of equipment for 12 soldiers (probably of 12 horsemen[706]) of the royal army: 12 garments, 12 *šir'am* armours[707], 12 *karballatu* headgears, 12 leather bags and 24 pairs of shoes[708]. Another, more explicit document drafted in the second year of Darius I (520 B.C.) in Nippur contains a loan to acquire military equipment for an armoured Babylonian cavalryman[709]:

6. 1-*en*(*ištēn*) ANŠE.KUR.RA(*sīsî*) *a-di ḫu-šu-ki-šu ù pu-gu-da-tu*$_4$ 1-*en*(*ištēn*) ^{túg}*su-ḫat-tu*$_4$
7. 1-*en*(*ištēn*) *ši-ir-'a-an-nu* AN.BAR(*parzillu*) 1-*en*(*ištēn*) *kar-bal-la-tu*$_4$ *šà ši-ir-'a-an-nu*
8. 1-*en*(*ištēn*) *ku-ú-ra-pa-nu šà su-ḫat-tu*$_4$ 1-*en*(*ištēn*) *kar-bal-la-tu*$_4$ *su-ḫat-tu*$_4$ 1-*en*(*ištēn*) ^{kuš}*šal-ṭu e-ru-ú*
9. 1 *me* 20 *ši-il-ta-aḫ šu-uš-ku-bu* 10 *ši-il-ta-aḫ gi-ir-ri* 1-*en*(*ištēn*) *ṭe-e-pu* AN.BAR(*parzillu*)
10. *šà* ^{kuš}*šal-ṭu* 2 ^{giš}*az-ma-ru-ú* AN.BAR(*parzillu*) *ù* 1 *ma-na* KÙ.BABBAR(*kaspu*)[710]

In that case *karballatu* would refer to a kind of helmet which belongs to the armour (or made from the same material, iron, as the scale-armour). L. Oppenheim interpreted this passage as which illustrates a leather helmet covered with metal scales[711]. But this type of helmet (leather base covered with metal scales) seems to be outdated at least from the end of the 2nd millennium B.C. E. Ebeling had probably right when he assumed that the meaning of *karballatu* in that case (line 7) is an iron cap[712], because of in line 8 the text itself specifies another type of *karballatu* made of probably linen (or felt?) belonging to the *suhattu* cloth (which was worn probably under the armour). The shape of the *karballatu* was high pointed as known from the Naks-i-Rustam inscription of Darius I[713]:

14. ... ^{kur}*gi-mir-ri*
15. *ša* ^{túg}*kar-bal-la-ti-šu-nu zaq-pa-'a*
("Cimmerians [Sakae] whose *karballatu*s are pointed")

---

[700] Dandamayev 1989, 564, note 8.
[701] Oppenheim *et al.* 1971, *s.v. karballatu*, 215.
[702] von Soden 1965—1981, *s.v. karballatu*, 448.
[703] Oppenheim *et al.* 1971, *s.v. karballatu*, 215. In these texts (lists of various kind of garments) the denotes probably a kind of garment, or headgear (?) made of GADA (*kitû*) = linen. *karballat kitî* = linen garment/headgear).
[704] Oppenheim *et al.* 1971, *s.v. karballatu*, 215.
[705] Strassmaier 1893, no. 253.
[706] Dandamayev 1989, 564.
[707] Oppenheim *et al.* 1971, *s.v. karballatu*, 215: "*šir'am*-coats"; Dandamayev 1989, 564: "breast-plates".
[708] Strassmaier 1893, no. 253, ll. 6-8.
[709] Lutz 1928, 275, pl. I; Ebeling 1952, 201-213, ll. 6-10.
[710] (6) 1 horse with its harness and "saddles", 1 *suḫattu* cloth garment (7) 1 iron scale armour and a *karballatu* cap which belongs to the armour (8) 1 neck protector which belongs to the *suḫattu* cloth, 1 *suḫattu* cap, 1 bow-and-arrow case made of bronze(?). Unfortunately for the other weapons of the cavalryman (lines 8-10) there is no agreement between the interpreters. Ebeling 1952, 206-208, 210 translated lines 8-10 as follows: "*einem Nackenschutz (bestehend aus) ein(em) Schweisstuch, einer Kappe, (bestehend aus) ein(em) Schweisstuch, einem Schild aus bronze, 120 Pfeilen, auflegbar, 10 Pfeilen, gimirräische(?), einer Keule aus Eisen für den Schild, 2 Lanzen aus Eisen und einer Mine Silber ...*". The translation of *šalṭu* for Schield seems wrong. Reiner *et al.* 1989, *s.v. šalṭu*, 271, translates the same lines as follows: "one bow-and-arrow case with ..., 120 mounted arrows, ten unmounted(?) arrows", "one *ṭēpu* weapon of iron with case, two lances [iron]".
[711] Oppenheim 1950, 192-193, note 18.
[712] Ebeling 1952, 206: "eine zum Panzer gehörig *karballatu*, also wohl aus Eisen, eine Eisenkappe", while Oppenheim *et al.* 1971, s.v. *karballatu*, 215 suggested a cowe attached to the *šir'anu* coat of mail.
[713] Weissbach 1911, 89, 3, lines 14-15 (cf. Weissbach-Markkleeberg 1938, 163, lines 17-18; Herzfeld 1938, 48, line 15 (Artaxerxes II?)).

This high pointed headgear of the Scythians (Sakae) is shown on the Behistun reliefs of Darius I[714] and on the reliefs of the Apadana at Persepolis[715] as well, and was mentioned by Herodotus (*Historiae*, VII:64)[716].

Concluding the evidence of the written sources it can be assumed that at the end of 6th century B.C. (just hundred years after the fall of Assyria and twenty years after the fall of Babylon) a Babylonian armoured cavalryman of the Persian royal army could wear iron scale armour[717] and probably a conical (iron or bronze) helmet as did the Assyrians and Babylonians a few decades earlier. There is no reason to suppose that the Persian army forgot the Assyrian (and probably Babylonian) art of armoured heavy infantry and cavalry, and did not use the helmet types (conical and crested) which were widespread from Syria to Iran, from Urartu to Assyria for more than two hundred years. Pointed bronze helmet **Cat. no. 141**, the Median booty found at Olympia[718] proves the continuity of the earlier Near Eastern helmet tradition to the Persian (Achaemenid) Period, as known from the description of the Mesopotamian (Assyrian and Babylonian) soldiers in the army of Xerxes by Herodotus (*Historiae* VII:63):

"The Assyrians were equipped with bronye helmets made in a complicated outlandish way which is hard to describe, shields, spears, daggers (like Egyptian ones), wooden clubs studded with iron, and linen corslets. These people used to be called Syrians by the Greeks, Assyrians being the name for them elsewhere. With them were the Chaldeans."[719]

There are five helmets, which for different reasons can be arranged into two groups of Persian helmets:

Group P.1: Persian conical bronze helmets (Cat. nos. 139-141)

The main characteristic of this group is that helmets belonging to it do not follow the decorative systems of the earlier Near Eastern conical helmet traditions (*e.g.* curves ("*Krummstab*"), arcs at the front, horizontal and vertical ribs and figurative decoration). In consequence of the total lack of its decoration, helmet **Cat. no. 139** (*Pl. 124*)[720] can not be incorporated into any of the known groups of ancient Near eastern helmets, so it may be a late (Persian?) example of the Near Eastern conical helmet tradition. The main characteristic of the second helmet (**Cat. no. 140**, *Pl. 125*)[721] is the conical knob, which is unknown in the earlier Near Eastern helmet tradition. The construction of the helmet is extraordinary as well. The skull-piece of the helmet was riveted together from two halves, onto which the long (massive or tubular) point of the helmet was fastened (riveted?).

Helmet **Cat. no. 141**[722] found at Olympia bears an inscription[723] along the edge mentioning that this helmet was taken as booty from the Medes. This helmet proves the survival of the conical type of helmets in the early 6th century B.C., in spite of the fact that there are no contemporary representations of Persian (conical) helmets. The history of the Iron Age Near Eastern helmet is unknown after the fall of Nineveh (Assyria) or Babylon (Neo-Babylonian Empire).

139. Provenance unknown (said to come from "Northwest Iran"), Group no. P.1.1 – H.: 27 cm, Diam.: 23.2-20.4 cm. Bronze helmet. It has a conical shape with pointed top. There is a series of holes around the base of the helmet for the attachment probably of a lining. Undecorated. (*Pl. 124*) – 7th–6th century B.C. – Mainz, Römisch-Germanisches Zentralmuseum, Inv. no. O.39606 – Schaaff 1973, pl. 49; Kellner 1976, 78, Cat. no. 143.

---

[714] Luschey 1968, pls. 31, 32, 38, 41.

[715] Herzfeld 1941, pl. 79.

[716] de Sélincourt – Burn 1983, 467 (Herodotus VII:64): "The Sacae (a Scythian people) wore trousers and tall pointed hats set upright on their heads...."

[717] For the scale armour of the Achaemenid Period *see* the descriptions of Xerxes' army by Herodotus (*Historiae* VII:61: the iron scale armour of the Persians; IX:22: the death of Masistius at Plataiai 479 B.C. Actual pieces of iron and bronze scales from Persian scale armours were found by F. Petrie at Memphis, Palace of Apries, now in the Ashmolean Museum, Oxford under the museum numbers: 1966.1207, 1266.1209, 1933.1435, 1910.537, 1909.1084, 1909.1085, 1966.1208. Similar pieces of armour scales were found in Cyprus at Amathus, Tomb 2, Cat. no. 57 (Cypro-Geom. III – Cypro-Archaic I. Period: Westholm 1935, 13-14, pls. V:57, CL; Westholm 1938, 163-173, fig. 1-2; Gjerstad 1948, 132, fig. 20) and at Idalion, Western Acropolis, Squares G-H:6-7 (Period 6A = Cypro-Archaic II, around 470 B.C.: Gjerstad 1935, 538-539, fig. 247-249, pl. CLXXII; Westholm 1938, 163-173, figs. 3-4; Gjerstad 1948, 132, fig. 20). For Achaemenian iron scale armour found at Pasargadae (Tall-i Takht, Room 94, 63/102; now MMA 1978.93.15) and from Persepolis *see* Muscarella 1988, 212, no. 321.

[718] Which bears a dedicatory inscription: Kunze 1961, 130, fig. 75.

[719] de Sélincourt – Burn 1983, 466-467 (Herodotus VII:63).

[720] Schaaff 1973, pl. 49; Kellner 1976, 78, Cat. no. 143.

[721] Schaaff 1973, pl. 48; Kellner 1976, 79, Cat. no. 151.

[722] Kunze 1961, 129-137, pls. 56-57; Schaaff 1973, pl. 50.

[723] Kunze 1961, 130, fig. 75.

140. Provenance unknown (said to come from "Northwest Iran"), Group no. P.1.2 – H.: 34 cm, Diam.: 22.8-19.1 cm. High pointed bronze helmet with conical knob. The helmet was made from three parts: the skull-piece was made from two similar halves riveted together; and a high, massive cast (?) point (or tube) ending in a conical knob holds the two halves together. There is a series of holes around the base of the helmet for the attachment probably of a lining. Undecorated. (*Pl. 125*) – 7th–6th century B.C. – Mainz, Römisch-Germanisches Zentralmuseum, O.39585 – Schaaff 1973, pl. 48; Kellner 1976, 79, Cat. no. 151.

141. Olympia, Group no. P.1.3 – H.: 23.1 cm, Diam.: 20.5-20.7 cm. Plain conical bronze helmet. The point of the helmet ends in a tube, which has a finished edge. The diameter of the hole at the top is 2.9-3.0 cm. It is most probable that th helmet once had a separate point or crest, which was fixed into this hole. There are two pairs of holes on the sides of the helmet along the edge for fastening the chin-strap loop. There is a loop made of bronze wire in one of the pairs of holes for fastening the chin-strap. There is a Greek inscription along the edge of the helmet: ΔΙΙ ἈΘΕΝΑΙΟΙ ΜΕΔΟΝ ΛΑΒΟΝΤΕΣ. – late 6th – early 5th century B.C. – Olympia, Olympia Museum, Inv. no. B 5100 – Kunze 1961, 129-137, pls. 56-57; Schaaff 1973, pl. 50.

Group P.2: Composite iron and bronze helmets (Cat. nos. 142-143)

The type of two composite iron and bronze helmets which were furnished with bronze knob, earflaps, neckguard and eyebrow cutouts found in western Anatolia was previously unknown. The two examples came from Sardis (**Cat. no. 142**, *Pls. 126-130*) and Old-Smyrna (Bayrakli) (**Cat. no. 143**). Unfortunately the Old-Smyrna helmet no longer exists[724], the only information is a short note which mentioned the finding of the helmet[725]. Fragments from the front and earflaps of probably a further example of this type were found in the excavations of the Lindos Acropolis at Rhodes[726]. The examination of the Sardis helmet (**Cat. no. 142**) probably of the same type however revealed the structure and construction of this type of helmet. The helmet was found in the destruction level (Brick Fall deposit of Colossal Lydian Structure) which can be associated with the siege and partial sack of Sardis by Cyrus the Great of Persia[727]. So the helmet should belong to one of the troops of Croesus or Cyrus. As Greenewalt has argued[728], the cultural identity of the helmet is obscure, and since the armies of Croesus and Cyrus included soldiers from other eastern Mediterranean and Near Eastern territories[729], Persia and Lydia are not the only cultural homes possible for this type of helmet.

The representational evidence for both Persian and Lydian helmets is very slight. A Graeco-Persian stele[730] shows a horseman wearing a helmet, but his helmet is crested with a tassel. A limestone statue from Golgoi, Cyprus[731] shows a headgear with radial structure at the front. But this headgear can not be considered as a helmet. Two iron earflaps with bronze cord decoration similar to that of the Sardis helmet were found at Idalion, Cyprus[732].

Various features of this type of helmet can be traced back to three different Near Eastern helmet types and/or traditions:

1. The earliest known examples of the eyebrow cutouts on helmets in the ancient Near East are Middle Elamite. This feature appears on Scythian cast bronze helmets of "Kuban" type[733] (late 7th – early 6th centuries B.C.), where the eyebrow cutout is bordered by a rib and follows the line of the pair of curves on the front of the earlier Near Eastern conical helmets, to which it harks back.

2. The closest parallels for the shape are the Scythian bronze helmets of "Kuban" type as well. These helmets derived probably from the Near Eastern (Assyrian) conical helmet tradition: the eyebrow cutouts and the vertical rib on the front of the Scythian helmets may be traced back to the pair(s) of curve(s) and the vertical rib between them on the front of Near Eastern conical helmets (Assyria, North Syria, Urartu, North-West Iran). Most of these Scythian helmets bear a knob of conical shape or in the form of a ring. The vertical rib of the Near Eastern helmets is similarly changed on the Marlik helmets (**Cat. nos. 135-137**) as well. So these Scythian helmets can be derived from various

---

724 C.H. Greenewalt, personal communication; Greenewalt – Heywood 1992, 12, note 19.

725 Cook – Akurgal 1952, 23. *See* Greenewalt – Heywood 1992, 12, note 19; Greenewalt 1992, 265, note 31.

726 Blinkenberg 1931, 189, no. 569: "*569 GD. Trois fragments d'un casque en fer, comprenant le sommet avec la partie frontale et une partie des couvre-joues fixes. L'état incomplet ne permet pas de déterminer précisément la forme du casque, mais ce qui en reste fait penser qu'il a été du type corinthien. La partie frontale est reinforcée par une morceau de fer fix, ... l'aide de rivets; au couvre-joue sont appliquées par dehors des bandes de bronze (larg. 0.008).*"

727 Greenewalt 1991, 12. Greenewalt dated the destruction of the Colossal Lydian Structure *ca.* 550 B.C.

728 Greenewalt – Heywood 1992, 16-17.

729 Greenewalt – Heywood 1992, 16-17, notes 37-40: Xenophon (*Cyroupaideia* 6.2.10) listed 12 non-Lydian auxilliary troops: Thrace, Egypt, Cyprus, Cilicia, Phrygia, Lycaonia, Paphlagonia, Cappadocia, Arabia, Phoenicia, Assyria, Ionia and Aeolis; and Egyptians (6.3.19-20; 7.1.30-45) Indians and Chaldaeans (6.2.9, 11; 7.2.2-8) in Cyrus's army.

730 Manisa, Archaeological Museum, Inv. no. 3389, Greenewalt – Heywood 1992, figs. 25-26.

731 New York, Metropolitan Museum of Art, MMA 74.51.2466; Greenewalt – Heywood 1992, fig. 23.

732 Gjerstad 1935, pl. CLXXVIII:14 (no. 505), 15 (no. 1071); Gjerstad 1948, 132-133, fig. 20:8. A further cheek-piece of probably the same type made of bronze was found in Idalion (Waldbaum 1989, 334-336). Similar shape of earflaps is known from the "Tamassos type" of Cypriote conical helmets (*see* pp. 60-64, *Pls. 51-54*), from Greek "*Kegelhelme*" of Geometric Age (pp. 65-67, *Pls. 55-56*) and from Illyrian helmets.

733 *See* p. 106, notes 698-699.

Near Eastern helmet traditions (Assyrian – Urartian tradition and North-West Iranian – Transcaucasian tradition) and played an important role through direct connections in the evolution of the Sardis type of composite helmets.

3. The closest parallel or archetype of the structure of the Sardis helmet would be a Transcaucasian helmet found at Safar-Haraba, Trialeti, Mogila 13 (**Cat. no. 117**), which was made from triangular bronze plates placed in radial arrangement(?)[734]and furnished with long earflaps. Unfortunately the possible inner structure of this helmet is unknown.

All these three features mentioned above prove the theory already suggested by Greenewalt[735] that the possible cultural home for the Sardis type of composite helmets would be the late 7th – early 6th century B.C. northern Iran (eyebrow cutouts) and the Transcaucasian – Caucasian region (radial structure). The role of the Scythians in the emergence of this type of helmet may have been very important since the shape of Scythian helmets of "Kuban" type is the closest parallel to the shape of the Sardis helmet[736]. This Iranian-Caucasian(-Scythian) inspiration does not signify that Lydia or Persia was the most likely place for it to have been made. This type of helmet could be considered as the product of the region of Anatolia and Transcaucasia under Scytho-Iranian (Median and/or Persian) influence, and in the general principle of construction may be presumed a prototype of the later Sassanian composite *Bandhelm* and the late Roman, early medieval *Strebenhelm* and *Spangenhelm* types[737].

142. Sardis, Colossal Lydian Structure, Group no. P.2.1 – H.: 22 cm (reconstructed), Diam.: 19.3 cm (reconstructed), W.: 1138.3 gr. Fragmentary helmet made of iron and bronze. "The skull-piece was flat or flattish on top (as was the bottom plate of the finial). Its vertical and horizontal curvetures are uncertain. It was formed of eight triangular plates of iron, placed in radial arrangement and fastened to an interior armature of eight iron bands or ribs. Each rib was a flat band of metal, *ca.* 1.5 cm wide, and supported the sides of two adjacent plates. Plates and ribs converged at the top, the points of the former meeting; the latter terminated somewhat short of center, their ends masked underneath by a thin iron disk (*ca.* 4.3 cm. in diameter). The brow of the helmet has a pair of scallops, one scallop in each of the two adjacent plates that form the front. The scallops evidently met in some kind of point, sharp or blunt, above the nose. There is no evidence for a nose guard; the central seam between the plates would have weakened the attachment place, but a nose guard could have been formed or supported by an extension of the rib to which the plates were attached. The edges of the two front plates are perforated by a series of small holes (*ca.* 0.1 cm. in diameter, *ca.* 1.1 cm. apart), for the attachment of a lining. Other edges are poorly preserved. The back plates extended at least 2.6 cm. below the hinges.

The cheek-pieces have a triangular semilunate shape. One is essentially complete, the other is attested by at least a few fragments, perhaps by many more among the miscellaneous fragments that have not been reconstructed. Of the complete cheek-piece, all edges are perforated by tiny holes (*ca.* 0.1 cm. in diameter, *ca.* 1 cm. apart), for the attachment of a lining. Near the tip is a larger hole (*ca.* 0.25 cm. in diameter) to secure the chin-strap.

The (cheek-piece) hinges are knuckle-and-pintle type. In each hinge the cheek-piece had two knuckles; the skull-piece had four. The knuckles of the skull-piece turned in, those of the cheek-pieces turned out. The pintles were iron; only segments of their shafts survive. The neck guard is a narrow strip of iron sheet (3.5 cm. wide/high; at least 28 cm long, including joining fragments 20 cm. long and nonjoining fragments), with an everted lip (0.3 cm. wide) at bottom. The vertical edges do not survive, or have not been recognized in the miscellaneous fragments that have not been reconstructed, and the profile restored in Figs. 21 and 22 is conjectural. The top edge was evidently contiguous to the bottom edge of the skull-piece because the former preserves patches of bronze corrosion that must be residue from the decorative cords on the latter. The method of attachment to the skull-piece is unclear. The lower edge is perforated by tiny holes (*ca.* 0.1 cm. in diameter, *ca.* 1.0-1.4 cm. apart.

Decoration: The finial is made of bronze, except for an iron cotter pin, which secured it to the skull-piece; it had four parts, from bottom to top as follows. The bottom plate had an everted lip and central hole framed by a "curb", the upper edge of which is broken. Above was a bell-shaped part with outflaring lip. The next part had the form of an inverted bowl with an outer edge cut to create a series of radiating petals or tongues (their alignment with the radial cords on the skull-piece is conjectural). Each tongue had a central groove and flanged edges. The method by which the tongues were attached to the bottom plate is unclear. The fourth part of the finial, the cap, has a shallow, conical top and straight lip. The iron cotter pin has a spherical head and a shaft that is square in section. It penetrates through the centers of the first three parts, between the points, and through the disk on the underside of the skull-piece, thus securing the component parts of the finial together and to the skull-piece; its two ends are splayed out against the under surface of the disk. The cords decorate skull-piece and cheek-pieces, contouring the edges of all and on the skull-piece radiating from the rim of the finial plate to the contour cords. Although the position of some segments has shifted, the original position is clear. On the skull-piece the cords occur in groups of three; on each cheek-piece the top edge is contoured by two cords, the side edges by one.

The cords are bronze except for the central cords of the skull-piece triplets, which are iron. The bronze cords are solid rods, the iron cords may be either solid rods or folded sheet. All bronze cords are articulated with bead-and-reel pattern; whether the iron cords had the same pattern is unclear (it is assumed for them in the reconstruction, figs. 21, 22). The cords were attached to the skull-piece with small rivets, 0.1 cm. in diameter, as the X-radiographs reveal.

---

[734] Kuftin 1941, 224: "... a cone-shaped bronze helmet sewn from bronze leaf, with long earflaps at the sides ...".
[735] Greenewalt – Heywood 1992, 12-14, esp. 17.
[736] The Kuban type of Scythian helmets itself is the latest example of the canonical decorative system of 9th and early 8th centuries B.C. Assyrian, 8th–7th centuries B.C. North-West Iranian conical helmets. The conical shape became mostly hemispherical (early Iranian local tradition), only the knob-like projection on the top of this Kuban type of Scythian helmets reminds us for the earlier shape. The vertical rib on the front of these helmets still exists, but the pair of curves bending to the centre on the front of the helmets lost its original function and became simply the border of two eyebrow cutouts (which was similarly an early Iranian feature: *see* pp. 99-101, **Cat. nos. 122-127**).
[737] Greenewalt – Heywood 1992, 12-14, notes, 22-23.

Perforations along the edges of skull-piece, cheek-piece, and neck-guard secured a lining. Adhering to the inside of the neck guard were pseudomorphs of leather, tentatively identified as goat leather, in three small patches." (Greenewalt – Heywood 1992, 5-7). (*Pls. 126-130*) – middle of the 6th century B.C. – Excav. no. M87.7/9444 – Greenewalt 1989, 183-184; Greenewalt 1991, 11-12, figs. 14-17; Greenewalt – Heywood 1992, 1-31, figs. 1-27; Greenewalt 1992, 260-269, figs. 12-19.

143. Old Smyrna (Bayrakli), Group no. P.2.2 – "... between the pithoi and the floor of the seventh century pylon, a large cache of iron weapons was uncovered; this cache also consisted largely of spearheads, some of considerable length, but also contained an iron helmet with a bronze plume-knob." (Cook – Akurgal 1952, 23) – middle of the 6th century B.C. – Cook 1952, 106, fig. 12; Cook – Akurgal 1952, 23; Greenewalt 1991, 12; Greenewalt – Heywood 1992, 12, note 19; Greenewalt 1992, 265, note 31.

LIST OF HELMET REPRESENTATIONS

1
"Stele of Vultures"
Telloh (Lagaš)
Eannatum
Paris, Louvre, AO 50, 2346, 2347, 2348, 16109
1.88 m
Strommenger 1962, pls. 66, 68
Börker-Klähn 1982, p. 124, no. 17
Parrot 1983, pl. 160

2
"Stele of Vultures"
Telloh (Lagaš)
Eannatum
Paris, Louvre, AO 50, 2346, 2347, 2348, 16109
1.88 m
Strommenger 1962, pls. 66, 68
Börker-Klähn 1982, p. 124, no. 17
Parrot 1983, pl. 160

3
"Standard of Ur"
Ur, PG 779
First half of 3rd millennium B.C.
London, British Museum
0.203 x 0.483 m
Woolley 1934, pls. 90-93
Parrot 1983, pls. 174-175
Strommenger 1962, pl. XI

4
Shell inlay (L' "Étendard" de Mari)
Mari, Temple of Ishtar
Fara / Ur I Periode (2900—2460 B.C.)
Paris, Louvre, AO 18215
9.9 cm
Parrot 1956, pl. LV
Strommenger 1962, pl. 74

5
Shell inlay (L' "Étendard" de Mari)
Mari, Temple of Ishtar
Fara / Ur I Periode (2900—2460 B.C.)
Paris, Louvre, AO 19820
9.9 cm
Parrot 1956, pl. LVI

Shell inlay
Mari, Nini-Zaza temple, Room/Court 13
Early Dynastic IIIa
Damascus, Archaeological Museum
Bänder 1995, pls. XIX, XX

Shell inlay
Mari, Nini-Zaza temple, Room 12
Early Dynastic IIIa
Damascus, Archaeological Museum
Bänder 1995, pl. XXI

Incised slab
Mari, Presargonic Palace, Room 46
Early Dynastic IIIb
Damascus, Archaeological Museum
Bänder 1995, pl. XXIV:a

Shell inlay
Khafajeh
Early Dynastic IIIa
Chicago, Oriental Institute Museum, A11461
Bänder 1995, pl. XIV:c

6
Shell inlay
Mari, Presargonic palace, Rooms 48/49 and 49/52
First half of 3rd millennium B.C.
Damascus, Archaeological Museum
10 cm
Parrot 1983, pl. 169
Bänder 1995, pls. XV, XVI

Shell inlay
Mari, Shamash temple
Early Dynastic IIIb
Damascus, Archaeological Museum
Bänder 1995, pls. XIVa

7
"Stele of Naram-Sin"
Susa, Acropolis
Narām-Sîn (2254—2218 B.C.)
Paris, Louvre, Sb 4
2 m
Strommenger 1962, pls. 122-123
Börker-Klähn 1982, pp. 134-136, no. 26
Parrot 1983, pl. 192

8
"Stele of Naram-Sin"
Susa, Acropolis
Narām-Sîn (2254—2218 B.C.)
Paris, Louvre, Sb 4
2 m
Parrot 1983, pl. 192
Strommenger 1962, pls. 122-123
Börker-Klähn 1982, pp. 134-136, no. 26

9
Stele
"Nasiriya-Gebiet"
2334—2154 B.C. (Dynasty of Agade)
Baghdad, Iraq Museum, IM 55639, IM 59205
Boston, Museum of Fine Arts, 66.893
Strommenger 1962, pl. 119
Börker-Klähn 1982, pp. 131-133, no. 22
Parrot 1983, pl. 197

Fragments of a stele
Susa
2334—2154 B.C. (Dynasty of Agade)
Paris, Louvre, Sb 6641, 6641bis
Börker-Klähn 1982, p. 133, no. 23.24

10
Stele
Telloh, Tell K
2415—2290 B.C.
Paris, Louvre, AO 2678, 2679, 1678
34 cm
Strommenger 1962, pl. 117
Börker-Klähn 1982, p. 130, no. 21
Bänder 1995, pls. XLII:a-b, XLIII

11
Shell inlay *tropaion*
Mari
18th century B.C.
Damascus, Archaeological Museum, Inv. no. 1915
Borchhardt 1972, p. 96, fig. 8, Kat. 21 I 4

12
Cylinder seal
"Anatolia"
18th–17th century B.C.

Paris, Louvre, AO 7223
Delaporte 1923, pl. 94, fig. 15
Rittig 1983, fig. 1

13
Cylinder seal
"Anatolia"
18th–17th century B.C.
Paris, Louvre, AO 7223
Delaporte 1923, pl. 94, fig. 15
Rittig 1983, fig. 1

14
Seal impression
Kültepe, Karum level II
18th–17th century B.C.
Özgüç 1965, pl. VIII, no. 24
Littauer - Crouwel 1979, fig. 28

15
Seal impression
"Anatolia"
18th–17th century B.C.
Paris, Louvre, AO 8306
Amiet 1969, figs. 1-2
Littauer - Crouwel 1979, fig. 29

16
Fragments of clay vessel with incised picture of a warrior
Boğazköy
15th–14th century B.C.
Bittel 1976a, fig. 2
Rittig 1983, fig. 3
Macqueen 1986^2, fig. 34

17
Vase painting (hydria)
Ugarit (Ras Shamra), South Palace, Room 219 W
LH III B/C (*c*. 1200 B.C.)
Inv. RS 64, 27.319
Schaeffer 1968, pl. VII

18
Wall-painting
Thebes TT 86 (Tomb of Menkheperre-Seneb)
Thutmose III (1479—1425 B.C.)
de G. Davies 1933, pls. V, VIII
Kendall 1981, fig. 2a

19
Wall-painting
Thebes TT 86 (Tomb of Menkheperre-Seneb)
Thutmose III (1479—1425 B.C.)
de G. Davies 1933, pls. V, VIII
Kendall 1981, fig. 2b

20
Wall-painting
Thebes TT 112 (Tomb of Amenmose)
Amenhotep II (1427—1401 B.C.)
de G. Davies 1933, pls. XXXIV, XXXV
Kendall 1981, fig. 3a

21
Wall-painting
Thebes TT 112 (Tomb of Amenmose)
Amenhotep II (1427—1401 B.C.)
de G. Davies 1933, pls. XXXIV, XXXV
Kendall 1981, fig. 3b

22
Wall-painting
Thebes TT 112 (Tomb of Amenmose)
Amenhotep II (1427—1401 B.C.)
de G. Davies 1933, pls. XXXIV, XXXV
Kendall 1981, fig. 3b

23
Wall-painting
Thebes TT 112 (Tomb of Amenmose)
Amenhotep II (1427—1401 B.C.)
de G. Davies 1933, pls. XXXIV, XXXV
Kendall 1981, fig. 3b

24
Wall-painting
Thebes TT 100 (Tomb of Rekhmire)
Amenhotep II (1427—1401 B.C.)
Davies 1953, vol. I, p. 38, vol. II, pl. XXXVII
Kendall 1981, fig. 5

25
Wall-painting
Thebes TT 96 (Tomb of Sennufer)
Amenhotep II (1427—1401 B.C.)
Kendall 1981, fig. 6

26
Wall-painting
Thebes TT 93 (Tomb of Kenamun)
Amenhotep II (1427—1401 B.C.)
Davies 1930, p. 27, pls. XVI, XXIV
Kendall 1981, fig. 7

27
Bronze relief on the chariot of Thutmose IV
Thebes
Thutmose IV (1401—1391 B.C.)
Cairo, National Museum
Yadin 1963, pp. 192-193
Kendall 1981, fig. 8c

28
Bronze relief on the chariot of Thutmose IV
Thebes
Thutmose IV (1401—1391 B.C.)
Cairo, National Museum
Yadin 1963, pp. 192-193
Kendall 1981, fig. 8c

29
Bronze relief on the chariot of Thutmose IV
Thebes
Thutmose IV (1401—1391 B.C.)
Cairo, National Museum
Yadin 1963, pp. 192-193
Kendall 1981, fig. 8c

30
Bronze relief on the chariot of Thutmose IV
Thebes
Thutmose IV (1401—1391 B.C.)
Cairo, National Museum
Yadin 1963, pp. 192-193
Kendall 1981, fig. 8c

31
Bronze relief on the chariot of Thutmose IV
Thebes
Thutmose IV (1401—1391 B.C.)
Cairo, National Museum
Yadin 1963, pp. 192-193
Kendall 1981, fig. 8d

32
Bronze relief on the chariot of Thutmose IV
Thebes
Thutmose IV (1401—1391 B.C.)
Cairo, National Museum
Yadin 1963, pp. 192-193

Kendall 1981, fig. 8d

33
Bronze relief on the chariot of Thutmose IV
Thebes
Thutmose IV (1401—1391 B.C.)
Cairo, National Museum
Yadin 1963, pp. 192-193
Kendall 1981, fig. 8b

34
Bronze relief on the chariot of Thutmose IV
Thebes
Thutmose IV (1401—1391 B.C.)
Cairo, National Museum
Yadin 1963, pp. 192-193
Kendall 1981, fig. 8a

35
Bronze relief on the chariot of Thutmose IV
Thebes
Thutmose IV (1401—1391 B.C.)
Cairo, National Museum
Yadin 1963, pp. 192-193
Kendall 1981, fig. 8e

36
Bronze relief on the chariot of Thutmose IV
Thebes
Thutmose IV (1401—1391 B.C.)
Cairo, National Museum
Yadin 1963, pp. 192-193
Kendall 1981, fig. 8e

37
Sculpture from the "King's Gate" at Boğazköy
Boğazköy
14th century B.C.
Akurgal 1961, Taf. 64-65

38
Cylinder seal
Anatolia
14th century B.C.
Boston, Tyszkiewicz Collection
4.5 inch
Terrace 1962, fig. 12

39
Cylinder seal
Anatolia
14th century B.C.
Boston, Tyszkiewicz Collection
Terrace 1962, fig. 12

40
Bronze statuette of "Hittite style" representing a warrior
Syria (?)
Second half of the 2nd millennium B.C.(?)
Damascus, National Museum, S.O. 112
11.6 cm
Joundi 1980, p. 36, fig. 14
Seeden 1980, p. 115, no. 1750, pl. 107

41
Cylinder seal impression
Ashur
Ashur-nerari II (1424—1418 B.C.)
Berlin, VAT 8951 = Ass 14446n = KAJ 177
Beran 1957, fig. 1

42
Stele of Ba'al
Ugarit (Ras Shamra)
15th–13th century B.C.
Paris, Louvre, AO 15775
144 cm
Schaeffer 1949, pl. 23
Frankfort 1963^{3}, pl. 141
Börker-Klähn 1982, pp. 238-239, no. 284

43
Bronze statuette of Ba'al
Ugarit (Ras Shamra)
14th century B.C.
Brochhardt 1972, p. 91, pl. 12:2
Seeden 1980, no. 1694, pl. 50

44
Faience rhyton
Kition
13th century B.C. (Late Cypriot II)
Sandars 1978, fig. 19

45
Bronze tripod (helmet of a sphinx)
Cyprus
12th century B.C.
Hencken 1971, fig. 5b

46
Bronze model of a chariot
Ltshashen (tumulus)
14th–13th century B.C.
Esaian 1986, p. 18, pl. 9:1

47
Bronze model of a chariot
Ltshashen (tumulus)
14th–13th century B.C.
Esaian 1986, p. 18, pl. 9:2

48
Bronze model of a chariot
Ltshashen (tumulus)
14th–13th century B.C.
Esaian 1986, p. 18, pl. 9:3

49
Bronze model of a chariot
Ltshashen (tumulus)
14th–13th century B.C.
Esaian 1986, p. 18, pl. 9:4

50
Bronze model of a chariot
Lori-Berd (tumulus)
14th–13th century B.C.
Esaian 1986, p. 20, pl. 11

51
Bronze model of a chariot
Lori-Berd (tumulus)
14th–13th century B.C.
Esaian 1986, p. 21, pl. 9:5

52
Bronze model of a chariot
Shirakavan
13th–11th century B.C.
Esaian 1986, p. 21, pl. 9:6

53
Bronze statuette
Khurvin
11th century B.C.(?)
Vanden Berghe 1959, p. 124, pls. 156b-c

54
Bronze belt

Asthi-Blur
11th–10th century B.C.
Esaian 1976, fig. 157

55
Relief representation of a Shardana
Thebes, Luxor, 1st pylon
Ramesses II (1290—1224 B.C.)
Borchhardt 1972, p. 113, Kat. 25 I 1

56
Relief representation of a Shardana
Thebes, Luxor, 1st pylon
Ramesses II (1290—1224 B.C.)
Borchhardt 1972, p. 113, Kat. 25 I 2
Abydos
Ramesses II (1290—1224 B.C.)
Borchhardt 1972, p. 113, Kat. 25 I 3
Medinet-Habu
Ramesses III (1194—1163 B.C.)
Medinet-Habu II, pls. 70, 72
Borchhardt 1972, p. 113, Kat. 25 I 6

57
Relief representation of a Shardana
Medinet-Habu
Ramesses III (1194—1163 B.C.)
Medinet-Habu II, pl. 94
Borchhardt 1972, p. 113, pl. 13:2, Kat. 25 I 4

58
Relief representation of a Shardana
Medinet-Habu
Ramesses III (1194—1163 B.C.)
Medinet-Habu I, pl. 17
Borchhardt 1972, p. 113, Kat. 25 I 5

59
Relief representation of a Shardana
Medinet-Habu
Ramesses III (1194—1163 B.C.)
Medinet-Habu I, pl. 34
Borchhardt 1972, p. 113, Kat. 25 I 7

60
Relief representation of a Shardana
Medinet-Habu
Ramesses III (1194—1163 B.C.)
Medinet-Habu II, pl. 65c
Borchhardt 1972, p. 114, pl. 14:1, Kat. 25 II 1

61
Relief representation of a Shardana
Medinet-Habu
Ramesses III (1194—1163 B.C.)
Medinet-Habu I, pl. 35
Borchhardt 1972, p. 114, pl. 14:3, Kat. 25 II 2
Medinet-Habu
Ramesses III (1194—1163 B.C.)
Medinet-Habu I, pl. 34
Borchhardt 1972, p. 115, Kat. 25 III 7

62
Relief representation of a Shardana
Medinet-Habu
Ramesses III (1194—1163 B.C.)
Borchhardt 1972, p. 114, pl. 16:1, Kat. 25 III 1

63
Relief representation of a Shardana
Medinet-Habu
Ramesses III (1194—1163 B.C.)
Medinet-Habu I, pl. 39
Borchhardt 1972, p. 114, Kat. 25 III 2

64
Relief representation of a Shardana
Medinet-Habu
Ramesses III (1194—1163 B.C.)
Medinet-Habu I, pl. 39
Borchhardt 1972, p. 114, pl. 13:1, Kat. 25 III 3

65
Relief representation of a Shardana
Medinet-Habu
Ramesses III (1194—1163 B.C.)
Medinet-Habu I, pl. 39
Borchhardt 1972, p. 114, Kat. 25 III 4

66
Relief representation of a Shardana
Medinet-Habu
Ramesses III (1194—1163 B.C.)
Medinet-Habu I, pl. 39
Borchhardt 1972, p. 114, Kat. 25 III 5

67
Relief representation of a Shardana
Medinet-Habu
Ramesses III (1194—1163 B.C.)
Medinet-Habu I, pl. 38
Borchhardt 1972, p. 115, pl. 14:2, Kat. 25 III 6

68
Krater
Ugarit (Ras Shamra)
SH III A (1375—1300 B.C.)
Damascus, National Museum, Inv. RS. 60, 23.253
Borchhardt 1972, p. 118, pl. 16:4, Kat. 26 IV

69
Relief representation of a Philistine
Medinet-Habu
Ramesses III (1194—1163 B.C.)
Borchhardt 1972, p. 118, pl. 13:1, Kat. 26 I

70
Relief representation of a Philistine
Medinet-Habu
Ramesses III (1194—1163 B.C.)
Borchhardt 1972, p. 118, pl. 16:6, Kat. 26 I

71
Stamp-seal
Enkomi, Tempel, Raum X
12th century B.C.
Borchhardt 1972, p. 118, pl. 16:2, Kat. 26 II 1

72
Ivory game-box
Enkomi, Tomb 58
12th century B.C.
Borchhardt 1972, p. 118, pls. 17:1,3 Kat. 26 II 2

73
Anthropomorph sarkophag
Beth-Shan
12th century B.C.
Borchhardt 1972, p. 118, pl. 17:2, Kat. 26 IV

74
Bronze statuette of an ingot-god
Enkomi
12th century B.C.
H.: 35 cm
Sandars 1978, figs. 112, 129, pl. VIII
Seeden 1980, no. 1794, pl. 112

75
White Obelisk

Nineveh
Ashur-nasir-pal I (1050—1032 B.C.)
London, British Museum, WA 118807
H.: 290 cm
Börker-Klähn 1982, pp. 179-180, no. 132, pls. 132a-d

76
Palace relief
Kalhu (Nimrud)
Ashur-nasir-pal II (883—859 B.C.)
London, British Museum, N.G. no. 7b
Budge 1914, pl. XXI:1

77
Palace relief
Kalhu (Nimrud)
Ashur-nasir-pal II (883—859 B.C.)
London, British Museum, N.G. no. 4a
Budge 1914, pls. XII:2, XIII:1-2, XVIII:1, XX:2

78
Palace relief
Kalhu (Nimrud)
Ashur-nasir-pal II (883—859 B.C.)
London, British Museum, N.G. no. 10a
Budge 1914, pls. XV:2, XXIV:1-2

79
Palace relief
Kalhu (Nimrud)
Ashur-nasir-pal II (883—859 B.C.)
London, British Museum, N.G. no. 15b
Budge 1914, pl. XVIII:2

80
Palace relief
Kalhu (Nimrud)
Ashur-nasir-pal II (883—859 B.C.)
London, British Museum, N.G. no. 1b
Budge 1914, pl. XXIII:2

81
Palace relief
Kalhu (Nimrud)
Ashur-nasir-pal II (883—859 B.C.)
London, British Museum, N.G. no. 3a
Budge 1914, pls. XII:1, XIII:1, XVII

82
Palace relief
Kalhu (Nimrud)
Ashur-nasir-pal II (883—859 B.C.)
London, British Museum, N.G. no. 16b(2)
Budge 1914, pl. XXV:2

83
Palace relief
Kalhu (Nimrud)
Ashur-nasir-pal II (883—859 B.C.)
London, British Museum, N.G. no. 9a
Budge 1914, pls. XV:1, XXI:2

84
Palace relief
Kalhu (Nimrud)
Ashur-nasir-pal II (883—859 B.C.)
London, British Museum, N.G. no. 9a
Budge 1914, pls. XV:1, XVII:1-2, XVIII:2, XXI:1, XXIII:2, XXV:2

85
Palace relief
Kalhu (Nimrud)
Ashur-nasir-pal II (883—859 B.C.)
London, British Museum, N.G. no. 14b
Budge 1914, pl. XXIV:1

86
Palace relief
Kalhu (Nimrud)
Ashur-nasir-pal II (883—859 B.C.)
London, British Museum, N.G. no. 15b
Budge 1914, pls. XVIII:2, XXIV:1

87
Palace relief
Kalhu (Nimrud)
Ashur-nasir-pal II (883—859 B.)
London, British Museum, N.G. no. 9a
Budge 1914, pl. XV:1

88
Fragments of a coloured brick orthostat
Ashur
9th century B.C. (Shalmaneser III?)
Ass. 10756
Andrae 1925, pl. 9e

89
Throne base, panel f
Nimrud, Fort Shalmaneser, great throne room T1
Shalmaneser III (858—824 B.C.), Mushallim-Marduk
Baghdad, Iraq Museum, ND 11000
Mallowan 1966, vol. 2, pp. 446-448

90
Bronze statuette
Beqaa (Lebanon)
10th–9th (?) century B.C.
Beirut, National Museum
Seeden 1980, p. 109, no. 1723, pl. 102

91
Bronze model of a chariot
"Tartous"
10th–9th (?) century B.C.
Paris, Louvre, AO 22265
Collon – Crouwel – Littauer 1976, pls. XIA-B
Seeden 1980, pp. 109-110, no. 1725, pls. N and 103

92
Reliefs from bronze gate
Balawat
Shalmaneser III (858—824 B.C.)
London, British Museum
King 1915, pl. IX, band II.3, lower register

93
Reliefs from bronze gate
Balawat
Shalmaneser III (858—824 B.C.)
London, British Museum
King 1915, pl. III, band I.3, lower register

94
Reliefs from bronze gate
Balawat
Shalmaneser III (858—824 B.C.)
London, British Museum
King 1915, pl. III, band I.3, lower register

95
Reliefs from bronze gate
Balawat
Shalmaneser III (858—824 B.C.)
London, British Museum
King 1915, pls. XIV, band III.2, lower register
XXIV, band IV.6, lower register

96

Ivory
Kalhu (Nimrud), NW Palace, ZT. 25
Late 9th – early 8th century B.C.
ND 3266
Mallowan – Davies 1970, pl. VI, no. 14

97
Ivory
Kalhu (Nimrud), Nabu Temple, Throneroom (SEB II)
Late 9th – early 8th century B.C.
ND 5612
Mallowan – Davies 1970, pl. XVIII, no. 62

98
Ivory
Kalhu (Nimrud), Nabu Temple, Throneroom (SEB II)
Late 9th – early 8th century B.C.
ND 5612
Mallowan – Davies 1970, pl. XVIII, no. 62

99
Relief slab (basalt orthostat)
Zinçirli, "Südtor der Stadt"
10th–9th century B.C.
Berlin, Staatliche Museen zu Berlin
H.: 128 cm
Bittel 1976b, fig. 297

100
Relief slab (basalt orthostat)
Tell Halaf
9th century B.C.
Berlin, Staatliche Museen zu Berlin
Klengel-Brandt 1986, pl. on the back cover

101
Relief slab (limestone)
Carchemish, Back wall of the Processional Entry, (King's Gate), no. 12
Katuwash (*c.* 900—873 B.C.)
Ankara, Archaeological Museum, Inv. no. 9651
H.: 90 cm
Woolley 1921, pl. B.26c
Orthmann 1971, p. 512, H/12, pl. 34c

102
Relief slab (limestone)
Carchemish, King's Gate
Katuwash (*c.* 900—873 B.C.)
Ankara, Archaeological Museum, Inv. no. 115
H.: 130 cm
Hogarth 1914, pl. B.2a
Orthmann 1971, p. 505, F/1, pl. 28e

Relief slab (limestone)
Carchemish, King's Gate
Katuwash (*c.* 900—873 B.C.)
Ankara, Archaeological Museum, Inv. no. 117
H.: 130 cm
Hogarth 1914, pl. B.3a
Orthmann 1971, p. 507, F/3, pl. 29a

Relief slab
Carchemish, King's Gate
Katuwash (*c.* 900—873 B.C.)
Ankara, Archaeological Museum, Inv, no. 9664
H.: 135 cm
Hogarth 1914, pl. B.3b
Orthmann 1971, p. 507, F/4, pl. 29b

Relief slab (limestone)
Carchemish, Long wall of sculptures
Suhish II
H.: 157 cm
Woolley – Barnett 1952, pl. B.44b
Orthmann 1971, p. 503, C/11, pl. 25b

103
Relief slab
Carchemish, King's Gate
Katuwash (*c.* 900—873 B.C.)
Ankara, Archaeological Museum, Inv. no. 116
H.: 135 cm
Hogarth 1914, pl. B.2b
Orthmann 1971, pp. 505-507, F/2, pl. 28f, 71c

104
Relief slab
Carchemish, King's Gate
Katuwash (*c.* 900—873 B.C.)
Ankara, Archaeological Museum, Inv. no. 117
H.: 130 cm
Hogarth 1914, pl. B.3a
Orthmann 1971, p. 507, F/3, pl. 29a

Relief slab
Carchemish, King's Gate
Ankara, Archaeological Museum, Inv. no. 9664
H.: 135 cm
Hogarth 1914, pl. B.3b
Orthmann 1971, p. 507, F/4, pl. 29b

Relief slab
Carchemish, Long wall of sculptures
Suhish II
H.: 165 cm
Woolley – Barnett 1952, pl. B.46a
Orthmann 1971, p. 503, C/14, pl. 25d

105
Relief slab
Kanish (Kültepe)
9th century B.C.
Kayseri, Archaeological Museum
H.: 92 cm
Bittel 1976b, fig. 321

106
Ivory
Kalhu (Nimrud), NW Palace, Room B
9th century B.C.
New York, Metropolitan Museum of Art, ND 1715a
Mallowan 1966, p. 62, fig. 25
Mallowan – Davies 1970, p. 26, no. 55, pl. XVI

107
Relief from bronze gate
Balawat
Shalmaneser III (858—824 B.C.)
London, British Museum
King 1915, pl. III, band I.3, lower register

108
Relief from bronze gate
Balawat
Shalmaneser III (858—824 B.C.)
London, British Museum
King 1915, pls. IV, band I.4, lower register; VII, band II.1, lower register; IX, band II.3, upper register

109
Relief from bronze gate
Balawat
Shalmaneser III (858—824 B.C.)
London, British Museum
King 1915, pls. X, band II.4, upper and lower registers; XI, band II.5, upper registers; XII, band II.6, upper register

110
Relief from bronze gate

Balawat
Shalmaneser III (858—824 B.C.)
London, British Museum
King 1915, pls. XXXVII, band VII.1, upper register; XXXVIII, band VII.2, upper register; XXXIX, band VII.3, upper register; XL, band VII.4, upper register; XLI, band VII.5, upper register; XLII, band VII.6, upper register

111
Bronze pectoral of a horse
"Urartu" - "NW Iran"
Late 9th century B.C.
Berlin, Sammlung Axel Guttmann, AG 230
Seidl 1986, fig. 2
Born – Seidl 1995, Abbs. 45-53

112
Bronze pectoral of a horse
"Urartu" - "NW Iran"
Late 9th century B.C.
Berlin, Sammlung Axel Guttmann, AG 230
Seidl 1986, fig. 1
Born – Seidl 1995, Abbs. 45-53

113
Bronze statuette of a warrior
Toprak Kale
Late 9th century B.C.
London, British Museum, BM 121177
Barnett 1954, pl. II,A-C

114
Bronze statuette
Maraş
Late 9th century B.C.
Louvre (from Pereti, Collection)
Barnett 1954, fig. 7

115
Bronzeplate (belt fragment?)
"Urartu"
Early 8th century B.C.
Adana Museum, no. 1466
W.: 9.1 cm; L.: 8.2 cm
Kellner 1991, no. 447, pl. 88

116
Bronze pectoral of a horse
"Urartu" - "NW Iran"
Late 9th century B.C.
Berlin, Sammlung Axel Guttmann, AG 230
Seidl 1986, fig. 3
Born – Seidl 1995, Abbs. 45-53

117
Bronze pectoral of a horse
"Urartu" - "NW Iran"
Late 9th century B.C.
Berlin, Sammlung Axel Guttmann, AG 230
Seidl 1986, fig. 3
Born – Seidl 1995, Abbs. 45-53

118
Wall-painting
Erebuni
Argishti I (789—766 B.C.)
Hodjash *et al.* 1979, pl. 40
Esaian 1986, p. 22, pl. 9:9

119
Wall-painting
Erebuni
8th century B.C.
Hodjash *et al.* 1979, pl. 46

120
Bronze model of a chariot
Paravakar
8th century B.C.
Esaian 1976, p. 238, fig. 148:1
Esaian 1986, p. 23, pl. 9:10

121
Bronze statuette of a warrior
Paravakar
8th century B.C.
Esaian 1976, p. 238, fig. 148:2
Esaian 1986, p. 24, pls. 2:1, 9:11

122
Bronze statuette of a warrior
Paravakar
8th century B.C.
Esaian 1976, p. 238, fig. 148:3
Esaian 1986, p. 24, pls, 2:2, 9:12

123
Bronze statuette of a warrior
Paravakar
8th century B.C.
Esaian 1976, p. 238, fig. 149:2
Esaian 1986, p. 24, pl. 2:4

124
Bronze belt
Hatshbulag (Azerbaijan)
9th–8th century B.C.
Kesamanli 1966, fig. 4

125
Bronze breastplate of a horse
Hasanlu
Late 9th century B.C.
Tehran, Musée Iran Bastan, HAS 74-241
Winter 1980, figs. 1, 17, 18, 21, 24, foldout plate

126
Rock relief
Eski Doğubayazit
8th century B.C.
Huff 1968, p. 65, fig. 2

127
Basalt orthostat
Zinçirli, outer front of the Western Wall of the Acropolis
First half of 8th century B.C.
Istanbul, Archaeological Museum, Inv. no. 7725
H.: 140 cm
von Luschan *et al.* 1902, fig. 102, pl. XXXIX
Maxwell-Hyslop 1959, pl. III
Orthmann 1971, p. 538, pl. 57a

128
Basalt orthostat
Zinçirli, Eastern Wall of the Outer Gate of the Acropolis
First half of 8th century B.C.
Berlin, Vorderasiatisches Museum, VA 2711
H.: 100 cm
von Luschan *et al.* 1902, figs. 122-123, pl. XXXVIIIc
Orthmann 1971, p. 542, pl. 59g

129
Basalt orthostat
Zinçirli, Eastern Wall of the Outer Gate of the Acropolis
First half of 8th century B.C.
Berlin, Vorderasiatisches Museum, VA 2713
H.: 90 cm
von Luschan *et al.* 1902, fig. 135, pl. XXXVIIIb
Orthmann 1971, p. 543, pl. 61b

130
Basalt orthostat
Zinçirli, Southern Gate of the City
First half of 8th century B.C.
Berlin, Vorderasiatisches Museum
H.: 130 cm
von Luschan *et al.* 1902, fig. 96, pls. XXXIVc, XXXV
Orthmann 1971, pp. 537-538, pl. 55b

Basalt orthostat
Zinçirli, Western Walls of the Outer Walls of the Acropolis
First half of 8th century B.C.
Istanbul, Archaeological Museum, Inv. no. 7780
H.: 116 cm
von Luschan *et al.* 1902, fig. 130, pl. XLIV
Orthmann 1971, p. 543, pl. 60c

131
Basalt orthostat
Zinçirli, outer front of the Eastern Walls of the Gate of Acropolis
First half of 8th century B.C.
Berlin, Vorderasiatisches Museum, VA 2647
H.: 137 cm
von Luschan *et al.* 1902, fig. 103, pl. XL
Orthmann 1971, p. 541, pl. 58f

132
Ivory
Kalhu (Nimrud), Ezida: Temple of Nabu, N.T. 14
First half of 8th century B.C.
ND 5395
Mallowan – Davies 1970, pl. XVII, no. 61a

133
Ivory
Kalhu (Nimrud), The Burnt Palace
First half of 8th century B.C.?
ND 1605
Mallowan 1966, p. 216, fig. 165

134
Palace reliefs
Kalhu (Nimrud), SW Palace
Tiglath-Pileser III (745—727 B.C.)
London, British Museum, BM 118908
Smith 1938, pl. IX
Barnett – Falkner 1962, pls. LXIX, LXX
London, British Museum, BM 118934
Smith 1938, pl. X

Palace reliefs
Kalhu (Nimrud), Central Palace
Tiglath-Pileser III (745—727 B.C.)
London, British Museum
Barnett – Falkner 1962, pls. IX, X, XIII, XIV (BM 118878 = Smith 1938, pl. XIX), XXVII, XXVIII (Istanbul, Inv. no. 23), XXXI, XXXII (Bombay, Inv. no. 7), LII (Leningrad, Hermitage, Inv. no. 3943), LIV, LV (Paris, de Clerq Coll.),
with unknown provenance:
Barnett – Falkner 1962, pls. LXXII (Zürich, Inv. no. 1918), LXXIII (Zürich, Inv. no. 1917), LXXIV (Berlin, VA 984), LXXV (Louvre, AO 19854), LXXVI (Bristol, H.3175)

135
Palace relief
Kalhu (Nimrud), "Between Central Palace and Upper Chambers"
Tiglath-Pileser III (745—727 B.C.)
London, British Museum, Or. Dr. I, pl. XIII
Barnett – Falkner 1962, pl. LVIII

Palace relief
Kalhu (Nimrud)
Tiglath-Pileser III (745—727 B.C.)
Private collection
Barnett – Falkner 1962, pl. LXXVII

136
Palace relief
Kalhu (Nimrud), Central Palace
Tiglath-Pileser III (745—727 B.C.)
London, British Museum, BM 115634
Smith 1938, pl. XIII
Barnett – Falkner 1962, pl. XXXVIII

137
Palace relief
Kalhu (Nimrud), Central Palace
Tiglath-Pileser III (745—727 B.C.)
London, British Museum, BM 118903
Smith 1938, pl. XIV
Barnett – Falkner 1962, pls. XXXIX, XL, XLVIII, XLIX

138
Palace relief
Kalhu (Nimrud), SW Palace
Tiglath-Pileser III (745—727 B.C.)
London, British Museum, BM 118934
Smith 1938, pl. X

139
Palace reliefs
Kalhu (Nimrud), Central Palace
Tiglath-Pileser III (745—727 B.C.)
Bombay, V.A.M. 78
Barnett – Falkner 1962, pl. XVI (= pl. XV, Or. Dr. III, Central XI)
Paris, De Clerq Collection
Barnett – Falkner 1962, pl. LV (= pl. LIV, Or. Dr. III, Central XXIV)
London, British Museum, BM 118934+118931
Barnett – Falkner 1962, pls. XCII, XCIII

140
Palace relief
Kalhu (Nimrud), "Between Central Palace and Upper Chambers"
Tiglath-Pileser III (745—727 B.C.)
London, British Museum, Or. Dr. I, pl. IX
Barnett – Falkner 1962, pl. LIX

141
Palace relief
Unknown provenance
Tiglath-Pileser III (745—727 B.C.)
Private collection
Barnett – Falkner 1962, pl. LXXVII

142
Palace relief
Kalhu (Nimrud), SW Palace
Tiglath-Pileser III (745—727 B.C.)
London, British Museum, BM 118907
Smith 1938, pl. XVIII
Barnett -- Flakner 1962, pls. LXVI (= Or. Dr. III, S.W. XV), LXVII

143
Palace reliefs
Kalhu (Nimrud), Central Palace
Tiglath-Pileser III (745—727 B.C.)
London, British Museum, BM 118904
Barnett – Falkner 1962, pl. XII, XI (= Or. Dr. III, Central IV)
London, British Museum, Or. Dr. III, Central VIII
Barnett – Falkner 1962, pl. IX

144
Palace reliefs
Kalhu (Nimrud), Central Palace
Tiglath-Pileser III (745—727 B.C.)
London, British Museum, BM 118904
Smith 1938, pl. XV,
Paris, Louvre, AO 19854
(unknown provenance)
Barnett -- Flakner 1962, pl. LXXV

145
Palace reliefs
Kalhu (Nimrud), Central Palace
Tiglath-Pileser III (745—727 B.C.)
London, British Museum, BM 115634
Smith 1938, pl. XIII
Barnett – Falkner 1962, pls. XXXVII, XXXVIII

146
Palace relief
Kalhu (Nimrud), Central Palace
Tiglath-Pileser III (745—727 B.C.)
London, British Museum, BM 115634
Barnett – Falkner 1962, pls. XXXVII (Or. Dr. III. Central XXVII), XXXVIII

147
Palace relief
Kalhu (Nimrud), Central Palace
Tiglath-Pileser III (745—727 B.C.)
Zürich, Archäologisches Institut der Universität, Inv. no. 1920
Barnett – Falkner 1962, pls. L (Or. Dr. III. Central XXVIII), LI

148
Palace relief
Kalhu (Nimrud), Central Palace
Tiglath-Pileser III (745—727 B.C.)
Zürich, Archäologisches Institut der Universität, Inv. no. 1916
Barnett – Falkner 1962, pls. XXXV (Or. Dr. III. Central I), XXXVI

149
Palace relief
Unknown provenance
Tiglath-Pileser III (745—727 B.C.)
Minneapolis, Institute of Arts
Barnett – Falkner 1962, pl. LXI

150
Palace relief
Kalhu (Nimrud), Central Palace
Tiglath-Pileser III (745—727 B.C.)
Leyden, Rijksmuseum van Oudheden, Inv. no. A.1952/12.1
Barnett – Falkner 1962, pls. XLI (Or. Dr. III. Central XXI), XLII

151
Palace relief
Unknown provenance
Tiglath-Pileser III (745—727 B.C.)
London, British Museum, BM 119402
Barnett – Falkner 1962, pl. LXXIX

152
Palace reliefs
Kalhu (Nimrud), SW Palace
Tiglath-Pileser III (745—727 B.C.)
London, British Museum, BM 118934
Smith 1938, pl. X
Barnett – Falkner 1962, pls. LXXXVIII, XC (Or. Dr. III. S.W. XVIII), XCI

153
Palace relief
Kalhu (Nimrud), SW Palace
Tiglath-Pileser III (745—727 B.C.)
London, British Museum, Or. Dr. III. S.W. XIX
Barnett – Falkner 1962, pl. LXII

154
Palace relief
Unknown provenance
Tiglath-Pileser III (745—727 B.C.)
Zürich, Archäologisches Institut der Universität, Inv. no. 1917
Barnett – Falkner 1962, pl. LXXIII

155
Palace relief
Kalhu (Nimrud), Central Palace
Tiglath-Pileser III (745—727 B.C.)
London, British Museum, BM 118902
Smith 1938, pl. XVI
Barnett – Falkner 1962, pls. XXXIII (Or. Dr. Central XIX), XXXIV

156
Palace relief
Kalhu (Nimrud), SW Palace
Tiglath-Pileser III (745—727 B.C.)
London, British Museum, BM 118905
Smith 1938, pl. XVII
Barnett – Falkner 1962, pls. LXIV (Or. Dr. III. S.W. IV), LXV

157
Palace relief
Kalhu (Nimrud), Central Palace
Tiglath-Pileser III (745—727 B.C.)
Zürich, Archäologisches Institut der Universität, Inv. no. 1920
Barnett – Falkner 1962, pls. L (Or. Dr. III, Central XXVIII), LI

158
Palace relief
Unknown provenance
Tiglath-Pileser III (745—727 B.C.)
Minneapolis, Institute of Arts
Barnett – Falkner 1962, pl. LXI

159
Palace relief
Kalhu (Nimrud), Central Palace
Tiglath-Pileser III (745—727 B.C.)
Leyden, Rijksmuseum van Oudheden, Inv. no. A.1952/12.1
Barnett – Falkner 1962, pls. XLI (Or. Dr. III. Central XXI), XLII

160
Coloured brick
Ashur, façade of the Temple of Ashur
Tiglath-Pileser III (745—727 B.C.)
Andrae 1925, pl. 6

161
Wall-painting
Nimrud, "Zentrum des Hügels"
Tiglath-Pileser III (745—727 B.C.)
H.: 0.31 m
W.: 0.225 m
Nunn 1988, pl. 124

162
Wall-painting
Nimrud, "Acropolis Palace"
Tiglath-Pileser III (745—727 B.C.)
H.: 0.17 m
W.: 0.10 m
Nunn 1988, pl. 123

163
Wall-painting
Til-Barsip, Room XXIV
Tiglath-Pileser III (745—727 B.C.)
Thureau-Dangin – Dunand 1936, pl. XLIX, upper register

164
Wall-painting
Til-Barsip, Room XXIV
Tiglath-Pileser III (745—727 B.C.)
Thureau-Dangin – Dunand 1936, pl. LI

165
Wall-painting
Til-Barsip, Room XXIV
Tiglath-Pileser III (745—727 B.C.)
Thureau-Dangin – Dunand 1936, pl. L

166
Wall-painting
Til-Barsip
Tiglath-Pileser III (745—727 B.C.)
Parrot 1961, fig. 339

167
Wall-painting
Til-Barsip
Tiglath-Pileser III (745—727 B.C.)
Thureau-Dangin – Dunand 1936, pl. XLIX, lower register

168
Wall-painting
Til-Barsip, Room XXIV
Tiglath-Pileser III (745—727 B.C.)
Thureau-Dangin – Dunand 1936, pl. XLIX

169
Wall-painting
Til-Barsip, Room XXIV
Tiglath-Pileser III (745—727 B.C.)
Thureau-Dangin – Dunand 1936, pl. LI

170
Wall-painting
Til-Barsip
Tiglath-Pileser III (745—727 B.C.)
Aleppo Museum
Amiet 1980, pl. 105

171
Wall-painting
Til-Barsip
Tiglath-Pileser III (745—727 B.C.)
Aleppo Museum
Amiet 1980, pl. 105

172
Wall-painting
Til-Barsip, Room XXIV
Tiglath-Pileser III (745—727 B.C.)
Thureau-Dangin – Dunand 1936, pl. XLIX, lower register

173
Reliefs
Hadatu (Arslan-Tash), Western gate
Tiglath-Pileser III (745—727 B.C.)
Istanbul, Archaeological Museum, Inv. no. 7-9
Thureau-Dangin *et al.* 1931, pls. IX:1, XI:1-2

174
Basalt relief
Karatepe, "Obere Grabung"
First half of the 8th century B.C ("Sph III")
H.: 135 cm
Bossert *et al.* 1950, pl. XIII, no. 64
Orthmann 1971, p. 494, pl. 18f

175
Basalt relief
Karatepe, "Obere Grabung"
First half of the 8th century B.C. ("Sph III")
H.: 135 cm
Bossert *et al.* 1950, pl. XIII, no. 64
Orthmann 1971, p. 494, pl. 18f

176
Basalt reliefs
Karatepe, "Rampe der Untere Grabung"
First half of the 8th century B.C. ("Sph. IIIb")
Bossert *et al.* 1950, pls. 82, 83
Orthmann 1971, p. 493, pl. 18a

177
Basalt relief
Karatepe, "Westwand der westliche Nebenkammer,Untere Grabung"
First half of the 8th century B.C. ("Sph. IIIb")
H.: 110 cm
Bossert *et al.* 1950, pl. XX, no. 100
Orthmann 1971, p. 491, pl. 16c

178
Basalt relief
Karatepe, "Obere Grabung"
First half of the 8th century B.C.
Bossert *et al.* 1950, pl. XIII, no. 69

179
Basalt reliefs
Karatepe, "Rampe der Untere Grabung"
First half of the 8th century B.C. ("Sph. IIIb")
Bossert *et al.* 1950, pl. XVI, nos. 82-83
Orthmann 1971, p. 493, pls. 18a

180
Basalt relief
Karatepe, "Obere Grabung"
First half of the 8th century B.C. ("Sph. IIIB")
Bossert *et al.* 1950, pl. XIII, no. 64
Orthmann 1971, p. 494, pl. 18f

181
Relief
Tell Ta'yinat
Ca. 750—725 B.C. (Amuq Phase O)
Chicago, Oriental Institute Museum, A.27854
Marfoe *et al.* 1982, no. 32

182
Palace relief
Khorsabad
Sargon II (721—705 B.C.)
Botta – Flandin 1849, pl. 49 (Salle 1, slab 2)

183
Palace relief
Khorsabad
Sargon II (721—705 B.C.)
Botta – Flandin 1849, pl. 76 (Salle 2, slab 1) and *passim*

184
Palace relief

Khorsabad
Sargon II (721—705 B.C.)
Botta – Flandin 1849, pl. 58 (Salle 2, slab 10)

185
Palace relief
Khorsabad
Sargon II (721—705 B.C.)
Botta – Flandin 1849, pls. 62-67 (Salle 11, slabs 15-21) and *passim*

186
Palace relief
Khorsabad
Sargon II (721—705 B.C.)
Botta – Flandin 1849, pl. 118 (Salle 8, slab 12)

187
Palace relief
Khorsabad
Sargon II (721—705 B.C.)
Botta – Flandin 1849, pl. 55 (Salle 11, slab 7)

188
Palace relief
Khorsabad
Sargon II (721—705 B.C.)
Botta – Flandin 1849, pl. 70 (Salle 11, slab 28)

189
Palace relief
Khorsabad
Sargon II (721—705 B.C.)
Botta – Flandin 1849, pl. 145 (Salle 14, slab 2)

190
Palace relief
Khorsabad
Sargon II (721—705 B.C.)
Botta – Flandin 1849, pl. 98 (Salle 5, slab 25)

191
Palace relief
Khorsabad
Sargon II (721—705 B.C.)
Botta – Flandin 1849, pls. 61-62 (Salle 11, slabs 14, 22)

192
Palace relief
Khorsabad
Sargon II (721—705 B.C.)
Botta – Flandin 1849, pl. 90 (Salle 5, slab 6)

193
Palace relief
Nineveh (Quyunjik, SW Palace, Room XIV)
Sennacherib (705—681 B.C.)
London, British Museum, BM 124787
Smith 1938, pl. XLIII

194
Palace relief
Nineveh (Quyunjik, SW Palace, Room XII)
Sennacherib (705—681 B.C.)
London, British Museum, BM 124783
Smith 1938, pl. XLIV

195
Palace relief
Nineveh (Quyunjik, SW Palace, Court VI)
Sennacherib (705—681 B.C.)
London, British Museum, BM 124773
Smith 1938, pl. XXXVII

Palace relief
Nineveh (Quyunjik, SW Palace, Court XIX)
Sennacherib (705—681 B.C.)
London, British Museum, BM 124782
Smith 1938, pl. XLVII

196
Palace relief
Nineveh (Quyunjik)
Sennacherib (705—681 B.C.)
Boston, Museum of Fine Arts
Terrace 1962, no. 22

197
Palace relief
Nineveh (Quyunjik, SW Palace)
Sennacherib (705—681 B.C.)
London, British Museum, BM 124789
Smith 1938, pl. XLI

198
Palace relief
Nineveh (Quyunjik, SW Palace, Room XIV)
Sennacherib (705—681 B.C.)
London, British Museum, BM 124787
Smith 1938, pl. XLIII

199
Palace relief
Nineveh (Quyunjik, SW Palace, Room XII)
Sennacherib (705—681 B.C.)
London, British Museum, BM 124783
Smith 1938, pl. XLIV

200
Palace relief
Nineveh (Quyunjik, SW Palace, Court LXIV)
Sennacherib (705—681 B.C.)
Paterson 1912, pl. 91

201
Palace relief
Nineveh (Quyunjik, SW Palace, Court XIX)
Sennacherib (705—681 B.C.)
London, British Museum, BM 124782
Smith 1938, pl. XLVII

202
Palace relief
Nineveh (Quyunjik, SW Palace)
Sennacherib (705—681 B.C.)
London, British Museum, BM 124786
Smith 1938, pl. XLII

203
Palace relief
Nineveh (Quyunjik, SW Palace)
Sennacherib (705—681 B.C.)
London, British Museum, BM 124775
Smith 1938, pl. XXXVIII:1

Palace relief
Nineveh (Quyunjik, SW Palace, Room XLV)
Sennacherib (705—681 B.C.)
London, British Museum, BM 124777
Smith 1938, pl. XXXIX

204
Palace relief
Nineveh (Quyunjik, SW Palace, Room XIV)
Sennacherib (705—681 B.C.)
London, British Museum, BM 124784
Smith 1938, pl. LVII

205
Palace relief

Nineveh
Sennacherib (705—681 B.C.)
Madrid, Real Academia de la Historia
Reade 1972, pl. XXXVIIa

206
Palace relief
Nineveh (Quyunjik, SW Palace, Room XII)
Sennacherib (705—681 B.C.)
London, British Museum, BM 124780
Smith 1938, pl. XLVI

207
Palace relief
Nineveh (Quyunjik, SW Palace, Room V)
Sennacherib (705—681 B.C.)
Paterson 1912, pl. 12

208
Palace relief
Nineveh (Quyunjik, SW Palace, Room XIV)
Sennacherib (705—681 B.C.)
London, British Museum, BM 124784
Smith 1938, pl. LVII

209
Palace relief
Nineveh (Quyunjik, SW Palace, Room XII)
Sennacherib (705—681 B.C.)
London, British Museum, BM 124779
Smith 1938, pl. XLV

210
Palace relief
Nineveh
Sennacherib (705—681 B.C.)
London, British Museum, BM 124775
Smith 1938, pl. XXXVIII:1

Palace relief
Nineveh
Sennacherib (705—681 B.C.)
London, British Museum, BM 124776
Smith 1938, pl. XXXVIII:2

211
Palace relief
Nineveh (Quyunjik, SW Palace)
Sennacherib (705—681 B.C.)
London, British Museum, BM 124789
Smith 1938, pl. XLI

212
Palace relief
Nineveh (Quyunjik, SW Palace)
Sennacherib (705—681 B.C.)
Oxford, Ashmolean Museum, 1950.240
Moorey 1987, fig. 11

213
Palace relief
Nineveh (Quyunjik, SW Palace, Court XIX)
Sennacherib (705—681 B.C.)
London, British Museum, BM 124782
Smith 1938, pl. XLVII

214
Palace relief
Nineveh (Quyunjik, SW Palace, Room XIV)
Sennacherib (705—681 B.C.)
London, British Museum, BM 124786
Smith 1938, pl. XLII

215
Palace relief
Nineveh (Quyunjik, SW Palace, Room XIV)
Sennacherib (705-681 B.C.)
London, British Museum, BM 124784
Smith 1938, pl. LIX

Palace relief
Nineveh (Quyunjik, SW Palace, Room XIV)
Sennacherib (705—681 B.C.)
London, British Museum, BM 124785
Smith 1938, pls. LXIII-LXIV

216
Palace relief
Nineveh (Quyunjik, SW Palace, Room V)
Sennacherib (705—681 B.C.)
Paterson 1912, pl. 15

217
Palace relief
Nineveh (Quyunjik, SW Palace, Room XIV)
Sennacherib (705—681 B.C.)
Paterson 1912, pl. 39

218
Palace relief
Nineveh (Quyunjik, SW Palace, Room XIV)
Sennacherib (705—681 B.C.)
London, British Museum, BM 124784
Smith 1938, pl. LXII

Palace reliefs
Nineveh (Quyunjik, SW Palace, Room XXVIII)
Sennacherib (705—681 B.C.)
London, British Museum, BM 124774
Smith 1938, pls. L-LIII

219
Palace relief
Nineveh (Quyunjik, SW Palace, Room XIV)
Sennacherib (705—681 B.C.)
London, British Museum, BM 124786
Smith 1938, pl. XLII

220
Palace relief
Nineveh (Quyunjik, SW Palace)
Sennacherib (705—681 B.C.)
Oxford, Ashmolean Museum, 1933.1575
Moorey 1987, fig. 10

221
Palace relief
Nineveh (Quyunjik, SW Palace, Room V)
Sennacherib (705—681 B.C.)
Paterson 1912, pl. 14

222
Palace relief
Nineveh (Quyunjik, SW Palace, Great Hall)
Sennacherib (705—681 B.C.)
Paterson 1912, pl. 9

223
Palace relief
Nineveh
Sennacherib (705—681 B.C.)
Paterson 1912, pl. 7

224
Palace relief
Nineveh
Sennacherib (705—681 B.C.)
Paterson 1912, pl. 7

225

Palace relief
Nineveh (Quyunjik, SW Palace, Great Hall)
Sennacherib (705—681 B.C.)
Paterson 1912, pl. 7

226
Cylinder seal
Ashur-nadin-shumi (700—695 B.C.)
London, British Museum, BM 77611+2
Collon 1988, pp. 129-130, no. 555

227
Wall-painting
Nineveh
Sennacherib (705—681 B.C.) or Ashurbanipal (668—627 B.C.)
Excav. no. 1930-5.8-216
H.: 0.13 m
W.: 0.22 m
Nunn 1988, pl. 130

228
Stele A of Esarhaddon
Til-Barsip
Esarhaddon (680—669 B.C.)
Aleppo, Archaeological Museum
Thureau-Dangin - Dunand 1936, pp. 151-155 pl. XII

Stele B of Esarhaddon
Til-Barsip
Esarhaddon (680—669 B.C.)
Aleppo, Archaeological Museum, Inv. no. 31
Thureau-Dangin - Dunand 1936, pp. 155-156, pl. XIII

229
Palace relief
Nineveh
Esarhaddon (680—669 B.C.)
Venice, Museuo Archeologica, no. 46
Reade 1972, pl. XLb

230
Palace relief
Nineveh
Esarhaddon (680—669 B.C.)
Venice, Museo Archeologico, no. 46
Reade 1972, pl. XLb

231
Palace relief
Nineveh, North Palace, Room I
Ashurbanipal (668—627 B.C.)
London, British Museum, BM 124929-124930
Barnett 1976, pl. XVI and *passim*

Palace relief
Nineveh
Ashurbanipal (668—627 B.C.)
Oxford, Ashmolean Museum, 1940.202
Reade 1972, pl. XXXIXa

232
Palace relief
Nineveh, North Palace, Room V^1-T^1
Ashurbanipal (668—627 B.C.)
Paris, Louvre, AO 19909
Barnett 1976, pl. LXIX

233
Palace relief
Nineveh, North Palace, Room V^1-T^1
Ashurbanipal (668—627 B.C.)
Paris, Louvre, AO 19908, 19904
Barnett 1976, pl. LXVIII

234
Palace relief
Nineveh, North Palace
Ashurbanipal (668—627 B.C.)
Berlin, Vorderasiatisches Museum, VA 961
Barnett 1976, pl.LXXa

235
Palace relief
Nineveh, SW Palace, Room XXXIII
Ashurbanipal (668—627 B.C.)
London, British Museum
Amiet 1980, pl. 120

236
Palace relief
Nineveh, SW Palace, Room XXXIII
Ashurbanipal (668—627 B.C.)
London, British Museum
Barnett 1960, p. 126

237
Palace relief
Nineveh, SW Palace, Room XXXIII
Ashurbanipal (668—627 B.C.)
London, British Museum
Barnett 1960, p. 122

238
Palace relief
Nineveh, SW Palace, Room XXXIII
Ashurbanipal (668—627 B.C.)
London, British Museum
Amiet 1980, pl. 120

239
Palace relief
Nineveh, SW Palace, Room XXXIII
Ashurbanipal (668—627 B.C.)
London, British Museum
Barnett 1986, p. 123

240
Palace relief
Nineveh, SW Palace, Room XXXIII
Ashurbanipal (668—627 B.C.)
London, British Museum
Barnett 1960, p. 120

241
Palace relief
Nineveh, SW Palace, Room XXXIII
Ashurbanipal (668—627 B.C.)
London, British Museum
Amiet 1980, pl. 120

242
Palace relief
Nineveh, SW Palace, Room XXXIII
Ashurbanipal (668—627 B.C.)
London, British Museum
Amiet 1980, pl. 120

243
Palace relief
Nineveh, SW Palace, Room XXXIII
Ashurbanipal (668—627 B.C.)
London, British Museum
Barnett 1960, p. 131

244
Palace relief
Nineveh, North Palace
Ashurbanipal (668—627 B.C.)
London, Royal Geographical Society, Inv. no. 1

Barnett 1976, pl. LXXIm

245
Palace relief
Nineveh, North Palace
Ashurbanipal (668—627 B.C.)
London, British Museum, BM 124956

246
Palace relief
Nineveh, SW Palace, Room XXXIII
Ashurbanipal (668—627 B.C.)
London, British Museum
Barnett 1960, p. 124

247
Palace relief
Nineveh, North Palace, Room F
Ashurbanipal (668—627 B.C.)
London, British Museum, BM 124931-124932
Barnett 1976, pl. XVII

248
Palace relief
Nineveh, North Palace, Room C
Ashurbanipal (668—627 B.C.)
London, British Museum, BM 124860-124861
Barnett 1976, pl. VI

249
Palace relief
Nineveh, SW Palace, Room XXXIII
Ashurbanipal (668—627 B.C.)
London, British Museum
Barnett 1960, p. 119

250
Palace relief
Nineveh, North Palace, Room L
Ashurbanipal (668—627 B.C.)
London, British Museum, BM 124926
Barnett 1960, p. 110

251
Palace relief
Nineveh, North Palace, Room F
Ashurbanipal (668—627 B.C.)
London, British Museum, BM 124933-124934
Barnett 1976, pl. XVIII

252
Palace relief
Nineveh, SW Palace, Room XXXIII
Ashurbanipal (668—627 B.C.)
London, British Museum
Barnett 1960, p. 126

253
Palace relief
Nineveh, North Palace, Room C
Ashurbanipal (668—627 B.C.)
London, British Museum, BM 124861
Barnett 1976, pl. VI

254
Palace relief
Nineveh, SW Palace, Room XXXIII
Ashurbanipal (668—627 B.C.)
London, British Museum
Barnett 1960, p. 125

255
Palace relief
Nineveh, North Palace
Ashurbanipal (668—627 B.C.)
London, British Museum, BM 124957

256
Palace relief
Nineveh, North Palace, Room S^1
Ashurbanipal (668—627 B.C.)
London, British Museum, BM 124788
Barnett 1976, pl. LXVI

257
Rockrelief
Wadi Brisa (Lebanon)
18th year of Nebuchadrezzar II (586 B.C.)
Börker-Klähn 1982, pls. 259-260

258
Stele
Babylon
Nabonidus (555—539 B.C.)
London, British Museum, BM 90837
H.: 0.41 m
W.: 0.46 m
Börker-Klähn 1982, pl. 266

259
Stele
Harran
Nabonidus (555—539 B.C.)
Urfa Museum
H.: 1.98 m
W.: 0.97 m
Börker-Klähn 1982, pls. 263a-b

260
Stele
Harran
Nabonidus (555—539 B.C.)
Urfa Museum
H.: 1.87 m
W.: 1.01 m
Börker-Klähn 1982, pls. 264a-b

261
Stele
Tema-oasis
Nabonidus (555—539 B.C.)
Paris, Louvre, AO.1509
H.: 1.48 m
W.: 0.40-0.43 m
Börker-Klähn 1982, pl. 265

262
Rockrelief
Gebel Akrum (Lebanon)
Neo-Babylonian
Börker-Klähn 1982, pl. 269

263
Architectural terracotta
Gordion, 2801 A 129
First half of 7th century B.C.
Akerström 1966, pl. 79

264
Architectural terracotta
Pazarli
First half of 7th century B.C.
Ankara, Museum of Anatolian Civilizations
Akerström 1966, pl. 90

265
Architectural terracotta
Pazarli
First half of 7th century B.C.
Ankara, Museum of Anatolian Civilizations

Akerström 1966, pl. 90

266
Architectural terracotta
Pazarli
First half of 7th century B.C.
Ankara, Museum of Anatolian Civilizations
Akerström 1966, pl. 91

267
Architectural terracotta
Pazarli
First half of 7th century B.C.
Ankara, Museum of Anatolian Civilizations
Akerström 1966, pl. 91

M = Mediterranean
M.1 = Local Geometric helmets
M.2 = Conical helmets
M.3 = Conical crested helmets
M.4 = Conical/pointed helmets with crest curving forward
M.5 = Conical/pointed helmets with unstilted crest curving forward
M.6 = Conical/pointed helmetswith crescent-shaped crest
M.7 = Hemispherical crested helmets

M.1.1
Warrior with Dipylon shield and two spears
Vase painting from an Attic sherd
Late Geometric
Lorimer 1950, fig. 12

M.1.2
Warriors with or without Dipylon shield
Vase painting of an Athenian Geometric Vase
Late Geometric
Richter 1934, figs. 1-2

M.2.1
Terracotta chariot group
Ayia Irini, Cyprus
Nicosia, Cyprus Museum
Gjerstad *et al.* 1935, pl. CCXXXIV:5, no. 2000

M.2.2
Terracotta chariot group
Ayia Irini, Cyprus
Second half of Cypro-Archaic I
Gjerstad *et al.* 1935, pl. CCXXXIV:6, no. 1046

Terracotta chariot group
Ayia Irini, Cyprus
Second half of Cypro-Archaic I
Gjerstad *et al.* 1935, pl. CCXXXV:5, no. 1998

Terracotta warrior with shield
Ayia Irini, Cyprus
Second half of Cypro-Archaic I
Gjerstad *et al.* 1935, pl. CCXXXII:7, no. 1032

Terracotta head of a warrior
Ayia Irini, Cyprus
Second half of Cypro-Archaic I
Gjerstad *et al.* 1935, pl. CCXXXII:15, no. 2384

Terracotta horseman
"Cyprus"
Edinburgh, Royal Scottish Museum, 1921.354
Second half of Cypro-Archaic I
Crouwel – Tatton-Brown 1988, pl. XXVI:4

M.2.3
Terracotta chariot group
Ayia Irini, Cyprus
Second half of Cypro-Archaic I
Gjerstad *et al.* 1935, pl. CCXXXIV:6, no. 1046

Terracotta horseman
"Cyprus"
London, British Museum, 1876.4.9.92
Second half of Cypro-Archaic I
Crouwel – Tatton-Brown 1988, pl. XXVI:2

M.2.4
Terracotta chariot group
Ayia Irini, Cyprus
Nicosia, Cyprus Museum
Second half of Cypro-Archaic I
Gjerstad *et al.* 1935, pl. CCXXXV:3, no. 1781+798

Terracotta chariot group
Ayia Irini, Cyprus
Nicosia, Cyprus Museum
Second half of Cypro-Archaic I
Gjerstad *et al.* 1935, pl. CCXXXV:4, no. 1170

M.2.5
Terracotta warrior
"Cyprus"
Early 7th century B.C.
Heuzey 1880, fig. F

M.2.6
Terracotta warrior
"Cyprus"
Early 7th century B.C.
Heuzey 1880, fig. G

M.2.7
Terracotta warrior with spiked shield
Ayia Irini, Cyprus
Nicosia, Cyprus Museum
Second half of Cypro-Archaic I
Gjerstad *et al.* 1935, 721-722, pl. CXCIV, no. 1385+1530

M.2.8
Geryon terracotta
Pyrgia, Cyprus
H.: 24 cm, W.: 23 cm
London, British Museum, 1917.7-1.13
670/660—600 B.C.
Tatton-Brown 1979, pl. XXXII

M.2.9
Terracotta warrior
Tamassos, Cyprus
670/660—600 B.C.
Pflug 1988b, fig. 13

M.2.10
Terracotta chariot warrior
"Cyprus"
670/660—600 B.C.
Heuzey 1880, fig. H

M.2.11
Terracotta archer
Salamis, Cyprus
Nicosia, Cyprus Museum, Sal. 6471 = Tc. 2112
Second half of 7th century B.C.
Monloup 1984, 182-183, (fig.), pl. 33, no. 663

M.2.12
Terracotta archer
Salamis, Cyprus
Nicosia, Cyprus Museum, Sal. 3535 = Tc. 1038
Second half of 7th century B.C.
Monloup 1984, 184, (fig.), pl. 33, no. 665

M.2.13
Terracotta archer
Salamis, Cyprus
Nicosia, Cyprus Museum, Sal. 4992 = Tc. 1839
Second half of 7th century B.C.
Monloup 1984, 184, pl. 33, no. 666

M.2.14
Terracotta horseman
Salamis, Cyprus
Nicosia, Cyprus Museum, Sal. 5022 = Tc. 1909
Second half of 7th century B.C.
Monloup 1984, 184, no. 668, (fig.)

M.2.15
Terracotta warrior drawing a sword from scabbard
"Cyprus"
H.: 10.5 inch
Myres 1914, 157, no. 1049

M.2.16
Terracotta chariot group
Kourion, Sanctuary of Apollo Hylates
H.: 19.5 cm
University Museum, Philadelphia (?), T. 1700
"Close to the seventh century"
Young - Young 1955, pp. 55-56, no. 1055, pl. 60

M.2.17
Bronze statuette of Geryon
"Etruscan"?
de Chanot 1880, pl. 22

M.2.18
Terracotta head of a warrior
Idalion (?), Cyprus
Berlin, 499X
Middle of the 7th century B.C.
Schmidt 1968, pl. 44
Pflug 1988b, figs. 7-8

M.2.19
Terracotta head of a warrior
Samos
Samos Museum, T.2637
Middle of the 7th century B.C.
Schmidt 1968, pl. 44
Pflug 1988b, figs. 5-6

M.2.20
Terracotta head of a warrior
Samos
Samos Museum, T.2741
Middle of the7th century B.C.
Schmidt 1968, pls. 44, 46

M.2.21
Bronze shield
Crete, Zeus Cave, Mt. Ida
Late 8th – early 7th century B.C.
Iraklion, Archaeological Museum, Inv. no. 7
Kunze 1931, p. 8, fig. 1, pl. 14

M.2.22
Bronze bowl
Delphi
Late 8th century B.C.
Delphi, Archaeological Museum, Inv. no. 4463
Markoe 1985, pp. 205-206, 320-323, G4

M.2.23
Bronze bowl
Delphi
Late 8th century B.C.
Delphi, Archaeological Museum, Inv. no. 4463
Markoe 1985, pp. 205-206, 320-323, G4

M.2.24
Bronze bowl
"Olympia"
Oxford, Ashmolean Museum, G401
Late 8th century B.C.
Markoe 1985, pp. 207, 326-327, G7

M.2.25
Bronze statuette of a warrior
Olympia
Olympia Museum, Inv. B 4240
Late 8th – early 7th century B.C.

M.2.26
Bronze statuette of a warrior
Athens, Acropolis
Athens, National Museum, Inv. 6612
H.: 21.2 cm
Early 7th century B.C.
Schweitzer 1971, pls. 164-165
Lamb 1929, pl. 20a

M.2.27
Bronze statuette of a warrior
Delphi
Inv. 3649
H.: 13 cm
Early 7th century B.C.
Perdrizet 1908, 30, no. 16, pl. 1:4
Rolley 1969, 45, no. 28, pl. IX:28

Bronze statuette of a warrior
Delphi
H.: 18.5 cm
7th century B.C.
Perdrizet 1908, 32, no. 23, pl. 1:8

M.2.28
Bronze statuette
Tegea
H.: 6.5 cm
7th century B.C.
Dugas 1921, 355-356, no. 53, fig. 19

M.2.29
Bronze statuette
Thermon
Early 7th century B.C.
Lamb 1929, pl. 17a

M.2.30
Bronze statuette
Karditsa
Early 7th century B.C.
Lamb 1929, pl. 17b

M.2.31
Plate fibula: Herakles and Iolaios killing the Hydra
Philadelphia, private collection
H.: 14 cm
Early 7th century B.C.
Schweitzer 1969, fig. 119

M.2.32
Terracotta horseman with round shield and sword
"Cyprus"
Cesnola Collection
New York, MMA 74.51.1778
Myres 1914, 344, no. 2093

M.3.1
Bronze "belt"

Fortetsa, Tomb P, no. 1568
Middle of the 8th century B.C.
Iraklion, Archaeological Museum, Inv. no. 1568
Brock 1957, pl. 115
Borchhardt 1972, 70, Kat. 15 II 2

M.3.2
Bronze "belt"
Fortetsa, Tomb P, no. 1569
Middle of the 8th century B.C.
Iraklion, Archaeological Museum, Inv. no. 1569
Brock 1957, pl. 116
Borcchardt 1972, 69-70, Kat. 15 II 1, pl. 25:5

M.3.3
Bronze openwork
Kato-Symi (Symi-Biannou)
Middle of the 8th century B.C.
Lempesi 1985, pl. 57, G4

M.3.4
Gold band
Khaniale Tekke
Middle of the 8th century B.C.
Dunbabin 1944, 85
Borchhardt 1972, 70, Kat. 15 IV, pl. 25:3

M.3.5
Bronze plaque (lion hunter)
Kavousi
Middle of the 8th century B.C.
Boyd 1901, 147-148, figs. 10-11
Kunze 1931, 218, fig. 31, pl. 56e

M.4.1
Bronze belt
Fortetsa, Tomb P, no. 1569
Middle of the 8th century B.C.
Iraklion, Archaeological Museum, Inv. no. 1569
Brock 1957, pl. 116
Borchhardt 1972, 69-70, Kat. 15 II 1, pl. 27:1

M.4.2
Sphinx
Bronze plaque
Kavousi
Middle of the 8th century B.C.
Boyd 1901, 147-148, figs. 10-11
Kunze 1931, 218, fig. 31, pl. 56e

M.4.3
Warriors
Relief pithos (the Wooden Horse at Troy)
Mykonos
Around 675 B.C.
Ervin 1963, pls. 18, 20

M.4.4
Warriors
Relief pithos (the Wooden Horse at Troy)
Mykonos
Around 675 B.C.
Ervin 1963, pl. 18

M.4.5
Sphinx
Late Orientalizing alabastron
Fortetsa, Tomb P, no. 1299
Iraklion, Archaeological Museum, Inv. 1299
680—630 B.C.
Brock 1957, 111-112, pl. 101

M.4.6
Sphinx
Melian hydria
Rheneia
Middle of 7th century B.C.
Kunze 1930, pl. 55a

M.4.7
Warrior with small round shield
Bronze corslet
"Crete"
Hamburg, Museum für Kunst und Gewerbe, Inv. 1970, 26a
Middle of the 7th century B.C. / 640 B.C.
Hoffmann 1972, 7-8, pls. 19-23, esp. 21

M.4.8
Stele of a warrior (with hoplite equipment)
Prinias, Sidérospila
Iraklion, Archaeological Museum, ar. 399
H.: 1.06 m
670—660 B.C.
Lempesi 1976, B1, pls. 14-15

M.4.9
Stele of a warrior (with hoplite equipment)
Prinias, Sidérospila
Iraklion, Archaeological Museum, ar. 398
H.: 0.93 m
660—650 B.C.
Lempesi 1976, B2, pls. 16-17

M.4.10
Stele of a warrior (with hoplite equipment)
Prinias, Sidérospila
Iraklion, Archaeological Museum, ar. 402
H.: 0.4 m
660—650 B.C.
Lempesi 1976, B3, pls. 18-19

M.4.11
Stele of a warrior (with hoplite equipment)
Prinias, Sidérospila
Iraklion, Archaeological Museum, ar. 401
H.: 0.835 m
640 B.C.
Lempesi 1976, B4, pls. 20-21

M.4.12
Stele of a warrior (with hoplite equipment)
Prinias, Sidérospila
Iraklion, Archaeological Museum, ar. 481
H.: 0.725 m
640 B.C.
Lempesi 1976, B5, pls. 22-23

M.4.13
Stele of a warrior (with hoplite equipment)
Prinias, Sidérospila
Iraklion, Archaeological Museum, ar. 400
H.: 1.04 m
620—610 B.C.
Daux 1960, 840, fig. 2
Lempesi 1976, B7, pls. 26-27

M.4.14
Terracotta plaque of a warrior (with hoplite equipment)
Gortyn
Middle of the 7th century B.C.
Lempesi 1976, pl. 47b

M.4.15
Sphinx
Laconian Sub-Geometric sherd
First half of 7th century B.C.
Droop 1926—1927, 54, fig. 4

M.4.16
Amazon

Terracotta shield
Samos, Heraion
Early 7th century B.C.
Eilmann 1933, 120, fig. 66, pl. 37:1

M.4.17
Warrior
Terracotta shield
Tiryns
Early 7th century B.C.
Lorimer 1950, pl. IX:1

M.4.18
Amazon
Terracotta shield
Tiryns
Early 7th century B.C.
Lorimer 1950, pl. IX:1

M.4.19
Amazon
Terracotta shield
Tiryns
Early 7th century B.C.
Lorimer 1950, pl. IX:1

M.4.20
Terracotta head
Sparta, Amyklaion
Athen, National Museum, Tc. Inv. 4381
H.: 11.5 cm
Middle of the 7th century B.C.
Schweitzer 1971, pls. 161-162

M.4.21
Bronze statuette (Zeus?)
Munich, Antikensammlung, Inv. 4315
H.: 14.5 cm
Lullies 1962, 627-628, no. 14, figs. 25-26

M.4.22
Bronze plaquette (Athena)
Dreros
Ca. 600 B.C.
Marinatos 1936, fig. 42, pl. 30

M.4.23
Head of a terracotta warrior
Salamis, Cyprus
Nicosia, Cyprus Museum, Sal. 4923 = Tc. 1840
Late 7th – early 6th century B.C.
Monloup 1984, 184, no. 668, (fig.)

M.5.1
Bronze warrior (linch pin)
Salamis, Tomb 79
Nicosia, Cyprus Museum, T.79, 129/220/5+5a and 188/220/4+4a
End of the 8th century B.C.
Törnkvist 1973, figs. 9-10

M.5.2
Miniature votive bronze helmets (5 pieces)
Gortyn
Iraklion, Archaeological Museum
First half of 7th century B.C.
Hoffmann 1972, 2, pl. 41:4

M.5.3
Miniature votive bronze helmets (5 complete and fragments of others)
Praisos, Altar Hill
Candia Museum
First half of 7th century B.C.
Bosanquet 1901—1902, 258, pl. 10
Benton 1939—1940, pl. 31:17

M.5.4
Miniature votive bronze helmet
Palaikastro
First half of 7th century B.C.
Benton 1939—1940, pl. 28:31

M.5.5
Rock carving
Thera
Middle of the 7th century B.C.
von Gaertringen – Wilski 1904, 79, fig. 64

M.5.6
Bronze statuette of a warrior
Olympia, Heraion
Second quarter of the 7th century B.C.
Steiner 1906, pl. 18
Kunze 1944, 119-123, pl. 42

M.5.7
Bronze statuette of a warrior
Olympia, Inv. 5834
Athens, National Museum, Inv. 6094
Second quarter of the 7th century B.C.
Furtwängler 1890, 39, no. 245, pl. 16:245-245a

M.5.8
Bronze warrior cauldron attachement
Olympia
H.: 17 cm
Second quarter of the 7th century B.C.
Hampe – Jantzen 1936—1937, 66, fig. 29

M.5.9
Bronze statuette of a warrior
Olympia, Inv. B 2000
H.: 21 cm
Early 7th century B.C.
Kunze 1944, 120, pls. 38-39

M.5.10
Bronze statuette of a warrior
Olympia, Inv. B 1999
H.: 17.3 cm
Early 7th century B.C.
Kunze 1944, 122, pls. 43-44

M.5.11
Bronze statuette of a warrior
Olympia, Inv. B 1701
H.: 16.9 cm
Early 7th century B.C.
Kunze 1944, 122, pls. 45-46

M.5.12
Bronze statuette of a warrior
Olympia, Inv. B 5600
Early 7th century B.C.
Kunze 1967b, pls. 108-109

M.5.13
Bronze statuette of a warrior
Olympia
Early 7th century B.C.
Kunze 1944, 113, pl. 36:3

M.5.14
Bronze statuette of a warrior
Olympia, Inv. 1671
Early 7th century B.C.
Kunze 1944, 109, fig. 90, pl. 34

M.5.15

Bronze statuette of a warrior
Amorgos
Berlin, Inv. 10388
H.: 8.5 cm
Early 7th century B.C.
Neugebauer 1922, 61, no. 4
Kunze 1967b, fig. 78

M.5.16
Bronze statuette of a warrior
Delphi
Athens, National Museum, Inv. 7415
Early 7th century B.C.
Casson 1922, 213, fig. 7b

M.5.17
Bronze statuette of a warrior
Athens, Carapanos Collection
Early 7th century B.C.
Bequignon 1929, fig. 3

M.5.18
Warrior
Early Orientalizing pithos
Knossos, Inv. 6391
Second quarter of the 7th century B.C.
Payne 1927—1928, 240, no. 38, pls. XI:10-11, XII

M.5.19
Bronze Group
Samos
Formerly Samos Museum, Inv. B.190
H.: 9 cm
Early 7th century B.C
Karo 1930, fig. 27
Schweitzer 1971, pls. 186-187

M.5.20
Bronze statuette of an armourer
New York, MMA, Inv. 42.11.42
H.: 5.2 cm
Early 7th century B.C
Schweitzer 1971, pl. 200

M.5.21
Scene from a tripod cauldron leg
Olympia
Plympia Museum, Inv. B 1730
L.: 46.7 cm
Early 7th century B.C
Schweitzer 1971, pl. 213

M.5.22
Bronze statuette of a warrior
Athens, Acropolis
Athens, national Museum, Inv. 6613
H.: 20.5 cm
Lullies – Hirmer 1960, 53, no. 2, pl. 2
Schweitzer 1971, pls. 159-161

M.5.23
Horsemen on a bronze relief
Olympia, "Brunnen 17"
Olympia Museum, Inv. B 5085a
Kunze 1961—1962, 115-116, pl. 130b

M.5.24
Handle attachement from a bronze vessel
Olympia
Olympia Museum, Inv. B 4774

M.6.1
Warrior fighting on ship
Athenian Geometric (Dipylon) Ware
Late 8th century B.C.
Richter 1934, figs. 1-2

M.6.2
Horserider
Crater-fragment
Athens, "Piräusstrasse"
Athens, National Museum
Late 8th century B.C.
Pernice 1892, pl. 10
Borchhardt 1972, 65, Kat. 13 III 2, pl. 32:4

M.6.3
Foot soldier with round shield and two spears
Crater-fragment
Athens, "Piräusstrasse"
Athens, National Museum
Late 8th century B.C.
Pernice 1892, pl. 10
Borchhardt 1972, 65, Kat. 13 III 3, pl. 33:4

M.6.4
Pyxisfragment
Argos, Heraion
Athens, National Museum
Late 8th century B.C.
Lorimer 1950, 233, pl. 17:2

M.6.5
Warrior (hoplite)
Geometric amphora
Athens
Daux 1958, fig. 8

M.6.6
Warrior (Geometric gem)
Siphnos
Late 8th early 7th century B.C.
Lorimer 1950, pl. VI:5

M.6.7
Warrior (votive terracotta shield)
Tiryns
Late 8th early 7th century B.C.
Lorimer 1950, pl. X:1

M.6.8
Warriors (hoplites)
Reliefpithos
Tenos
700—675 B.C.
Kontoleon 1969, pls. 52-53

M.6.9
Warrior
Pottery sherd
Lindos, Acropolis
First half of 7th century B.C.
Blinkenberg 1931, pl. 37:867

M.6.10
Warrior (hoplite)
Late Geometric pottery sherd
Athens, Kerameikos, Inv. 112
First half of 7th century B.C.
Tölle 1963, 650, no. 7, fig. 5

M.6.11
Scene from a tripod cauldron leg
Olympia
Olympia Museum, Inv. B 1730
L.: 46.7 cm
Early 7th century B.C
Schweitzer 1971, pl. 213

M.6.12

Warrior with small round shield
Bronze helmet aplique
"Crete"
Hamburg, Museum für Kunst und Gewerbe
Middle of the 7th century B.C.
Hoffmann 1972, 6, fig. 1
Gröschel 1986, 72, fig. 5

M.6.13
Warrior (hoplite)
"Melian amphora"
Late 7th century B.C.
Conze 1862, pl. 3

M.6.14
Warrior
Proto-Corinthian ivory seal
Perachora, Sanctuary of Hera Akraia
First quarter of 7th century B.C.
Illustrated London News, 2 May 1931, 748

M.6.15
Bronze statuette of a warrior
Olympia, Inv. 2914
Early 7th century B.C.
Furtwängler 1890, 39, no. 243, pl. 16:243-243a

M.6.16
Rock carving
Thera
Middle of the 7th century B.C.
von Gaertringen – Wilski 1904, 79, fig. 63

M.6.17
Bronze statuette of a warrior
Delphi
Inv. 7733
Early 7th century B.C.
Rolley 1969, 38, no. 17, pl. VII:17

M.6.18
Terracotta horseman
"Cyprus"
Nicosia, Cyprus Museum
Second half of 7th century B.C.
Törnkvist 1973, 25, fig. 21

M.6.19
Terracotta warrior with shield
Kaloriziki, Cyprus
Museum of Episkopi, Cyprus
Second half of 7th century B.C.
Törnkvist 1973, 25, fig. 22-23

M.6.20
Terracotta warrior with shield
Kaloriziki, Cyprus, K.1144
Second half of 7th century B.C.
Benson 1973, pl. 41

M.6.21
Terracotta warrior with "figure-of-eight" shield
"Cyprus"
Pierides Foundation Museum
H.: 13.5 cm
End of 7th century B.C.
Karageorghis 1985, 216, no. 214 (pl.)

M.6.22
Warrior leading a horse
Argive Late geometric pottery sherd
Early 7th century B.C.
Waldstein 1905, pl. 57:4

M.6.23
Warrior
Argive Late geometric krater fragment
Early 7th century B.C.
Charitonidis 1954, 413, fig. 4

M.6.24
Warriors (hoplites)
Vase of Aristonothos
7th century B.C.
Lorimer 1950, pl. X:1

M.7.1
Bronze belt
Fortetsa, Tomb P, no. 1568
Middle of the 8th century B.C.
Iraklion, Archaeological Museum, Inv. no. 1568
Brock 1957, pl. 115
Borchhardt 1972, 70, Kat. 15 II 2

M.7.2
Warriors (hoplites)
Relief pithos (the Wooden Horse at Troy)
Mykonos
Around 675 B.C.
Ervin 1963, pls. 19, 21

M.7.3
Warriors (hoplites)
Relief pithos (the Wooden Horse at Troy)
Mykonos
Around 675 B.C.
Ervin 1963, pls. 19, 20, 21

M.7.4
Not warlike context
Lakonian sherd
Amyklaion
Late 8th – early 7th century B.C.
Wide 1899, fig. 42
Snodgrass 1964, 8

M.7.5
Bronze statuette of a warrior
Olympia, Inv. 9788
Early 7th century B.C.
Furtwängler 1890, 38, no. 242, pl. 16:242

M.7.6
Warriors
Terracotta plaques (2 pieces)
Praisos (?)
New York, Metropolitan Museum of Art
Dohan 1930—1931, figs. 1-2

M.7.7
Hoplite
Bone relief
Sparta
Late 7th century B.C.(?)
Richards 1891, pl. 11

M.7.8
Terracotta warrior
"Cyprus"
H.: 15.3 cm
Cypro-Archaic Period
Karageorghis 1985, 204, no. 199 (pl.)

BIBLIOGRAPHY

Aharoni, Y., 1975
Lachish V, Investigations at Lachish, The Sanctuary and the Residency, Tel Aviv, Tel Aviv University, Institute of Archaeology

Ahlberg, G., 1971
Fighting on Land and Sea in Greek Geometric Art, (Skrifter Utgivna av Svenska Institutet i Athen, 4°, XVI), Lund, Berlingska Boktryckeriet

Akerström, A., 1966
Die architektonischen Terrakotten Kleinasiens, (Skrifter Utgivna av Svenska Institutet i Athen, 4°, XI), Lund, C.W.K. Gleerup

Akurgal, E., 1961
Die Kunst der Hethiter, München, Hirmer

Albright, W.F., 1930—1931
"Mitannian *maryannu*, 'chariot warrior' and the Canaanite and Egyptian Equivalents", *Archiv für Orientforschung* 7, 217-221

Albright, W.F., 1934
The Vocalization of the Egyptian Syllabic Orthography, (American Oriental Series, 5), New Haven, American Oriental Society

Alexander, R.L., 1973—1976
"The Tyskiewicz Group of Stamp Cylinders", *Anatolica* 5, 141-215

Almagro, M., 1958
"Deposito de la Ría de Huelva", *Inventaria Arch. Espana* 1-4 E.1 39, 273ff.

Almagro-Gorbea, M., 1973
"Cascos del Bronze Final en la Peninsula Iberica", *Trabajos de Prehistorica* 30, 349ff.

Amandry, P., 1949
"Casques grecs à decor gravé", *Bulletin de correspondance hellénique* 73, 437-446, figs. 1-7, pls. 22-25

Amiet, P., 1969
"Quelques ancêtres du chasseur royal d'Ugarit", *Ugaritica* VI, Paris, P. Geuthner, 1-8

Amiet, P., 1971
Les Antiquités Orientales au Musée du Louvre, Paris

Amiet, P., 1977
"Appliqués iraniennes?", *La Revue du Louvre et des Musées de France* 27, 63ff.

Amiet, P., 1978
Musée du Louvre, Département des Antiquités Orientales, Guide du Visiteur, Paris

Amiet, P., 1980
Art of the Ancient Near East, New York, Abrams

Andrae, W., 1925
Coloured Ceramics from Ashur and Earlier Ancient Assyrian Wall Paintings. From Photographs and Water-Colours by Memebers of the Ashur Expedition Organized by the Deutsche Orient-Gesellschaft, London, Kegan Paul, Trench, Trubner & CO. LTD.

Andrae, W., 1938
Das wiedererstandene Assur, Leipzig, Hinrichs

Andrae, W., 1977^2
Das wiedererstandene Assur, B. Hrouda, ed., München, Beck

Anonymus, 1968
Guides to the National Museum, Egypt and Western Asia, Copenhagen

Anonymus, 1973
"Acquisitions du Musée en 1972", *Cahiers de Mariemont* 4, 59-60, No. 4 (fig.)

Anonymus, 1975
Münchner Jahrbuch der bildenden Kunst 26, p. 221, fig. 5

Archi, A. – Biga, M.G., 1982
Testi amministrativi di vario contenuto. Archivio L.2769: TM.75.G.3000-4101, (Archivi Reali di Ebla, Testi, 3), Roma, Missione Archeologica Italiana di Siria

Åström, P., *The Cuirass Tomb and Other Finds at Dendra. Part 1: The Chamber Tombs*, (Studies in Mediterranean Archaeology 4:1), Göteborg, 1977, Paul Åströms Förlag

Åström, P., *The Cuirass Tomb and Other Finds at Dendra. Part 2: Excavations in the Cemeteries, the Lower Town and the Citadel*, (Studies in Mediterranean Archaeology 4:2), Göteborg, 1983, Paul Åströms Förlag

Aubet, M.E., 1993
The Phoenicians and the West. Politics, Colonies and Trade, Cambridge, University Press

Azarpay, G., 1968
Urartian Art and Artifacts. A Chronological Study, Berkeley and Los Angeles, University of California Press

Bachmann, W., 1927
Felsreliefs in Assyrien, Bawian, Maltai und Gundük, (Wissenschaftliche Veröffentlichung der Deutschen Orient-Gesellschaft, 52), Leipzig, Hinrichs

Bänder, D., 1995
Die Siegesstele des Naramsîn und ihre Stellung in Kunst- und Kulturgeschichte, (Wissenschaftliche Schriften im Schulz-Kircher Verlag, Reihe 12, Beiträge zur Kunstgeschichte, 103), Idstein, Schulz-Kircher Verlag

Bär, J., 1996
Der assyrische Tribut und seine Darstellung. Eine Untersuchung zur imperialen Ideologie im neuassyrischen Reich, (Alter Orient und Altes Testament, 243), Neukirchen-Vluyn, Verlag Butzon & Bercker Kevelaer

Barag, D., 1985
Catalogue of Western Asiatic Glass in the British Museum, vol. I, London, The Trustees of the British Museum

Barnett, R.D., 1953
"An Assyrian Helmet", *The British Museum Quarterly* 18, 101-102, pls. 31-32

Barnett, R.D., 1954
"Excavations of the British Museum at Toprak Kale, Near Van – Addenda", *Iraq* 16, 3-22

Barnett, R.D., 1959
"Further Russian Excavations in Armenia (1949—1953)", *Iraq* 21, 1-19

Barnett, R.D., 1967
The British Museum Report of the Trustees 1966, London, The Trustees of the British Museum

Barnett, R.D., 1974
"The hieroglyphic Writing of Urartu", in: K. Bittel ,ed., *Anatolian Studies presented to Hans Gustav Güterbock on the Occasion of His 65th Birthday*, (Uitgaven van het Nederlands Historisch-Archaeologisch Instituut te Istanbul 35), Istanbul, Nederlands Historisch-Archaeologisch Instituut in het Nabije Osten, 43-56, pls. 11-13, 29c

Barnett, R.D., 1976
Sculptures from the North Palace of Ashurbanipal at Nineveh (668—627 B.C.), London, The Trustees of the British Museum

Barnett, R.D. – Curtis, J.E., 1973
"A Review of Acquisitions 1963-1970 of Western Asiatic Antiquities, 2", *The British Museum Quarterly* 37, 119-137

Barnett, R.D. – Falkner, M., 1962
The Sculptures of Assur-nasir-apli II (883—859 B.C.), Tiglath-pileser III (745—727 B.C.), Esarhaddon (681—669 B.C.) from the Central and South-West Palaces at Nimrud, London, The Trustees of the British Museum

Barnett, R.D. – Gökce, N., 1953
"The Find of Urartian Bronzes at Altin Tepe, Near Erzincan", *Anatolian Studies* 3, 121-129

Barnett, R.D. – Watson, W., 1952
"Russian Excavation in Armenia", *Iraq* 14, 132-147, pls. 32-33

Barrelet, M.-Th., 1955
"Les déesses armées et ailées", *Syria* 32, 222ff.

Barrelet, M.-Th., 1968
Figurines et reliefs en terre cuite de la Mésopotamie antique, Paris, Geuthner

Belinskiy, A.B., 1990
"K voprosu o vremeni poyavleniya shlemov assiriykogo tipa na Kavkaze" ("About the Date the Assyrian-type Helmets First Appeared in the Caucasus"), *Sovetskaya Archeologiya*, 1990/4, 190-195

Belli, O., 1993
"Der beschriftete Bronzehelm des Königs Menua aus der Festung Burmageçit bei Tunceli", in: M.J. Mellink – E. Porada – T. Özgüç, eds., *Nimet Özgüç'e Armagan. Aspects of Art and Iconography: Anatolia and its Neighbours. Studies in Honor of Nimet Özgüç*, Ankara, Türk Tarih Kurumu Basimevi, 61-67.

Belloni, G.G. – Dall'Asèn, L.F., 1969
Iranian Art, London, Pall Mall Press
Benson, J.L., 1973
The Necropolis of Kaloriziki, (Studies in Mediterranean Archaeology, 36), Göteborg, Paul Aströms Vörlag
Benton, S., 1939–40
"The Dating of Helmets and Corselets in Early Greece", *Annals of the British School of Archaeology at Athens* 40, 78-82
Béquignon, Y., 1929
"Études thessaliennes III: statuette de guerrier trouvée ... Phères", *Bulletin de correspondance hellénique* 53, 101-116
Beran, T., 1957
"Assyrische Glyptik des 14. Jahrhunderts", *Zeitschrift für Assyriologie und Vorderasiatischen Archäologie* 52, 141ff.
Bittel, K., 1976a
"Tonschale mit Ritzzeichnung von Boghazköy", *Revue Archéologique* NS 1, 9-14, figs. 1-3
Bittel, K., 1976b
Die Hethiter, (Universum der Kunst), München, C.H. Beck
Bittel, K. – Neve, P. – Fischer, F. – Beran, T. – Otten, H., 1958
"Vorläufiger Bericht über die Ausgrabungen in Boğazköy im Jahre 1957", *Mitteilungen der Deutschen Orient-Gesellschaft* 91, 1-84
Black, J. – Green, A., 1992
Gods, Demons and Symbols of Ancient Mesopotamia. An Illustrated Dictionary, London, British Museum Press
Blegen, C., 1953
Troy: Excavations Conducted by theUniversity of Cincinnati 1932—1938, vol. III, Princeton
Blinkenberg, Chr., 1931
Lindos. Fouilles de l'Acropole 1902-1914. I. Les petits objets, Berlin, Walter de Gruyter
Boardman, J., 1970
"Orientalen auf Kreta", in: Hoffmann, H., ed., *Dädalische Kunst auf Kreta*, Hamburg, 14-25
Boardman, J., 1980
The Greeks Overseas: Their Early Colonies and Trade, (new and enlarged edition), London
Boehmer, R.M., 1972
Die Kleinfunde von Boghazköy, (WVDOG 87), Berlin, Verlag Gebr. Mann
Boehmer, R.M., 1981
"Kopfbedeckung, B. Archäologisch", *Reallexikon der Assyriologie und Vorderasiatischen Archäologie VI*, 3-4, Berlin, Walter de Gruyter, 203-210
Bol, P.C., 1989
Argivische Schilde, (Olympische Forschungen, XVII), Berlin, Walter de Gruyter
Borchhardt, J., 1972
Homerische Helme. Helmformen der Ägäis in ihren Beziehungen zu orientalischen und europäischen Helmformen in der Bronze- und frühen Eisenzeit, Mainz am Rhein, Philipp von Zabern
Borchhardt, J., 1977
"Helme", in: H.G. Buchholz – J. Wiesner, eds., *Kriegswesen, vol. I: Schutzwaffen und Wehrbauten*, (Archaeologia Homerica, Band 1, Kapitel E), Göttingen, Vandenhoeck & Ruprecht, E57-E74
Borger, R., 1956
Die Inschriften Asarhaddons Königs von Assyrien, (Archiv für Orientforschung, Beiheft 9), Graz
Born, H. – Seidl, U., 1995
Schutzwaffen aus Assyrien und Urartu, (Sammlung Axel Guttmann, 4), Berlin, Verlag Sammlung Guttmann – Philipp von Zabern
Bosanquet, R.C., 1901—1902
"Excavations at Praisos, I.", *Annals of the British School of Archaeology at Athens* 8, 231-270
Bossert, H.Th., *et al.*, 1950
Karatepe Kazilari (Die Ausgrabungen auf dem Karatepe, Erster Vorbericht), (Türk Tarih Kurumu Yayinlarindan V. seri – No. 9), Ankara, Türk Tarih Kurumu Basimevi
Botta, M.P.E. – Flandin, M.E., 1849—1850
Monument de Ninive, vols. I-V, Paris, Imprimerie Nationale

Bottéro, J., 1956
Archives Royales de Mari, VII, Textes administratifs de la salle 110, (Musée du Louvre, Départment des Antiquités Orientales, Textes Cunéiformes, 28), Paris, Paul Geuthner
Bottini, A., 1988
"Apulisch-korinthische Helme", in: A. Bottini, *et al.*, Antike Helme. Sammlung Lipperheide und andere Bestände des Antikenmuseums Berlin, (Römisch-Germanisches Zentralmuseum, Monographien 14), Mainz, Verlag des Römisch-Germanisches Zentralmuseums, 107-136
Boyd, H.A., 1901
"Excavations at Kavousi, Crete, in 1900", *American Journal of Archaeology* 5, 125-158
Börker-Klähn, J., 1982
Altvorderasiatische Bildstelen und vergleichbare Felsreliefs, (Baghdader Forschungen, 4), Mainz am Rhein, Philipp von Zabern
Briquel-Chatonnet, F., 1992
Les relations entre les cités de la Cote Phénicienne et les Royaumes d'Israël et de Juda, (Studia Phoenicia 12, Orientalia Lovaniensia Analecta 46), Leuven, Peeters, esp. 141-226
British Museum, 1929[3]
British Museum. A Guide to the Exhition illustrating Greek and Roman Life, London
Brize, P., 1985
"Samos und Stesichoros. Zu einem früharchaischen Bronzeblech", *Mitteilungen des Deutschen Archäologischen Instituts, Athenische Abteilung* 100, 53-90
Brock, J.K., 1957
Fortetsa. Early Greek Tombs near Knossos, (BSA Supplementum, 2), Cambridge, Cambridge University Press
Bronzes et terres cuites, 1981
Bronzes et terres cuites Louristan – Amlash, et autres cites iraniens du 3e millénaire av. J.C. au 1er millénaire, Nouveau Drouot – Salle No. 14, Le lundi 30 mars 1981 à 21 heures
Brückner, A. – Pernice, E., 1893
"Ein attischer Friedhof", *Mitteilungen des Deutschen Archäologischen Instituts, Athenische Abteilung* 18, 73-191
Buchanan, B. – Moorey, P.R.S, 1988
Catalogue of Ancient Near Easter Seals in the Ashmolean Museum, vol. III, The Iron Age Stamp Seals (c. 1200-350 B.C.), Oxford, Clarendon Press
Buchholz, H.-G., 1972
"Politiko-Tamassos 1971", *Report of the Department of Antiquities Cyprus*, 183-186
Buchholz, H.-G., 1973
"Tamassos, Zypern, 1970-1972", *Archäologischer Anzeiger*, 295-388
Buchholz, H.-G., 1978
"Tamassos, Zypern, 1974 bis 1976. 3. Bericht", *Archäologischer Anzeiger* 93, 155-230
Budge, E.A.W., 1914
Assyrian Sculptures in the British Museum. Reign of Ashur-nasir-pal, 885—860 B.C., London
Budge, E.A.W. – King, L.W., 1902
Annals of the Kings of Assyria, London, The British Museum
Buhl, M.-L., 1962
"Bronzefund fra Persien", *Fra Nationalmuseets Arbejdsmark* 1962, 30-31, colour pl. on p. 31
Calmeyer, P., 1969
Datierbare Bronzen aus Luristan und Kirmanshah, (Untersuchungen zur Assyriologie und Vorderasiatischen Archäologie 5), Berlin, Walter de Gruyter
Calmeyer, P., 1975
"Helm", *Reallexikon der Assyriologie und Vorderasiatischen Archäologie* IV, Berlin, Walter de Gruyter, 313-317
Calmeyer, P., 1986
"Zu einem ziselierten Bronzehelm des 8. Jahrhunderts", *Archäologische Mitteilungen aus Iran* 19, 79-86, pls. 15-20
Calmeyer, P., 1991a
"Helmets and Quivers", in: R. Merhav, ed., *Urartu. A Metalworking Center in the First Millennium B.C.E.*, Jerusalem, Israel Museum, 123-133
Calmeyer, P., 1991b
"Some Remarks on Iconography", in: R. Merhav, ed., *Urartu. A Metalworking Center in the First Millennium B.C.E.*, Jerusalem, Israel Museum, 312-319
Carter, H. – Newberry, P.E., 1904

The Tomb of Thutmosis IV, Westminster
Casson, S., 1922
"Bronze Work of the Geometric Period", *Journal of Hellenic Studies* 42, 207-219
Catling, H.W., 1964
Cypriote Bronzework in the Mycenaean World, (Oxford Monographs on Classical Archaeology), Oxford, Clarendon Press
Catling, H.W., 1970
"A Bronze Plate from a Scale-Corslet found at Mycenae", *Archäologischer Anzeiger* 85, 441-449
Catling, H.W., 1977
"Panzer", in: H.G. Buchholz – J. Wiesner, eds., *Kriegswesen, vol. I: Schutzwaffen und Wehrbauten*, (Archaeologia Homerica, Band 1, Kapitel E), Göttingen, Vandenhoeck & Ruprecht
Chambon, A., 1984
Tell el-Far'ah I. L'âge du fer, ("Mémoire" n° 31), Paris, Éditions Recherche sur les Civilisations
Charitonidis, S., 1954
"Recherches dans le quartier Est d'Argos", *Bulletin de correspondance hellénique* 78, 410-426
Chiera, E., 1934
Joint Expedition with the Iraq Museum at Nuzi, vol. V, Mixed Texts, (American Schools of Oriental Research, Publications of the Baghdad School, 5), Philadelphia, University of Pennsylvania Press
Cialowicz, K.M., 1983
"Casques corinthiens dans les collections polonaises", *Études et Travaux* 13, 47-52
Civil, M., 1969
The Series lú = ša and Related Texts, (Materials for the Sumerian Lexikon XII), Roma, Pontificium Institutum Biblicum
Colbow, G., 1991
Die kriegerische Ištar. Zu den Erscheinungsformen bewaffneter Gottheiten zwischen der Mitte des 3. und der Mitte des 2. Jahrtausends, (Münchner Vorderasiatische Studien, 8), München, Profil Verlag
Coldstream, J.N., 1977
Geometric Greece, (A Benn Study: Archaeology), London
Coldstream, J.N., 1982
"Greeks and Phoenicians in the Aegean", in: Niemeyer, H.G., ed., *Phönizier im Westen. Die Beiträge des Internationalen Symposiums über die phönizische Expansion im Westlichen Mittelmeerraum in Köln vom 24. bis 27. April 1979*, (Madrider Beiträge, 8), Mainz am Rhein, 261-275
Coldstream, J.N., 1983
"Gift Exchange in the Eighth Century B.C.", in: Hägg, R., ed., *The Greek Renaissance in the Eighth Century B.C.: Tradition and Innovation*, (Skrifter Utgivna av Svenska Institutet i Athen 4, XXX), Stockholm, P. Aströms Vörlag, 201-206
Collon, D., 1988
First Impressions. Cylinder Seals in the Ancient Near East, Chicago, The University of Chicago Press
Collon, D. – Crouwel, J. – Littauer, M.A., 1976
"A Bronze Chariot Group from the Levant in Paris", *Levant* 8, 71-81
Conze, A., 1862
Melische Thongefässe, Leipzig, Breitkopf und Härtel
Cook, J.M., 1952
"Archaeology in Greece, 1951", *Journal of Hellenic Studies* 72, 92-112
Cook, J.M. – Akurgal, E., 1952
"Old Smyrna (Bayrakli)", (in: Archaeological Work in Turkey, 1951), *Anatolian Studies* 2, 23-24
Courbin, P., 1957
"Une tombe géométrique d'Argos", *Bulletin de correspondance hellénique* 81, 322-386
Crawford, V.E., 1954
Sumerian Economic Texts from the First Dynasty of Isin, (Babylonian Inscriptions in the Collection of J.B. Nies, 9), New Haven

Crouwel, J.H., 1987
"Chariots in Iron Age Cyprus", *RDAC* 1987, 101-118
Crouwel, J.H. - Tatton-Brown, V., 1988
"Ridden Horses in Iron Age Cyprus", *RDAC* 1988/II, 77-87
Curtis, J.E., 1979
An examination of Late Assyrian metalwork with special reference to material from Nimrud, vols. I-II, Unpublished Ph. D. Diss., University of London
Curtis, J.E., 1994
"Mesopotamian Bronzes from Greek Sites: the Workshops of Origin", *Iraq* 56, 1-25
Curtis, J.E. - Wheeler, T.S. - Muhly J.D. - Maddin, R. 1979
"Neo-Assyrian Ironworking Technology", *Proceedings of the American Philosophical Society* 123, 369-390
Dalley, S. - Postgate, J.N., 1984
The Tablets from Fort Shalmaneser, (Cuneiform Texts from Nimrud, III), London, The British School of Archaeology in Iraq
Dandamayev, M.A., 1989
"The Old Iranian Pasa'du", in: L. De Meyer - E. Haerinck, eds., *Archaeologia Iranica et Orientalis. Miscellanea in Honorem Louis Vanden Berghe*, Gent 1989, 563-566
Daux, G., 1958
"Chronique des fouilles et découvertes archéologiques en Grèce en 1957: Athènes", *Bulletin de correspondance hellénique* 82, 656-669, Tiryns, p. 707, fig. 26
Daux, G., 1960
"Chronique des fouilles et découvertes archéologiques en Grèce en 1959", *Bulletin de correspondance hellénique* 84, 617-868
Daux, G., 1961
"Chronique des fouilles et découvertes archéologiques en Grèce en 1960", *Bulletin de correspondance hellénique* 85, 671-675
Daux, G., 1968
"Chronique des fouilles et découvertes archéologiques en Grèce en 1967", *Bulletin de correspondance hellénique* 92, 996, fig. 5
Dawkins, R.M., ed., 1929
The Sanctuary of Artemis Orthia at Sparta, London, Macmillan
de Chanot, E., 1880
"Géryon. Bronze Étrusque", *Gazette Archéologique* 6, 136-138
de G. Davies, N., 1930
The Tomb of Ken-Amun at Thebes, vols. I-II, (Publications of the Metropolitan Museum of Art. Egyptian Expedition, 5), New York
de G. Davies, N., 1953
The Tomb of Rekh-mi-re at Thebes, vols. I-II, (Publications of the Metropolitan Museum of Art. Egyptian Expedition, 11), New York
de G. Davies, N. and N., 1933
The Tomb of Menkheperrasonb, Amenmose, and Another (Nos. 86, 112, 42, 226), (Theban Tomb Series, Fifth Memoir), London
de Genouillac, H. 1911
Inventaire des tablettes de Tello conservées au Musée Impérial Ottoman, vol. II: Textes de l'époque de Agadé et de l'époque d'Ur (Fouilles d'Ernest de Sarzec en 1894), Paris, Ernest Leroux
Delaporte, L., 1910
Catalogue des cylindres Orientaux et des cachets assyro-babyloniens, perses et syro-cappadociens de la Bibliothèque Nationale, Paris, E. Leroux
Delaporte, L., 1923
Musée du Louvre, Catalogue des cylindres, cachets et pierres gravée de style orientale. II. Acquisitions, Paris, Musée du Louvre
de Mertzenfeld, C.D., 1954
Inventaire commenté des Ivories Phéniciennes et apparentés découvertes dans le Proche-Orient, Paris, E. de Boccard
Deonna, W., 1938
Le mobilier Délien, (Délos XVIII), Paris, de Boccard
Deonna, W., 1954
"Casque gravé sur une hache minoenne", *Bulletin de correspondance hellénique* 78, 253-257

de Schauensee, M., 1988
"Northwest Iran as a Bronzeworking Centre: the View from Hasanlu", in: J. Curtis ed., *Bronzeworking Centres of Western Asia c. 1000-539 B.C.*, London, Kegan Paul, 45-62, pls. 13-56
de Sélincourt, A. – Burn, A.R., 1983
Herodotus, The Histories, Harmondsworth, Penguin Books
de Vaux, R., 1951
"La troisième campagne de fouilles à Tell el-Fàr'ah prés Naplouse. Rapport préliminaire", *Revue Biblique* 58, 393-430
Despini, K., 1981
"Korinthiako kranos apo te Khalkidike" ("Ein archaischer Helm aus Chalkidike"), *Athens Annals of Archaeology* 14, 246-250
Dewanckel, G., 1979
"Observations techniques concernant l'identification et la restauration", *Iranica Antiqua* 14, 64-65
Dezső, T., 1998
Oriental Influence in the Aegean and Eastern Mediterranean Helmet Traditions in the 9th–7th Centuries B.C.: The Patterns of Orientalization, (British Archaeological Reports, International Series, 691), Oxford, BAR
Dezső, T. – Curtis, J.E., 1991
"Assyrian Iron Helmets from Nimrud now in the British Museum", *Iraq* 53, 105-126, pls. 15-20
Dikaios, P., 1969
Enkomi. Excavations 1948–1958. Vol. I., Mainz, Philipp von Zabern
Dohan, E.H., 1930–1931
"Archaic Cretan Terracottas in America", *Metropolitan Museum Studies* 3, 209-228
Dörig, J., 1981
"Ein korinthischer Helm in Athener Privatbesitz", in: A. Mallwitz, *X. Bericht über die Ausgrabungen in Olympia*, (Fs. E. Kunze), Berlin, Walter de Gruyter, 109-110, pl. 12
Droop, J.P., 1926–1927
"Excavations at Sparta, 1927. The Native Pottery from the Acropolis", *Annals of the British School of Archaeology at Athens* 28, 49-81
Dugas, Ch., 1921
"Le Sanctuaire d'Aléa Athéna a Tégée", *Bulletin de correspondance hellénique* 45, 335-435
Dunbabin, T.J., 1944
"Archaeology in Greece 1939—1945", *Journal of Hellenic Studies* 64, 78-97
Dunbabin, T.J., 1957
The Greeks and Their Eastern Neighbors. Studies in the Relations between Greece and the Countriesof the Near East in the Eight and Seventh Centuries BC., (Society for the Promotion of Hellenic Studies, Supplementary Paper, 8), London
Durand, J.-M., *et al.*, 1988
Archives epistolaires de Mari, vols. I-II, (Archives Royales de Mari, 26), Paris, Éditions Recherche sur les Civilisations
Dyson, R.H., 1961
"Excavating the Mannaean Citadel of Hasanlu; And new Light on several Millennia of Persian Azerbaijan", *The Illustrated London News*, September 30, 1961, 534-537
Dyson, R.H. – Muscarella, O.W., 1989
"Constructing the Chronology and Historical Implications of Hasanlu IV", *Iran* 27, 1-27
Ebeling, E., 1927
Ausgrabungen der Deutschen Orient-Gesellschaft in Assur, E: Inschriften, IV, Keilschrifttexte Juristischen Inhalts, (50. Wissenschaftliche Veröffentlichung der Deutschen Orient-Gesellschaft), Leipzig, Hinrichs
Ebeling, E., 1952
"Die Rüstung eines babylonischen Panzerreiters nach einem Vertrage aus der Zeit Darius II.", *Zeitschrift für Assyriologie und Vorderasiatischen Archäologie* 50, 203-213
Edzard, D.O., 1981
Verwaltungstexte versciedenen Inhalts aus dem Archiv L.2769, (Archivi Reali di Ebla, Testi, 2), Roma, Missione Archeologica Italiana di Siria
Egg, M., 1984

"Neuerwerbungen für die Sammlungen", *Jahrbuch des Römisch-Germanischen Zentralmuseums Mainz* 31, 647-655

Egg, M., 1986
Italische Helme (Römisch-Germanisches Zentralmuseum Mainz, Monographien 11, 1-2), Mainz, Verlag des Römisch-Germanisches Zentralmuseum

Egg, M., 1988
"Die ältesten Helme der Hallstattzeit", in: Bottini, A., *et al.*, *Antike Helme. Sammlung Lipperheide und andere Bestände des Antikenmuseums Berlin. Handbuch mit Katalog*, (Römisch-Germanisches Zentralmuseum, Monographien 14), Mainz, Verlag des Römisch-Germanisches Zentralmuseums, 212-221

Egg, M. - Waurick, G., 1990
Antike Helme. Ausstellungskatalog, Speyer, Mainz, Verlag des Römisch-Germanisches Zentralmuseum

Eichler, S., 1984
Götter, Genien und Mischwesen in der urartäischen Kunst, (Archäologische Mitteilungen aus Iran, Ergänzungsband 12), Berlin, Dietrich Reimer

Eilmann, R., 1933
"Frühe Griechische Keramik im Samischen Heraion", *Mitteilungen des Deutschen Archäologischen Instituts, Athenische Abteilung* 58, 47-145

Elayi, J., 1983
"Les cités phéniciennes et l'empire assyrien à l'époque d'Assurbanipal", *Revue Assyriologique* 77, 45-58

Elayi, J., 1985a
"Les relations entre les cités phéniciennes et l'empire assyrien sous le régne de Sennacherib", *Semitica* 35, 19-26

Elayi, J., 1985b
"Byblos et la domination assyro-babylonienne", *Baghdader Mitteilungen* 16, 393-397

Ervin, M., 1963
"A Relief Pithos from Mykonos", *Archaiologikon Deltion* 18, 37-75

Erzen, A., 1973
"Die Grabungen von Cavustepe", in: *The Proceedings of the Xth International Congress for Classical Archaeology, Ankara - Izmir 23-30, IX, 1973*, Ankara

Esaian, S.A., 1966
Oruzhie i Voennovo Dyelo Drevnyey Armenii, Erevan, Izdatyelstvo Akademii Nauk Armyanskoy S.S.R.

Esaian, S.A., 1976
Drevnyaya Kultura Plemen Severo-Vostotshnoy Armenii (III-I Tis. do N.E.), Erevan, Izdatyelstvo Akademii Nauk Armyanskoy S.S.R.

Esaian, S.A., 1984
"Gürtelbleche der älteren Eisenzeit in Armenien", Beiträge zur allgemeinen und vergleichenden Archäologie 6, 97-198

Esaian, S.A., 1986
Dospeh Drevney Armenii, Erevan, Izdatelstvo Erevanskogo Universiteta

Evrard-Derriks, C., 1977—1978
"Les Antiquités du Proche-Orient au Musèe Royal de Mariemont. Les armes et les outils", *Les Cahiers de Mariemont* 8-9, 7-43

Fales, F.M., 1989
Prima dell'alfabeto. La storia della scrittura attraverso testi cuneiformi inediti, (Studi e Documenti 4), Venezia, Erizzo Editrice

Fales, F.M., 2000
"Preparing for War in Assyria", in: Andreau, J. - Briant, P. - Descat, R., Eds., *La guerre dans les économies antiques*, (Entretiens d'Archéologie et d'Histoire, 5; Économie Antique, 3), Saint-Bertrand-de-Comminges, Musée Départmental de Saint-Bertrand-de-Comminges, 35-62

Falsone, G., 1988
"Phoenicia as a Bronzeworking Centre in the Iron Age", in: Curtis, J.E., ed., *Bronzeworking Centres of Western Asia c. 1000–539 BC*, London, Keegan and Paul, 227-250

Felsch, R.C.S., 1987
"Kalapodi", *Archäologischer Anzeiger* 1987, 1-25

Frankel, D., 1979
The Ancient Kingdom of Urartu, London

Franken, H.G., 1961
"The Excavations at Deir 'Alla in Jordan, 2nd Season", *Vetus Testamentum* 11, 361-372
Frankenstein, S., 1979
"The Phoenicians in the Far West: A Function of Neo-Assyrian Imperialism", in: Larsen, T., ed., *Power and Propaganda*, (Mesopotamica, Copenhagen Studies in Assyriology, 7), Copenhagen, 263-294
Frankfort, H., 1963[3]
The Art and Architecture of the Ancient Orient, (The Pelican History of Art), London, Penguin Books
Friedrich, J. – Meyer, R. – Ungnad, A. – Weidner, E., 1940
Die Inschriften vom Tell Halaf. Keilschrifttexte und aramäische Urkunden aus einer assyrischen Provinzhauptstadt, (Archiv für Orientforschung, Beiheft 6), Graz
Fuchs, H. – Kellner, H.-J., 1978
"Von den Aufgaben des Museums", in: B. Hrouda ed., *Methoden der Archäologie. Eine Einführung in ihre naturwissenschaftlichen Techniken*, München, C.H. Beck, 372-390
Furtwängler, A., 1890
Olympia IV, Die Bronzen, Berlin, A. Asher
Galanina, L.K., 1985
"Shlemi Kubanskogo tipa. Voprosi, hronologii i proishozhdeniya", in: *Kulturnoe nacledie vostoka. Problemi, poicki, suzhdeniya*, Leningrad, Nauka, 169-183
Galanina, L.K., 1987
"Die Kelermes-Kurgane als Quelle für die Erforschung der Kontakte zwischen den Skythen und den altorientalischen Kulturen", in: H. Franke ed., *Skythika*, (Bayerische Akademie der Wissenschaften, Philosophisch-Historische Klasse, Abhandlungen, NF. 98), 7-18
Galanina, L.K., 1997
"Die Kurgane von Kelermes", in: Ivantchik, A. – Parzinger, H., eds., *Steppenvölker Eurasiens*, Moscow
Gamber, O., 1978
Waffe und Rüstung Eurasiens, Frühzeit und Antike, (Bibliothek für Kunst- und Antiquitätenfreunde 51), Braunschweig, Klinkhard and Biermann
Garelli, P., 1983
"Remarques sur les rapports entre l'Assyrie et les cités phéniciennes", in: *Atti del I Congresso Internazionale di Studi Fenici e Punici, vol. I*, Rome, 61-66
Ghirshman, R., 1961
7000 Ans d'Art en Iran, (Exposition, Paris, Petit Palais, octobre 1961 – janvier 1962), Paris
Ghirshman, R., 1963
Perse, Proto-Iraniens, Mèdes, Achéménides, Paris, Edition Gallimard
Ghirshman, R., 1964
Persia from the Origins to Alexander the Great, (The Arts of Mankind), London, Thames and Hudson
Ghirshman, R., 1966
Tchoga Zanbil I, (Mémoires de la Mission Archéologique en Iran, 39), Paris, Presses Universitaires de France
Gjerstad, E., 1935
"Idalion", in: E. Gjerstad *et al.*, eds., *The Swedish Cyprus Expedition. Finds and Result of the Excavations in Cyprus, 1927-1931, vol. II*, Stockholm, The Swedish Cyprus Expedition, 460-642
Gjerstad, E., 1948
The Swedish Cyprus Expedition IV:2, Stockholm, The Swedish Cyprus Expedition
Gordon, C.H., 1965
Ugaritic Textbook. Grammar, texts in translitteration, cuneiform selections, glossary, indices, Roma
Greenewalt, C.H., Jr., 1989
"Sardis, 1988", (in: Recent Archaeological Research in Turkey), *Anatolian Studies* 39, 182-184
Greenewalt, C.H., Jr., 1991
"The Sardis Campaign of 1987", in: W.E. Rast ed., *Preliminary Reports of ASOR-Sponsored Excavations 1982-1989*, Baltimore, The Johns Hopkins University Press, 1-28
Greenewalt, C.H., Jr., 1992

"When a Mighty Empire was Destroyed: The Common Man at the Fall of Sardis, Ca. 546 B.C.", *Proceedings of the American Philosophical Society* 136, II, 247-271
Greenewalt, C.H., Jr. – A.M. Heywood, 1992
"A Helmet of the Sixth Century B.C. from Sardis", *Bulletin of the American Schools of Oriental Research* 285, 1-31
Gröschel, S.-G., 1986
"Der goldene Helm der Athena (Ilias V, 743/44)", *Archäologische Mitteilungen aus Iran* 19, 43-78, pls. 13-20
Guida, P.C., 1973
Le armi difensive dei Micenei nelle figurazione, (Incunabula Greca, 56), Roma
Guralnick, E., 1988
"Greece and the Near East: Art and Archaeology", in: Sutton, R.F., ed., *Daidalikon: Studies in Memory of Raymond V. Schoder, S.J.*, Wauconda, Illinois, 151-176
Guralnick, E., 1992
"East to West: Near Eastern Artifacts from Greek Sites", in: Charpin, D. – Joannès, F., eds., *La circulation des biens, des personnes et des idées dans le Proche-Orient ancien, XXXVIIIe R.A.I.*, (Editions Recherche sur les Civilisations), Paris, 327-340
Hampe, R. – Jantzen, U., 1936—1937
"Eisengeräte und Waffen", in: *Bericht über die Ausgrabungen in Olympia*, Berlin, W. de Gruyter, 49-64
Hančar, F., 1966
"Das urartäische Lebensbaummotiv – Eine neue Bedeutungstradition?", *Archaeologia Iranica* 6, 92-108
Hawkins, J.D., 1976—1980
"Karkamis", *Reallexikon der Assyriologie und Vorderasiatischen Archäologie* V, Berlin, Walter de Gruyter, 426-446
Heinrich, E., 1984
Die Paläste im Alten Mesopotamien, (Denkmäler Antiker Architektur, Band 15), Berlin, Walter de Gruyter
Hencken, H., 1971
The Earliest European Helmets. Bronze Age and Early Iron Age, (American School of Oriental Research, Peabody Museum, Harvard University, Bulletin 28), Cambridge, Mass., Harvard University Press
Herrmann, G., 1986
Ivories from SW 37 Fort Shalmaneser, (Ivories from Nimrud (1949—1963), Fasc.IV, 1-2), London, British School of Archaeology in Iraq
Herrmann, H.-V., 1966
"Urartu und Griechenland", *Jahrbuch des Deutschen Archäologischen Instituts* 81, 79-141
Herzfeld, E.E., 1938
Altpersische Inschriften, (Archäologische Mitteilungen aus Iran, Ergänzungsband 1), Berlin, Reimer
Herzfeld, E.E., 1941
Iran in the Ancient East, New York, *etc.*, Oxford University Press
Heuzey, L., 1880
"Sur un petit vase en forme de téte casquée prtant une inscription hiéroglyphique", *Gazette Archéologique* 6, 145-160
Hodjash, S.I. – Trukhtanova, N.S. – Hovhannissian, K.L., 1979
Erebuni. Pamyatnik Urartskogo zodtshestva VIII—VI v. do n.e., Moskva, Iskusstvo
Hoffmann, H., 1972
Early Cretan Armorers, Mainz, Zabern
Hogarth, D.G., 1914
Charchemish I, Introductory, London
Hood, M.S.F. – de Jong, P., 1952
"Late Minoan Warrior Graves from Ayios Ioannis and the New Hospital Site at Knossos", *Annual of the British School of Archaeology at Athens* 47, 243-277, pls. 50-56
Hori A. – Miyashita S. – Ishida K., 1982
"Studies in the Urartian Bronze Objects from Japanese Collections, 1", *Bulletin of the Ancient Orient Museum* 4, 68-76, fig. 2, pl. 7
Horn, H.G. – Rüger, C.B., 1979
Die Numider, Bonn, Rheinisches Landesmuseum

Hovhannissian, C., 1973
Erebooni, Erevan, Hayastan
Hrouda, B., 1965
Die Kulturgeschichte des assyrischen Flachbildes, (Saarbrücker Beiträge zur Altertumskunde, 2), Bonn, Rudolf Habelt
Huff, D., 1968
"Das Felsgrab von Eski Dogubayazit", *Istanbuler Mitteilungen* 18, 58-86, pls. 17-19
Ilinskaya, V.A. – T'erenozhkin, A.I., 1983
Skifiya VII—IV vv. do n.e., Kiev, Naukova Dumka
Jackson, A.H., 1987
"An Early Corinthian Helmet in the Museum of the British School at Athens", *Annals of the British School of Archaeology at Athens* 82, 107-114, pl. 17
Jakob-Rost, L., 1975
Die Stempelsiegel im Vorderasiatischen Museum, Berlin, Akademie-Verlag
Jantzen, U., 1962
"Phrygishe Fibeln", in: Hanfman, G.M.A., *et al.*, *Festschrift für Friedrich Matz*, Mainz, Ph. von Zabern, 39-43
Joundi, A., 1980
"Department of Syrio-Oriental Antiquities", in: M. Abu-l-Faraj Al-Ush, *et al.*, *A Concise Guide to the National Museum of Damascus*, Damascus, General Directorate of Antiquities and Museums
Karageorghis, V., 1966
"Chronique des fouilles et découvertes archéologiques a Chypre en 1965", *Bulletin de correspondance hellénique* 90, 287-389
Karageorghis, V., 1967
"Nouvelles tombes de guerriers a Palaepaphos", *Bulletin de correspondance hellénique* 91, 202-245
Karageorghis, V., 1973
"Chronique des fouilles et découvertes archéologiques à Chypre en 1972", *Bulletin de correspondance hellénique* 97, 601-689
Karageorghis, V., 1985
Ancient Cypriote Art in the Pierides Foundation Museum, Larnaca
Karageorghis, V., 1988
"Chronique des fouilles et découvertes archéologiques à Chypre en 1987", *Bulletin de correspondance hellénique* 112, 793-855
Karageorghis, V. – Masson, E., 1975
"A propos de la découvertes d'écailles d'armure en bronze à Gastria Alaas (Chypre)", *Archäologischer Anzeiger* 90, 209-222
Karo, G., 1930
"Archäologische Funde aus dem Jahre 1929 und der Ersten Hälfte von 1930: Inseln", *Archäologischer Anzeiger* 45, 128-165
Kellner, H.-J., 1976
Urartu, ein wiederentdeckter Rivale Assyriens, Katalog der Ausstellung, München, Prähistorische Staatssammlung
Kellner, H.-J., 1979
"Eisen in Urartu", in: *Akten des VII. Internationalen Kongresses für Iranische Kunst und Archäologie, München 7—10. September 1976*, (Archäologische Mitteilungen aus Iran, Ergänzungsband 6), Berlin, Dietrich Reimer Verlag, 151-156
Kellner, H.-J., 1980
"Bemerkungen zu den Helmen in Urartu", *Anadolu Arastirmalari* 8, 205-213, pls. 1-16
Kellner, H.-J., 1987
"Datierungsfragen zu Urartu", in: H. Franke ed., *Skythika*, (Bayerische Akademie der Wissenschaften, Philosophisch-Historische Klasse, Abhandlungen, NF. 98), 19-23, pls. 13-25
Kellner, H.-J., 1991
Gürtelbleche aus Urartu, (Prähistorische Bronzefunde, Abteilung XII, Band 3), Stuttgart, F. Steiner Verlag
Kellner, H.-J., 1993
"Ein Assyrische Beutehelm?", in: M.J. Mellink – E. Porada – T. Özgüç, eds., *Nimet Özgüçe Armagan. Aspects of Art and Iconography: Anatolia and its Neighbours. Studies in Honor of Nimet Özgüç*, Ankara, Türk Tarih Kurumu Basimevi, 325-331.

Kendall, T., 1979
Warfare and Military Matters in the Nuzi Tablets, Ph.D. Diss., Brandeis University, Ann Arbor, University Microfilms
Kendall, T., 1981
"*Gurpisu ša awēli*: The Helmets of the Warriors at Nuzi", in: M.A. Morrison – D.I. Owen, eds., *Studies on the Civilization and Culture of Nuzi and the Hurrians, In Honor of Ernest R. Lacheman on his Seventy-Fifth Birthday, April 29, 1981*, Winona Lake, Eisenbrauns, 201-231
Kesamanli, G.P., 1966
"Pogrebenie s bronzovim poyasom iz Hatschbulaga (Azerbayjanskaya SSR)", *Sovetskaya Archeologiya* 3, 221-227
Kestemont, G., 1983
"Tyr et les Assyriens", *Studia Phoenicia* 1-2, 53-78
Khondkarian, M. – Amirian, S. – Hovhannissian, J., 1968
Guidebook to the State Historical Museum of Armenia, Erevan
Kilian-Dirlmeyer, I., 1985a
"Fremde Weihungen in griechischen Heiligtümern vom 8. bis zum Beginn des 7. Jahrhunderts v. Chr.", *Jahrbuch des Römisch-Germanischen Zentralmuseums Mainz* 32, 215-254
Kilian-Dirlmeyer, I., 1985b
"Noch einmal zu den 'Kriegergräbern' von Knossos", *Jahrbuch des Römisch-Germanischen Zentralmuseums Mainz* 32, 196-214
King, L.W., 1909
Cuneiform Texts from Babylonian Tablets in the British Museum, Part XXV, London, The Trustees of the British Museum
King, L.W., 1915
Bronze Reliefs from the Gates of Shalmaneser, London
Klengel-Brandt, E., 1986
Syrien, Kleinasien, (Vorderasiatisches Museum, Kleine Schriften, 6), Berlin, Staatlichen Museen zu Berlin
Knudtzon, J.A., 1915
Die el-Amarna Tafeln, I-II, (Vorderasiatische Bibliothek, 2), Leipzig, Hinrichs
Kohlmeyer, K. – Saherwala, G., 1984
Frühe Bergvölker in Armenien und Kaukasus, (Berliner Forschungen des 19. Jahrhunderts), Berlin, Berliner Gesellschaft für Anthropologie
Kontoleon, N.M., 1969
"Die frühgriechische Reliefkunst", *Archaiologike Ephemeris* 1969, 215-236
Kopcke, G., 1992
"What Role for Phoenicians?", in: Kopcke, G. – Tokumaru, I., eds., *Greece between East and West: 10th – 8th Centuries BC*, (Papers of the Meeting at the Institute of Fine Arts, New York University, March 15–16th, 1990), Mainz, von Zabern, 103-113
König, F.W., 1955
Handbuch der chaldischen Inschriften, (Archiv für Orientforschung, Beiheft 8), Graz, Im Selbsverlage des Herausgebers
Kuftin, B.A., 1936
Materiali k istorii kulturnogo i etnicheskogo oformleniya geofeodalnoy Abhazii. Otchet o rabote otryada Abhazskoy arkheologicheskoy ekspeditsii 1934 g., Tbilisi, Akademii Nauk Gruzinskoy S.S.R.
Kuftin, B.A., 1941
Arkheologogitcheskiye raskopki v Trialeti I, Opit periodizatsii pamyatnikov, Tbilisi, Akademii Nauk Gruzinskoy S.S.R.
Kuftin, B.A., 1946
"Prehistoric Culture Sequence in Transcaucasia", *Southwestern Journal of Anthropology* 2, 340-360
Kuftin, B.A., 1949
Materiali k arkheologii Kolhidi, vol. I, Tbilisi, Akademii Nauk Gruzinskoy S.S.R.
Kukahn, E., 1936
Der griechische Helm, Diss., Marburg-Lahn, H. Bauer
Kunst im Handwerk früher Völker, 1961
Kunst im Handwerk früher Völker, Ausstellungskatalog, Museum für Vor- und Frühgeschichte, Berlin
Kunze, E., 1931

Kretische Bronzereliefs, Stuttgart
Kunze, E., 1938a
"Helme", in: Kunze, E. – Schleif, H., *II. Bericht über die Ausgrabungen in Olympia*, Berlin, Walter de Gruyter, 93-96
Kunze, E., 1938b
"Waffenfunde", in: Kunze, E. – Schleif, H., *II. Bericht über die Ausgrabungen in Olympia*, Berlin, Walter de Gruyter, 67-103
Kunze, E., 1941
"Helme", in: Kunze, E. – Schleif, H., *III. Bericht über die Ausgrabungen in Olympia*, Berlin, Walter de Gruyter, 106-114
Kunze, E., 1944
"Bronzestatuetten", in: Kunze, E. – Schleif, H., *IV. Bericht über die Ausgrabungen in Olympia*, Berlin, Walter de Gruyter, 105-142
Kunze, E., 1958a
"Der frühgriechische Kegelhelm", in: Kunze, E., *VI. Bericht über die Ausgrabungen in Olympia*, "Helme", Berlin, W. de Gruyter, 118-151
Kunze, E., 1958b
"Helme: Der frühgriechische Kegelhelm", in: E. Kunze, *VI. Bericht über die Ausgrabungen in Olympia*, Berlin, Walter de Gruyter, 118-125, pls. 33-38
Kunze, E., 1961a
"Die Ausgrabungen in den frühjahren 1956 bis 1958", in: Kunze, E., *VII. Bericht über die Ausgrabungen in Olympia*, Berlin, Walter de Gruyter, 1-28
Kunze, E., 1961b
"Korinthische Helme", in: Kunze, E., *VII. Bericht über die Ausgrabungen in Olympia*, Berlin, Walter de Gruyter, 56-128, pls. 13-55
Kunze, E., 1961c
"Ein Bronzehelm aus der Perserbeute", in: Kunze, E., *VII. Bericht über die Ausgrabungen in Olympia*, Berlin, Walter de Gruyter, 129-137, pls. 56-57
Kunze, E., 1961–1962
"Die Ausgrabungen in Olympia. Laufbahn und Nordwall des Stadions", *Archaiologikon .Deltion* 17, 107-124
Kunze, E., 1967a
"Helme. Der frühgriechische Kegelhelm", in: Kunze, E., *VIII. Bericht über die Ausgrabungen in Olympia*, Berlin, W. de Gruyter, 111-116
Kunze, E., 1967b
"Kleinplastik aus Bronze", in: Kunze, E., *VIII. Bericht über die Ausgrabungen in Olympia*, Berlin, W. de Gruyter, 213-250, pls. 106-119
Kunze, E., 1967c
"Waffenweihungen", in: Kunze, E., *VIII. Bericht über die Ausgrabungen in Olympia*, Berlin, W. de Gruyter, 83-110, figs. 28-37, pls. 30-49
Lacheman, E.R., 1939
"Appendix D", in: R.F.S. Starr, *Nuzi I*, Cambridge, Mass., Harvard University Press, 540-541
Lacheman, E.R., 1950
Excavations at Nuzi, vol. V: Miscellaneous Texts from Nuzi, II, Cambridge, Mass., Harvard University Press
Lacheman, E.R., 1955
Excavations at Nuzi, vol. VI: The Administrative Archives, (Harvard Semitic Series, 15), Cambridge, Mass., Harvard University Press
Lamb, W., 1929
Greek and Roman Bronzes, London, Methuen
Lamon, R.S. – Shipton, G.M., 1939
Megiddo, vol. I. Seasons of 1925—34, Strata I—V, (Oriental Institute Publications, 42), Chicago
Landsberger, B., 1955
Materialen zum Sumerischen Lexikon, Vol. 3, Roma, Pontificium Institutum Biblicum
Layard, A.H., 1850
Nineveh and its Remains, vols. I-II, London, John Murray
Layard, A.H., 1853
The Monuments of Nineveh, vol. I, London, John Murray
Legrain, L., 1937

Business Documents of the Third Dynasty of Ur, Plates, (Ur Excavation Texts, 3), London
Legrain, L., 1947
Business Documents of the Third Dynasty of Ur, Indexes, Vocabulary, Catalogue, List, (Ur Excavation Texts, 3), London
Lempesi, A.K., 1976
Hoi steles tou Prinia, Athinai
Lempesi, A.K., 1985
To Hiero tou Herme kai tes Aphrodites ste Symi Biannou, vol. I, 1. Chalkina Cretica Toreumata, Athen
Lepore, L., 1984
"Un elmo bronzeo della collezione Ceccanti", in: *Studi di antichità in onore di Guglielmo Maetzke*, vol. 2, (Archaeologica, 49), Roma, Bretschneider, 329-332, pl. 1
Ligabue, G. – Salvatori, S., 1977
"Oriental Bronzes in Private Collections in Venice", *Rivista di Archeologia* 1, 7-15, figs. 1-18
Limet, H., 1960
Le travail du métal au pays de Sumer au temps de la IIIe dynastie d'Ur, (Bibliothèque de la Faculté de Philosophie et Lettres de l'Université de Liège, fasc. CLV), Paris, Société d'Édition "Les Belles Lettres"
Liverani, M., 1988
"The Growth of the Assyrian Empire in the Habur/Middle Euphrates Area: A New Paradigm", *State Archives of Assyria Bulletin* 2, 81-97
Littauer, M.A. – Crouwel, J.H., 1979
Wheeled Vehicles and Ridden Animals in the Ancient Near East, (Handbook der Orientalistik VII/I/II/B/1), Leiden – Köln, Brill
Livingstone, A., ed., 1989
Court Poetry and Literary Miscellanea, (State Archives of Assyria, 3), Helsinki, Helsinki University Press
Lloyd, S., 1967
Early Highland Peoples of Anatolia, London, Thames and Hudson
Lloyd, S. – Müller, H.W., 1986
Ancient Architecture, (History of World Architecture), London, Faber and Faber
Lorimer, H.L., 1950
Homer and the Monuments, London, Macmillan
Loud, G., 1948
Megiddo, vol. II. Seasons of 1935—39, (Oriental Institute Publications, 42), Chicago
Luckenbill, D.D., 1927
Ancient Records of Assyria and Babylonia II. Historical Records of Assyria from Sargon to the End, Chicago, U.P.
Lullies, R., 1962
"Neuerwerbungen der Antikensammlungen in München 1958–1961", *Archäologischer Anzeiger* 77, 594-631
Lullies, R. – Hirmer, M., 1960
Greek Sculpture, London, Thames and Hudson
Luschey, H., 1968
"Studien zu dem Darius-Relief in Bisutun", *Archäologische Mitteilungen aus Iran* (NF) 1, 63-94
Lutz, H.F., 1928
An Agreement between a Babylonian Feudal Lord and His Retainer in the Reign of Darius II., (University of California Publications in Semitic Philology 9/III), Berkeley
Maass, M., 1987
"Helme, Zubehör von Wagen und Pferdegeschirr aus Urartu", *Archäologische Mitteilungen aus Iran* 20, 65-92, pls. 1-10
Maass, M., 1988
Badisches Landesmuseum. 150 Jahre Antikensammlungen Karlsruhe 1838—1988, Karlsruhe, Badisches Landesmuseum
Maass, M., 1989
"Badisches Landesmuseum, Neuerwerbungen 1988, Antike", *Jahrbuch der Staatlichen Kunstsammlungen in Baden-Württemberg* 26, 194-197
Macqueen, J.G., 1986

The Hittites, and Their Contemporaries in Asia Minor, (Revised and Enlarged edition), London, Thames and Hudson
Madhloom, T.A., 1970
The Chronology of Neo-Assyrian Art, London, University of London, The Athlone Press
Magen, U., 1986
Assyrische Königsdarstellungen – Aspekte der Herrschaft, (Baghdader Forschungen 9), Mainz am Rhein, Philipp von Zabern
Magou, E. – Pernot, M. – Rolley, C., 1991
"Bronzes orientaux et orientalisants analyses complémentaires", *Bulletin de correspondance hellénique* 115, 561-577
Mahboubian, H., 1997
Art of Ancient Iran. Copper and Bronze. The Houshang Mahboubian Family Collection, London, 1997, Philip Wilson
Mahmoud, Y., 1970
"Tell al-Fakhar. Report on the First Season's Excavations", *Sumer* 26, 109-126, pls. 2-27
Maier, F.G. – Karageorghis, V., 1984
Paphos. History and Archaeology, Nicosia, Leventis Foundation
Mallowan, M.E.L., 1966
Nimrud and its Remains, vols. I-II, London, Collins
Mallowan, M.E.L. – Davies, L.G., 1970
Ivories in Assyrian Style, (Ivories from Nimrud (1949-1963), vol. 2), London, British School of Archaeology in Iraq
Marfoe, E., *et al.*, 1982
A Guide to The Oriental Institute Museum, Chicago, The Oriental Institute
Margueron, J., 1977
"Ras Shamra 1975 et 1976. Rapport préliminaire sur les campagnes d'automne", *Syria* 54, 151-188
Marinatos, Sp., 1936
"Le temple géométrique de Dréros", *Bulletin de correspondance hellénique* 60, 214-285
Markoe, G., 1985
Phoenician Bronze Bowls from Cyprus and the Mediterranean, (University of California Publications, Classical Studies, 26), Berkeley – Los Angeles, University of California Press
Martirosyan, A.A., 1974
Argishtichinili, (Arkheologicheskie pamyatniki Armenii, 8), Erevan, Akademii Nauk Armyanskoy S.S.R.
Mastrokostas, E., 1961–1962
"Chronika: Achaia", *Archaiologikon Deltion* 17, 126-132
Maxwell-Hyslop, K.R., 1959
"An Urartian Archer on the Zinjirli Chariot Relief", *Bulletin of the Institute of Archaeology* 2, 65-66, pl. 3
Maxwell-Hyslop, K.R., 1974
"Assyrian Sources of Iron", *Iraq* 36, 139-154
Mayer, W., 1983
"Sargons Feldzug gegen Urartu – 714 v. Chr. Text und Übersetzung", *Mitteilungen der Deutschen Orient-Gesellschaft* 115, 65-132
Medinet-Habu I
Medinet-Habu, vol. I. Earlier Historical Records of Ramses III. By the Epigraphic Survey. Field dir. Harold Hayden Nelson, (Oriental Institute Publications, 8), Chicago 1930, The University of Chicago
Medinet-Habu II
Medinet-Habu, vol. II. Later Historical Records of Ramses III. By the Epigraphic Survey. Field dir. Harold Hayden Nelson, (Oriental Institute Publications, 9), Chicago 1932, The University of Chicago
Medvedskaya, I., 1988
"Who Destroyed Hasanlu IV?", *Iran* 26, 1-15
Medvedskaya, I., 1991
"Once More on the Destruction of Hasanlu IV: Problems of Dating", *Iranica Antiqua* 26, 149-161
Meissner, B., 1920
Babylonien und Assyrien, vol. I, Heidelberg, Carl Winters Universitätsbuchhandlung

Melikishvili, G.A., 1960
Urartskie Klinoobraznie Nadpisi, Moskva, Akademii Nauk S.S.S.R.
Mellink, M.J., 1976
"Local, Phrygian and Greek Traits in Northern Lycia", *Revue d'Assyriologie* 1976, 21-34
Menzel, B., 1981
Assyrische Tempel, vols. I-II, (Studia Pohl: Series Major, 10/1), Rome, Biblical Institute Press
Merhav, R., 1991
"Everyday and Ceremonial Utensils", in: R. Merhav, ed., *Urartu. A Metalworking Center in the First Millennium B.C.E.*, Jerusalem, Israel Museum, 200-241
Miron, R., 1990
Kāmid el-Lōz. 10. Das 'Schatzhaus' im Palastbereich. Die Funde, (Saarbrücker Beiträge zur Altertumskunde, Band 46), Bonn, Habelt
Mongait, A.L., 1959
Archaeology in the U.S.S.R., Moscow, Academy of Sciences of the U.S.S.R.
Monloup, Th., 1984
Les figurines de terre cuite de tradition archaique, (Salamine de Chypre XII), Paris, de Boccard
Moorey, P.R.S., 1971
Catalogue of the Ancient Persian Bronzes in the Ashmolean Museum, Oxford, Clarendon Press
Moorey, P.R.S., 1975
"The Terracotta Plaques from Kish and Hursagkalama, c. 1850 to 1650 B.C.", *Iraq* 37, 84ff.
Moorey, P.R.S., 1987
The Ancient Near East, (Ashmolean Museum Publications), Oxford, Ashmolean Museum
Moorey, P.R.S., 1988
"The Technique of Gold-Figure Decoration on Achaemenid Silver Vessels and its Antecedents", *Iranica Antiqua* 23, 231-246
Moorey, P.R.S., 1999
Ancient Mesopotamian Materials and Industries. The Archaeological Evidence, Winona Lake, Eisenbrauns
Moortgat, A., 1959
Alt-Vorderasiatische Malerei, (Meisterwerke Aussereuropäischer Malerei), Berlin, Safari Verlag
Moortgat, A., 1969
The Art of Ancient Mesopotamia: The Classical Art of the Near East, London and New York, Phaidon
Muscarella, O.W., 1977
"Book Review of P. Calmeyer, Reliefbronzen in babylonischem Stil", *Journal of the American Oriental Society* 97, p. 80
Muscarella, O.W., 1980
The Catalogue of Ivories from Hasanlu, Iran, (University Museum Monograph, 40, Hasanlu Special Studies, 2), Philadelphia, The University Museum
Muscarella, O.W., 1981
"Urartäische Gürtel", in: *Archäologie zur Bibel. Kunstsätze aus den biblischen Länden*, Mainz, Philipp von Zabern, 191-197
Muscarella, O.W., 1984
"Ancient Near Eastern Art", *The Metropolitan Museum of Art Bulletin*, Spring 1984, 21, pl. 20
Muscarella, O.W., 1988
Bronze and Iron. Ancient Near Eastern Artifacts in the Metropolitan Museum of Art, New York
Muscarella, O.W., 1989
"King Midas of Phrygia and the Greeks", in Emre, K., *et al.*, *Anatolia and the Anient Near East: Studies in Honour of Tahsin Özgüç*, Ankara, 333-344
Muscarella, O.W., 1992
"Greek and Oriental Cauldron Attachments: A Review", in: Kopcke, G. – Tokumaru, I., eds., *Greece between East and West: 10th – 8th Centuries BC*, (Papers of the Meeting at the Institute of Fine Arts, New York University, March 15—16th, 1990), Mainz, von Zabern, 16-45
Müller-Karpe, H., 1962
"Zur spätbronzezeitlichen Bewaffnung in Mitteleuropa und Griechenland", *Germania* 40, 255-287

Myres, J.L., 1914
Handbook of the Cesnola Collection of Antiquities from Cyprus, New York, The Metropolitan Museum of Art
Nagel, W., 1959—1960
"Ein urartäischer Helm aus dem Argisti-Magazin", *Archiv für Orientforschung* 19, 144-147
Nagel, W., 1963
Altorientalisches Kunsthandwerk, (Berliner Beiträge zur Vor und Frühgeschichte, 5), Berlin, Walter de Gruyter
Nagoev, A.H., 1979
"Raskopki kurganov skifskogo vremeni u s. Nartan", *Arkheologitcheskie Otkritiya* 1979, 119ff.
Neeft, C.W., 1977
"Two Corinthian Helmets in the Allard Pierson Musuem, Amsterdam", in: Boersma, J.S., *et al.*, eds., *Festoen. Opgedragen aan A.N. Zadoks-Josephus Jitta bij haar zeventigste verjaardag*, (Scripta Archaeologica Groningana, 6), Groningen, H.D. Tjeenk Willink, 443-449
Negahban, E.O., 1964
A Preliminary Report on Marlik Excavation (Gohar Rud Expedition, Rudbar 1961-1962), Tehran
Negahban, E.O., 1995
Weapons from Marlik, (Archaeologische Mitteilungen aus Iran, Ergänzungsband 16), Berlin, Dietrich Reimer
Neugebauer, K.A., 1922
"Neuerwerbungen der Antikensammlungen in Deutschland, Berlin", Archäologischer Anzeiger 37, 59-119
Nougayrol, J., 1968
"Textes suméro-accadiens des archives et bibliothèques privées d'Ugarit", *Ugaritica* 5, Paris, Geuthner, 1-371
Nunn, A., 1988
Die Wandmalerei und der glasierte Wandschmuck in Alten Orient, (Handbuch der Orientalistik, I/2/B/6), Leiden, Brill
Oded, B., 1974
"The Phoenician Cities and the Assyrian Empire in the Time of Tiglath-Pileser III", *Zeitschrift des Deutschen Palästinavereins* 90, 38-49
Opificius, R., 1961
Das Altbabylonische Terrakottarelief, Berlin
Oppenheim, A.L., 1950
"Review of H.H. Figulla, Business Documents of the New-Babylonian Period (UET IV), London 1949", *Journal of Cuneiform Studies* 4, 188-195
Oppenheim, A.L., *et al.*, 1956a
The Assyrian Dictionary of the Oriental Institute of the University of Chicago, Vol. 5, G, Chicago, Oriental Institute
Oppenheim, A.L., *et. al.*, 1956b
The Assyrian Dictionary of the Oriental Institute of the University of Chicago, Vol. 6, H, Chicago, Oriental Institute
Oppenheim, A.L., *et al.*, 1964
The Assyrian Dictionary of the Oriental Institute of the University of Chicago, Vol. 1, A, Chicago, Oriental Institute
Oppenheim, A.L., *et al.*, 1971
The Assyrian Dictionary of the Oriental Institute of the University of Chicago, Vol. 8, K, Chicago, Oriental Institute
Orthmann, W., 1971
Untersuchungen zur Späthethitischen Kunst, (Saarbrücker Beiträge zur Altertumskunde, 8), Bonn, R. Habelt
Orthmann, W., 1975
Der Alte Orient, (Propyläen Kunstgeschichte, 14), Berlin, Propyläen Verlag
Otcheti 1896
Otcheti arkheologitcheskoy komissii za 1894 g., S.-Peterburg
Overlaet, B.J., 1979
"Pointed Helmets of the Iron Age from Iran", *Iranica Antiqua* 14, 51-63, pls. 1-4
Özgen, E., 1982

The Urartian Bronze Collection at the University Museum: The Urartian Armor, Ph. D. Diss. University of Pennsylvania, Ann Arbor, University Microfilms
Özgen, E., 1983
"The Urartian Chariot Reconsidered: I. Representational Evidence, 9th – 7th centuries B.C.", *Anatolica* 10, 111-131
Özgüç, N., 1965
Kültepe Mühür Baskilarinda Anadolu Grubu (The Anatolian Group of Cylinder seal Impressions from Kültepe), (Türk Tarih Kurumu Yayinlarindan – V. Seri, No. 22), Ankara, Türk Tarih Kurumu Basimevi
Özgüç, T., 1966
Altintepe I, (Türk Tarih Kurumu Yayinlarindan – V. Seri, No. 24), Ankara, Türk Tarih Kurumu Basimevi
Özgüç, T., 1986
Kültepe – Kanis II. Eski yakindogu'nun ticaret merkezinde yeni arastirmalar, (New Researches at the Trading Center of the Ancient Near East), (Türk Tarih Kurumu Yayinlari, V. Dizi – Sa. 41), Ankara, Türk Tarih Kurumu Basimevi
Pallottino, M., 1955
"Gli scavi di Karmir Blur in Armenia e il problema delle connessioni tra l'Urartu, la Grecia e l'Etruria", *Archeologia Classica* 7, 109-123, pls. 42-49
Papathanasopoulos, G.A., 1969
"Archaiotetes kai mnemeia Eleias: 1. Mouseion Olympias", *Archaiologikon Deltion* 24, 146-150
Paris – Hotel Drouot, 1991
Archéologie (Bronzes du Luristan, Important masque funéraire Égyptien, Rare casque Ourartéen, Terres cuites Iraniennes, Verrerie Antique, Textiles coptes) Vente aux enchères publiques. La Vendredi 4 Octobre 1991 à 14 heures 30, Paris – Hotel Drouot – Salle 4
Parrot, A., 1948
Tello. Vingt campagnes des fouilles (1877—1933), Paris, Éditions Albin Michel
Parrot, A., 1956
Mission archéologique de Mari, vol. I: Le Temple d'Ishtar, (Institut Français d'Archéologie de Beyrouth. Bibliothèque Archéologique et Historique, 65), Paris, Paul Geuthner
Parrot, A., 1958
Mission archéologique de Mari, vol. II: Le palais, vol. 2: Peintures murales, (Institut Français d'Archéologie de Beyrouth. Bibliothèque Archéologique et Historique, 69), Paris, Paul Geuthner
Parrot, A., 1959
Mission archéologique de Mari, vol. II: Le palais, vol. 3: Documents et monuments, (Institut Français d'Archéologie de Beyrouth. Bibliothèque Archéologique et Historique, 70), Paris, Paul Geuthner
Parrot, A., 1961
Ninieveh and Babylon, London
Parrot, A., 1983
Sumer und Akkad, (Universum der Kunst), München, C.H. Beck
Paterson, A., 1912
Assyrian Sculptures, Palace of Sinacherib, Hague, M. Nijhoff
Payne, H.G.G., 1927-1928
"Early Greek Vases from Knossos", *Annals of the British School of Archaeology at Athens* 29, 224-298
Perdrizet, P., 1908
Monuments figurés. Petits bronzes, terres-cuites, antiquités diverses, (Fouilles de Delphes V), Paris, A. Fontemoing
Pernice, E., 1892
"Geometrische Vase aus Athen", *Mitteilungen des Deutschen Archäologischen Instituts, Athenische Abteilung* 17, pp.
Petrie, W.M.F., 1897
Six Temples at Thebes, London, Bernhard Quatritch
Petrie, W.M.F., 1917
Tools and Weapons, (British School of Archaeology in Egypt, 30), London, British School of Archaeology
Petrie, W.M.F., 1930

Beth-Pelet (Tell Fara), vol. I, (British School of Archaeology in Egypt, 48), London, British School of Archaeology in Egypt – B. Quatrich

Petrie, W.M.F., 1933
Ancient Gaza, vol. III, (British School of Archaeology in Egypt, 55), London, British School of Archaeology – B. Quatrich

Petrie, W.M.F., 1934
Ancient Gaza, vol. IV, (British School of Archaeology in Egypt, 56), London, British School of Archaeology

Pettinato, G., 1979
Catalogo dei testi cuneiformi di Tell Mardikh – Ebla, (Materiali Epigrafici di Ebla, 1), Napoli, Istituto Universitario Orientale di Napoli

Pettinato, G., 1980
Testi amministrativi della Biblioteca L.2769, (Materiali Epigrafici di Ebla, 2), Napoli, Istituto Universitario Orientale di Napoli

Pfeiffer, R.H. – Lacheman, E.R. 1942
Excavations at Nuzi, vol IV: Miscellaneous Texts from Nuzi I, (Harvard Semitic Series, 13), Cambridge, Mass., Harvard University Press

Pflug, H., 1988a
"Griechische Helme Geometrischer Zeit", in: A. Bottini, *et al.*, *Antike Helme. Sammlung Lipperheide und andere Bestände des Antikenmuseums Berlin*, (Römisch-Germanisches Zentralmuseum, Monographien, 14), Mainz, Verlag des Römisch-Germanisches Zentralmuseums, 11-26

Pflug, H., 1988b
"Kyprische Helme", in: A. Bottini, *et al.*, *Antike Helme. Sammlung Lipperheide und andere Bestände des Antikenmuseums Berlin*, (Römisch-Germanisches Zentralmuseum, Monographien, 14), Mainz, Verlag des Römisch-Germanisches Zentralmuseums, 27-41

Pflug, H., 1988c
"Korinthische Helme", in: A. Bottini, *et al.*, *Antike Helme. Sammlung Lipperheide und andere Bestände des Antikenmuseums Berlin*, (Römisch-Germanisches Zentralmuseum, Monographien, 14), Mainz, Verlag des Römisch-Germanisches Zentralmuseums, 65-106

Pflug, H., 1989
"Kyprische Helme", in: H. Pflug, *Schutz und Zier. Helme aus dem Antikenmuseum Berlin und Waffen anderer Sammlungen*, (Antikenmuseum Basel und Sammlung Ludwig), Basel, 17-18

Piaskowski, J. – Wartke, R.-B., 1989
"Technologische Untersuchungen an einigen urartaischen Eisenobjekten aus Toprakkale", *Forschungen und Berichte* 27, 89-114

Pigott, V.C., 1982
The Adoption of Iron in Western Iran in the Early First Millennium B.C.: An Archaeo-Metallurgical Study, Ann Arbor, University Microfilms International

Piotrovsky, B.B., 1944
Istoriya i kultura Urartu, Erevan, Akademii Nauk Armyanskoy S.S.R.

Piotrovsky, B.B., 1949
"Tri urartskie nadpisi na bronzovih predmetah iz Teishebaini (Karmir-blur)", *Epigrafika Vostoka* 3, 88-89

Piotrovsky, B.B., 1950
Karmir Blur I, Rezultaty Raskopok 1939—1949, Erevan, Akademii Nauk Armyanskoy S.S.R.

Piotrovsky, B.B., 1952
Karmir Blur II, Rezultaty Raskopok 1949—1950, Erevan, Akademii Nauk Armyanskoy S.S.R.

Piotrovsky, B.B., 1955
Karmir Blur III. Rezultaty Raskopok 1950—1953, Erevan, Akademii Nauk Armyanskoy S.S.R.

Piotrovsky, B.B., 1959
Vanskoe Tsarstvo. Urartu, Moskva, Izdatelstvo Vostotshnoy Literaturi

Piotrovsky, B.B., 1962
Iskusstvo Urartu VIII-VII vekov do n.e., Leningrad, Izdatyelstvo Gosudarstvennogo Ermitaza

Piotrovsky, B.B., 1965
"Teichebaini, centre ourartien du VIIe siècle avant notre ère et ses rapports avec les autres pays du monde antique", in: *Le rayonnement des civilisations grecque et romaine sur les cultures périphériques. Huitième congrès international d'archéologie classique (Paris, 1963)*, Paris, E. de Boccard, 411-419, pls. 96-99

Piotrovsky, B.B., 1966a

Il regno di Van. Urartu, (Incunabula Greca, 12), Roma, Edizioni dell'Ateneo

Piotrovsky, B.B., 1966b

"Urartu", in: *Enciclopedia dell'Arte Antica Classica e Orientale, vol. VII*, Roma, Istituto della Enciclopedia Italiana, 1060-1062

Piotrovsky, B.B., 1967

Urartu. The Kingdom of Van and its Art , London, Evelyn Adams and Mackay

Piotrovsky, B.B., 1968

Urartu , (Archaeologia Mundi 26), München, Wilhelm Heine Verlag

Piotrovsky, B.B., 1969

Urartu, London, Barrie & Rockliff: The Cresset Press

Piotrovsky, B.B., 1970

Karmir Blur, Leningrad, Aurora

Place, V., 1867—1870

Ninive et l'Assyrie, vol. I, Paris

Pleiner, R. – Bjorkman, J.K. 1974

"The Assyrian Iron Age: The History of Iron in Assyrian Civilization", *Proceedings of the American Philosophical Society* 118, 283-313

Porada, E., 1948

Corpus of Ancient Near Eastern Seals in North American Collections. The Collection of the Pierpont Morgan Library, (The Bollingen Series, 14), Washington, Pantheon Books

Postgate, J.N., 2000

"The Assyrian Army in Zamua", *Iraq* 62 (2000), 89-108

Protonotariou-Deilaki, E., 1971

"Chronika: Argos", *Archaiologikon Deltion* 26, 74-82

Protonotariou-Deilaki, E., 1973

"Chronica, Argos", *Archaiologikon Deltion* 28, 97-99

Protonotariou-Deilaki, E., 1982

"Απο το Αργος του 8ου και 7ου ΑΙ Π. Χ.", *Annuario della Scuola Archaeologica di Atene* 60, 33-48

Rabinovich, B.Z., 1941

"Shlem skifskogo perioda", *Gosudarstvenniy Ermitage, Trudi otdela istorii pervobitnoy kulturi*, vol. I, 100-128

Radan, G., 1981

"Assyrian Helmet found off the Coast of Atlit", *Sefunim* 6, 49-57

Rawlinson, H.C., 1870

The Cuneiform Inscriptions of Western Asia, vol. III, London

Rawlinson, H.C., 1909

The Cuneiform Inscriptions of Western Asia , vol. V, London

Reade, J.E., 1972

"The Neo-Assyrian Court and Army: Evidence from the Sculptures", *Iraq* 34, 87-112

Reiner, E., *et al.*, 1984

The Assyrian Dictionary of the Oriental Institute of the University of Chicago, Vol. 15, S, Chicago, Oriental Institute

Reiner, E., *et al.*, 1989

The Assyrian Dictionary of the Oriental Institute of the University of Chicago, Vol. 17, Š, Part I, Chicago, Oriental Institute

Renfrew, C., 1983

Before Civilization: The Radiocarbon Revolution and Prehistoric Europe, Pelican Books

Richards, G.C., 1891

"Archaic Reliefs at Dhimitzana", *Journal of Hellenic Studies* 12, 41-45

Richter, G.M.A., 1934

"A Colossal Dipylon Vase", *Bulletin of the Metropolitan Museum of Art* 29, 169-172

Riemschneider, M., 1965

Das Reich am Ararat, Leipzig, Koehler & Amelang

Riis, P.J., 1948

Hama. Fouilles et recherches 1931—1938, vol. II/3: Les cimitères à crémation, Kobenhavn, Nordisk Forlag

Rittig, D., 1983

"Ein kleinasiatisches Rollsiegel mit Reitermotiv", *Orientalia* 52, 156-160

Rittig, D., 1994
"Zum Ursprung des Raupenhelmes", in: Calmeyer, P. – Hecker, K. – Jakob-Rost, L. – Walker, C.B.F., Hrsg., *Beiträge zur Altorientalischen Archäologie und Altertumskunde. Festschrift für Barthel Hrouda zum 65. Geburtstag*, Wiesbaden, Harrassowitz, 247-254
Roaf, M., 1990
Cultural Atlas of Mesopotamia and the Ancient Near East, Oxford, Equinox
Rostoker, W. – Bronson, B., 1990
Pre-Industrial Iron. Its Technology and Ethnology, (Archeomaterials Monograph, 1), Philadelphia
Rowe, A., 1940
The Four Canaanite Temples of Beth-Shan, Philadelphia
Röllig, W., 1992
"Asia Minor as a Bridge between East and West: the Role of the Phoenicians and Arameans in the Transfer of Culture", in: Kopcke, G. – Tokumaru, I., eds., *Greece between East and West: 10th – 8th Centuries BC*, (Papers of the Meeting at the Institute of Fine Arts, New York University, March 15—16th, 1990), Mainz, von Zabern, 93-102
Rubinstein, R.I., 1975
U sten Teishebaini, Moskva, Sovetskiy Hudozshnik
Salonen, A., 1951
Die Landfahrzeuge des alten Mesopotamien, Helsinki
Salonen, A., 1955
Hippologica Accadica, Helsinki
Salonen, E., 1966
Die Waffen der alten Mesopotamier, (Studia Orientalia, 33), Helsinki
Salvatori, S., 1975
"Un elmo assiro figurato proveniente dal Luristan", *Oriens Antiquus* 14, 255-264, figs. 1-2, pls. 36-38
Salvini, M., 1973
"Urartea, Arte", in: *Enciclopedia dell'Arte Antica Classica e Orientale, Supplement 1970*, Roma, Istituto della Enciclopedia Italiana, 881-888
Sandars, N.K., 1978
The Sea Peoples. Warriors of the ancient Mediterranean 1250—1150 B.C., London, Thames & Hudson
Schaaff, U., 1973
"Frühlatènezeitliche Grabfunde mit Helmen vom Typ Berru", *Jahrbuch des Römisch-Germanischen Zentralmuseums Mainz* 20, 81-106
Schaeffer, C.F.A., 1937
"Les fouilles de Ras Shamra – Ugarit. Neuvième campagne (Printemps 1937). Rapport sommaire", *Syria* 18, 144ff.
Schaeffer, C.F.A., 1949
"La grande stele du Baal au foudre de Ras Shamra", *Ugaritica II*, (Mission de Ras Shamra, 5), Paris, P. Geuthner, 121-130
Schaeffer, C.F.A., 1951
"Reprise de recherches archéologiques à Ras Shamra – Ugarit. *Syria* 28, 1-21, pls. I-IV
Schaeffer, C.F.A., 1962
"Fouillles et découvertes des XVIIIe et XIXe campagnes, 1954—1955, *Ugaritica* IV, (Mission de Ras Shamra, 15, Institut Français d'Archéologique de Beyrouth, Bibliothèque Archéologique et Historique, 74), Paris, Paul Geuthner, 1-147
Schaeffer, C.F.A., 1968
"Commentaires sur les lettres et documents trouvés dans les bibliothèques privées d'Ugarit", *Ugaritica V*, (Mission de Ras Shamra, 16), Paris, P. Geuthner, 607-760
Schauer, P., 1979—1980
"Urnenfelderzeitliche Helmformen und ihre Vorbilder", *Fundberichte aus Hessen* 19-20, 521-543
Schauer, P., 1983
"Orient im Spätbronze- und Früheisenzeitlichen Occident. Kulturbeziehungen zwischen der Iberischen Halbinsel und dem Vorderen Orient während des späten 2. und des ersten Drittels des 1. Jahrtausends v. Chr.", *Jahrbuch des Römisch-Germanischen Zentralmuseums Mainz* 30, 175-194
Schauer, P., 1988

"Die Kegel- und Glockenförmigen Helme mit Gegossenem Scheitelknauf der Jüngeren Bronzezeit Alteuropas", in: A. Bottini, *et al.*, *Antike Helme. Sammlung Lipperheide und andere Bestände des Antikenmuseums Berlin*, (Römisch-Germanisches Zentralmuseum, Monographien, 14), Mainz, Verlag des Römisch-Germanisches Zentralmuseums, 181-194

Scheil, V., 1913
Textes Élamites-Semitiques, (Mémoires de la Mission archéologique de Susiane, 14), Paris, Ernest Leroux

Schmidt, G., 1968
Kyprische Bildwerke aus dem Heraion von Samos, (Samos, 7), Bonn, R. Habelt

Schroeder, O., 1915
Die Thontafeln von el-Amarna, vols. I-II, (Vorderasiatische Schriftdenkmäler, 11-12), Leipzig, Hinrichs

Schumacher, E., 1986
"Luristan", in: *Museum (Altenessen Museum)*, München, Magazinpresse, 58-69

Schweitzer, B., 1971
Greek Geometric Art, London, Phaidon

Seeden, H., 1980
The Standing Armed Figurines in the Levant, (Prähistorische Bronzefunde, Abteilung I, Band 1), München, C.H. Beck

Seidl, U., 1976—1980
"Inanna/Istar (Mesopotamien)", *Reallexikon der Assyriologie und Vorderasiatische Archäologie V*, Berlin, de Gruyter, 87-89

Seidl, U., 1986
"Ein Pferde-Pektorale", in: H. Kelly-Bucellati ed., *Insight through Images. Studies in Honor of Edith Porada*, (Bibliotheca Mesopotamica, 21), Malibu, Undena, 229-236

Semionov, P.P., 1955
"Shlemi iz Severnoy Osetii", *Kratkie soobsheniya instituta istorii materialnoy kulturi* 52, pp.

Sethe, K. 1907
Urkunden der 18. Dynastie, vol. III, Leipzig

Shanks, H., 1984
"Destruction of Judean Fortress Portrayed in Dramatic Eighth-Century B.C. Pictures", (Book review), *Biblical Archaeology Review* 10/2, 48-65

Shaw, J.W., 1980
"Excavations at Kommos (Crete) During 1979", *Hesperia* 49, 207-250

Shaw, J.W., 1989
"Phoenicians in Southern Crete", *American Journal of Archaeology* 93, 165-183

Singer, S.F., 1985
"Treasures from the Lands of the Bible", *Biblical Archaeology Review* 11/2, 26-38

Smith, S., 1938
Sculptures in the British Museum, from Shalmaneser III to Sennacherib, London, The Trustees of the British Museum

Snodgrass, A.M., 1964
Early Greek Armour and Weapons from the End of the Bronze Age to 600 B.C., Edinburgh, University Press

Sollberger, E. 1956
Corpus des inscription "royales" présargonics de Lagas, Genève

Sotheby's 1976
Sotheby's Catalogue of Antiquities and Tribal Art. Tuesday, 7th December, London, London

Speiser, E.A., 1950
"On Some Articles of Armour and Their Names", *Journal of the American Oriental Society* 70, 47-49

Starr, R.F.S., 1939
Nuzi, vol. I, Cambridge, Mass., Harvard University Press

Steiner, P., 1906
"Bronze-Statuette aus Olympia", *Mitteilungen des Deutschen Archäologischen Instituts, Athenische Abteilung* 31, 219-227

Strassmaier, J.N., 1893
Inschriften von Darius, König von Babylon (521—485 v. Chr.) von den Thontafeln des Britischen Museums , vol. II, (Babylonische Texte, 11), Leipzig, E. Pfeiffer

Strommenger, E., 1962

Fünf Jahrtausende Mesopotamien. Die Kunst von den Anfängen um 5000 v Chr. bis zu Alexander dem Großen, München, Hirmer
Stronach, D., 1958
"Metal Objects from the 1957 Excavations at Nimrud", *Iraq* 20, 169-181, pls. XXXII-XXXVI
Sulimirski, T., 1970
Prehistoric Russia: An Outline, London, John Baker
Tadmor, H., 1975
"Assyria and the West. The Ninth Century and its Aftermath", in: Goedicke, H. – Roberts, J.J., eds., *Unity and Diversity*, Baltimore, 36-48
Tallqvist, K., 1938
Akkadische Götterepitheta. Mit einem Götterverzeichnis und einer Liste der prädikativen Elemente der sumerischen Götternamen, (Studia Orientalia, 7), Helsingfors, Societas Orientalis Fennica
Tasyürek, O.A., 1974
"The Bronze Urartian Helmet in the Gaziantep Museum", *Türk Arkeoloji Dergisi* 21, 177-181, fig. 1, pl. 1
Tasyürek, O.A., 1975
"Some Inscribed Urartian Bronze Armour", *Iraq* 37, 154ff.
Tasyürek, O.A., 1987
"Darstellungen des urartäischen Gottes Haldi", in: S. Sahin *et al.*, eds., *Studien zur Religion und Kultur Kleinasiens, vol. II*, (Fs. F.K. Dörner), Leiden, Brill, 940-955
Tatton-Brown, V., 1979a
"A Terracotta 'Geryon' in the British Museum", *RDAC* 1979, 281-288
Tatton-Brown, V., ed., 1979b
Cyprus B.C. 7000 Years of History, London, The Trustees of the British Museum
Tehov, B.V., 1977
Centralniy Kavkaz v XVI—X vekov do n.e., Moskwa, Nauka
Terrace, E.L.B., 1962
The Art of the Ancient Near East in Boston, Museum of Fine Arts, Boston
Thompson, R.C., 1904
Cuneiform Texts from the Babylonian Tablets in the British Museum, vol. XVIII, London
Thureau-Dangin, F., 1903
Recueil de tablettes chaldéennes, Paris
Thureau-Dangin, F., 1910
Inventaire des tablettes de Tello conservées au Musée Impérial Ottoman, vol. I: Textes de l'époque de Agadé (Fouilles d'Ernest de Sarzec en 1895), Paris, Ernest Leroux
Thureau-Dangin, F. – Dunand, M., 1936
Til-Barsib, (Bibliothèque Archéologique et Historique, XXIII), Paris, Geuthner
Thureau-Dangin, F. – Barrois, A. – Dossin, G. – Dunand, M., 1931
Arslan-Tash, (Bibliothèque Archéologique et Historique, XVI), Paris, Geuthner
Tölle, R., 1963
"Figürliche bemalte Fragmente der Geometrischen Zeit vom Kerameikos", *Archäologischer Anzeiger* 78, 642-665
Törnkvist, S., 1973
"Arms, Armour and Dress of the Terracotta Sculpture from Ajia Irini, Cyprus", *Medelhavsmuseet Bulletin* 6, 7-55
Trésors de l'ancien Iran, 1966
Trésors de l'ancien Iran, (Musée Rath – Genève, 8 juin – 25 septembre 1966)
Tshernenko, E.V., 1968
Skifskiy Dospeh, Kiev, Naukova Dumka
Tshernenko, E.V., 1984
Skifo-persidskaya voyna, Kiev, Naukova Dumka
Tufnell, O., *et al.*, 1953
Lachish III. The Iron Age, Oxford, Oxford University Press
Untracht, O., 1975
Metal Techniques for Craftsmen. A Basic Manual for Craftsmen on the Methods of Forming and Decorating Metals, New York, Doubleday
Ussishkin, D., 1969
"The Date of the Neo-Hittite Enclosure in Karatepe", *Anatolian Studies* 19, 121-137
Ussishkin, D., 1982

The Conquest of Lachish by Sennacherib, (Tel Aviv University. Publications of the Institute of Archaeology , 6) Tel Aviv, Tel Aviv University, Institute of Archaeology
Uvarova, P.S., 1900
Mogilniki Severnogo Kavkaza, (Materiali po arkheologii Kavkaza 8), Moskva, Isledovaniya Moskovskogo Arkheologicheskogo Obshchestva
Vanden Berghe, L., 1959
Archéologie de L'Iran Ancien, Leiden, Brill
Vanden Berghe, L., 1967
Musées Royaux d'Art et d'Histoire, Galerie de l'Asie Anterieure et de l'Iran Anciens, Bruxelles
Vanden Berghe, L., 1968
Het Archeologisch Onderzoek naar de Bronskultuur van Luristan: Opgravingen in Pusht-i Kuh, vol. I: Kalwali en War Kabud (1965 en 1966), Brussel, Paleis der Academiën
Vanden Berghe, L., 1982
"Un carquois Urartéen", in: J. Quaegebeur, ed., *Studia Paulo Naster Oblata, II: Orientalia Antiqua*, (Orientalia Lovaniensia Analecta, 13), Leuven, Department Oriëntalistiek – Uitgeverij Peeters, 245-257, pls. XVI-XIX
Vanden Berghe, L. – De Meyer, L., 1982
Urartu een vergeten cultuur uit het bergland Armenië, Sint-Pietersabdij – Gent, 9 oktober 1982 – 30 januari 1983
van Loon, M.N., 1966
Urartian Art. Its distinctive Traits in the Light of New Excavations, (Uitgaven van het Nederlands Historisch-Archaeologisch Instituut te Istanbul, 20), Istanbul, Nederlands Historisch-Archaeologisch Instituut
van Loon, M.N., 1975
"Die Kunst von Urartu", in: W. Orthmann, *Der Alte Orient*, (Propyläen Kunstgeschichte, 14), Berlin, Propyläen Verlag, 463
van Pée, M., 1965
"Note sur le traitment d'un casque en fer orné de bronze provenant du Louristan', *Institut Royal du Patrimoine Artistique, Bulletin* 8, 196-198, pl. on p. 196
Ventzke, W., 1983
"Zur Rekonstruktion eines bronzenen Schuppenpanzers", in: R. Hachmann, ed., *Kāmid el-Lōz 1977—81*, (Saarbrücker Beiträge zur Altertumskunde, 36), Bonn, Rudolf Habelt, 162-166, pls. 21-22
Ventzke, W., 1986
"Der Schuppenpanzer von Kāmid el-Lōz", in: R. Hachmann, ed., *Frühe Phöniker im Libanon. 20. Jahre deutsche Ausgrabungen in Kāmid el-Lōz*, Mainz am Rhein, Philipp von Zabern, 94-100
Verdelis, N.M., 1963
"Neue geometrische Gräber in Tiryns", *Mitteilungen des Deutschen Archäologischen Instituts, Athenische Abteilung* 78, 1-62
Verdelis, N.M., 1967
"Neue Funde von Dendra", *Mitteilungen des Deutschen Archäologischen Instituts, Athenische Abteilung* 82, 1-53, figs. 1-34, pls. 1-2
Verdelis, N.M., 1977
"Weapons", in: P. Åström, *The Cuirass Tomb and Other Finds at Dendra. Part 1: The Chamber Tombs*, (Studies in Mediterranean Archaeology, 4/1), Göteborg, Paul Åströms Ferlag
Vingt ans d'acquisitions, 1967
Vingt ans d'acquisitions au Musée du Louvre, 1947—1967, (Orangerie des Tuileries, 16 déc. 1967 – mars 1968), Paris
Vinogradov, V.B., 1973
"Kubanskiy shlem iz Tshetsheno-Ingushetii", in: *Skifskie drevnosti*, Kiev
von Brandenstein, C.-G., 1940
"Zum churritischen Lexikon. 5. Die Waffen der Sauska des Feldzuges", *Zeitschrift für Assyriologie und Vorderasiatischen Archäologie* 46, 103-105
von Gaertringen, H. – Wilski, P., 1904
Stadtgeschichte von Thera, (Thera III), Berlin, Reimer
von Haller, A., 1954

Die Gräber und Grüfte von Assur, (Wissenschaftliche Veröffentlichung der Deutschen Orient-Gesellschaft, 65), Berlin, Mann
von Lipperheide, F., 1896
Antike Helme, Munich
von Luschan, F., *et al.*, 1893
Ausgrabungen in Sendschirli, vol. I, (Mitteilungen aus den Orientalischen Sammlungen, XI), Berlin
von Luschan, F., *et al.*, 1902
Ausgrabungen in Sendschirli, vol. III, (Mitteilungen aus den Orientalischen Sammlungen, XIII), Berlin
von Luschan, F. – Andrae, W., 1943
Die Kleinfunde von Sendschirli, Ausgrabungen in Sendschirli , vol. V, (Mitteilungen aus den Orientalischen Sammlungen, XV), Berlin, Walter de Gruyter
von Soden, W., 1965—1981
Akkadisches Handwörterbuch, I-III, Wiesbaden, Harrassowitz
Voth, S.M., 1981
Analysis of Military Titles and Functions in the Old-Babylonian Period, (Diss., Hebrew Union College – Jewish Institute of Religion, Ohio)
Völcker-Janssen, W., 1992
"Helm, Spitzhelm", in: D. Hopp – H. Schaaf – W. Völcker-Janssen, *Iranische Metallfunde im Museum Altenessen*, (Saarbrücker Beiträge zur Altertumskunde, 57), Bonn, Rudolf Habelt, 51-52.
Waetzold, H., 1981
"Kopfbedeckung, A. Philologisch", *Reallexikon der Assyriologie und Vorderasiatischen Archäologie* VI, 3-4. Berlin, Walter de Gruyter, 197-203
Waldbaum, J.C., 1989
"B. Metalwork from Idalion, 1971—1980", in: L.E. Stager – A.M. Walker, eds., *American Expedition to Idalion, Cyprus 1973—1980*, (Oriental Institute Communications, 24), Chicago, The Oriental Institute of the University of Chicago, 328-351, pls. 1-4
Waldstein, Ch., 1905
The Argive Heraeum, vol. II, Boston and New York, Houghton, Mifflin and Co.
Walters, H.B., 1899
Catalogue of the Bronzes, Greek, Roman, and Etruscan, in the Department of Greek and Roman Antiquities, British Museum, London
Wartke, R.-B., 1991
"Production of Iron Artifacts", in: R. Merhav, ed., *Urartu. A Metalworking Center in the First Millennium B.C.E.*, Jerusalem, Israel Museum, 322-331
Wäfler, M., 1975
Nicht-Assyrer neuassyrischer Darstellungen, (Alter Orient und Altes Testament, 26), Neukirchen-Vluyn, Butzon & Bercker Kevelaer
Weidner, E.F., 1959
Die Inschriften Tukulti-Ninurtas I. und seiner Nachfolger, (Archiv für Orientforschung, Beiheft 12), Graz
Weidner, E.F., 1922
Keilschrifturkunden aus Boghazköy, vol. III, Berlin, Staatliche Museen zu Berlin, Vorderasiatische Abteilung
Weiss, C., 1977
"An Unusual Corinthian Helmet", *California Studies in Classical Antiquity* 10, 195-207
Weissbach, F.H., 1911
Die Keilinschriften der Achéémeniden, (Vorderasiatische Bibliothek, 3), Leipzig, Hinrichs
Weissbach-Markkleeberg, F.H., 1938
"Die dreisprachige Inschrift Darius Suse e.", *Zeitschrift für Assyriologie und Vorderasiatischen Archäologie* 44, 150-169
Westholm, A.W., 1935
"Amathus", in: E. Gjerstad *et al.*, eds., *The Swedish Cyprus Expedition II.*, Stockholm, 1-142
Westholm, A.W., 1938
"Cypro-Archaic Splint Armour", *Acta Archaeologica* 9, 163-173
Wide, S., 1899
"Geometrische Vasen aus Griechenland", *Mitteilungen des Deutschen Archäologischen Instituts, Athenische Abteilung* 14, 78-86

Wilcke, C., 1969
Das Lugalbandaepos, Wiesbaden, Harrassowitz
Wilcke, C., 1975
"Helm. A. Philologisch", in: *Reallexikon der Assyriologie und Vorderasiatischen Archäologie IV*, Berlin, Walter de Gruyter, 311-313
Wilkinson, C.K., 1965
"Art of the Marlik Culture", *Bulletin of the Metropolitan Museum of Art*, Nov. 1965 , 101-109
Williams, A.R. – Maxwell-Hyslop, K.R., 1976
"Ancient Steel from Egypt", *Journal of Archaeological Science* 3, 283-305
Winckler, H., 1896
Die Thontafeln von Tell el-Amarna, (Keilinschriftliche Bibliothek, 5), Berlin
Winckler, H. – Abel, L., 1889—1890
Der Thontafelfund von el Amarna, (Mitteilungen aus dem Orientalischen Sammlungen der Königlichen Museen zu Berlin, 1-3), Berlin
Winlock, H.E., 1947
Rise and Fall of the Middle Kingdom in Egypt, London
Winter, I.J., 1980
A Decorated Breastplate from Hasanlu, Iran, (University Museum Monograph, 39, Hasanlu Special Studies, 1), Philadelphia, The University Museum
Winter, I., 1988
"North Syria as a Bronzeworking Centre in the Early First Millennium BC: Luxury Commodities at Home and Abroad", in: Curtis, J.E., ed., *Bronzeworking Centres of Western Asia c. 1000—539 BC*, London, Keegan and Paul, 193-225
Woolley, C.L., 1921
Charchemish II, The Town Defences, London
Woolley, C.L., 1934
Ur Excavations, vol. II: The Royal Cemetery, Oxford, University Press
Woolley, C.L., 1955
Alalakh, Oxford, Oxford University Press
Woolley, C.L. – Barnett, R.D., 1952
Charchemish III, London
Wright, G.E., 1958
Biblische Archäologie, Göttingen, Vandenhoeck & Ruprecht
Xenaki-Sakellariou, A., 1953
"La représentation du casque en dents de sanglier", *Bulletin de correspondance hellénique* 77, 46-58
Yadin, Y., 1963
The Art of Warfare in Biblical Lands in the Light of Archaeological Discovery, London, Weidenfeld and Nicolson
Yalouri, A., 1971
"A Hero's Departure", *American Journal of Archaeology* 75, 269-275
Yalouris, N., 1960
"Mykenische Bronzeschutzwaffen", *Mitteilungen des Deutschen Archäologischen Instituts, Athenische Abteilung* 75, 42-67
Young, J.H. – Young, S.H., 1955
Terracotta Figurines from Kourion in Cyprus, (Museum Monographs), Philadelphia, University Museum
Zaccagnini, C., 1979
"Les rapports entre Nuzi and Hanigalbat", *Assur* 2, 1-27
Zaccagnini, C., 1983
"Pattern of Mobility among Ancient Near Eastern Craftsmen", *Journal of Near Eastern Studies* 42, 245ff.
Zahlhaas, G., 1993
"Zwei neue urartäische Inschriften", *Archäologische Mitteilungen aus Iran* 26, 47-54
Zervos, Ch., 1946
L'Art en Grèce du troisieme millénaire au IVe siècle avant notre ère, Paris, Cahiers d'Art

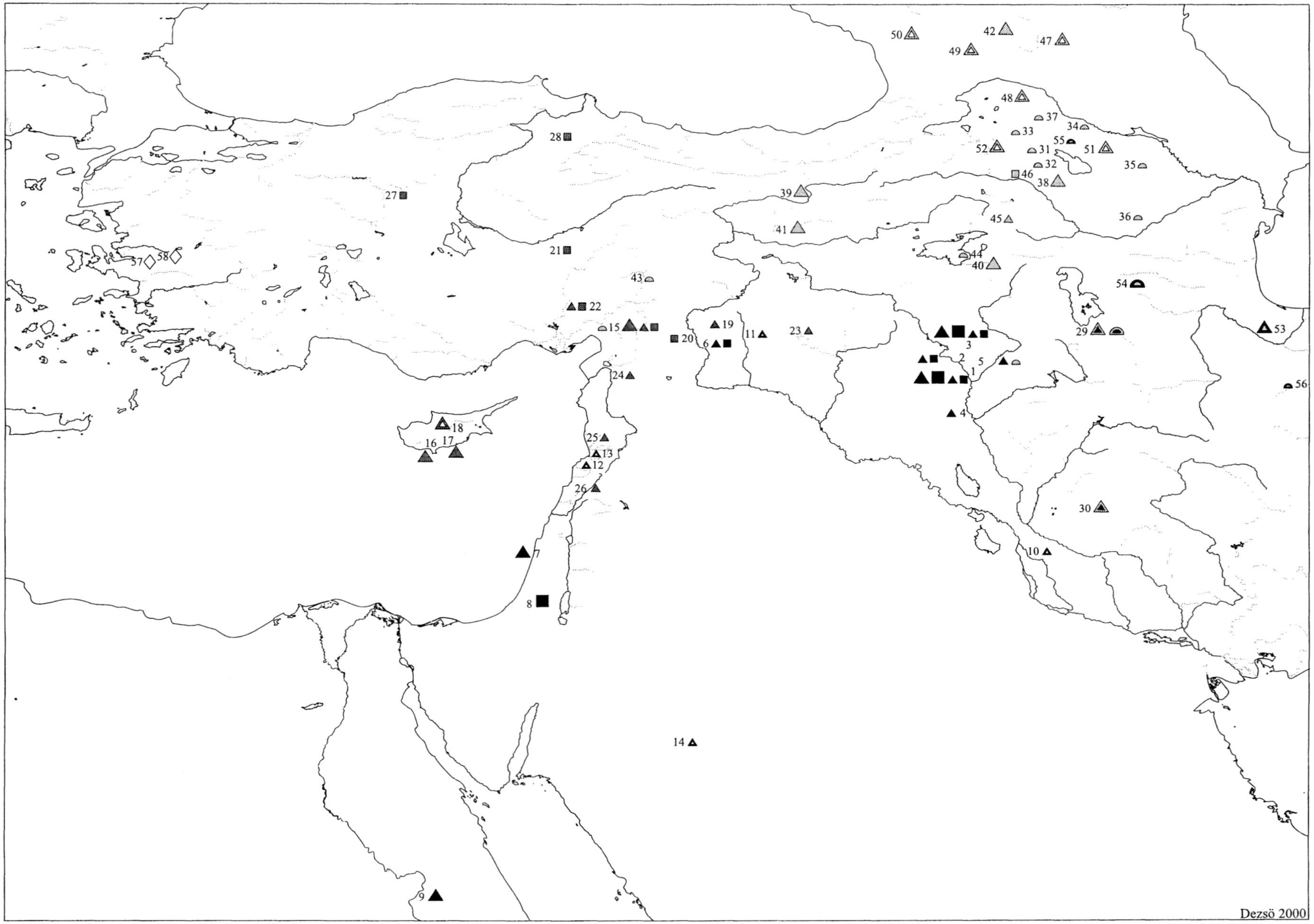

Helmet finds and representations of the Iron Age Near East

LIST OF PLACES
(listed according to their helmet groups)

ASSYRIAN

1. Nimrud
2. *Nineveh*
3. Khorsabad
4. *Ashur*
5. *Balawat*
6. *Til-Barsip*
7. Atlit
8. Lachish
9. Thebes

BABYLONIAN
(royal helmets)

10. Babylon
11. Harran
12. Wadi Brisa
13. Gebel Akrum
14. Tema Oasis

NORTH SYRIAN
(with Anatolian)

15. Zinçirli
16. Palaepaphos
17. Alassa
18. Tamassos
19. *Arslan-Tash*
20. *Carchemish*
21. *Kültepe*
22. *Karatepe*
23. *Tell Halaf*
24. *Tel Ta'yinat*
25. *Tartous*
26. *Beqaa*

27. *Gordion*
28. *Pazarli*

NORTHWEST IRANIAN
(with Hasanlu)

29. Hasanlu
30. Djub-i Gauhar

URARTIAN

31. *Ltshashen*
32. *Lori Berd*
33. *Shirakavan*
34. *Paravakar*
35. *Hatshbulag*
36. *Artsvanik*
37. *Ayrum*
38. Karmir Blur
39. Altintepe
40. Çavustepe
41. Burmageçit
42. Rutchi-Tig
43. *Marash*
44. *Toprak Kale*
45. *Eski Dogubayazit*
46. *Erebuni*

CAUCASIAN

47. Faskau
48. Trialeti
49. Kolhid
50. Primorskoye
51. Hurjin Hoger
52. Artchadzorskoy Kurgan

IRANIAN

53. Marlik
54. Meslekh-i Sahr
55. *Asthi Blur*
56. *Khurvin*

PERSIAN
(Anatolian)

57. Sardis
58. Bayrakli

Conical/pointed helmet — Conical crested helmet — Hemispherical crested helmet — Composit helmet

Large icons indicate helmet finds, small icons indicate helmet representations.
Colours/shades indicate possible relations.

CHART 1

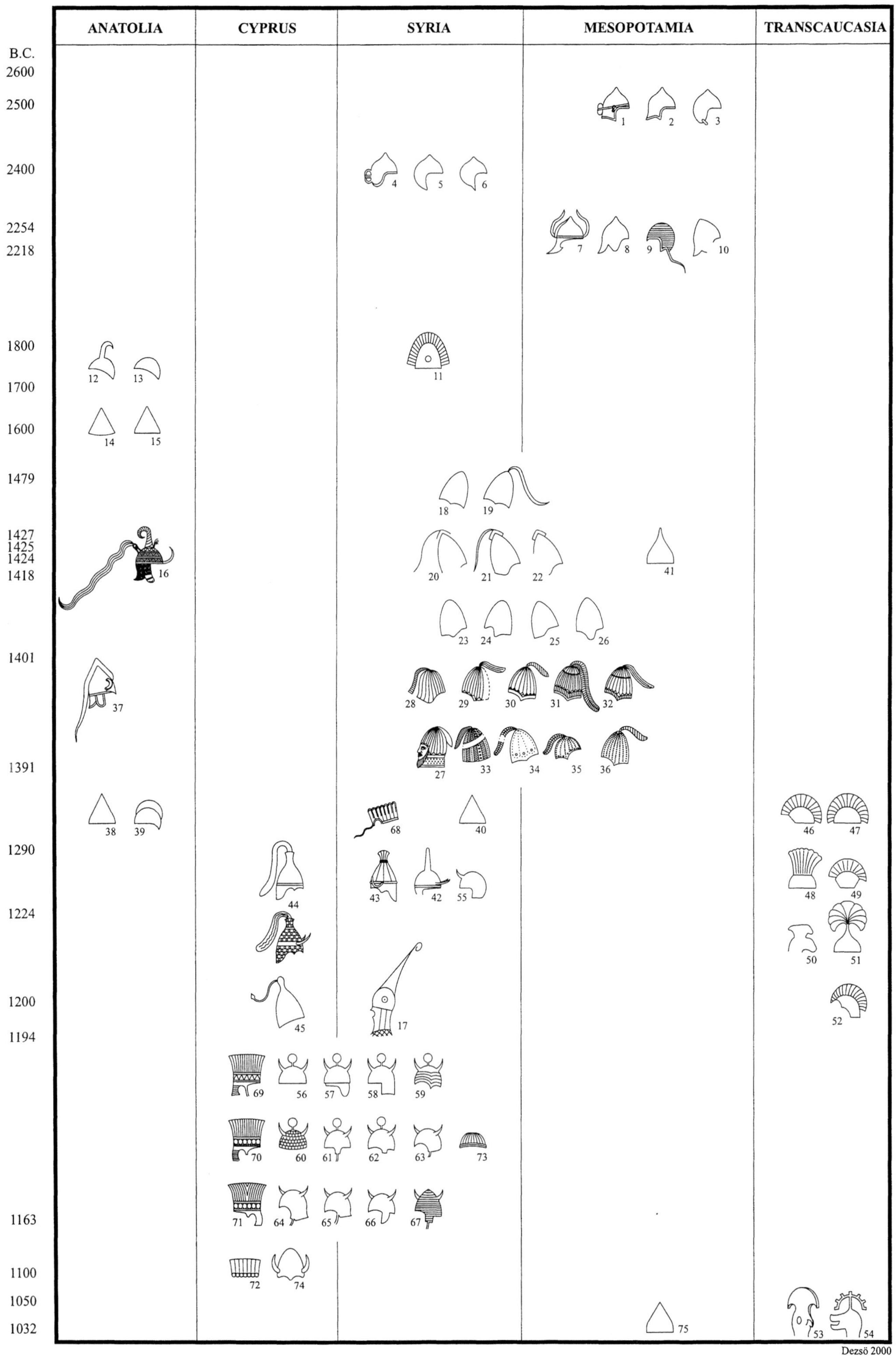

Dezsö 2000

CHART 2

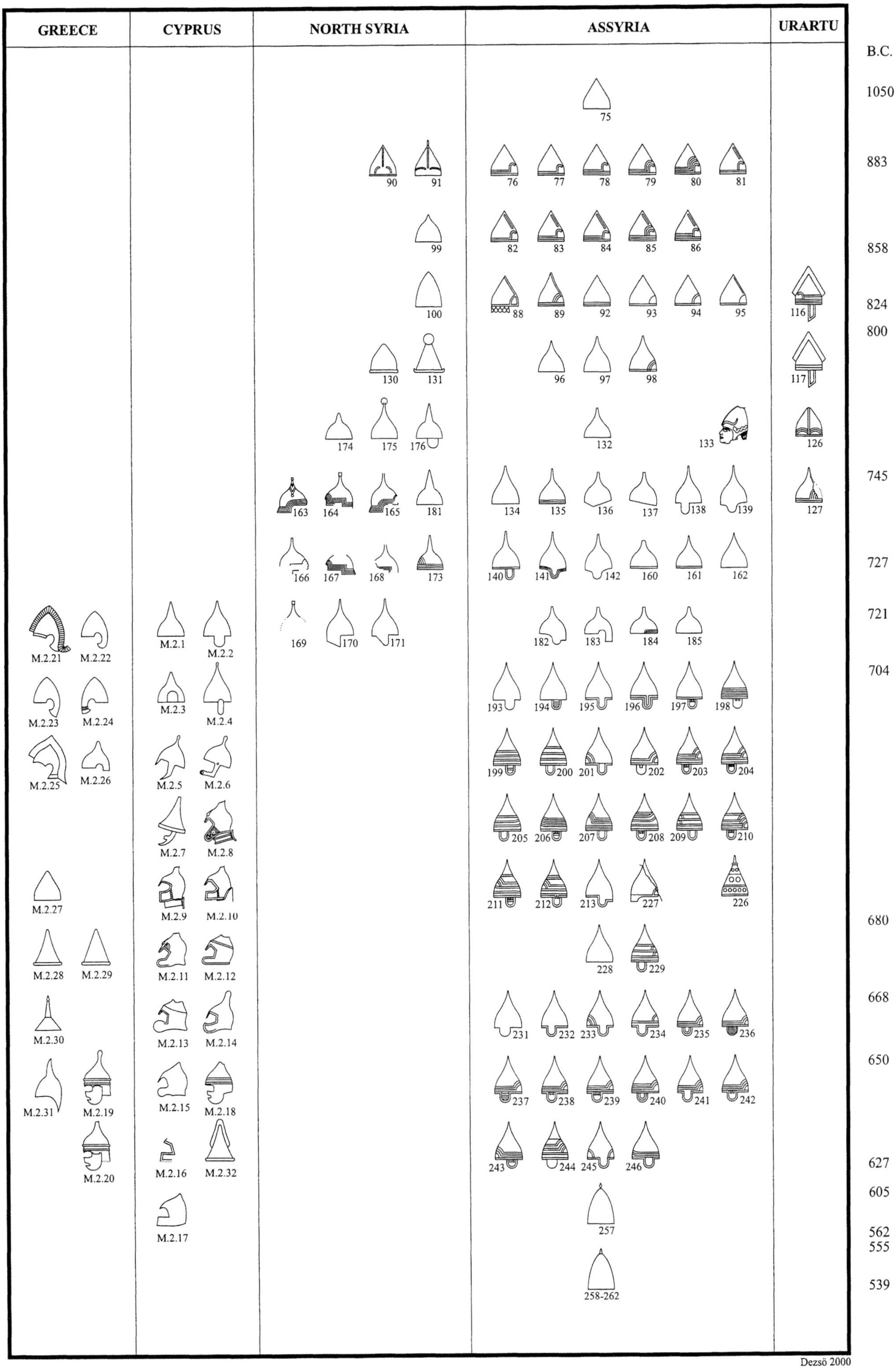

Dezsö 2000

CHART 3

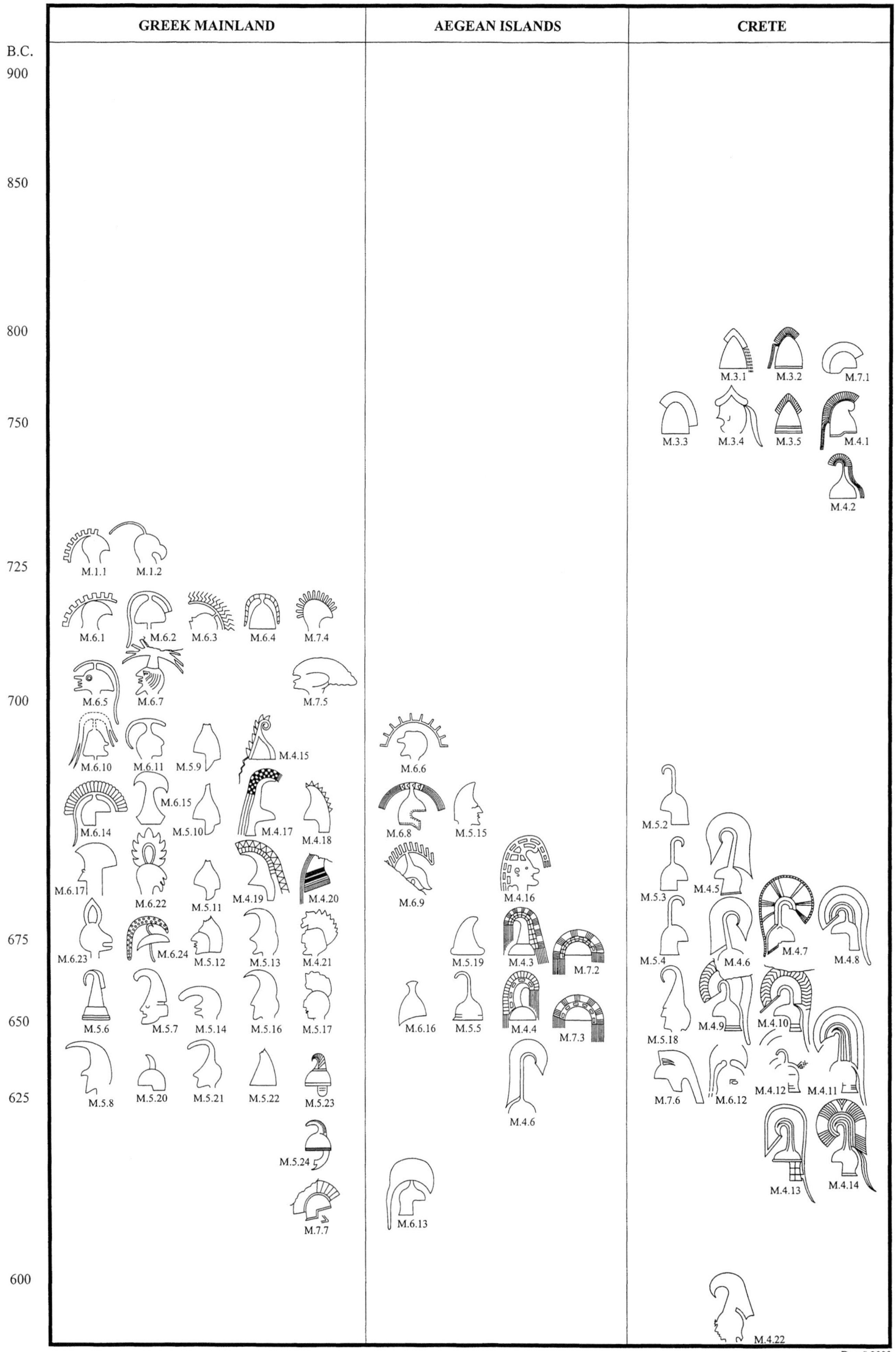

Dezsö 2000

CHART 3

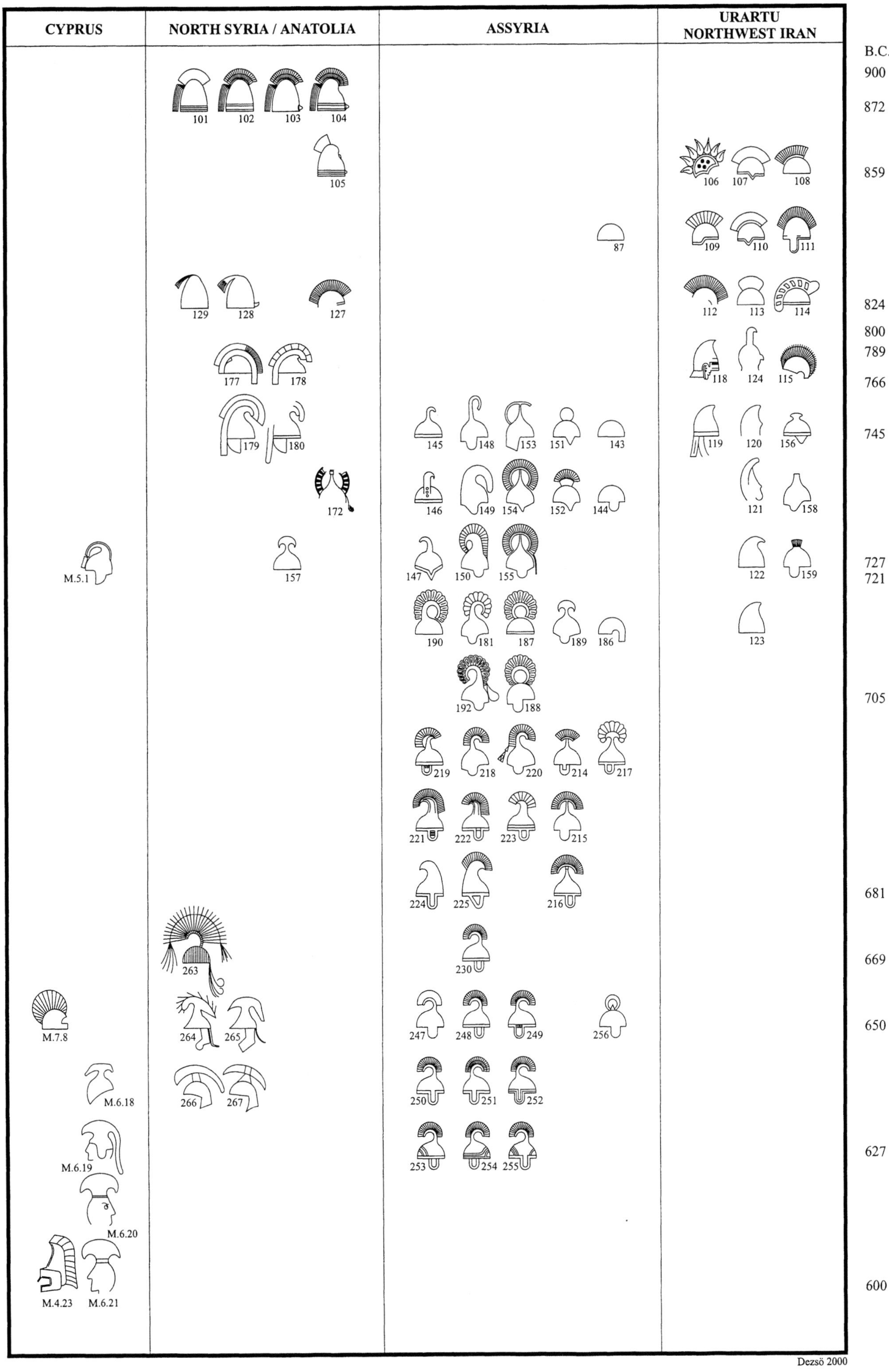

Dezsö 2000

CHART 4

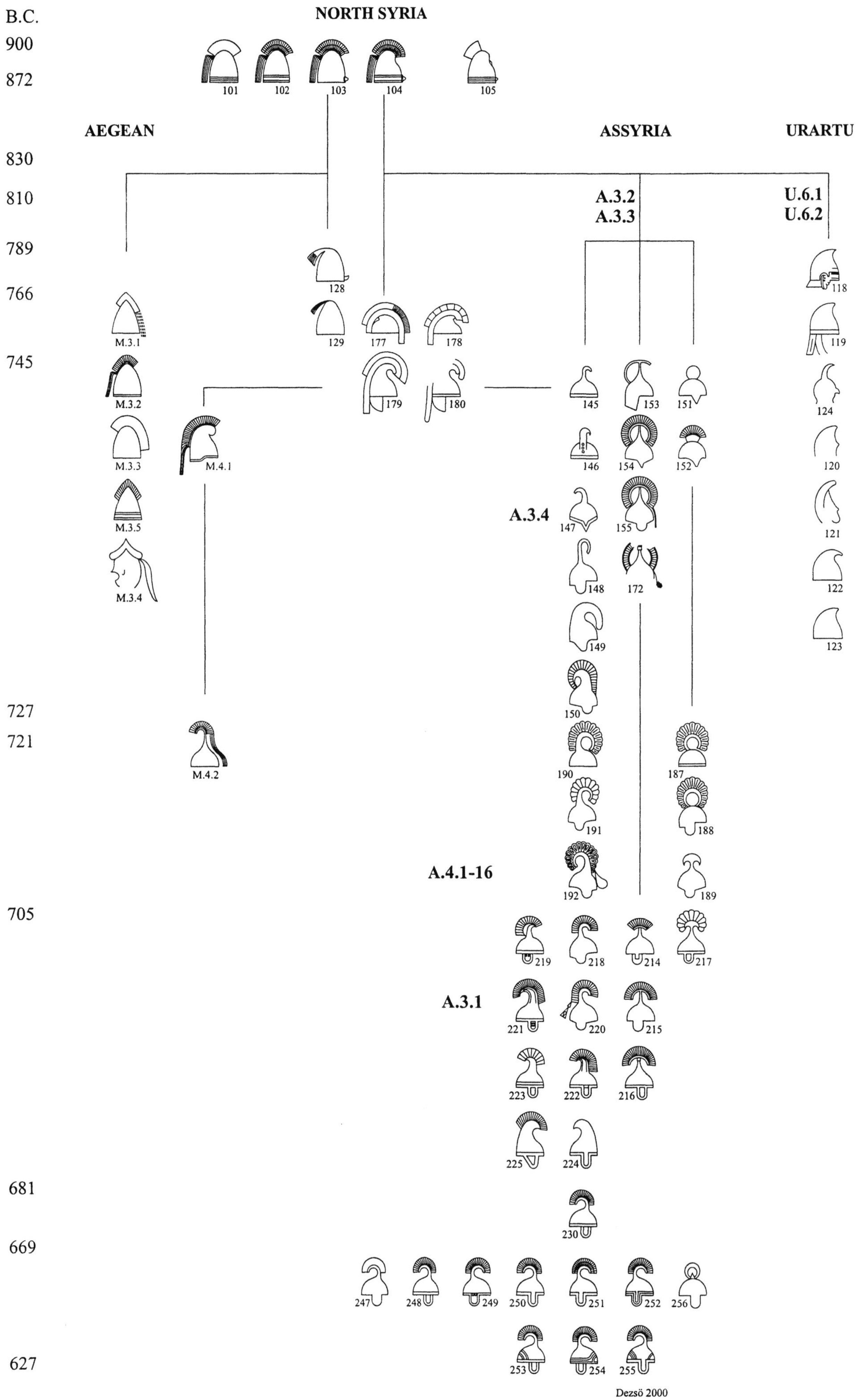
B.C.
900
872
830
810
789
766
745
727
721
705
681
669
627
NORTH SYRIA
AEGEAN
ASSYRIA
URARTU
101
102
103
104
105
A.3.2
A.3.3
U.6.1
U.6.2
128
118
M.3.1
129
177
178
119
M.3.2
179
180
145
153
151
124
M.3.3
M.4.1
146
154
152
120
M.3.5
A.3.4
147
155
121
M.3.4
148
172
122
149
123
150
M.4.2
190
187
191
188
A.4.1-16
192
189
219
218
214
217
A.3.1
221
220
215
223
222
216
225
224
230
247
248
249
250
251
252
256
253
254
255
Dezsö 2000

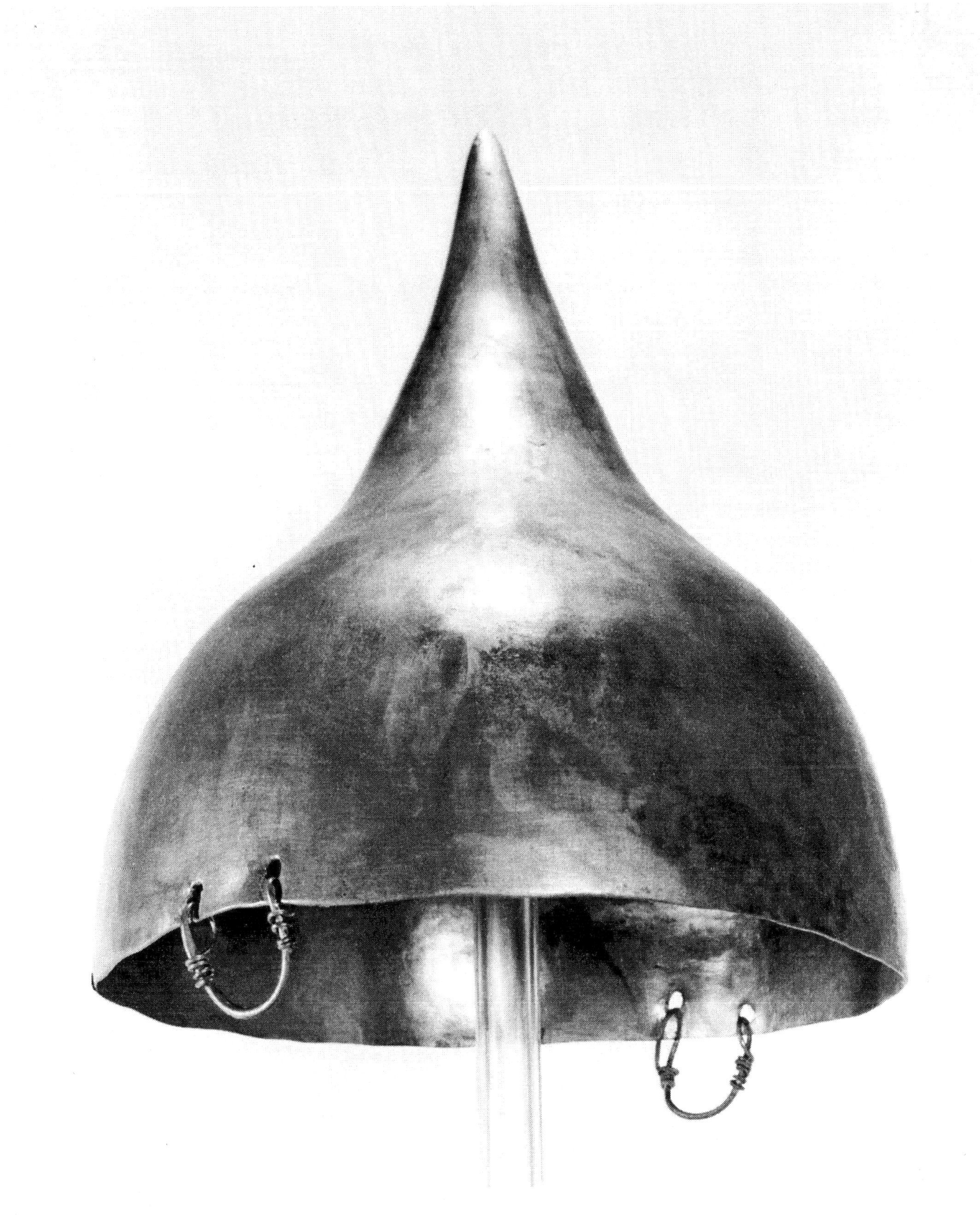

Cat. no. 1. Manchester, University of Manchester Museum, Inv. 1618

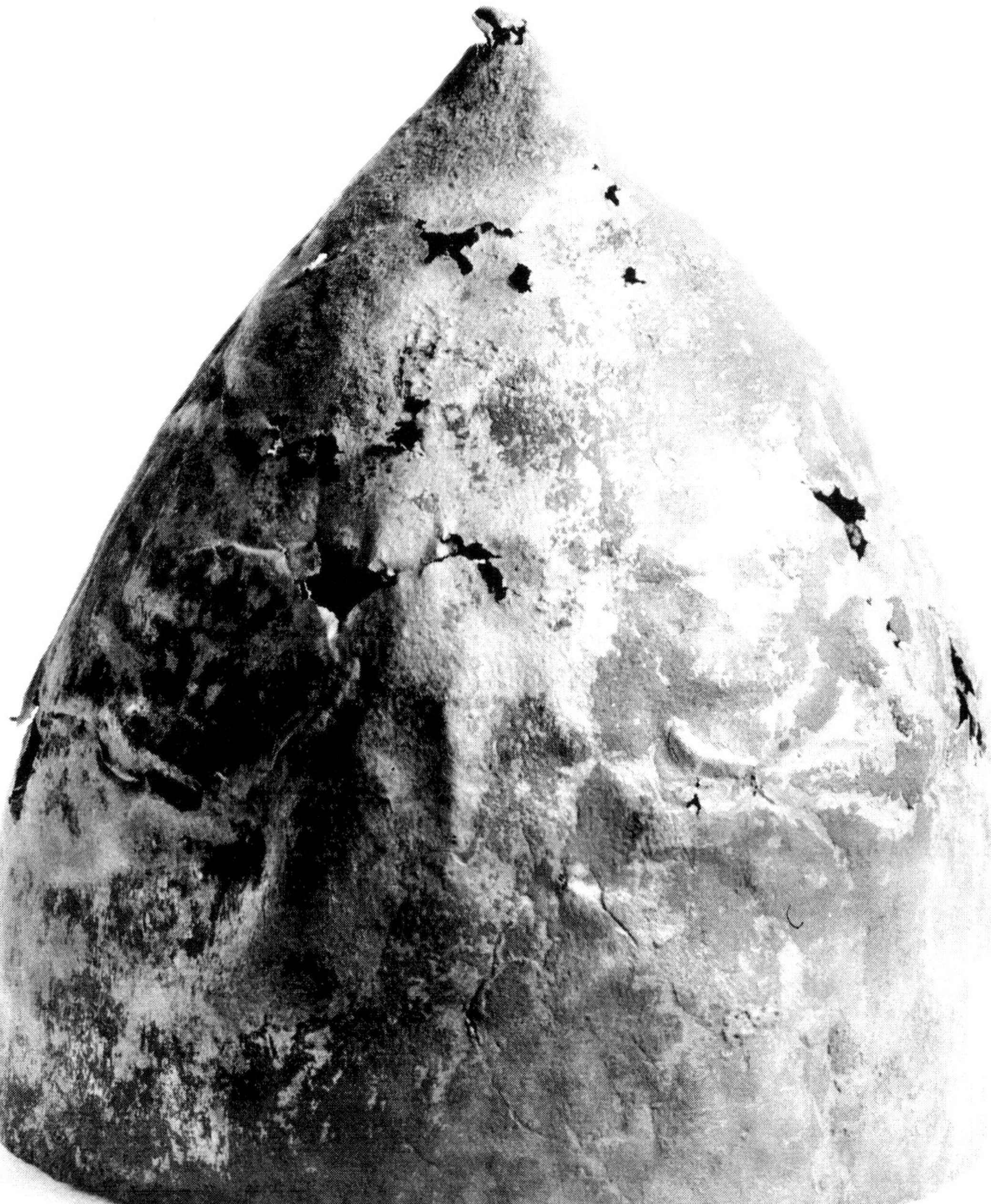

Cat. no. 2. Haifa, National Maritime Museum

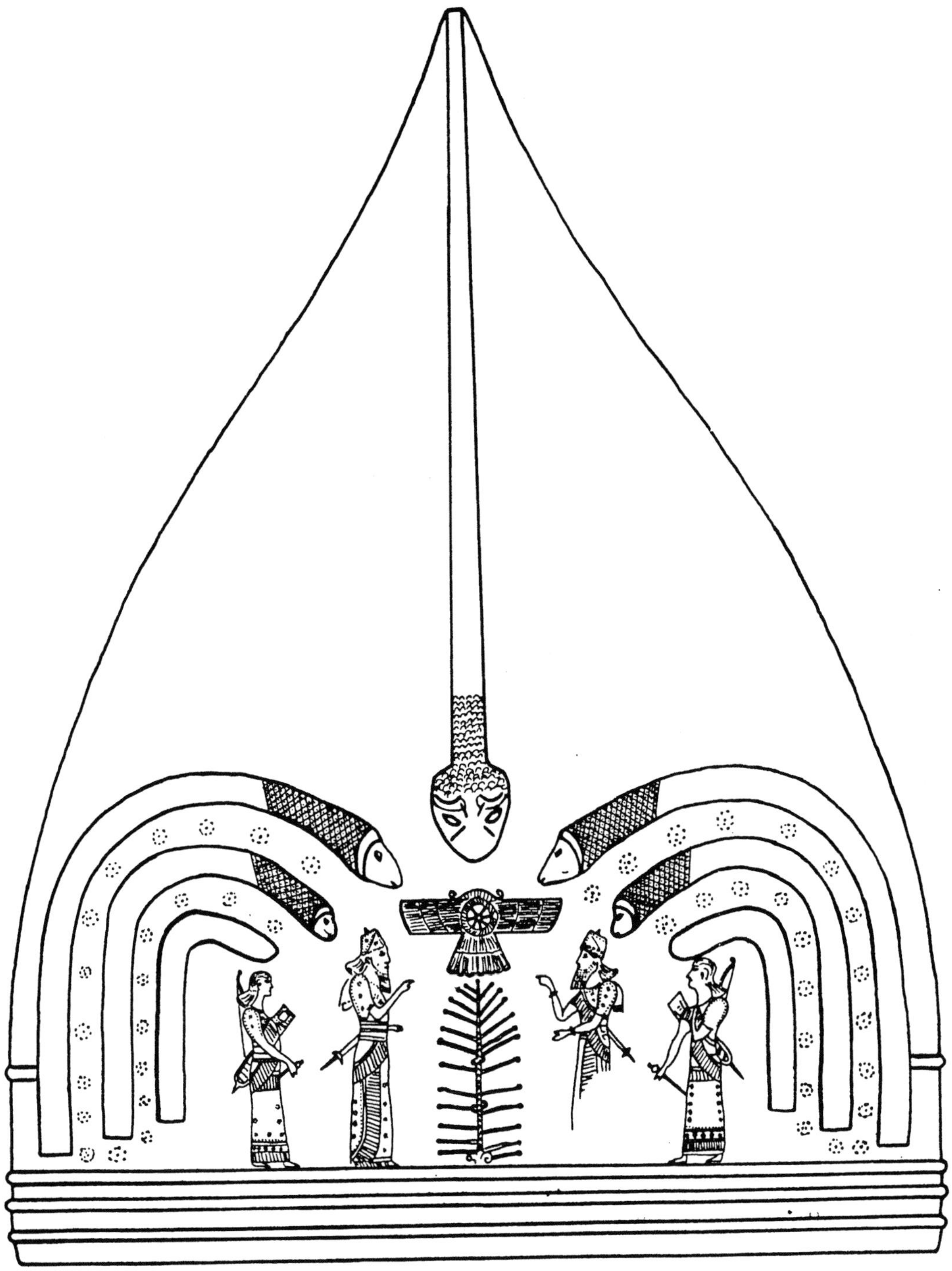

Cat. no. 3. Venice, Maria and Giancarlo Ligabue Collection.
Drawing of helmet.

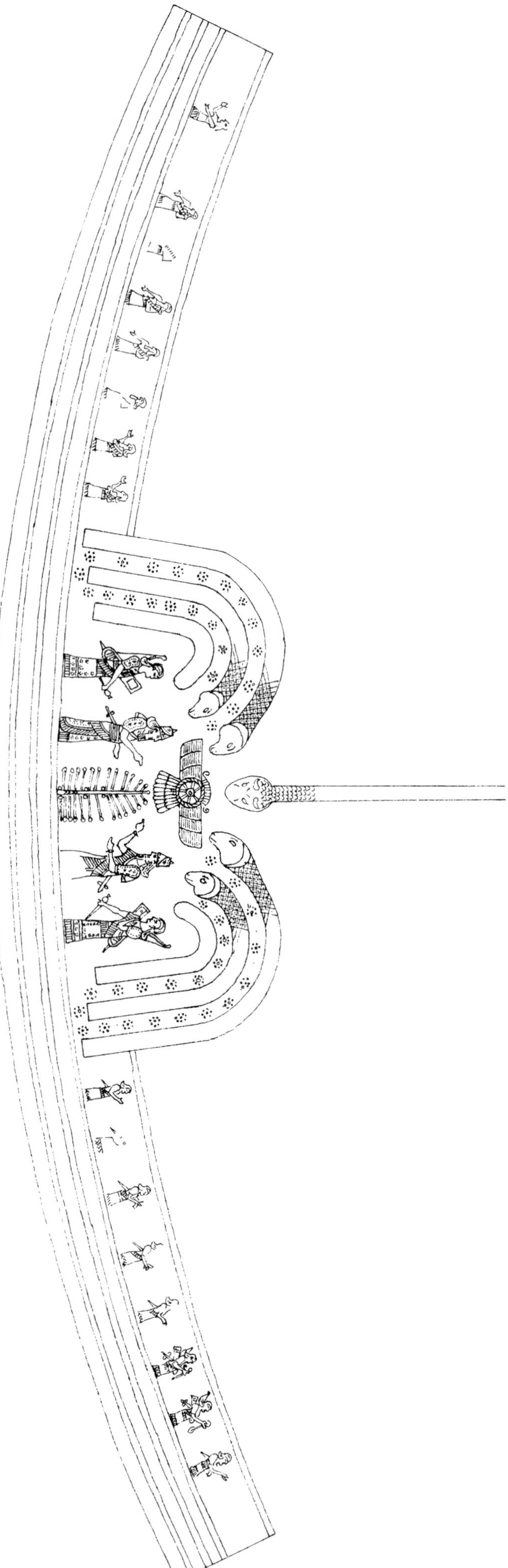

Cat. no. 3. Venice, Maria and Giancarlo Ligabue Collection.
Drawing of helmet decoration.

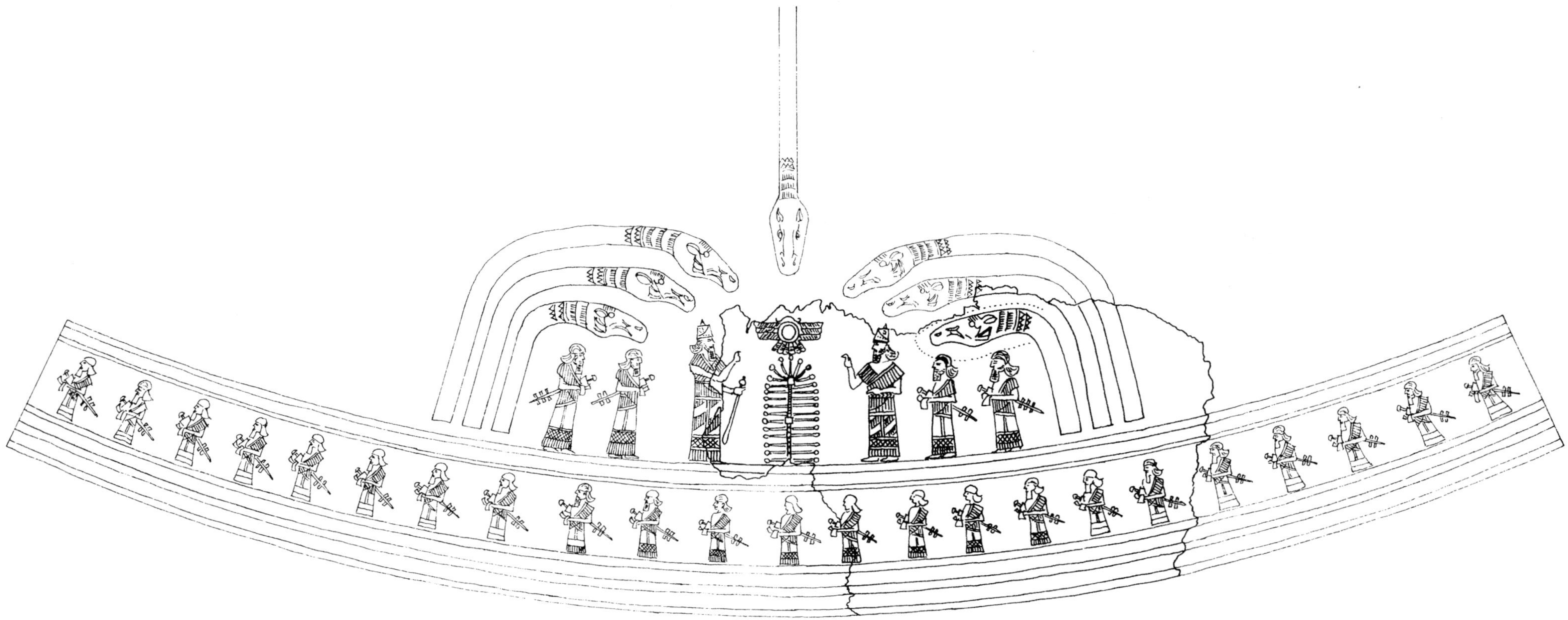

Cat. no. 4. Venice, private collection.
Drawing of helmet decoration (reconstruction).

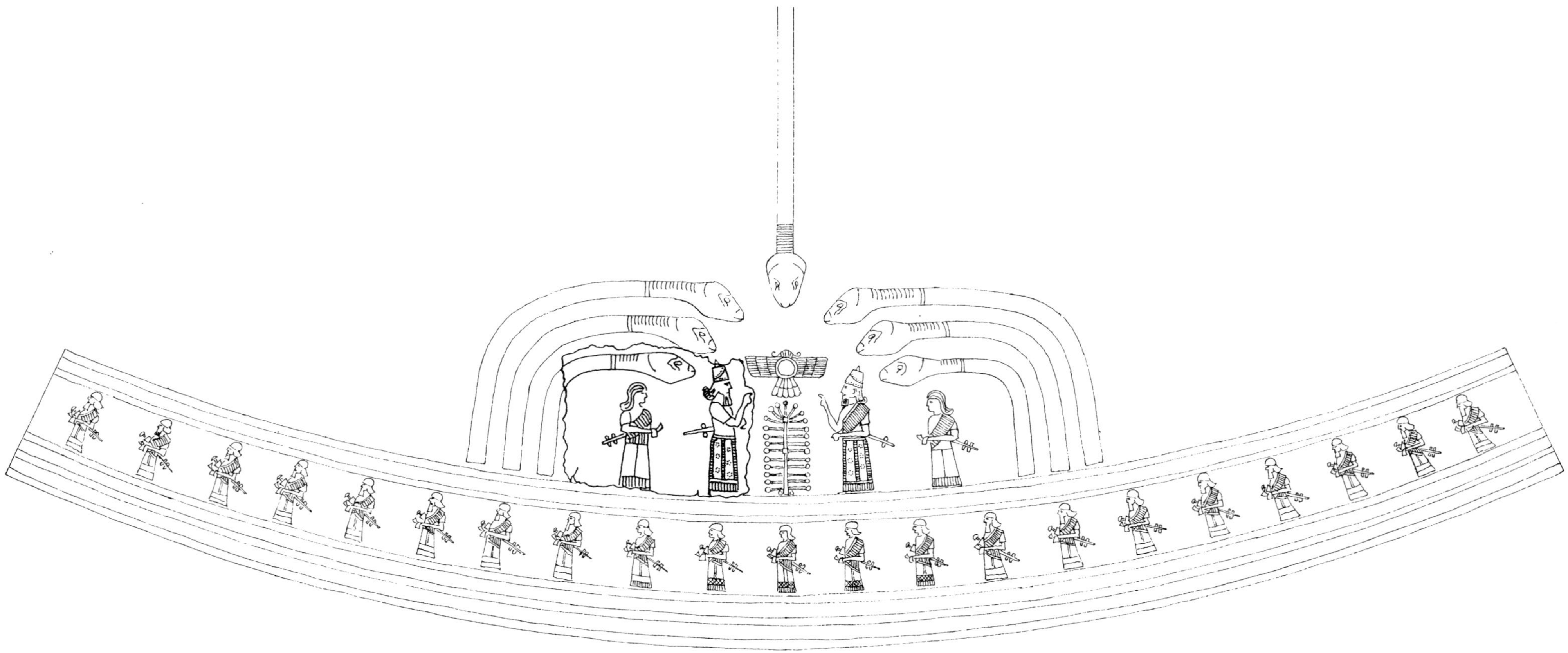

Cat. no. 5. Tehran, private collection.
Drawing of helmet decoration (reconstruction).

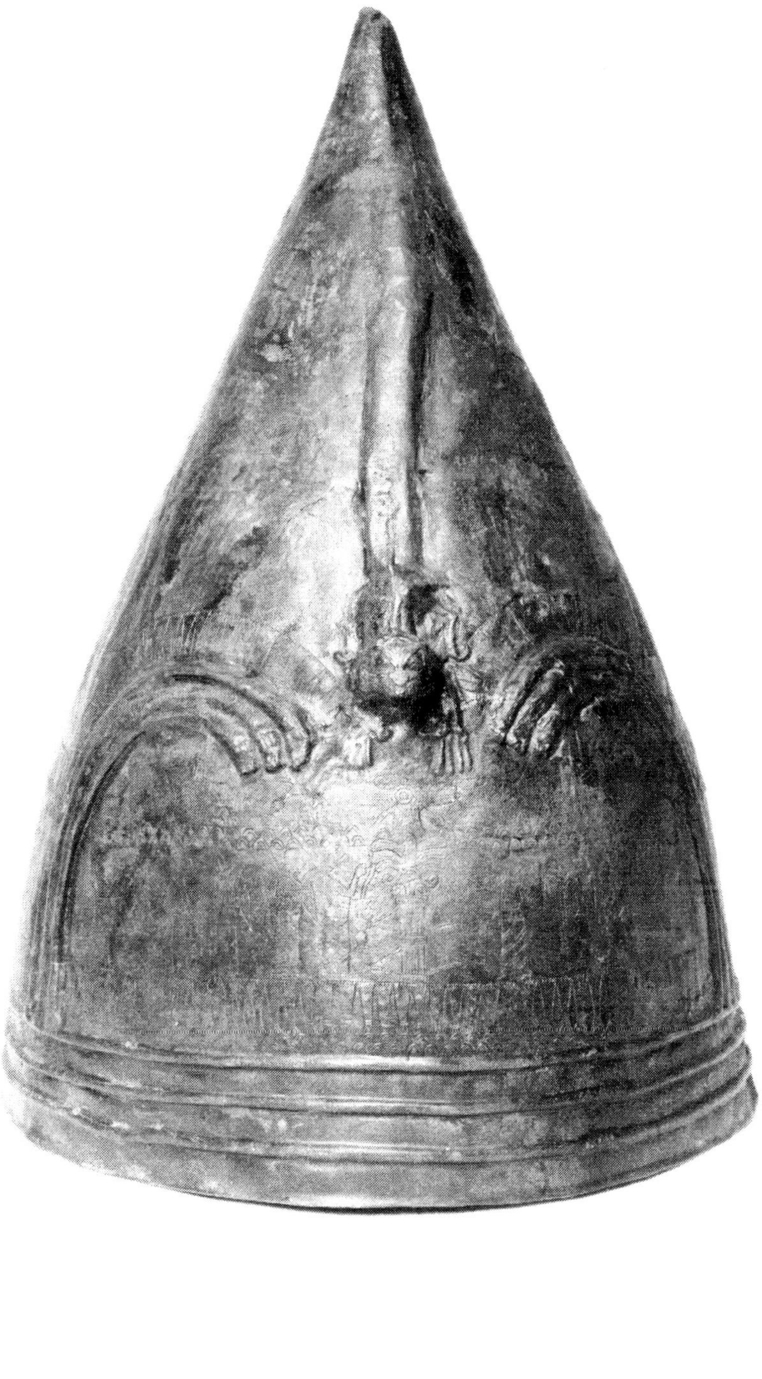

Cat. no. 8. Schaffhausen, Museum Allerheiligen.
(Galvano Copy: Munich, Archäologische Staatssammlung, Museum für Vor- und Frühgeschichte, Neg. no. 26-27)

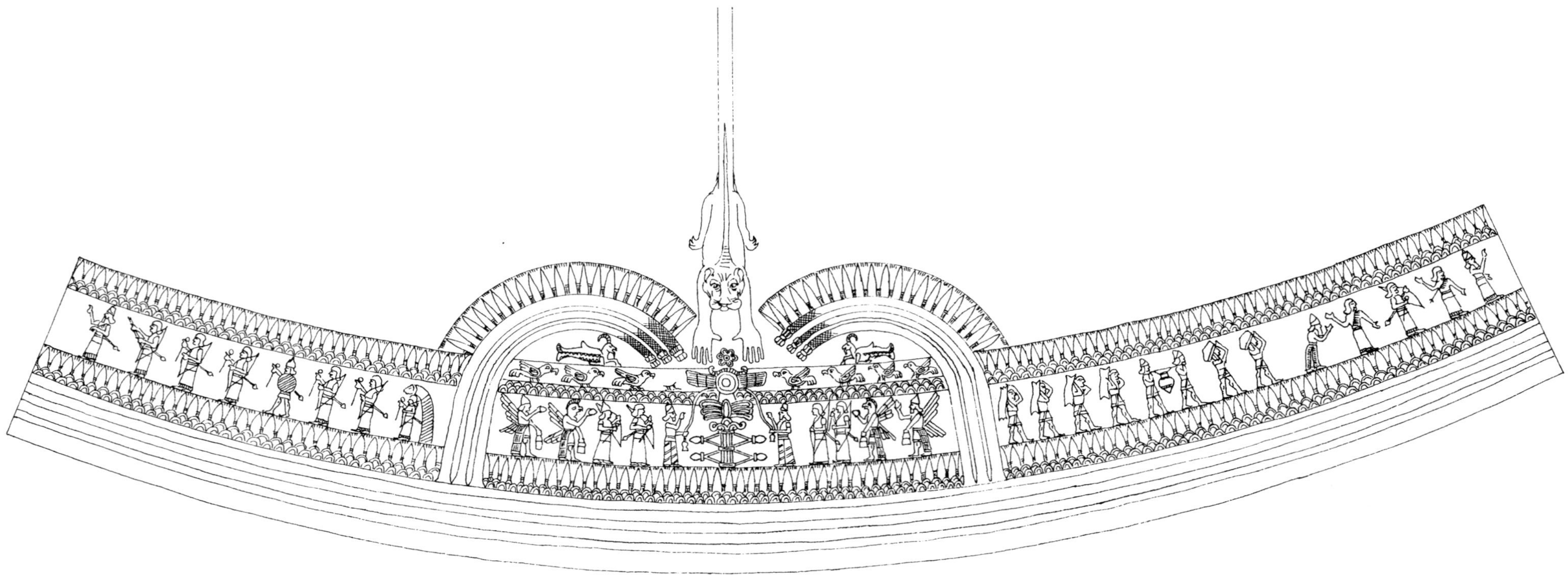

Cat. no. 8. Schaffhausen, Museum Allerheiligen.
(Galvano Copy: Munich, Archäologische Staatssammlung, Museum für Vor- und Frühgeschichte Neg. no. 26-27).
Drawing of helmet decoration.

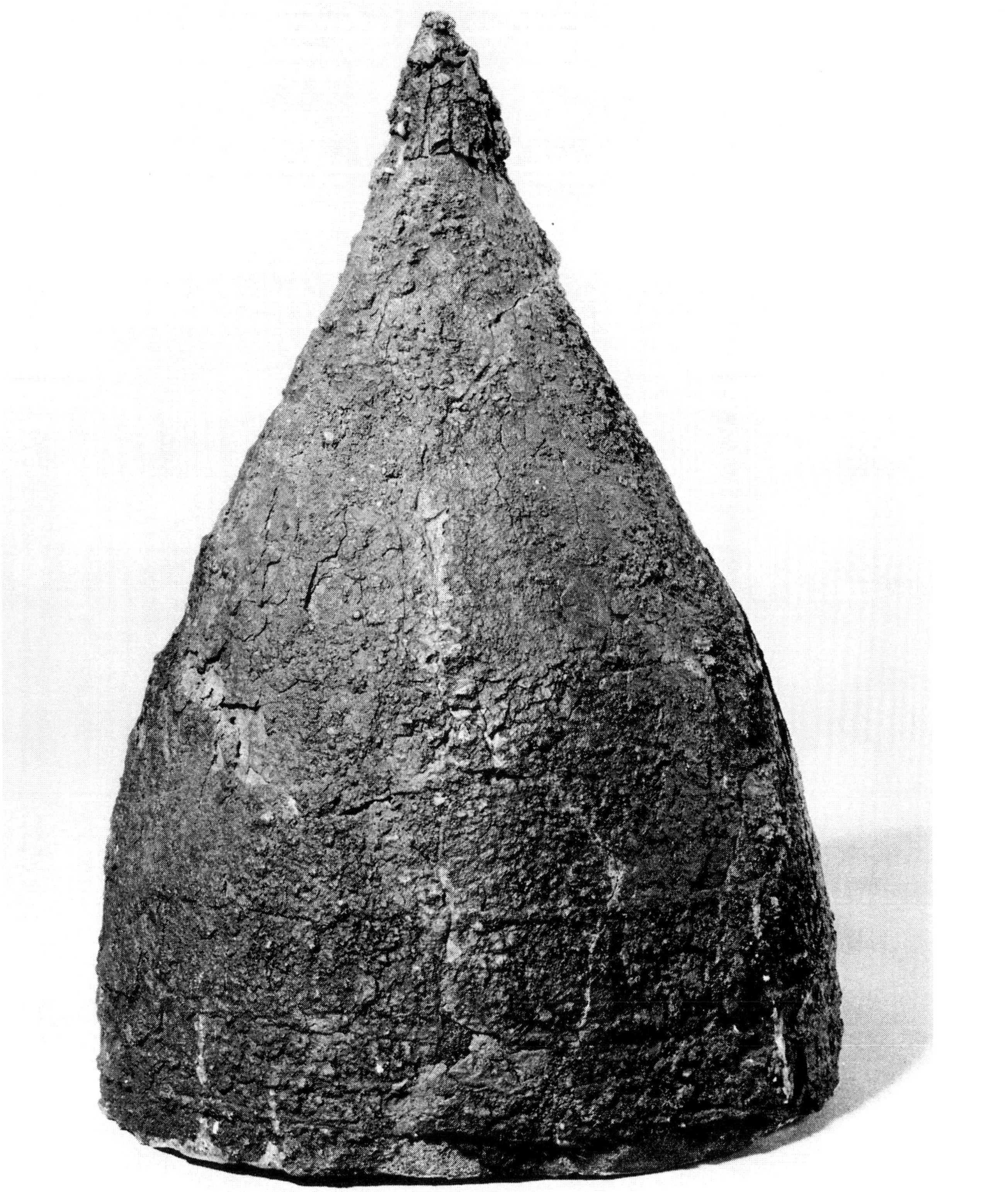

Cat. no. 10. London, British Museum, WA 22496.

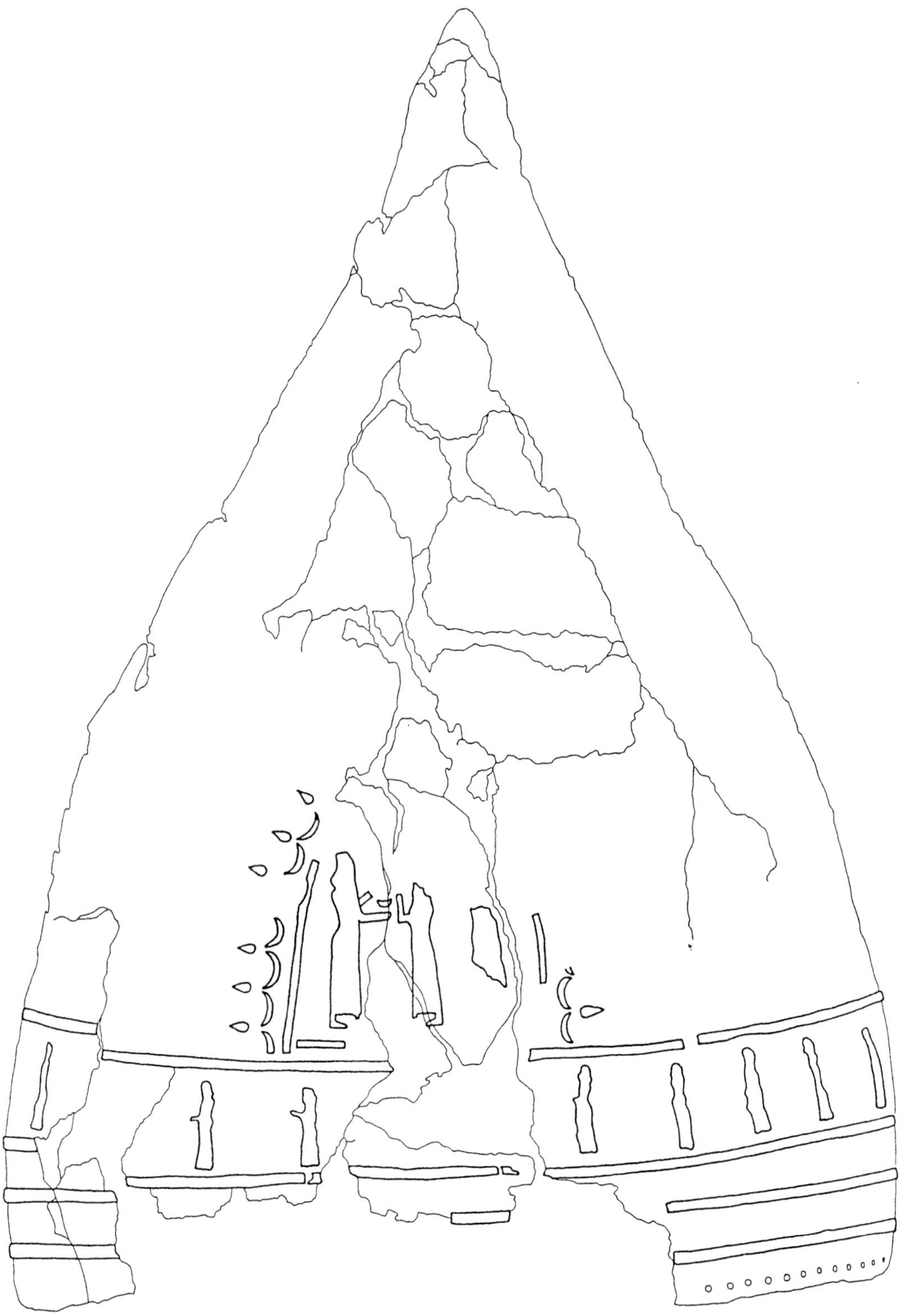

Cat. no. 10. London, British Museum, WA 22496.
Drawing of X-ray photograph.

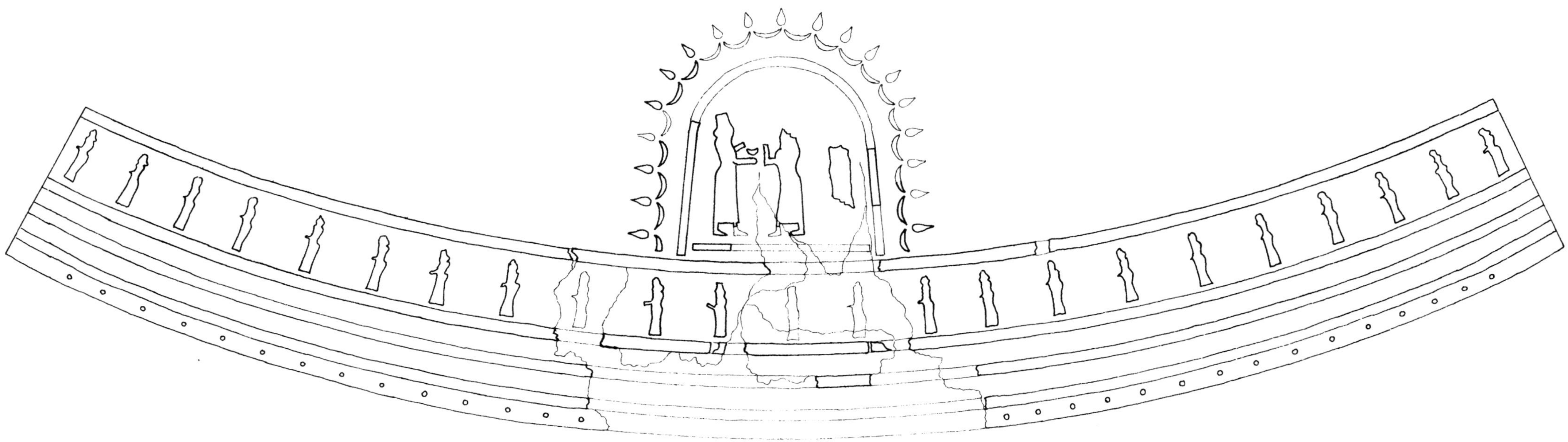

Cat. no. 10. London, British Museum, WA 22496.
Drawing of helmet decoration.

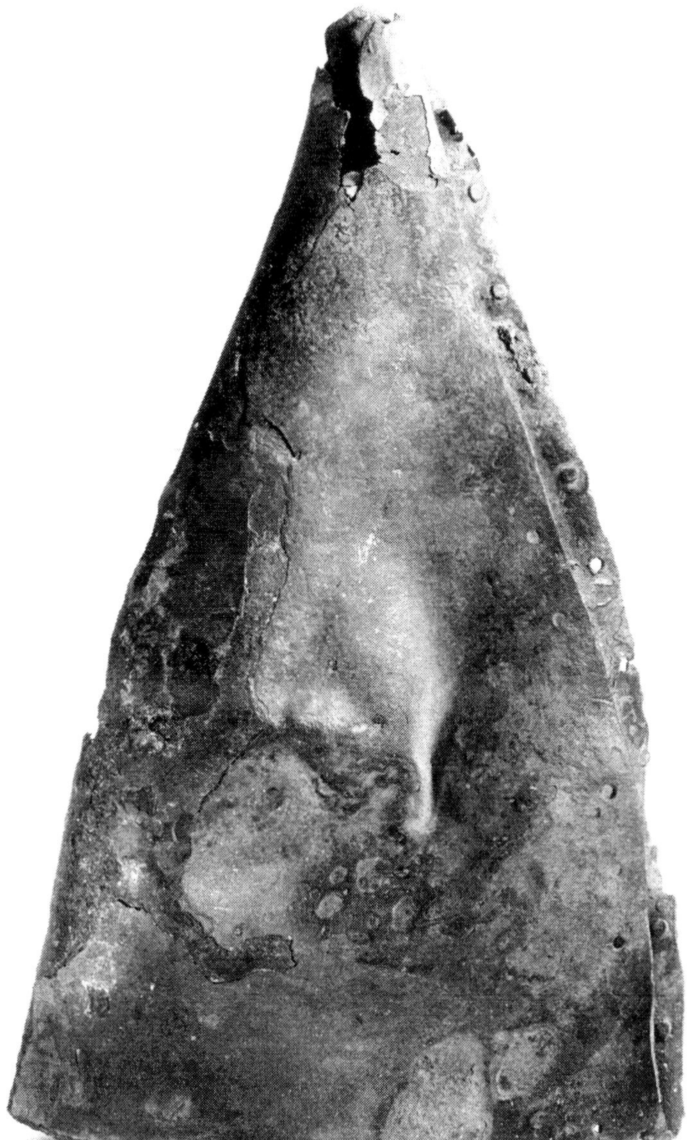

Cat. no. 11. Munich, Archäologische Staatssammlung, Museum für Vor- und Frühgeschichte, PS 1980/6227.

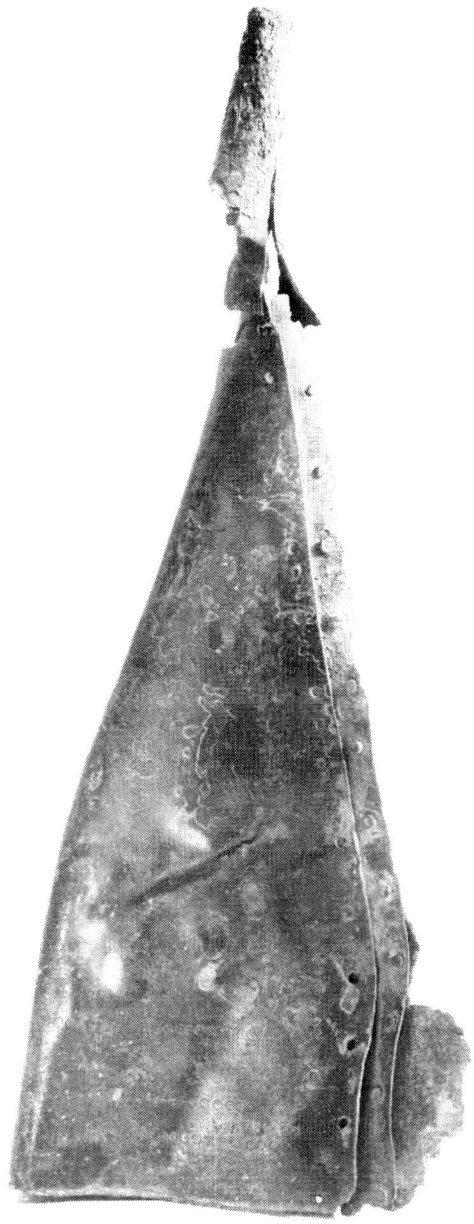

Cat. no. 11. Munich, Archäologische Staatssammlung, Museum für Vor- und Frühgeschichte, PS 1980/6228.

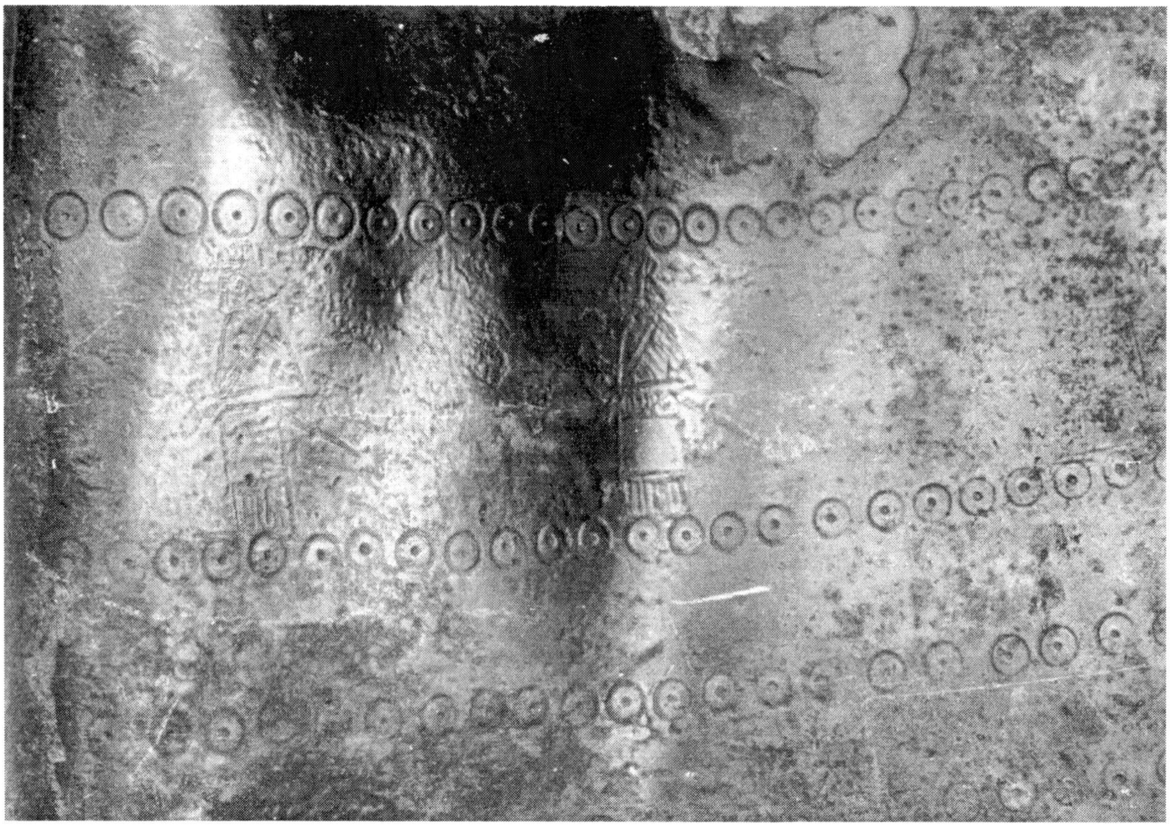

Cat. no. 11. Munich, Archäologische Staatssammlung, Museum für Vor- und Frühgeschichte, PS 1980/6227-6228. Details.

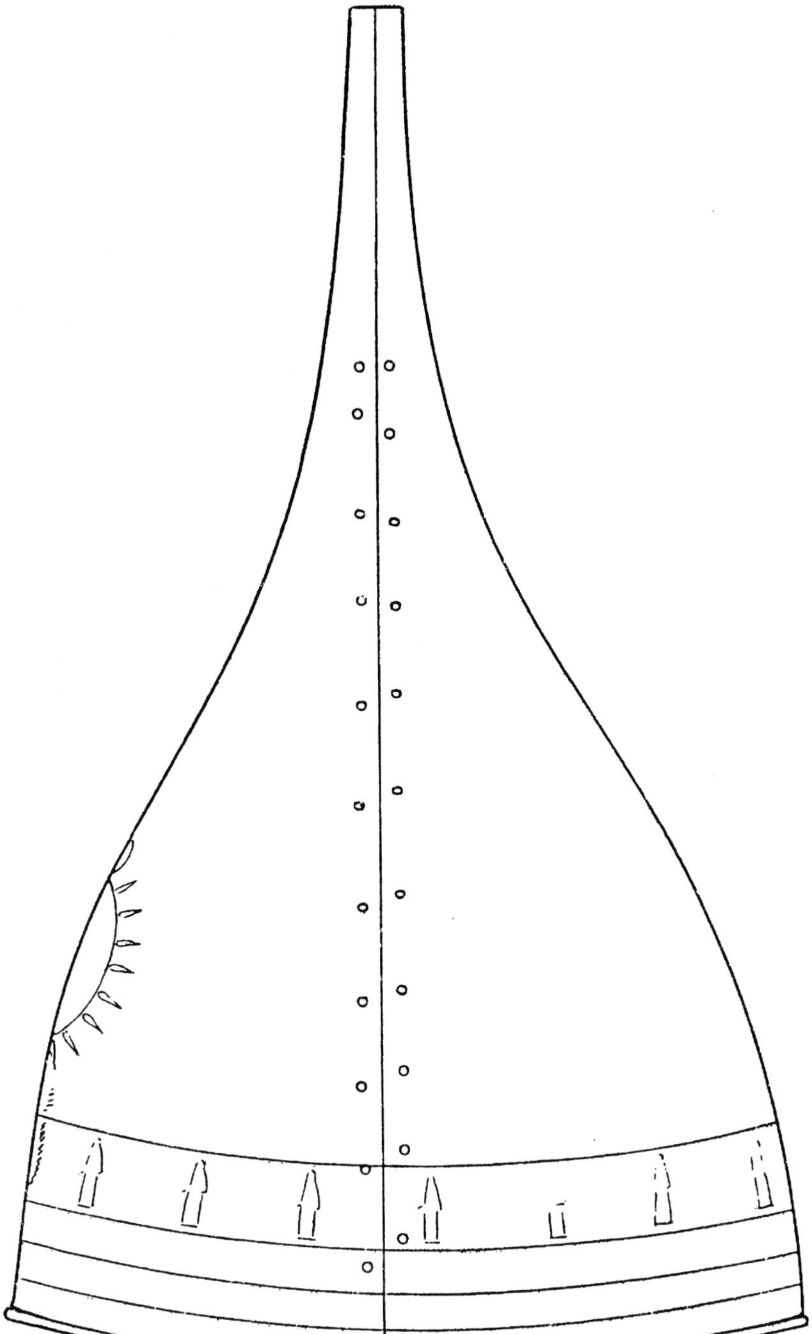

Cat. no. 11. Munich, Archäologische Staatssammlung, Museum für Vor- und Frühgeschichte, PS 1980/6227-6228.
Drawing of construction.

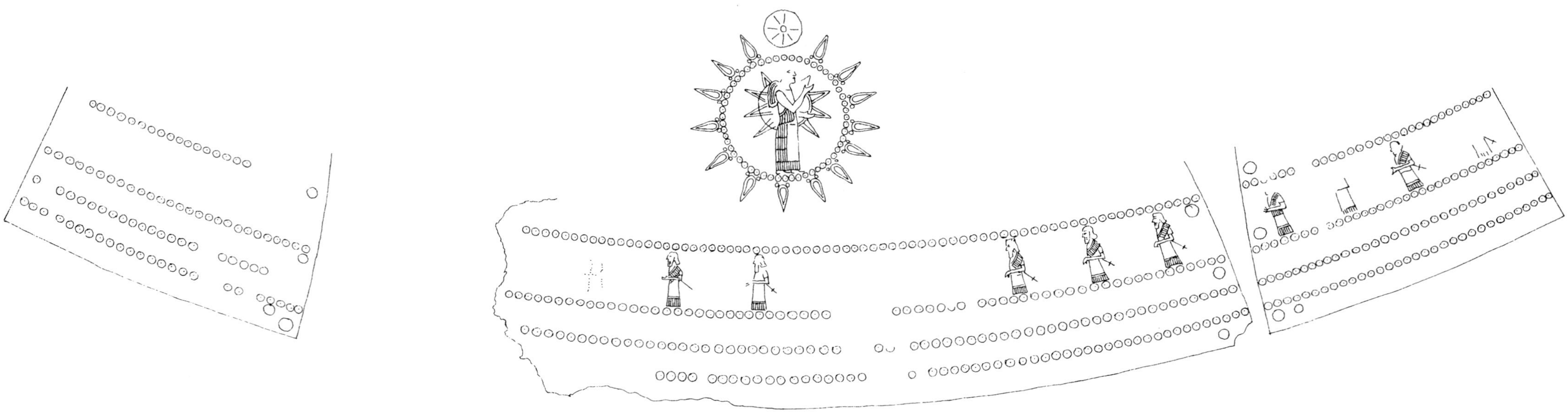

Cat. no. 11. Munich, Archäologische Staatssammlung, Museum für Vor- und Frühgeschichte, PS 1980/6227-6228.
Drawing of helmet decoration.

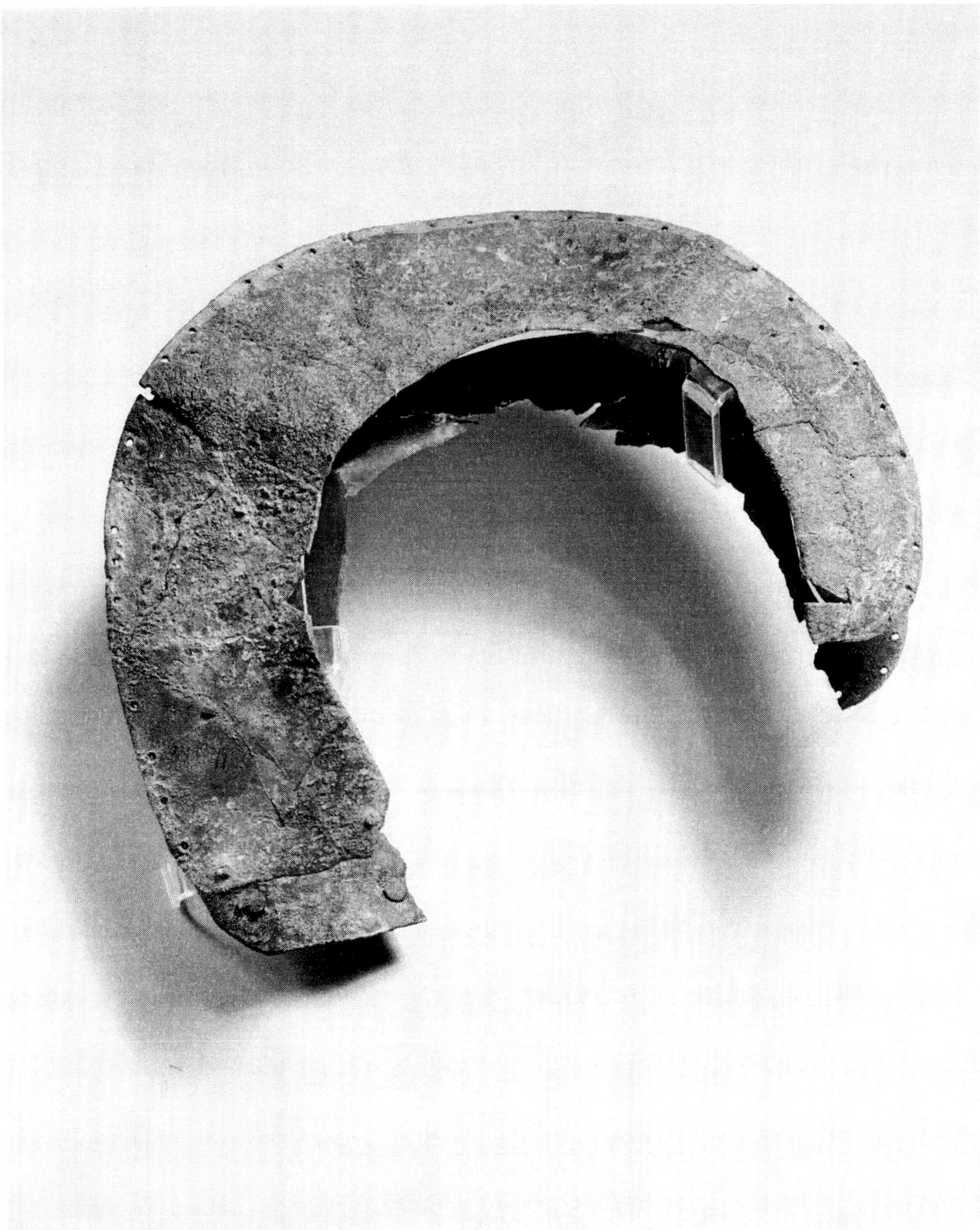

Cat. no. 12. Jerusalem, The Israel Museum, 34.133/1-2.

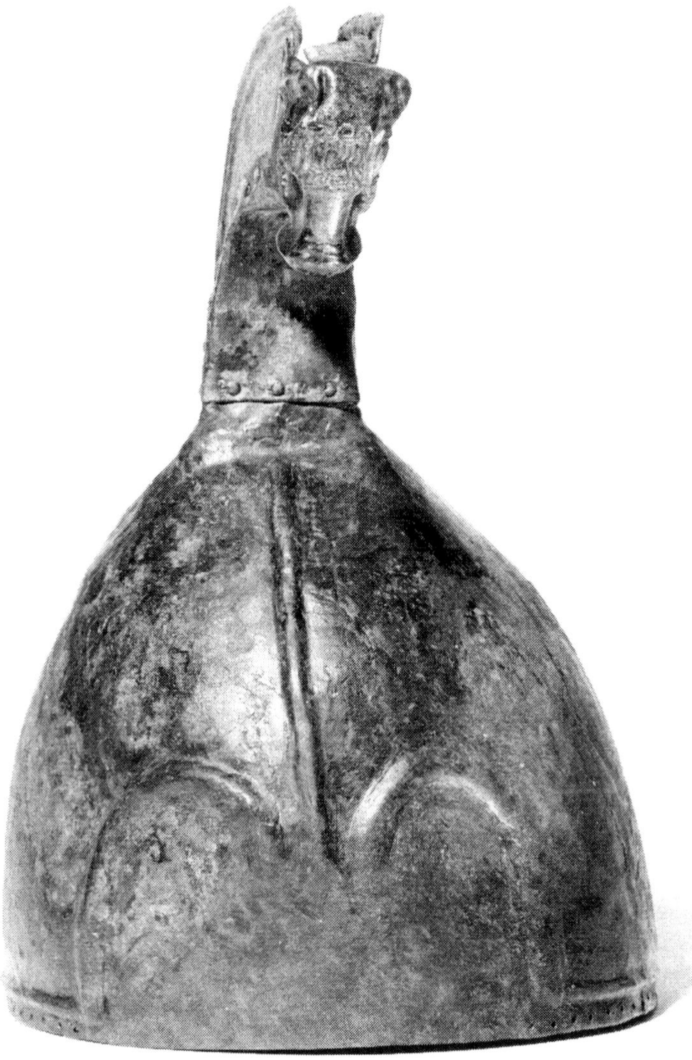

Cat. no. 13. Munich, Archäologische Staatssammlung, Museum für Vor- und Frühgeschichte, PS 1979/1181.
Front view.

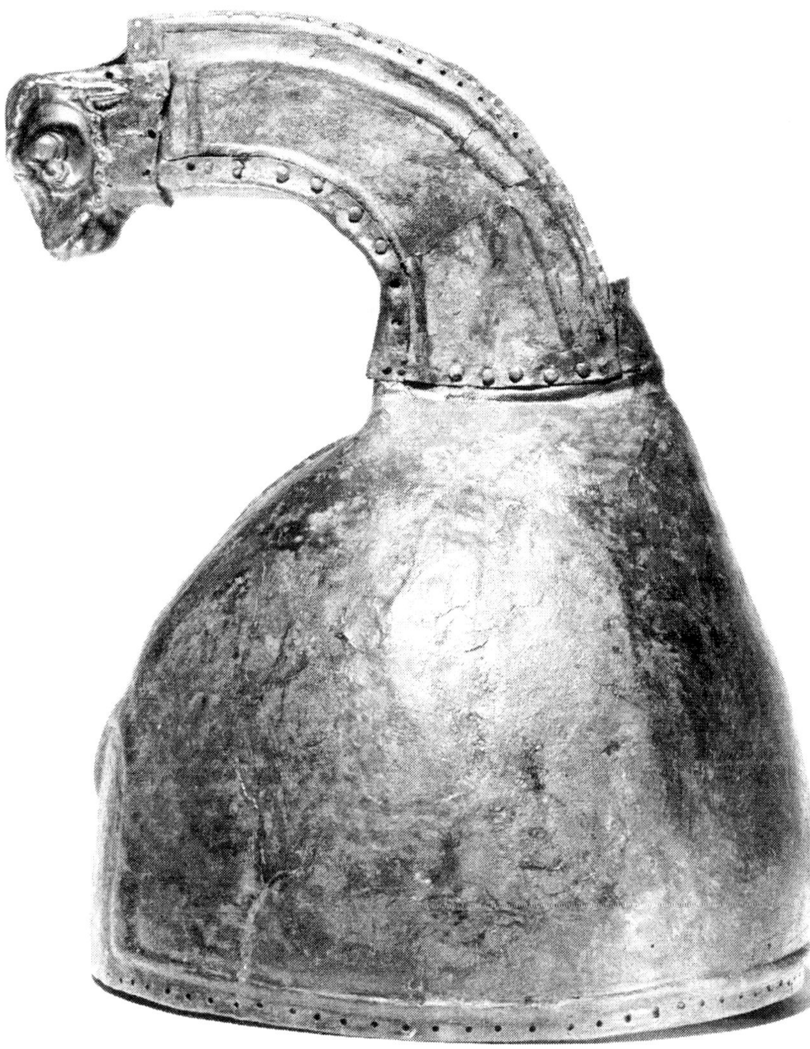

Cat. no. 13. Munich, Archäologische Staatssammlung, Museum für Vor- und Frühgeschichte, PS 1979/1181.
Side view.

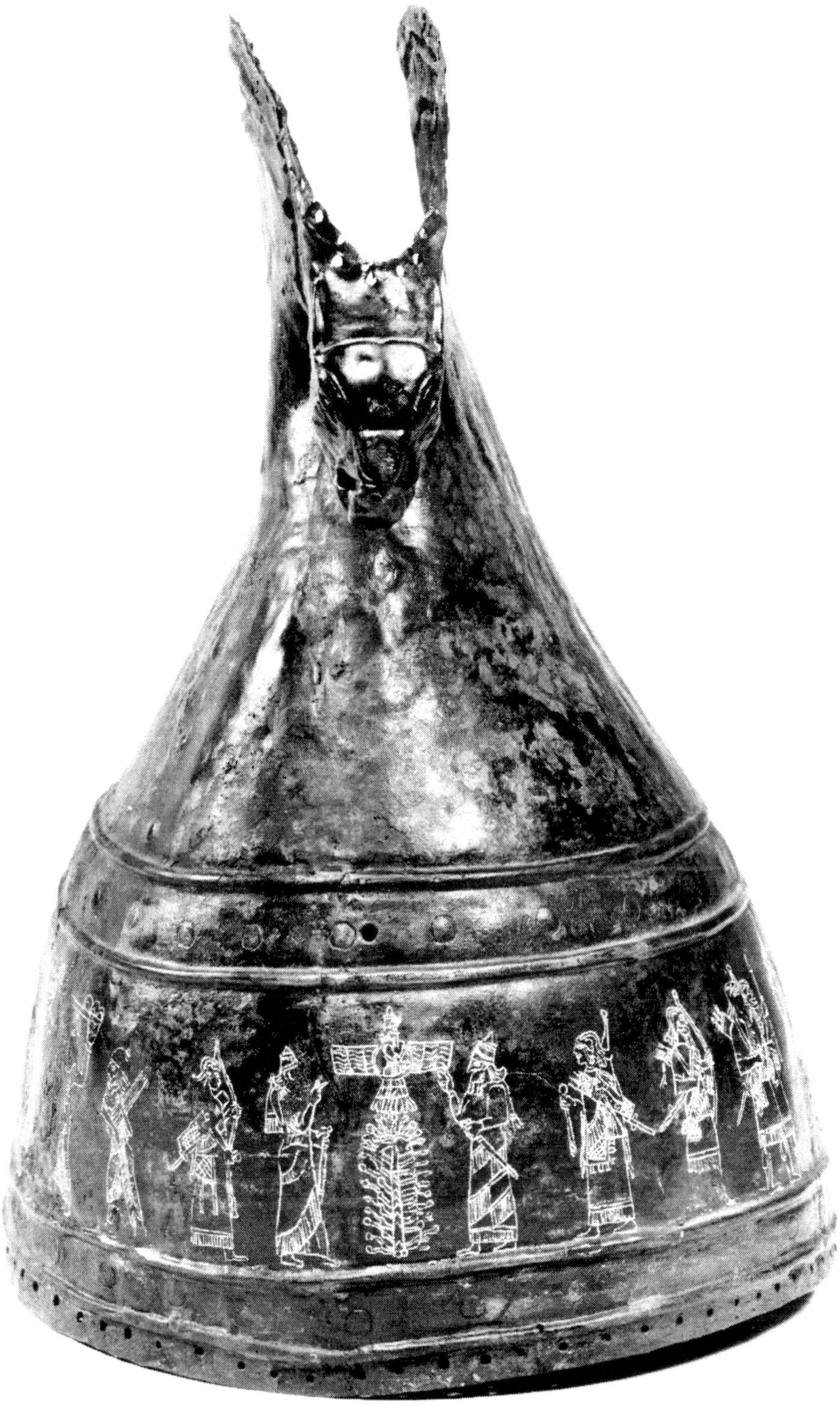

Cat. no. 14. Karlsruhe, Badisches Landesmuseum, 89/12.
Front view.

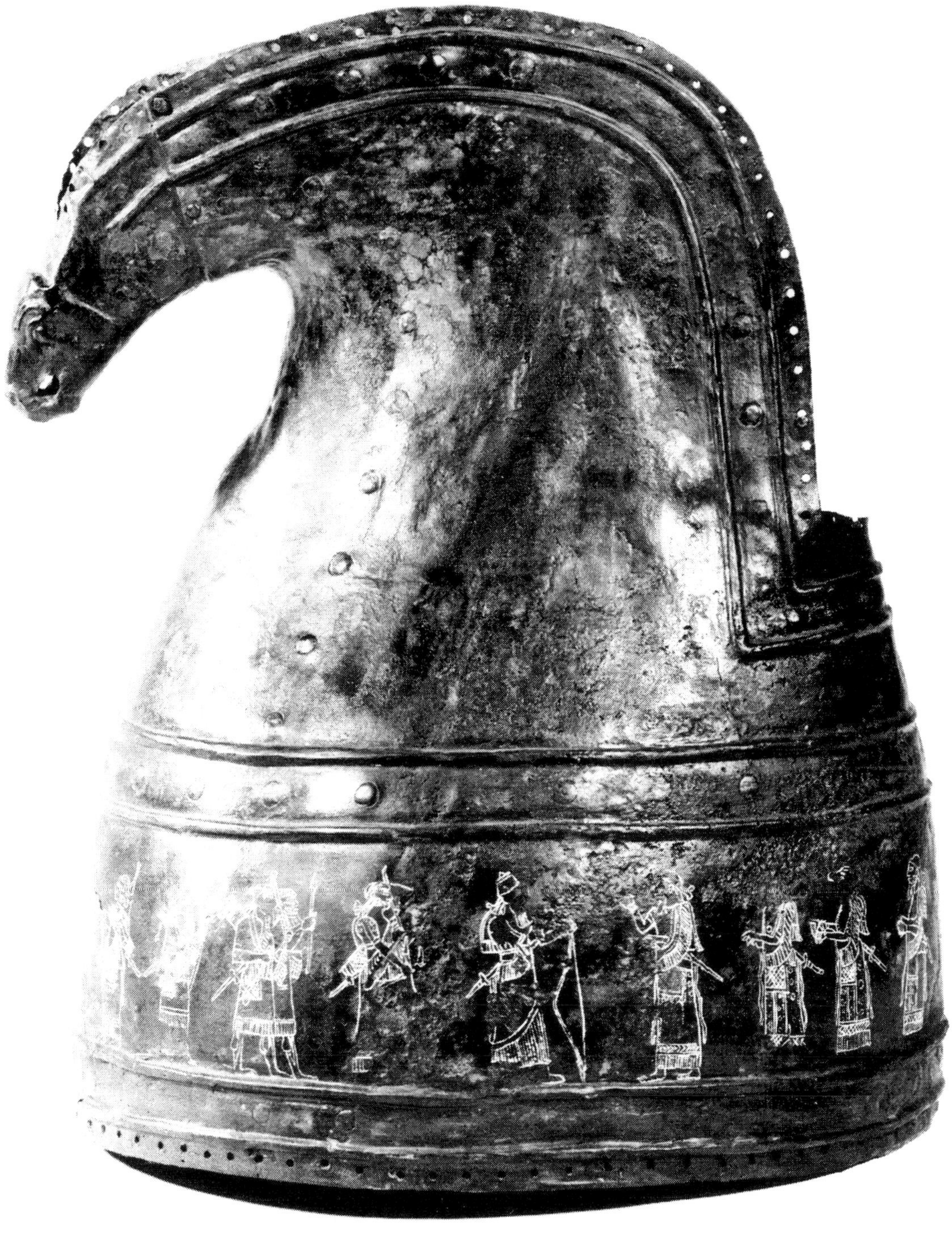

Cat. no. 14. Karlsruhe, Badisches Landesmuseum, 89/12.
Side view.

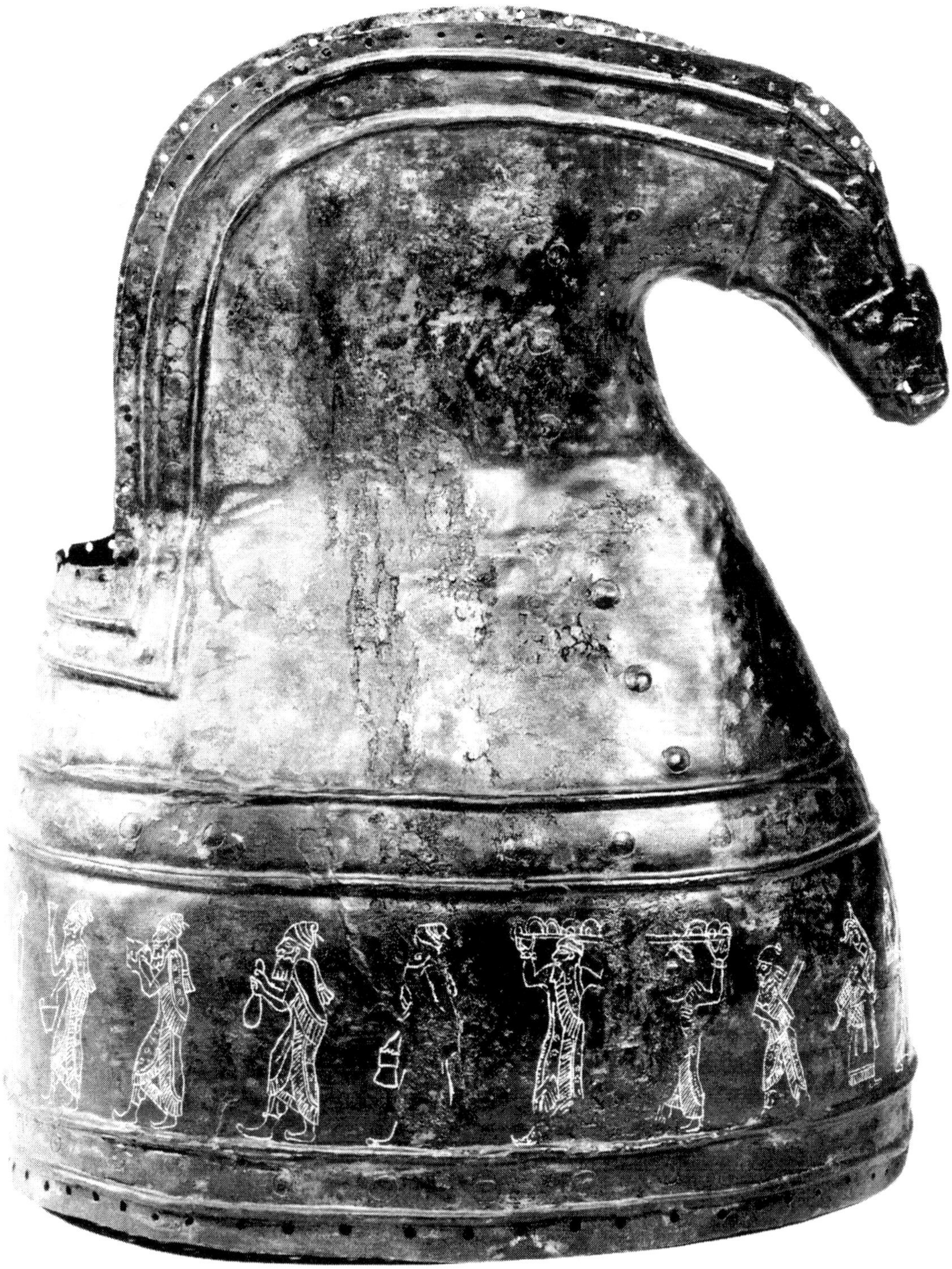

Cat. no. 14. Karlsruhe, Badisches Landesmuseum, 89/12.
Side view.

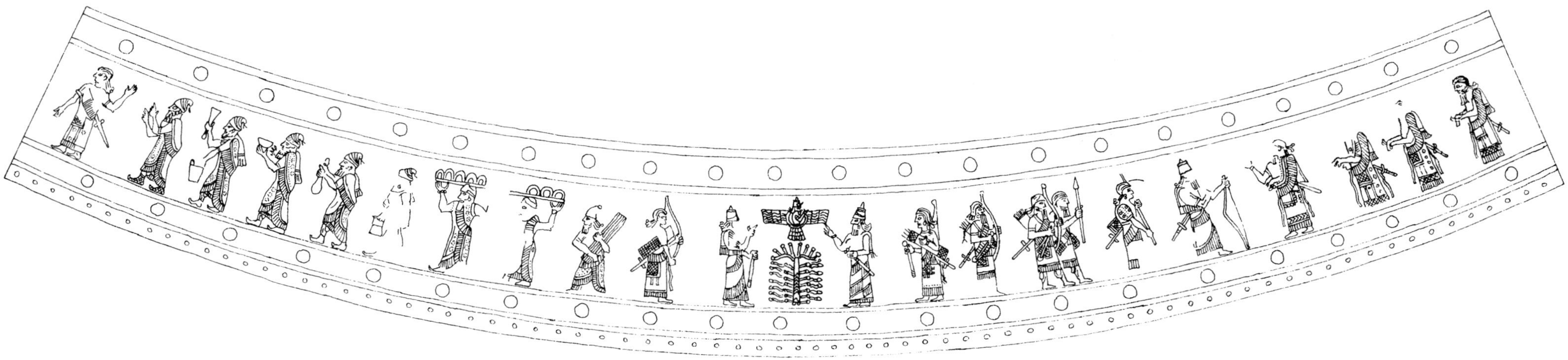

Cat. no. 14. Karlsruhe, Badisches Landesmuseum, 89/12.
Drawing of helmet decoration.

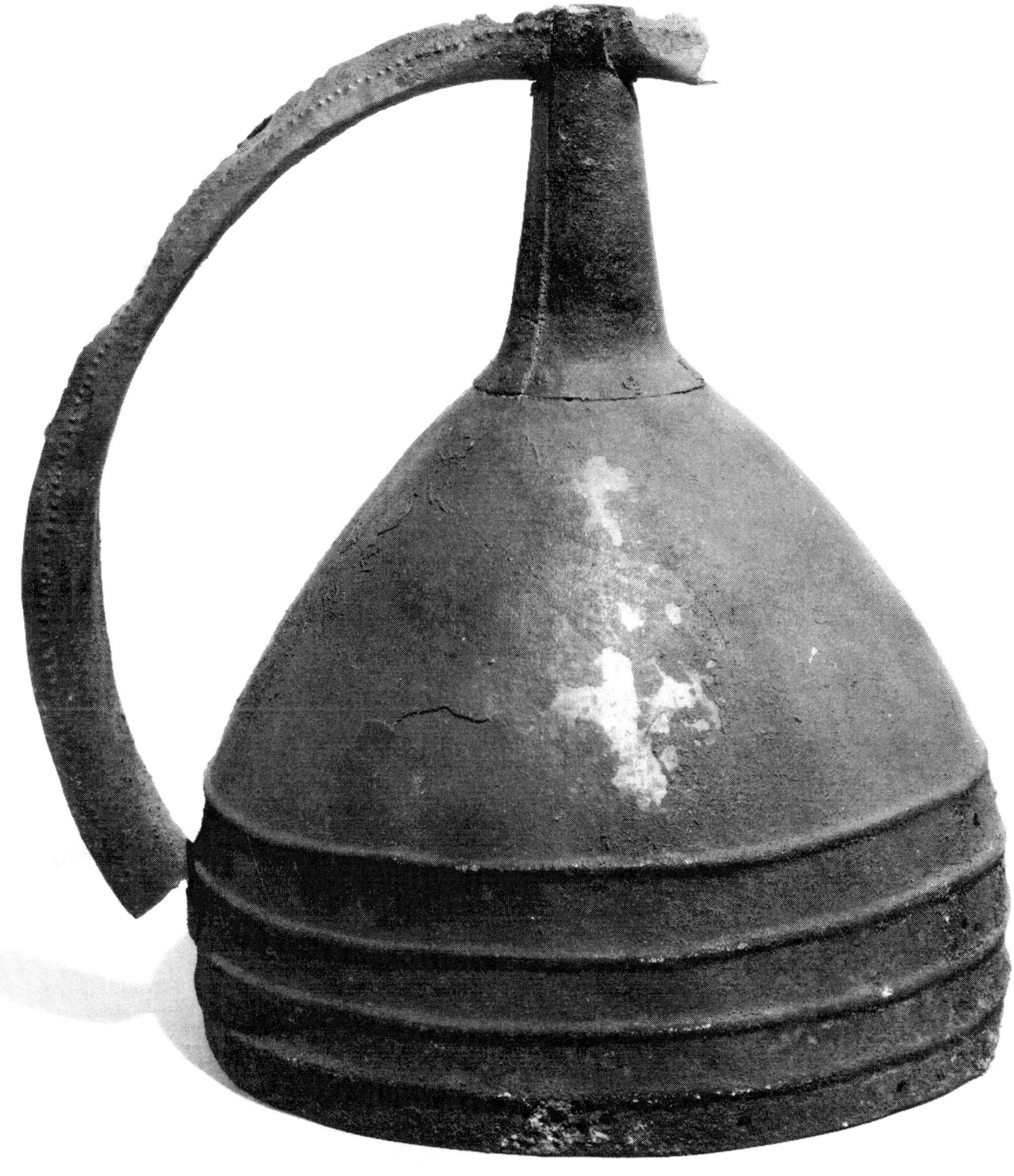

Cat. no. 15. London, British Museum, Richard Payne Knight Bequest, 2841.

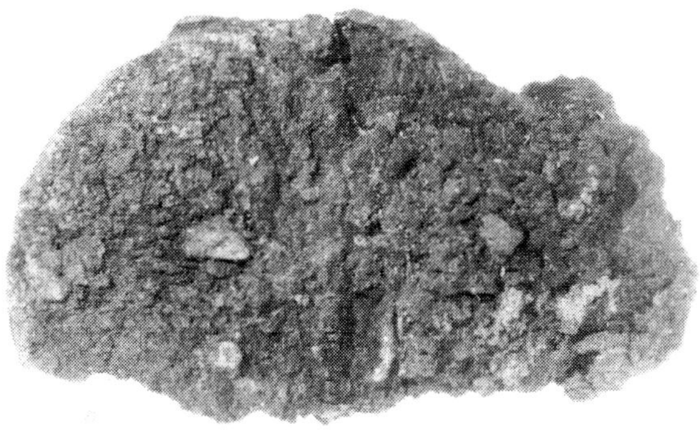

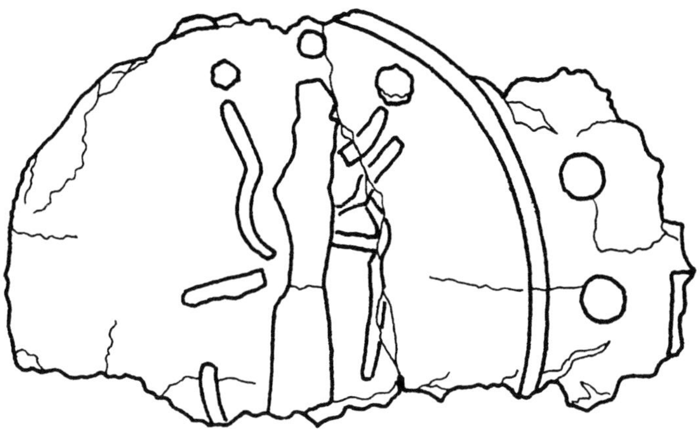

Cat. no. 16. London, British Museum, 48-11-4, 115.
Photograph and drawing of X-ray photograph.

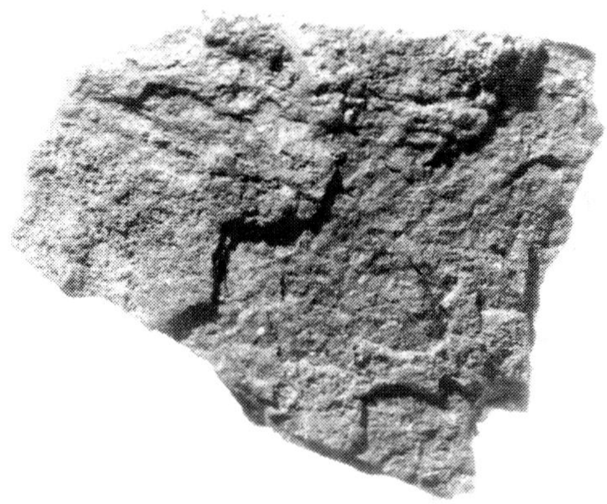

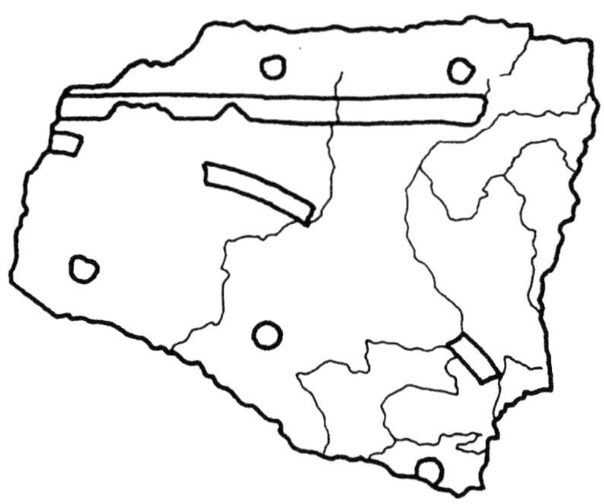

Cat. no. 17. London, British Museum, 48-11-4, 115.
Photograph and drawing of X-ray photograph.

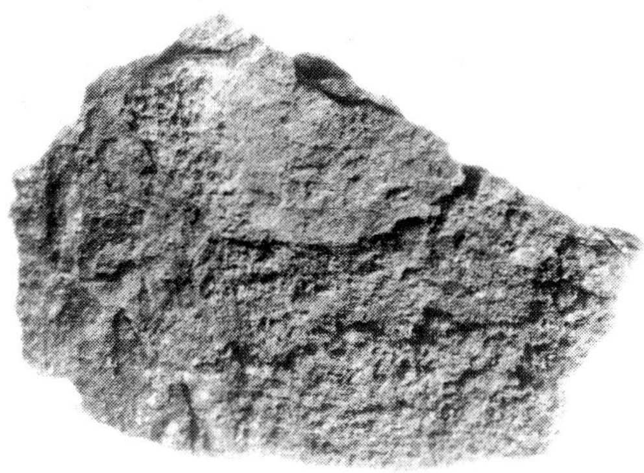

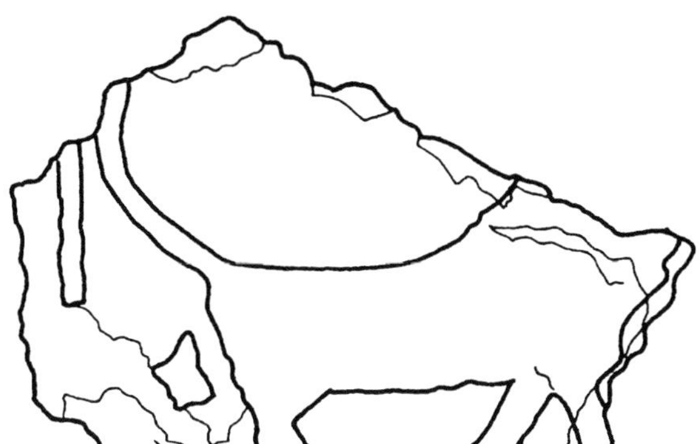

Cat. no. 18. London, British Museum, 48-11-4, 115.
Photograph and drawing of X-ray photograph.

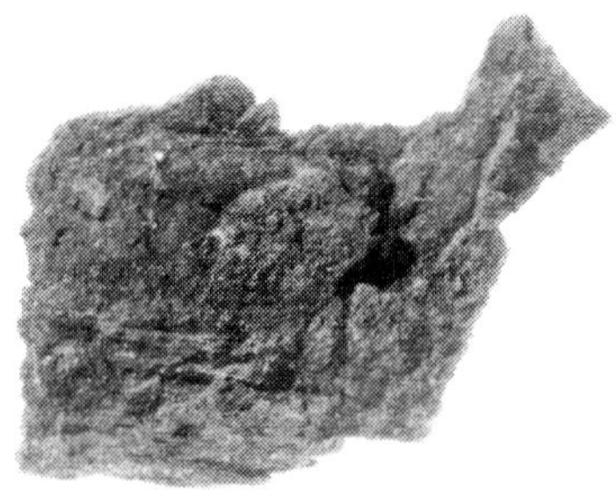

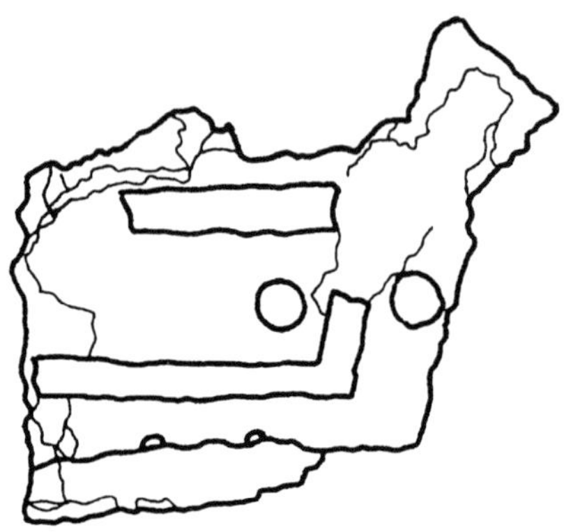

Cat. no. 19. London, British Museum, 48-11-4, 115.
Photograph and drawing of X-ray photograph.

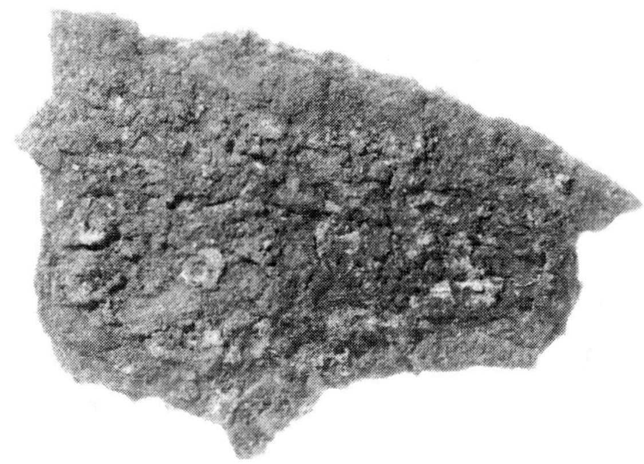

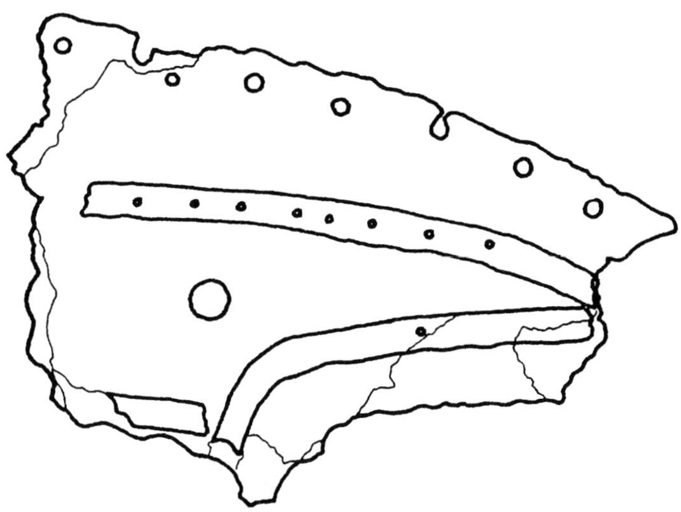

Cat. no. 20. London, British Museum, 48-11-4, 115.
Photograph and drawing of X-ray photograph.

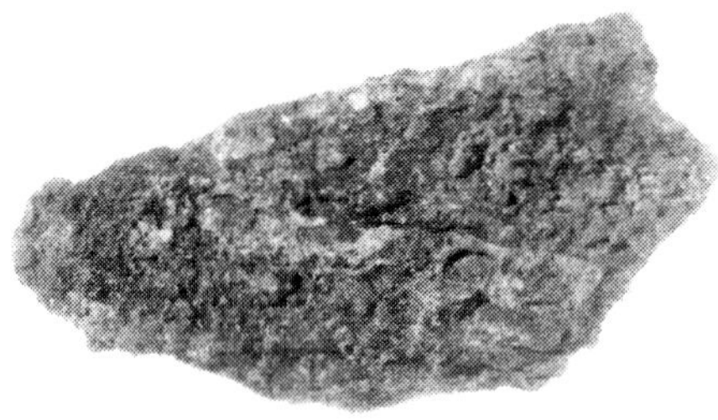

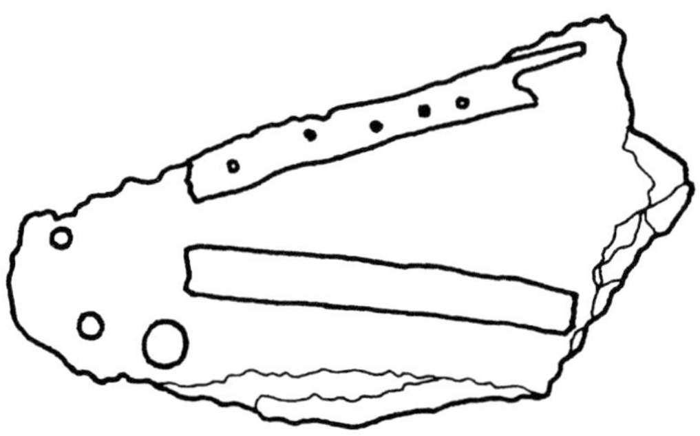

Cat. no. 21. London, British Museum, 48-11-4, 115.
Photograph and drawing of X-ray photograph.

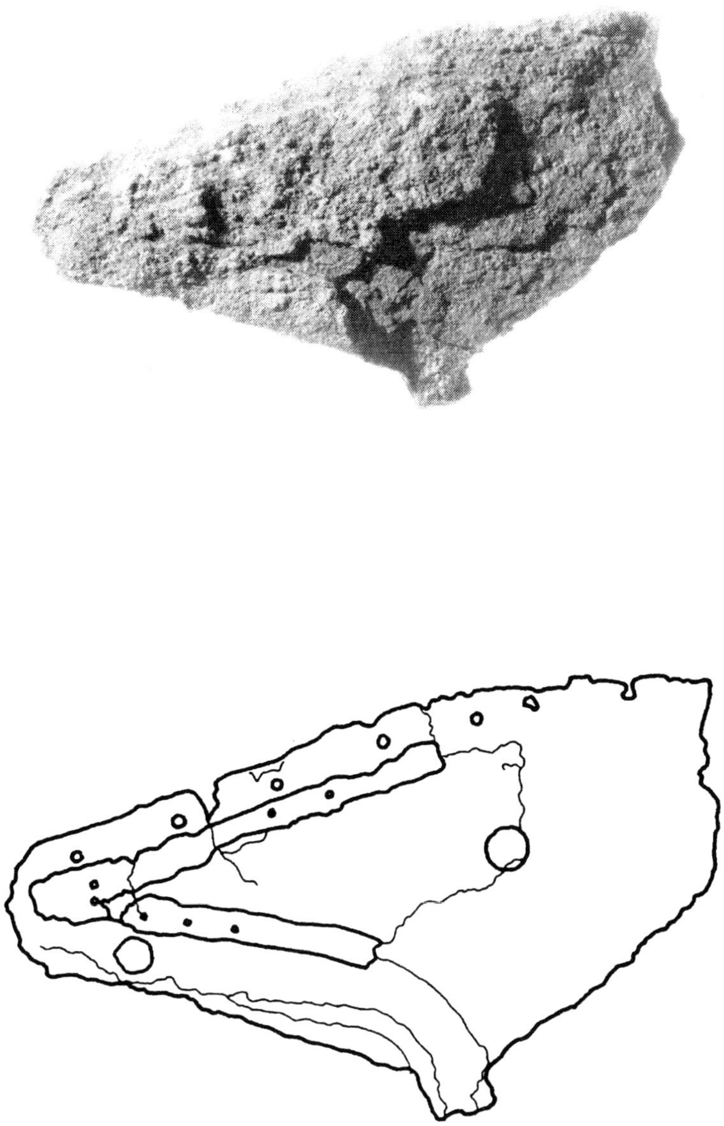

Cat. no. 22. London, British Museum, 48-11-4, 115.
Photograph and drawing of X-ray photograph.

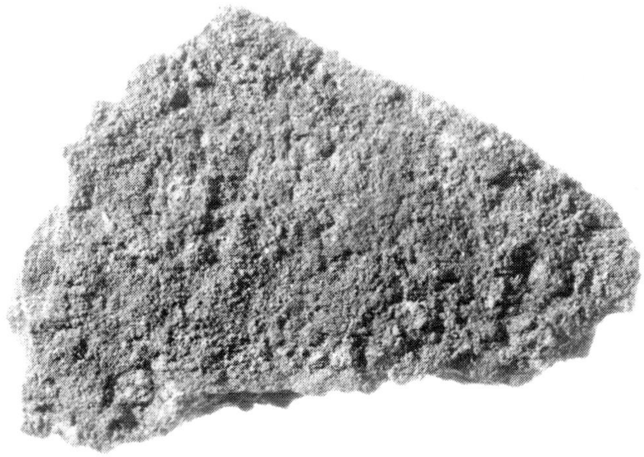

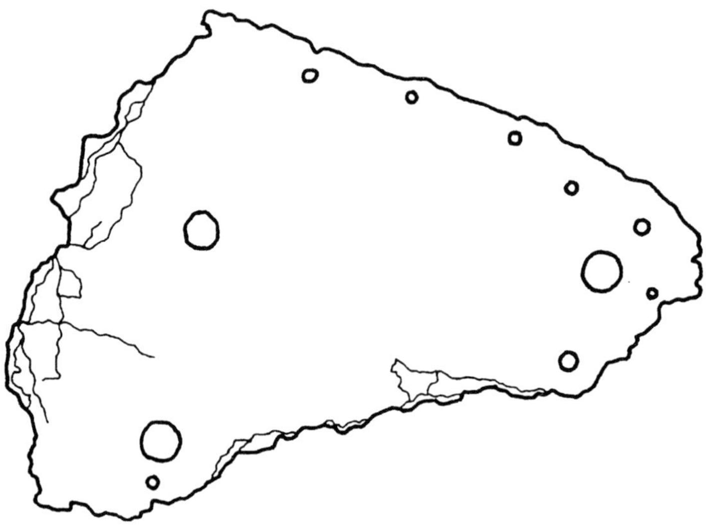

Cat. no. 23. London, British Museum, 48-11-4, 115.
Photograph and drawing of X-ray photograph.

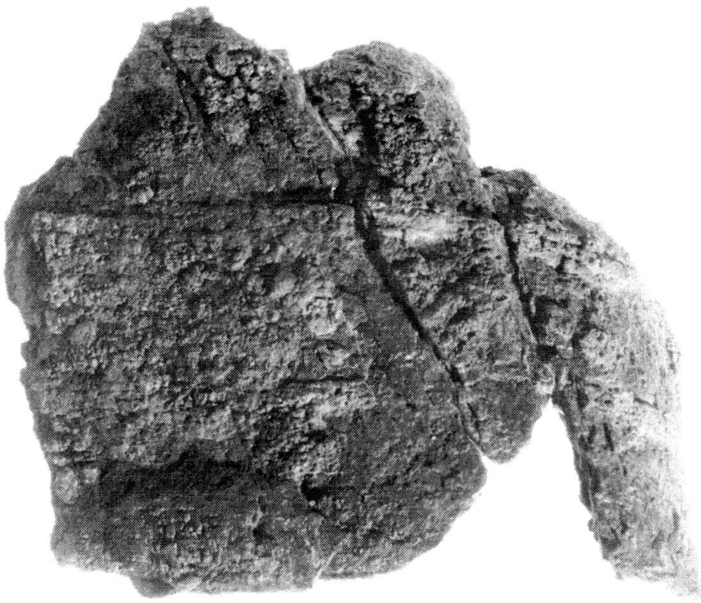

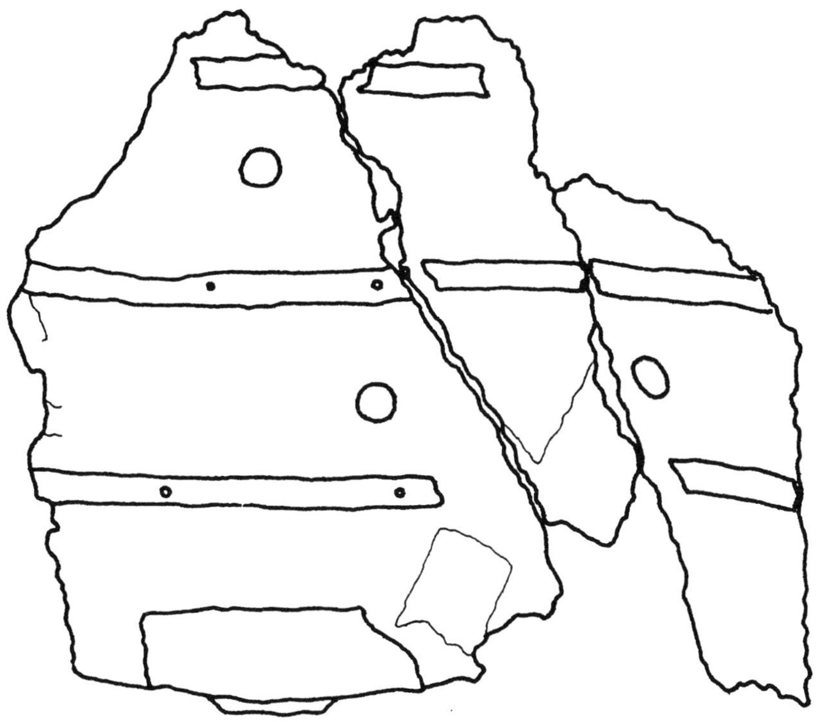

Cat. nos. 24-25-26. London, British Museum, 48-11-4, 115.
Photograph and drawing of X-ray photograph.

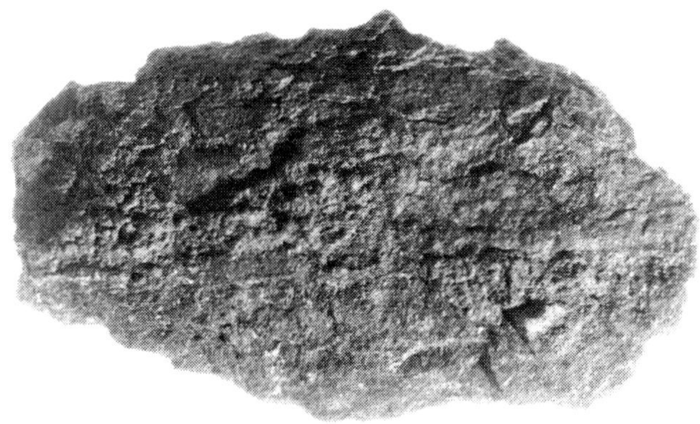

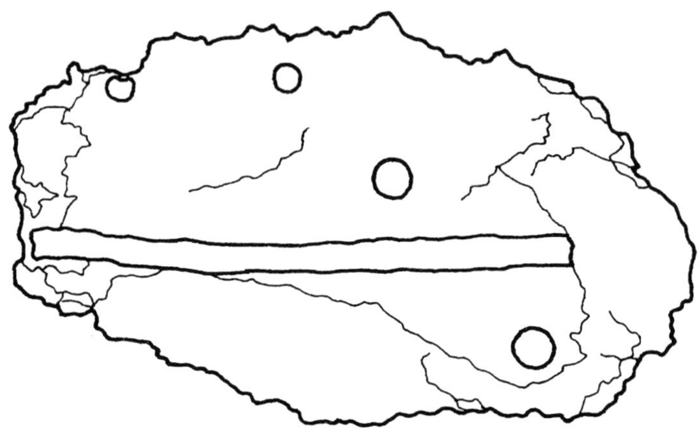

Cat. no. 27. London, British Museum, 48-11-4, 115.
Photograph and drawing of X-ray photograph.

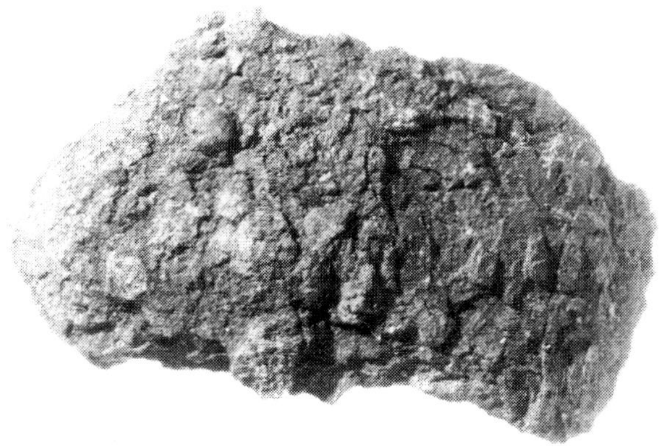

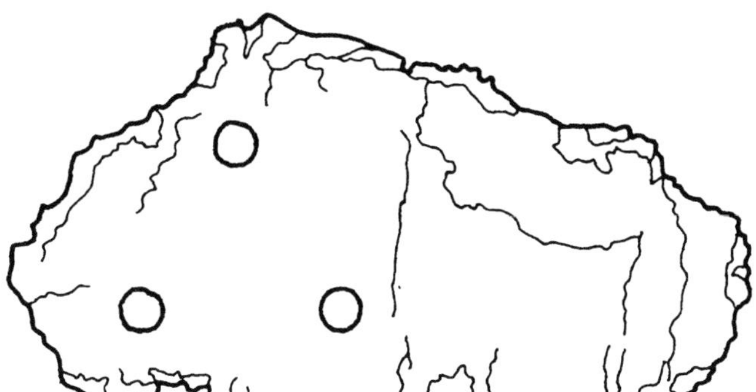

Cat. no. 28. London, British Museum, 48-11-4, 115.
Photograph and drawing of X-ray photograph.

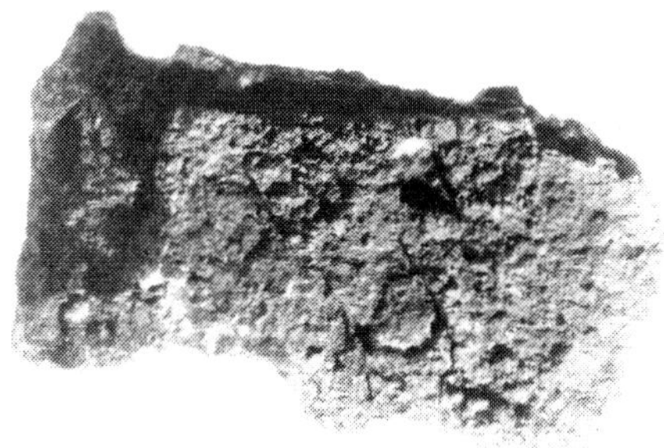

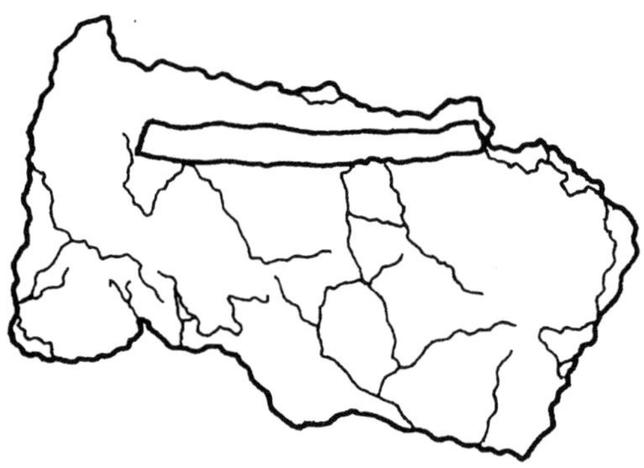

Cat. no. 29. London, British Museum, 48-11-4, 115.
Photograph and drawing of X-ray photograph.

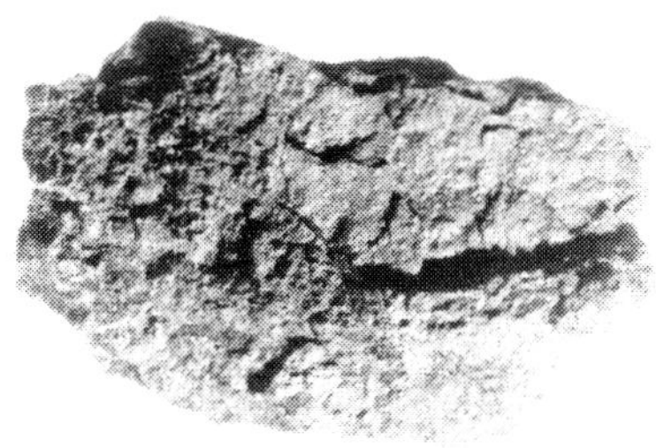

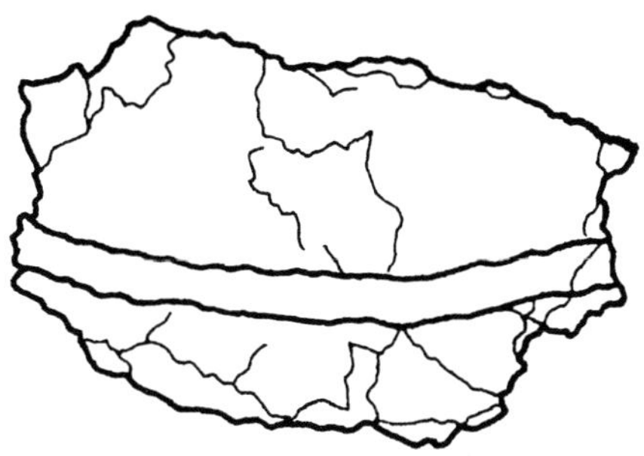

Cat. no. 30. London, British Museum, 48-11-4, 115.
Photograph and drawing of X-ray photograph.

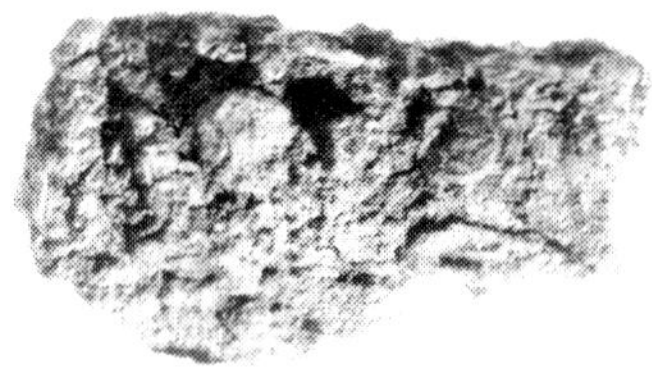

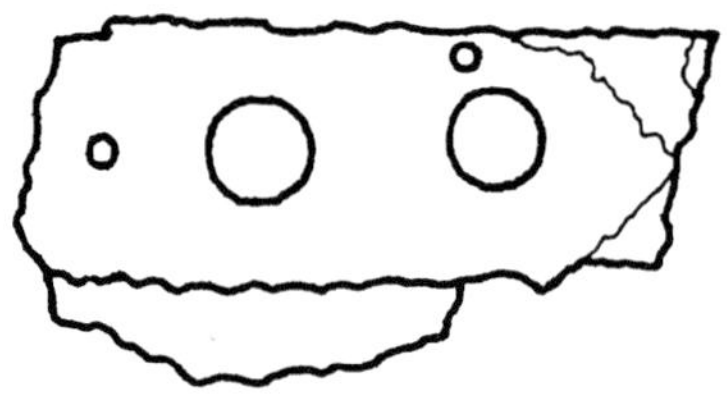

Cat. no. 31. London, British Museum, 48-11-4, 115.
Photograph and drawing of X-ray photograph.

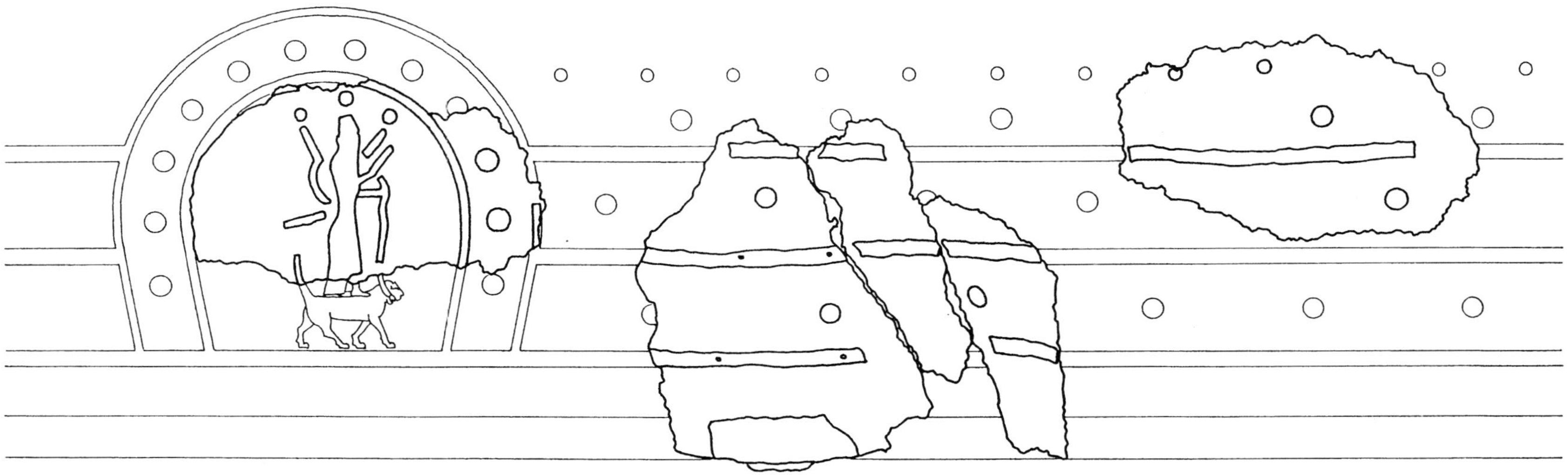

Cat. no. 16, 24-25-26, 27. London, British Museum, 48-11-4, 115.
Reconstruction: front decoration.

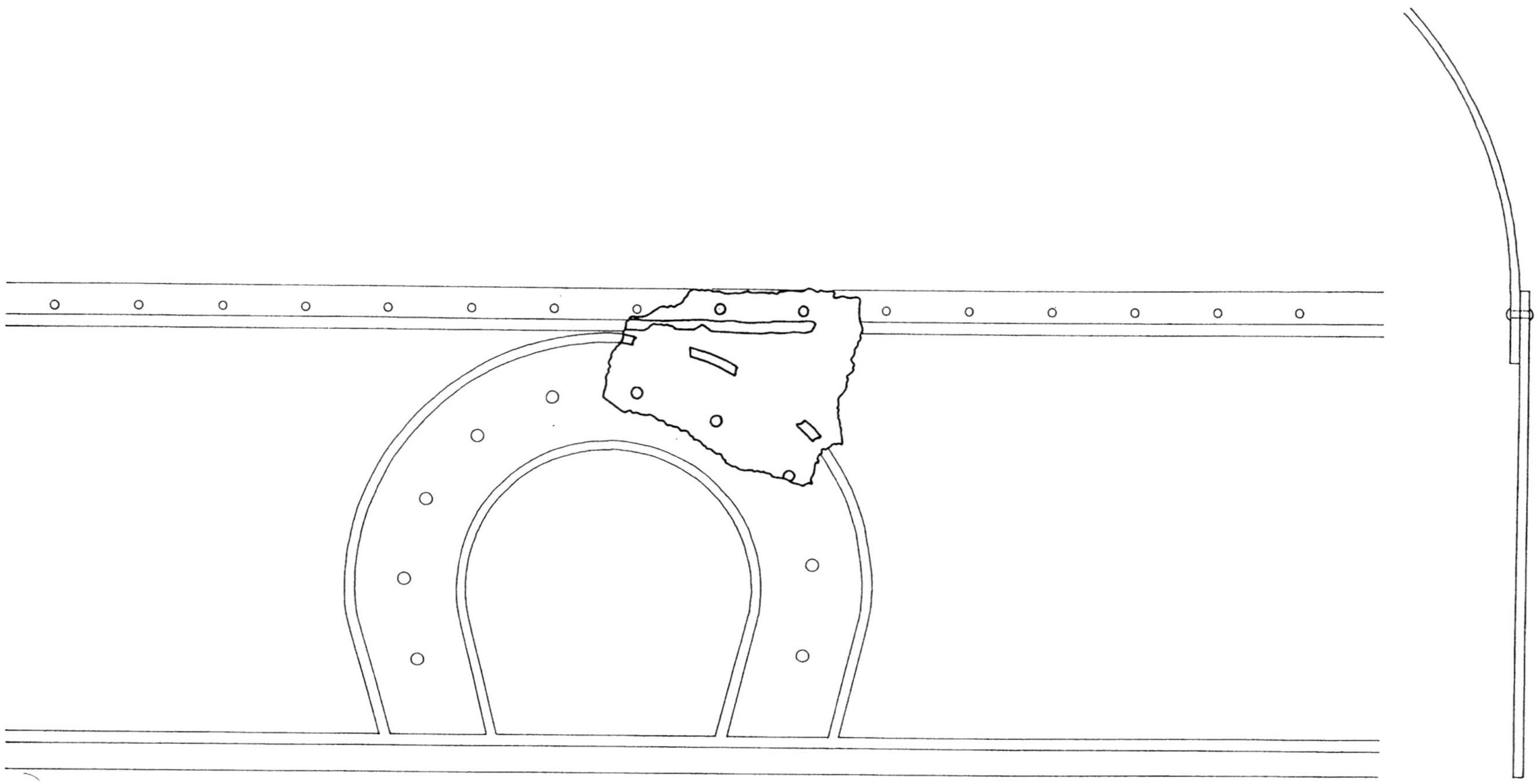

Cat. no. 17. London, British Museum, 48-11-4, 115.
Reconstruction: front decoration and construction.

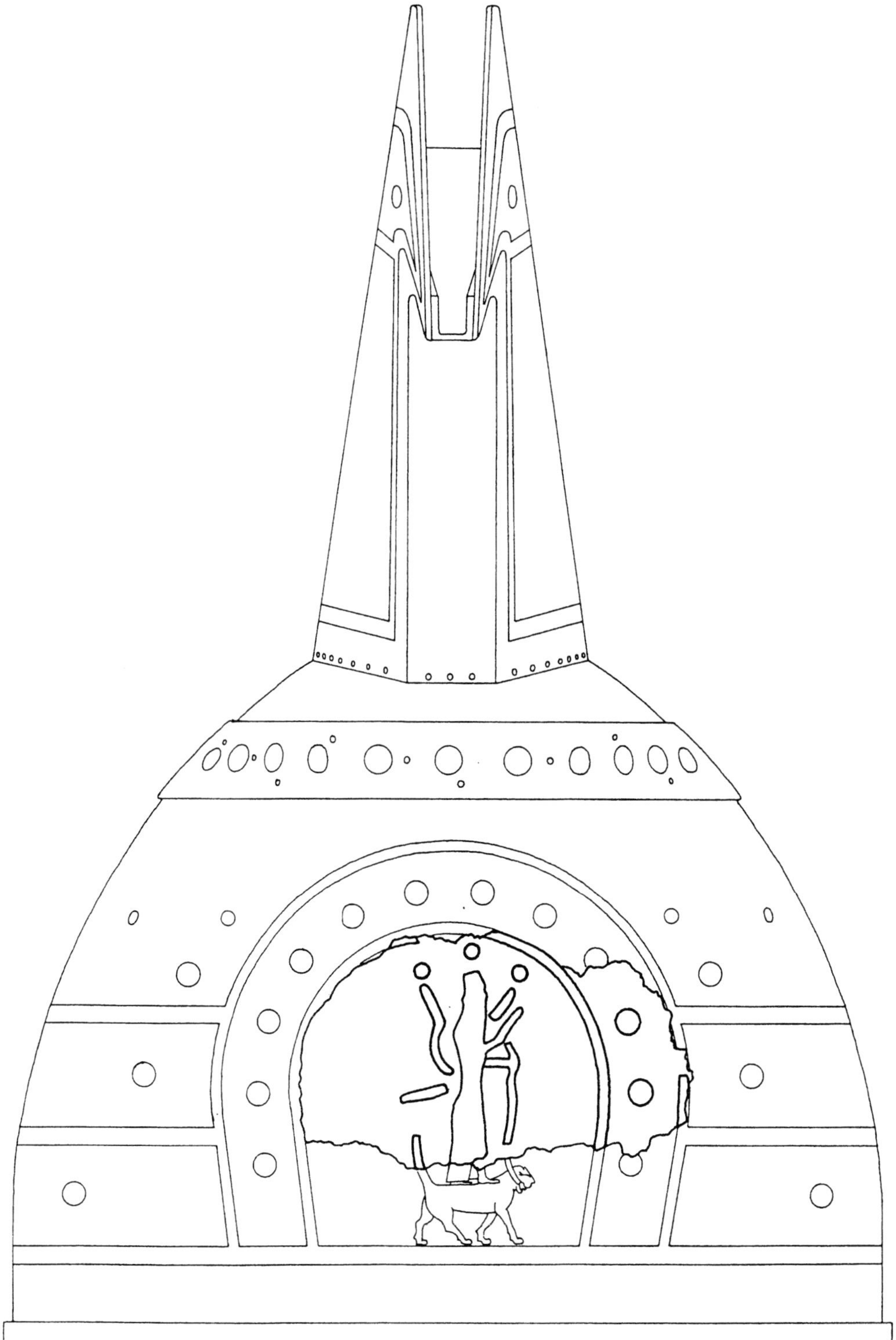

Cat. no. 16. London, British Museum, 48-11-4, 115.
Reconstruction: front view.

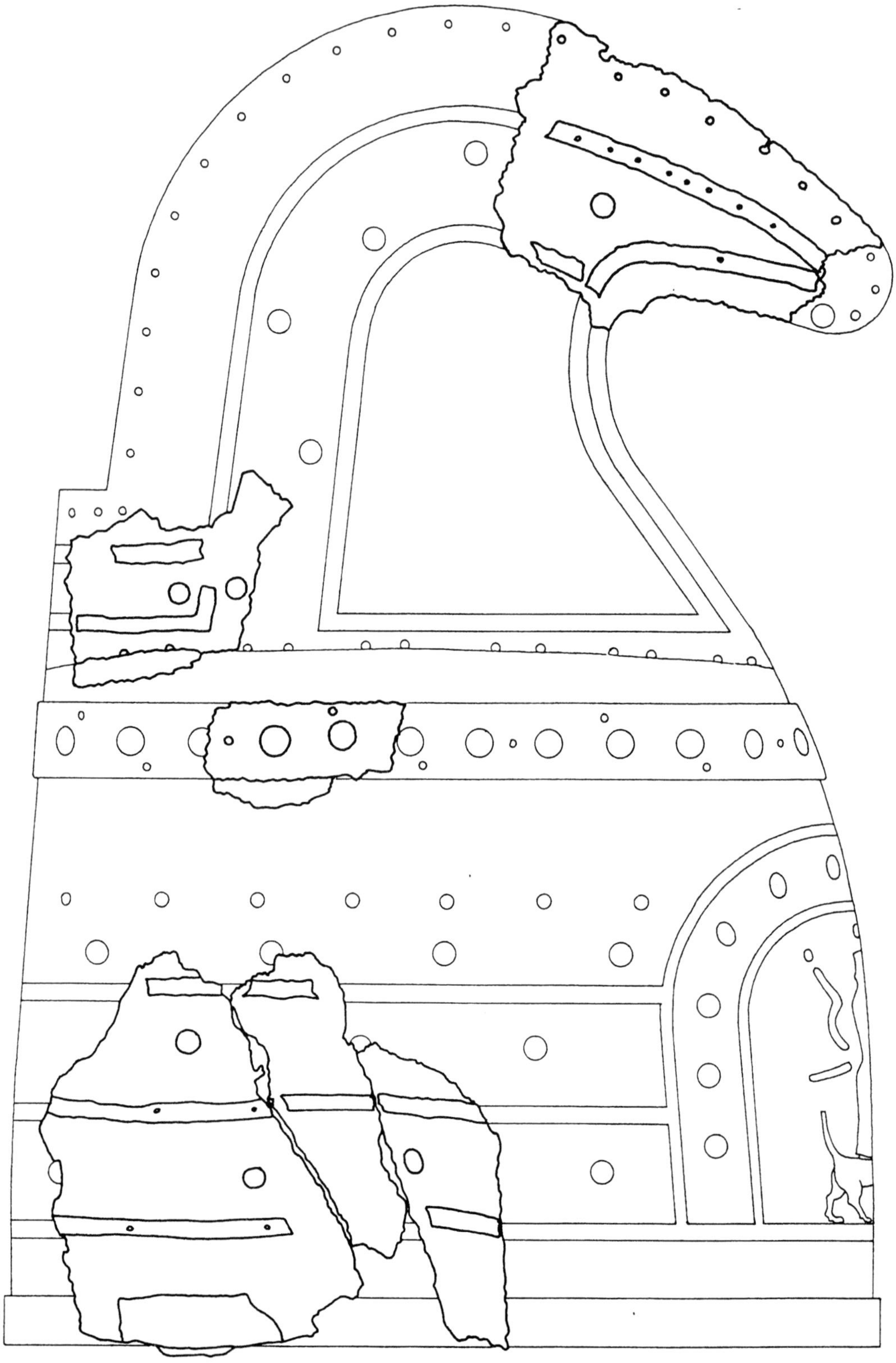

Cat. no. 19, 20, 24-25-26-27, 31. London, British Museum, 48-11-4, 115.
Reconstruction: side view.

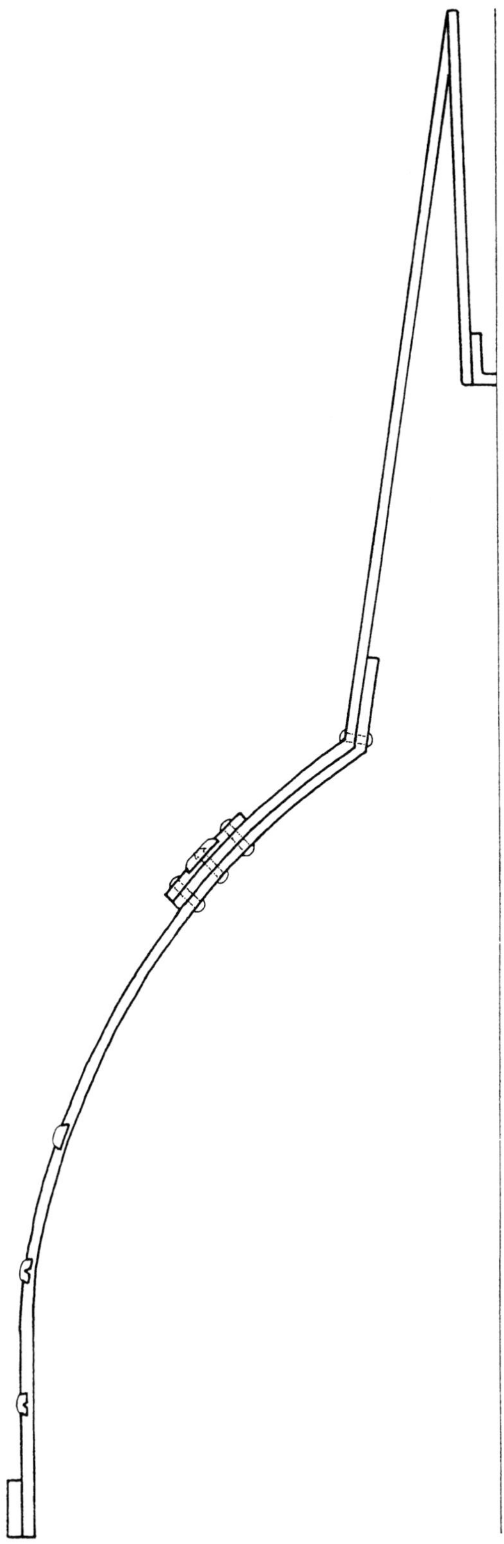

Cat. no. 16-31. London, British Museum, 48-11-4, 115.
Reconstruction: section of a crested helmet.

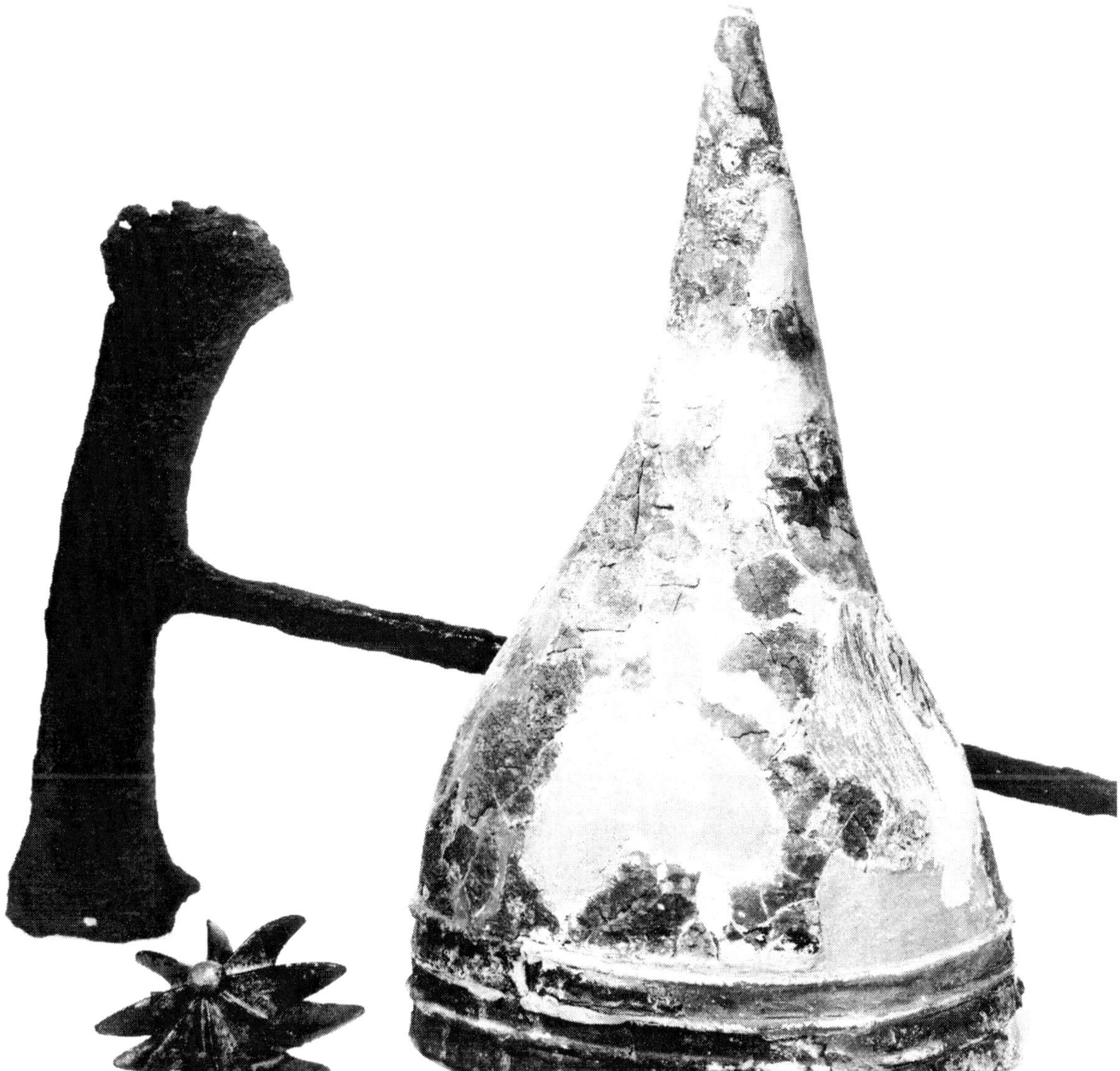

Cat. no. 32. Berlin, Staatliche Museen zu Berlin, Vorderasiatisches Museum, S.3976.

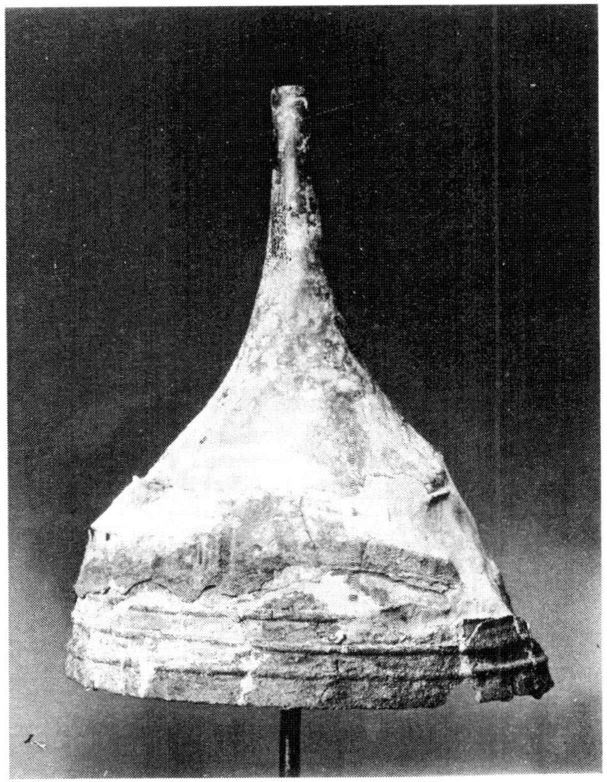

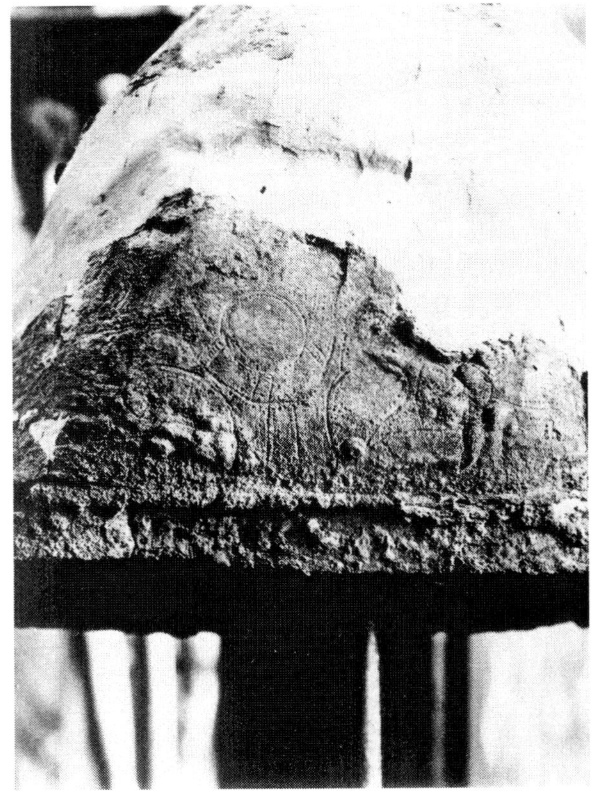

Cat. no. 33. Berlin, Staatliche Museen zu Berlin,Vorderasiatisches Museum, S.3695.
Helmet and detail.

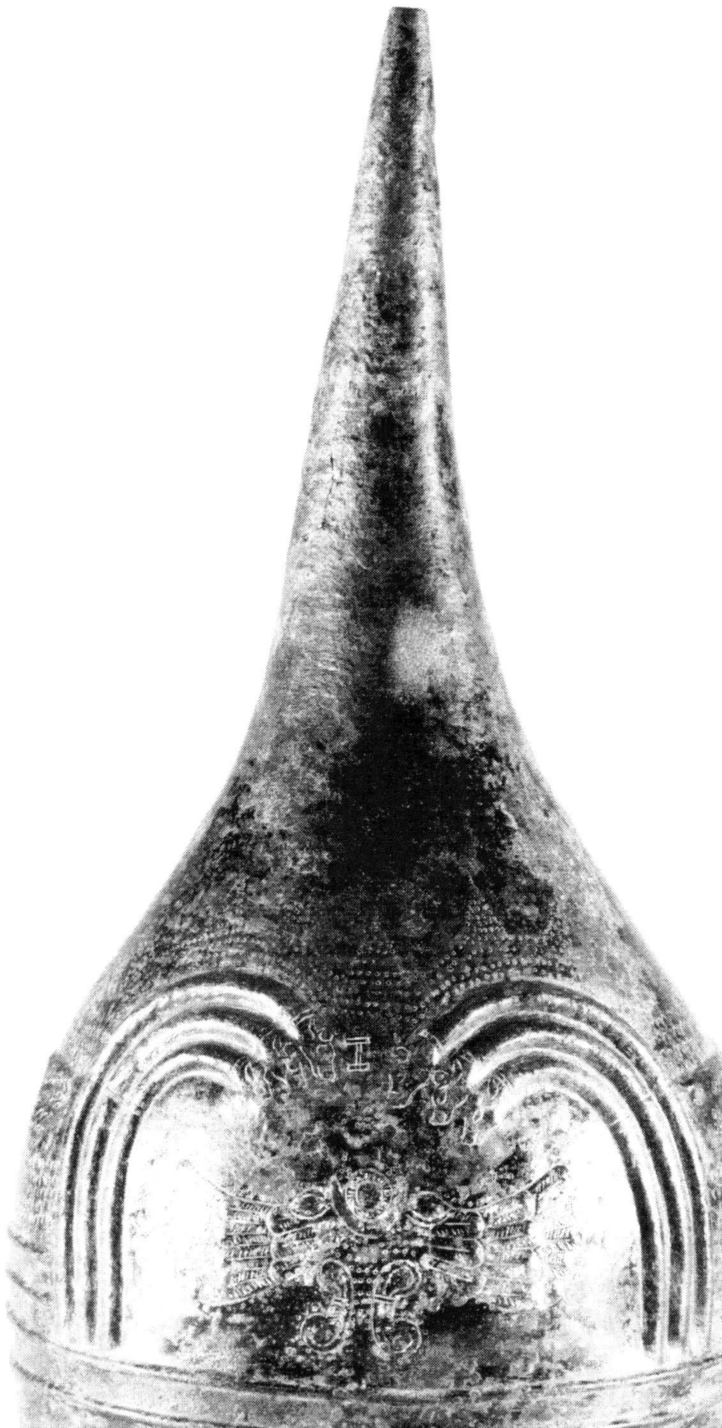

Cat. no. 34. London, British Museum, WA 134611.

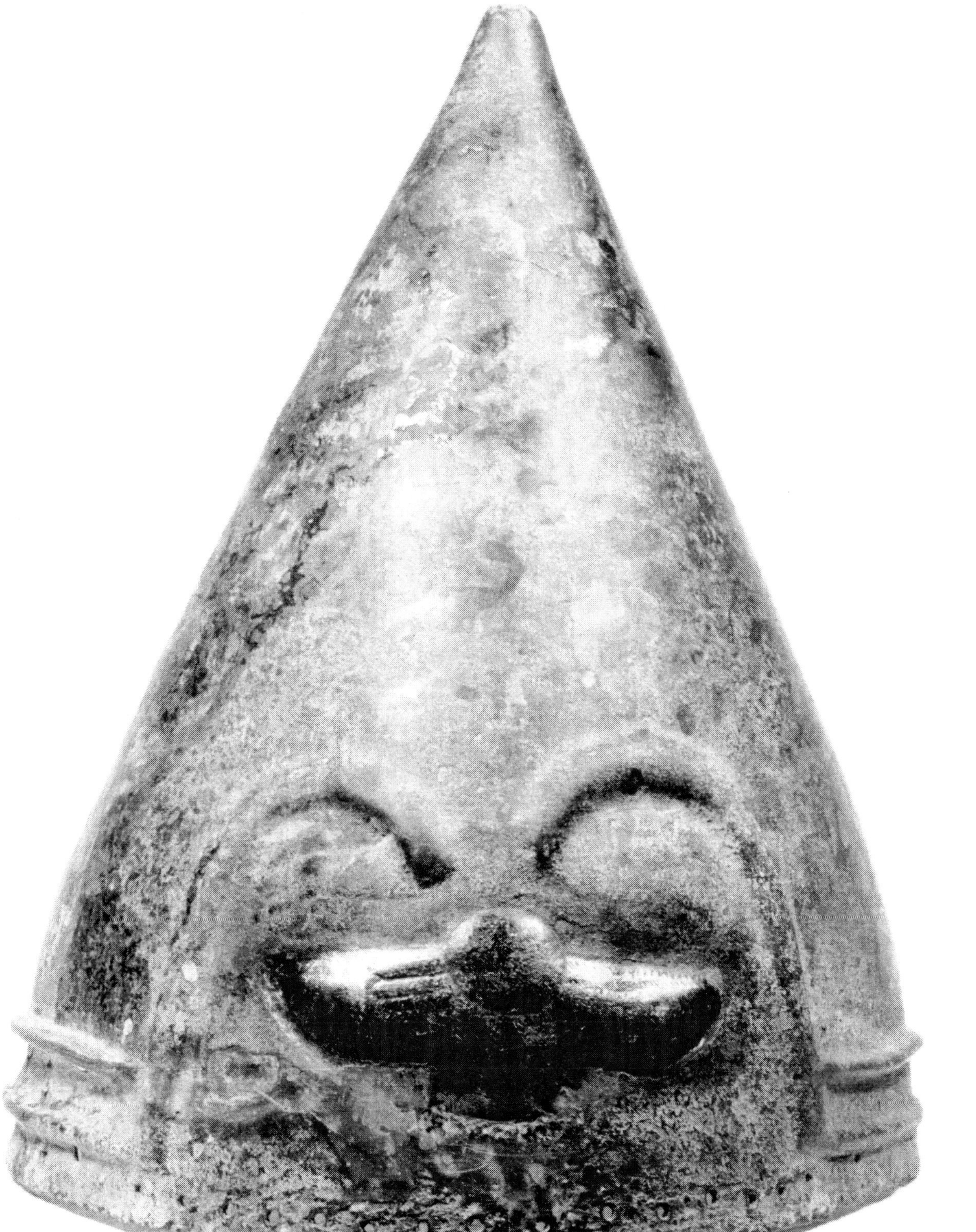

Cat. no. 35. Nicosia, Cyprus Museum, 1965/XI-29/62.

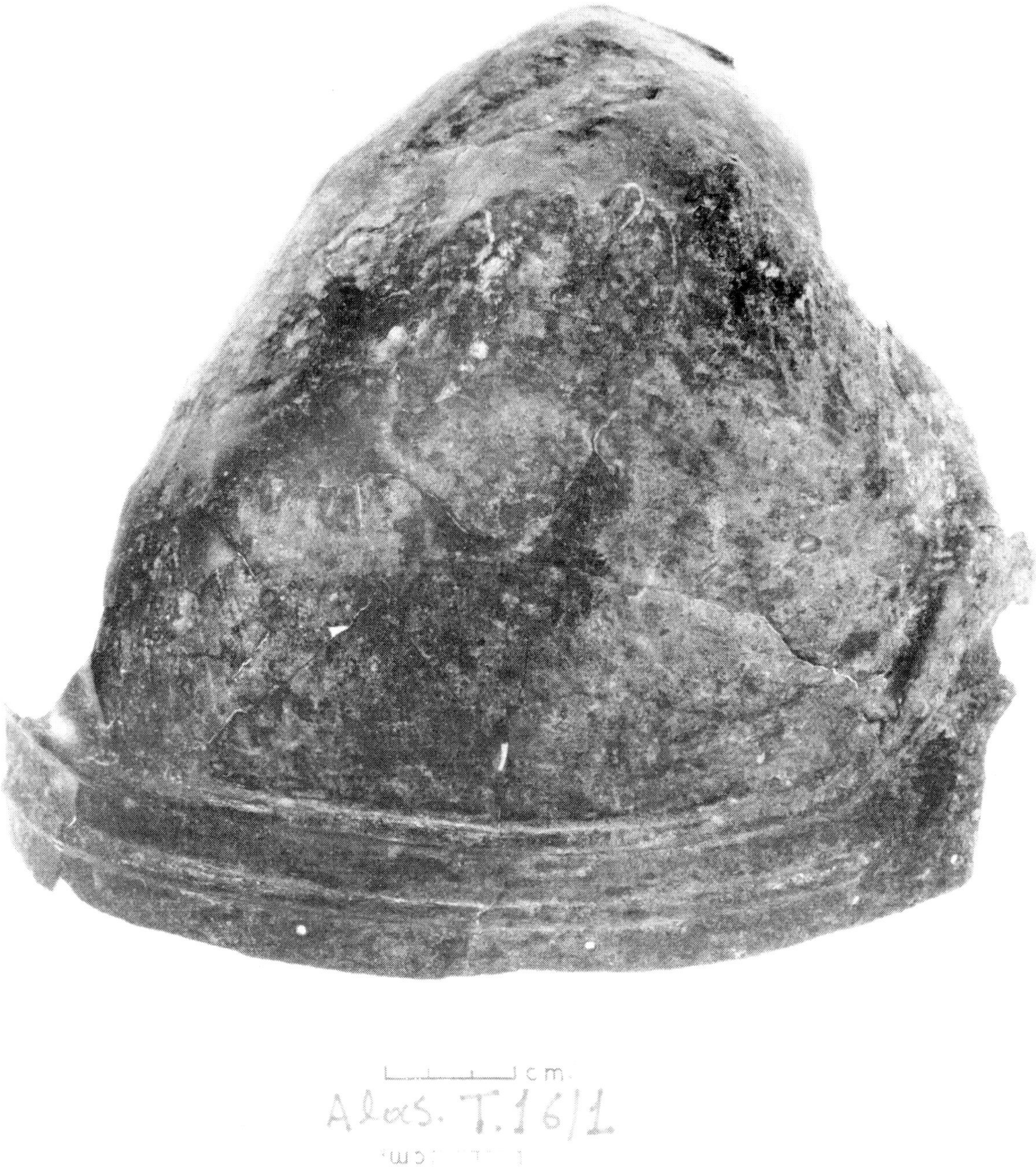

Cat. no. 36. Nicosia, Cyprus Museum, T16/1.

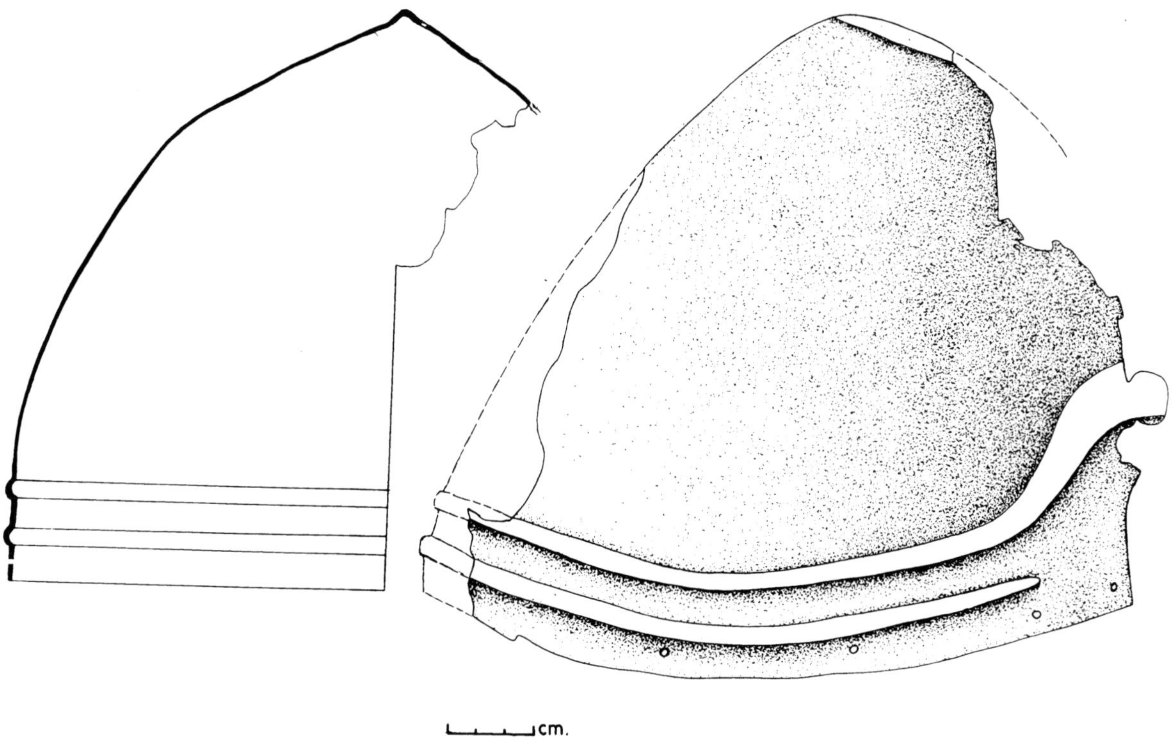

Cat. no. 36. Nicosia, Cyprus Museum, T16/1.
Drawing of helmet.

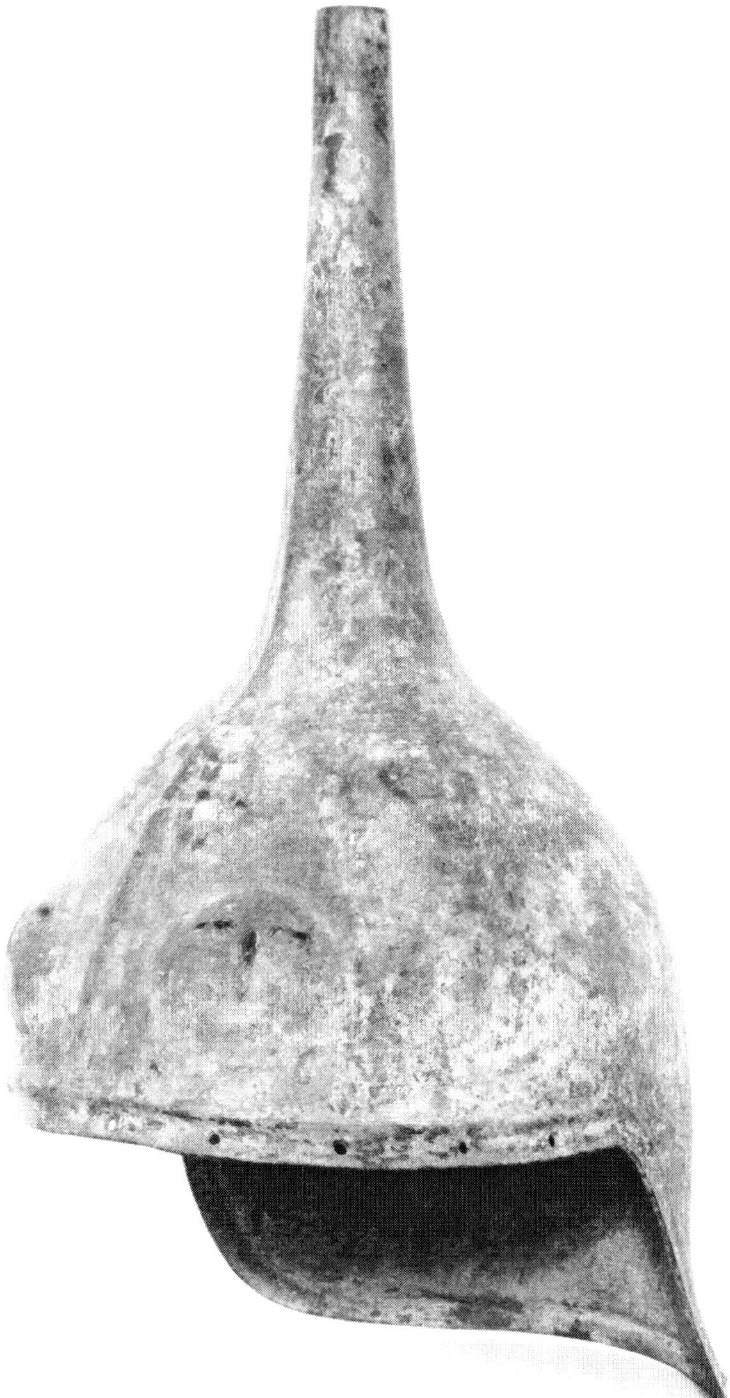

Cat. no. 37. Sotheby's, Private Collection.

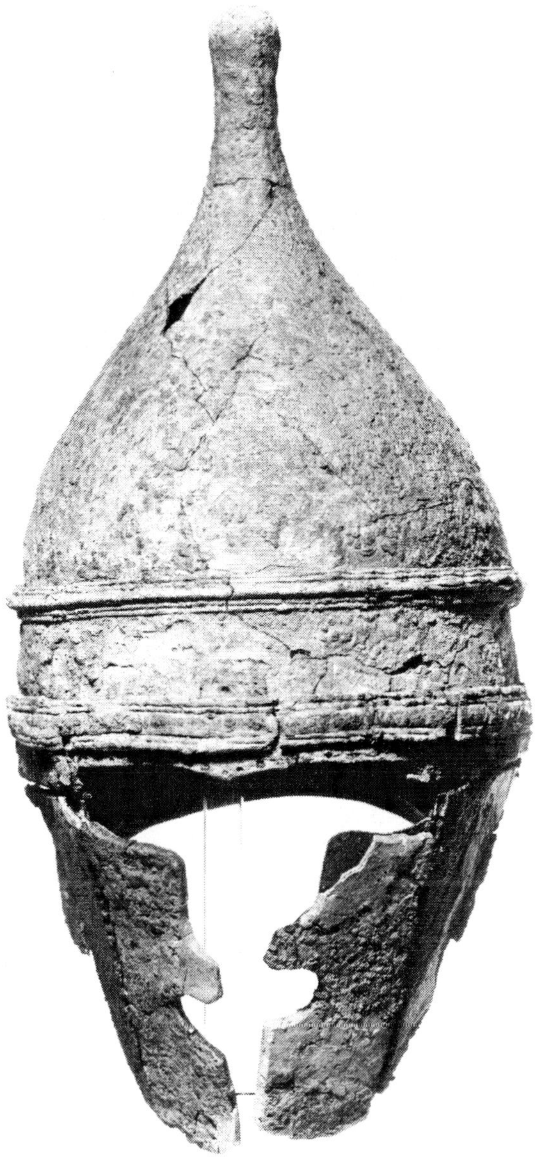

Helmet from Tamassos, Berlin, Staatliche Museen zu Berlin, Antikensammlung, Misc. 8142,620.
Front view.

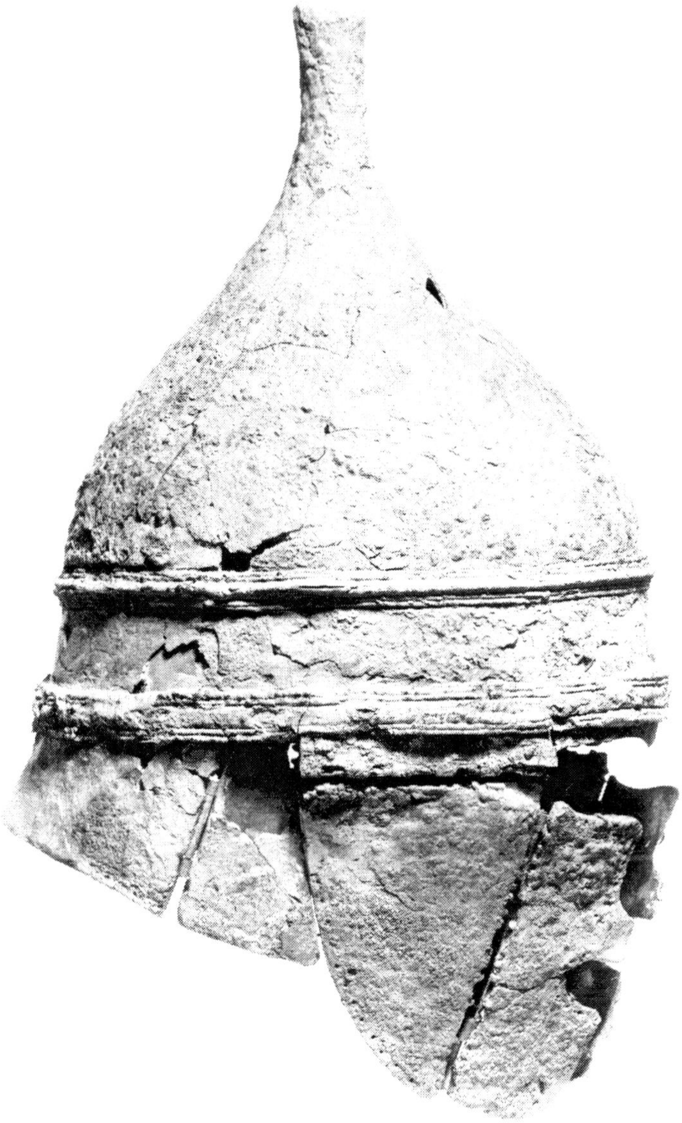

Helmet from Tamassos, Berlin, Staatliche Museen zu Berlin, Antikensammlung, Misc. 8142,620.
Side view.

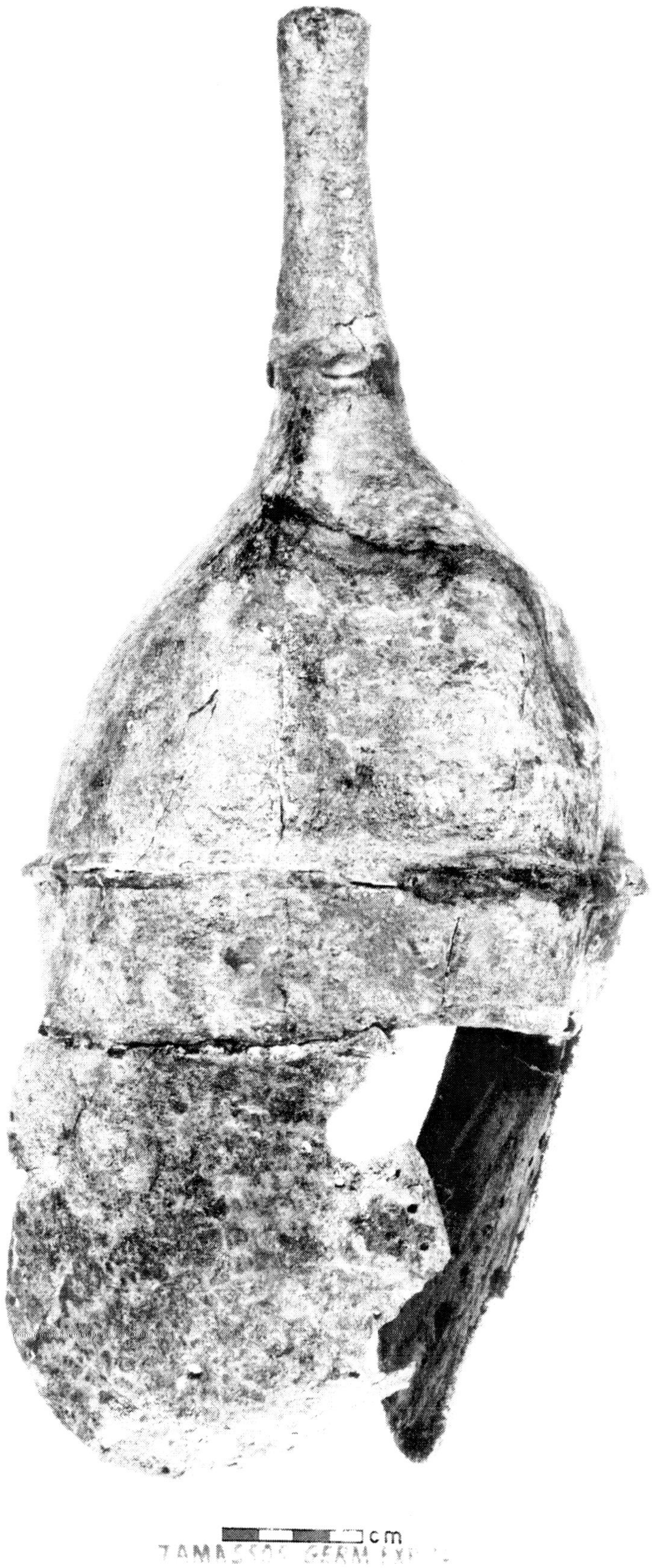

Helmet from Tamassos, Nicosia, Cyprus Museum, 98/1971.
Front view.

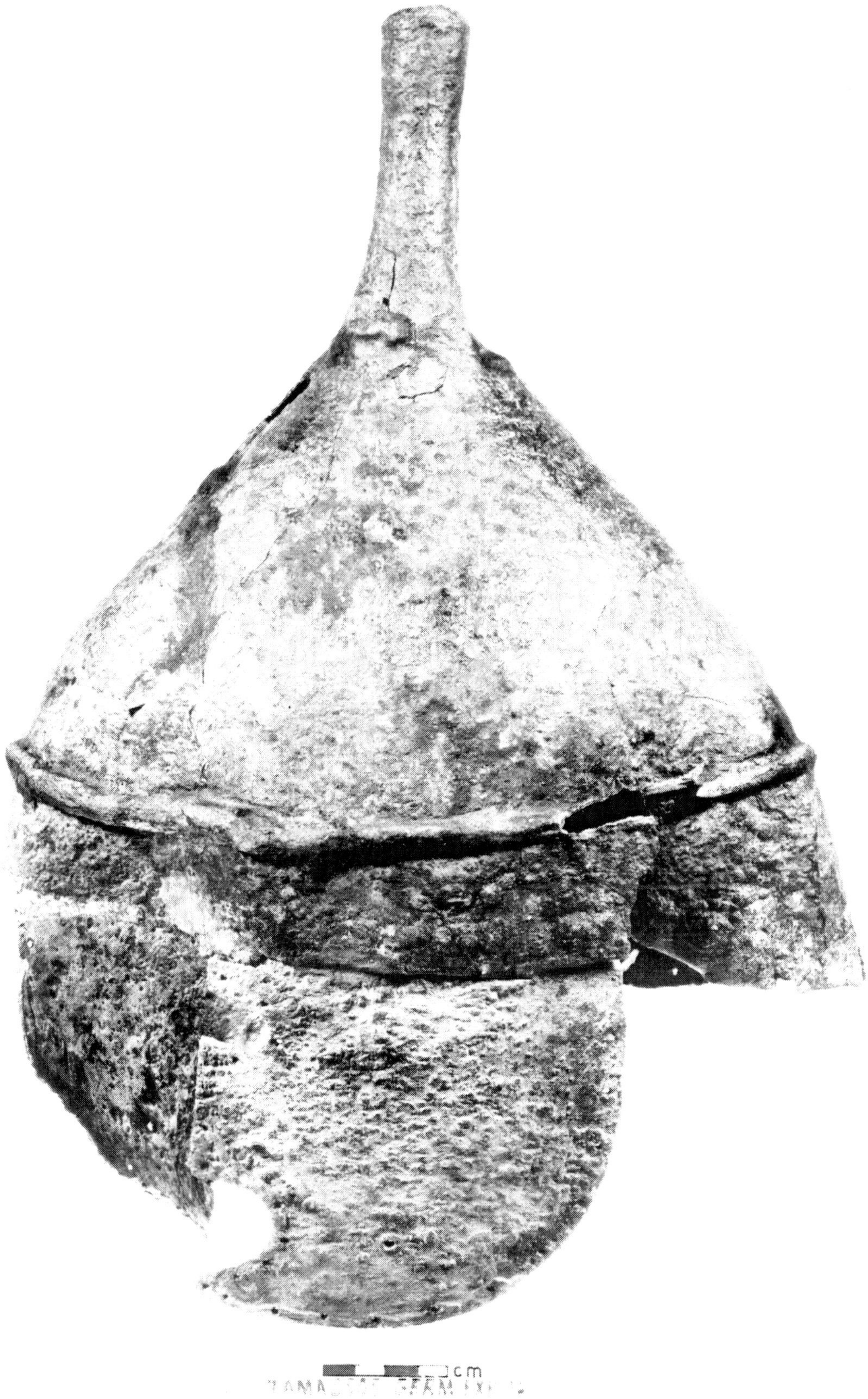

Helmet from Tamassos, Nicosia, Cyprus Museum, 98/1971.
Side view.

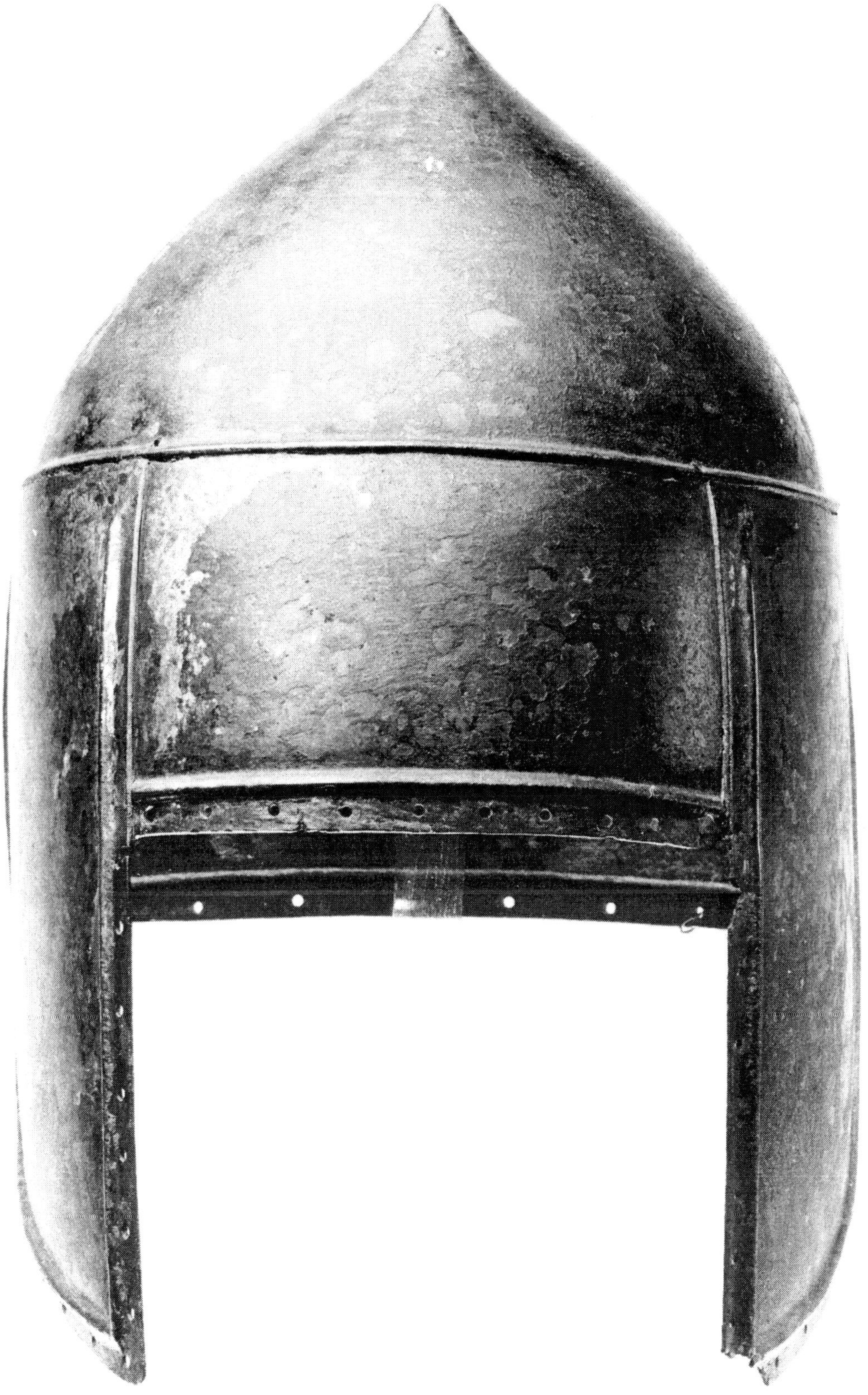

Greek Geometric helmet, Budapest, Museum of Fine Arts, 8442.
Front view.

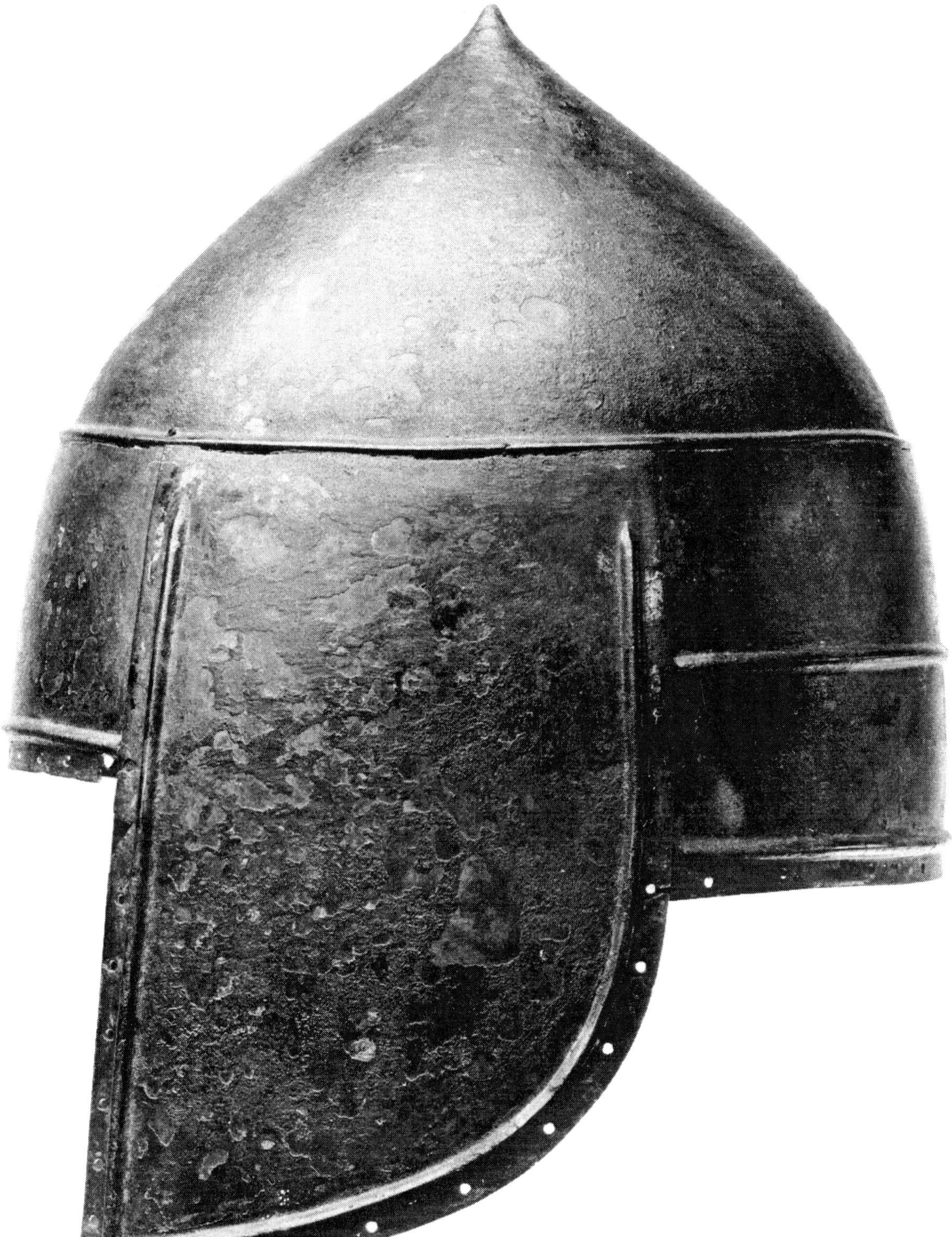

Greek Geometric helmet, Budapest, Museum of Fine Arts, 8442.
Side view.

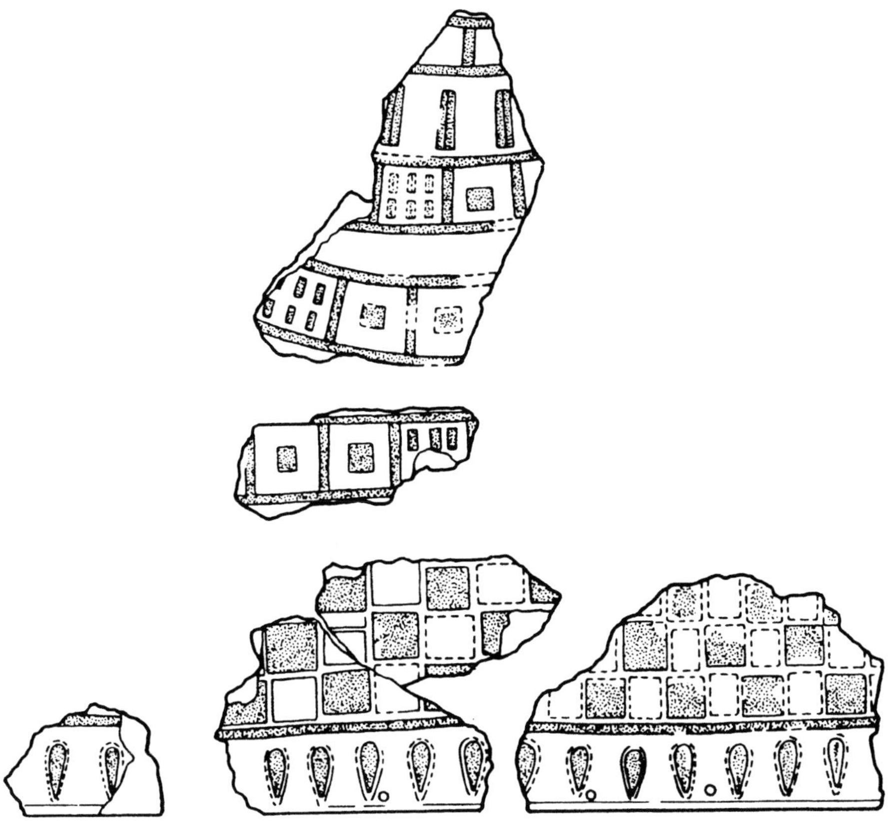

Cat. no. 38. Berlin, Staatliche Museen zu Berlin, Vorderasiatisches Museum, S.3964.
Drawing of fragment.

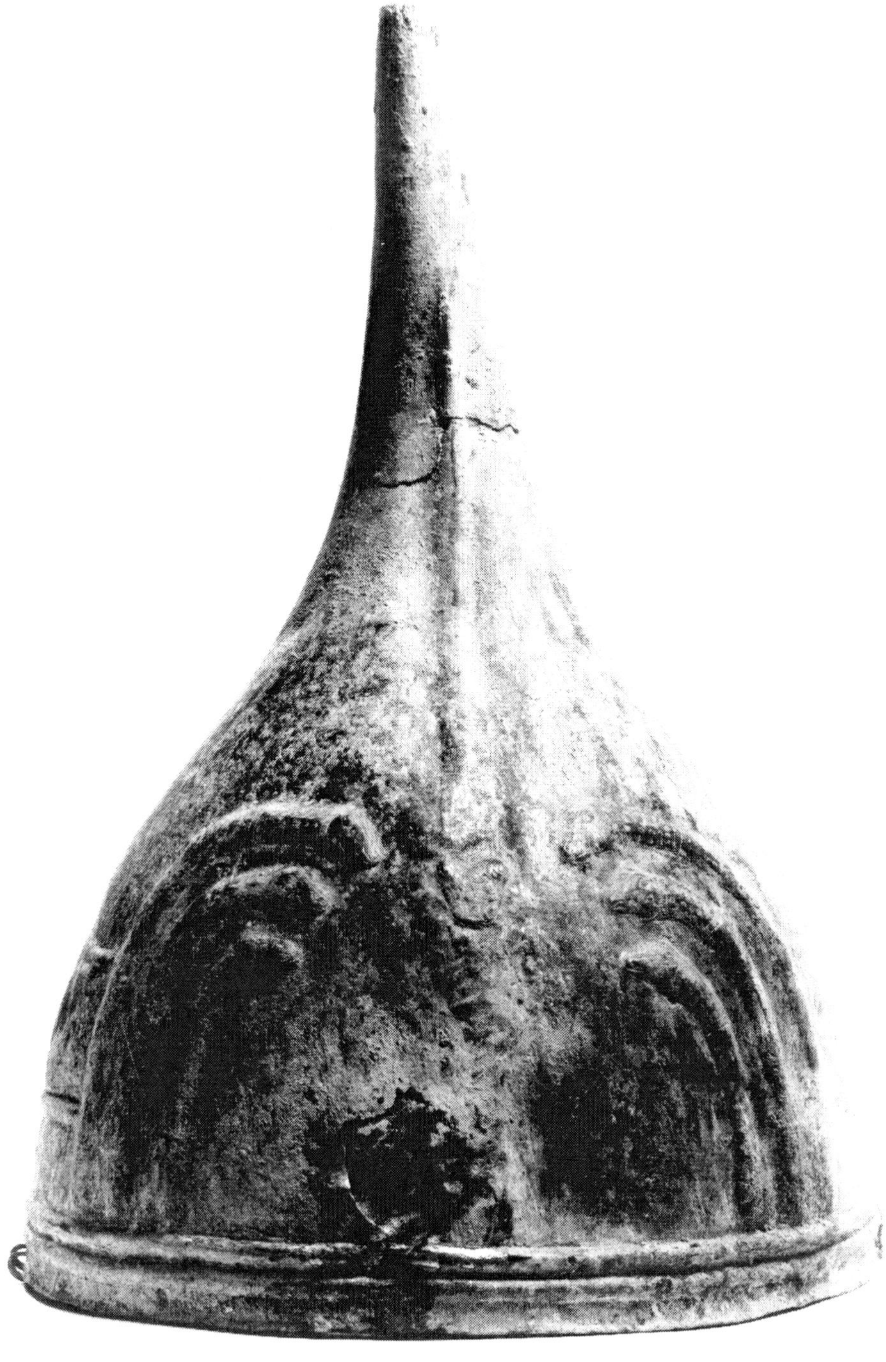

Cat. no. 39. Mariemont, Musée Royal de Mariemont, Ac. 72/27.

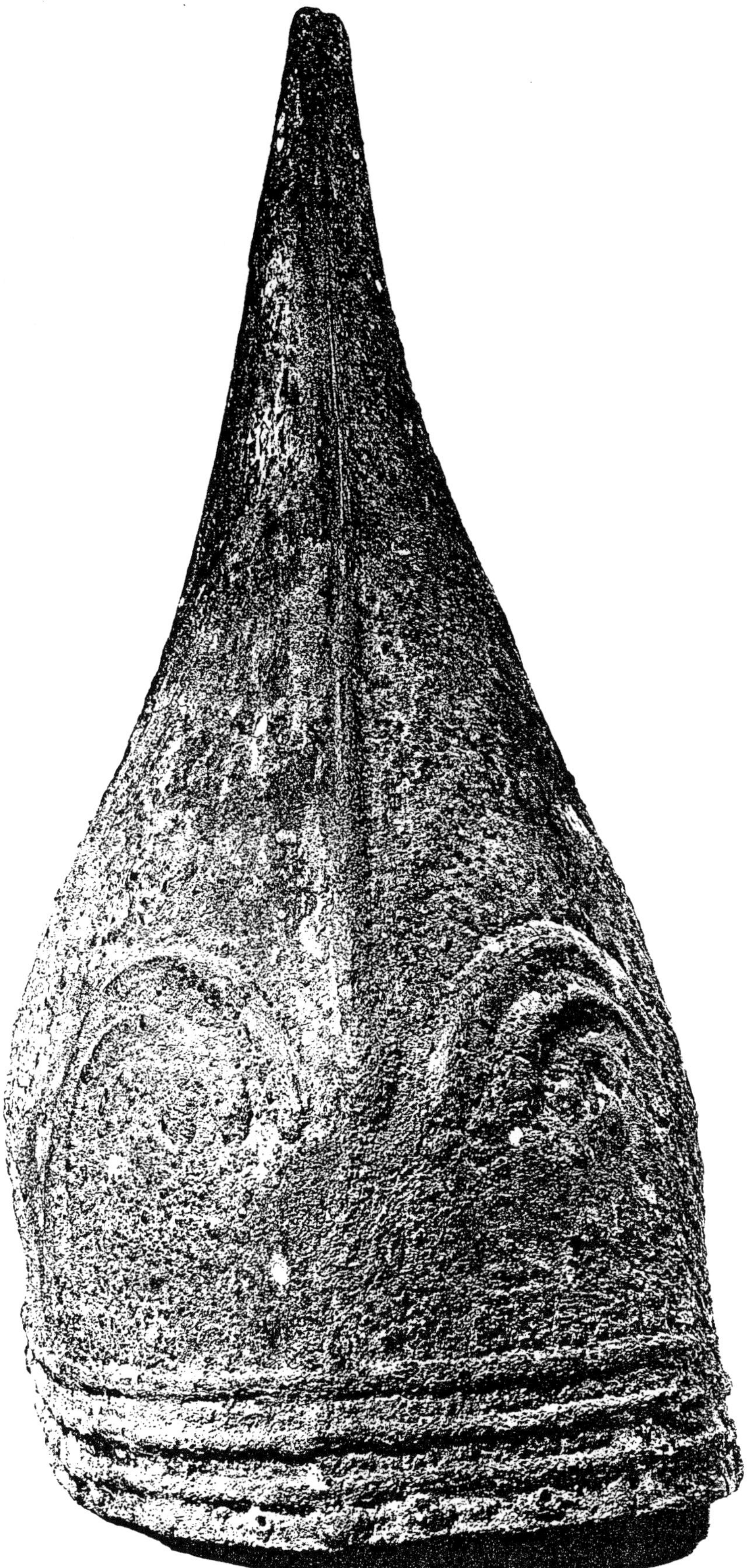

Cat. no. 40. Paris, Musée du Louvre, AO 22123.

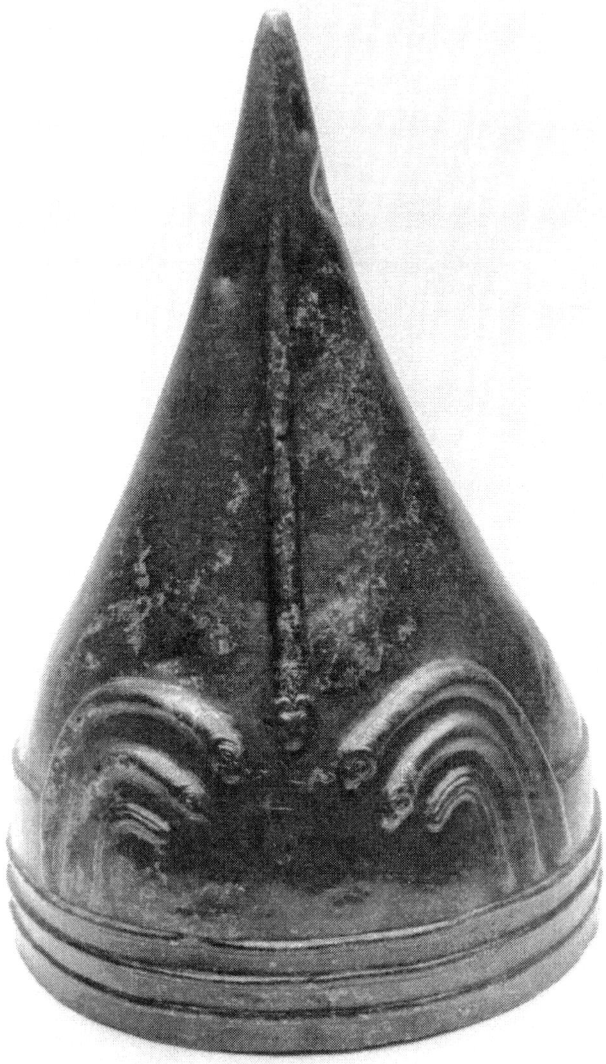

Cat. no. 41. Altenessen, Museum Altenessen, 72/246A.

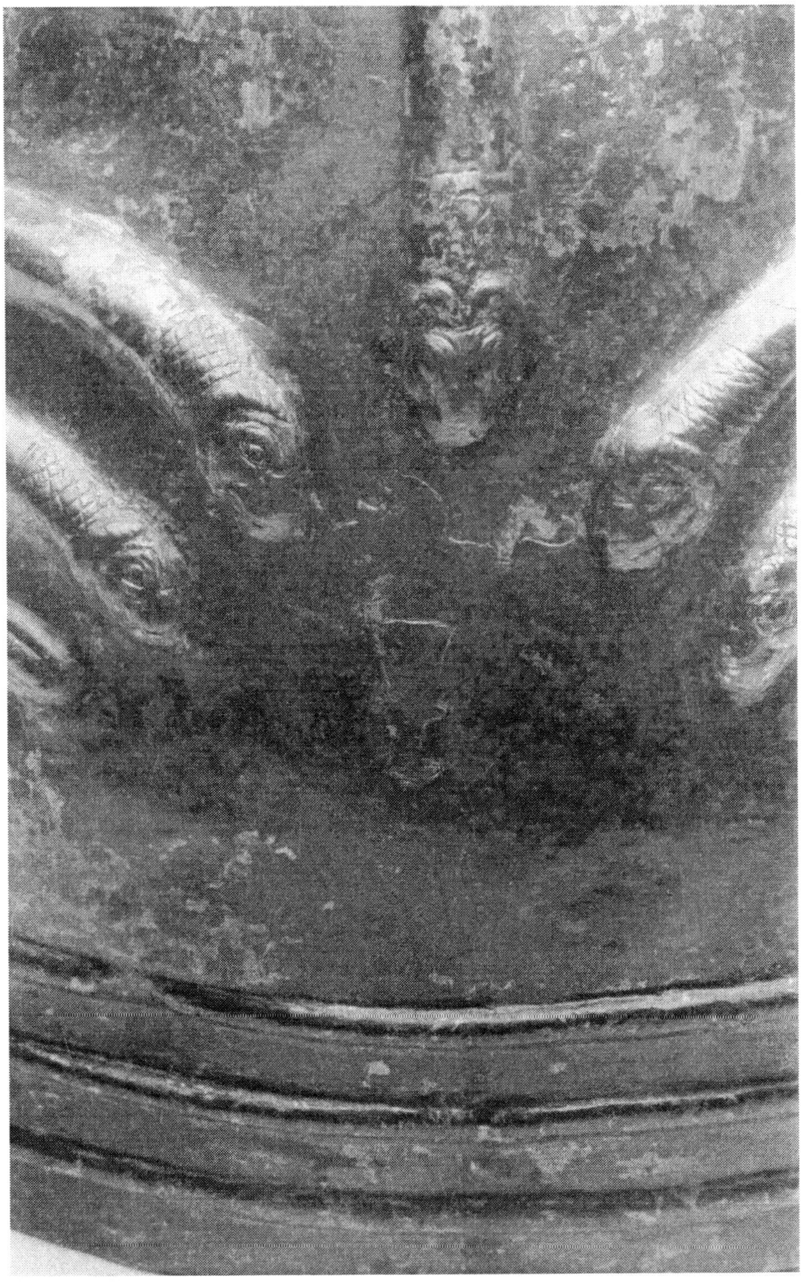

Cat. no. 41. Altenessen, Museum Altenessen, 72/246A.
Detail.

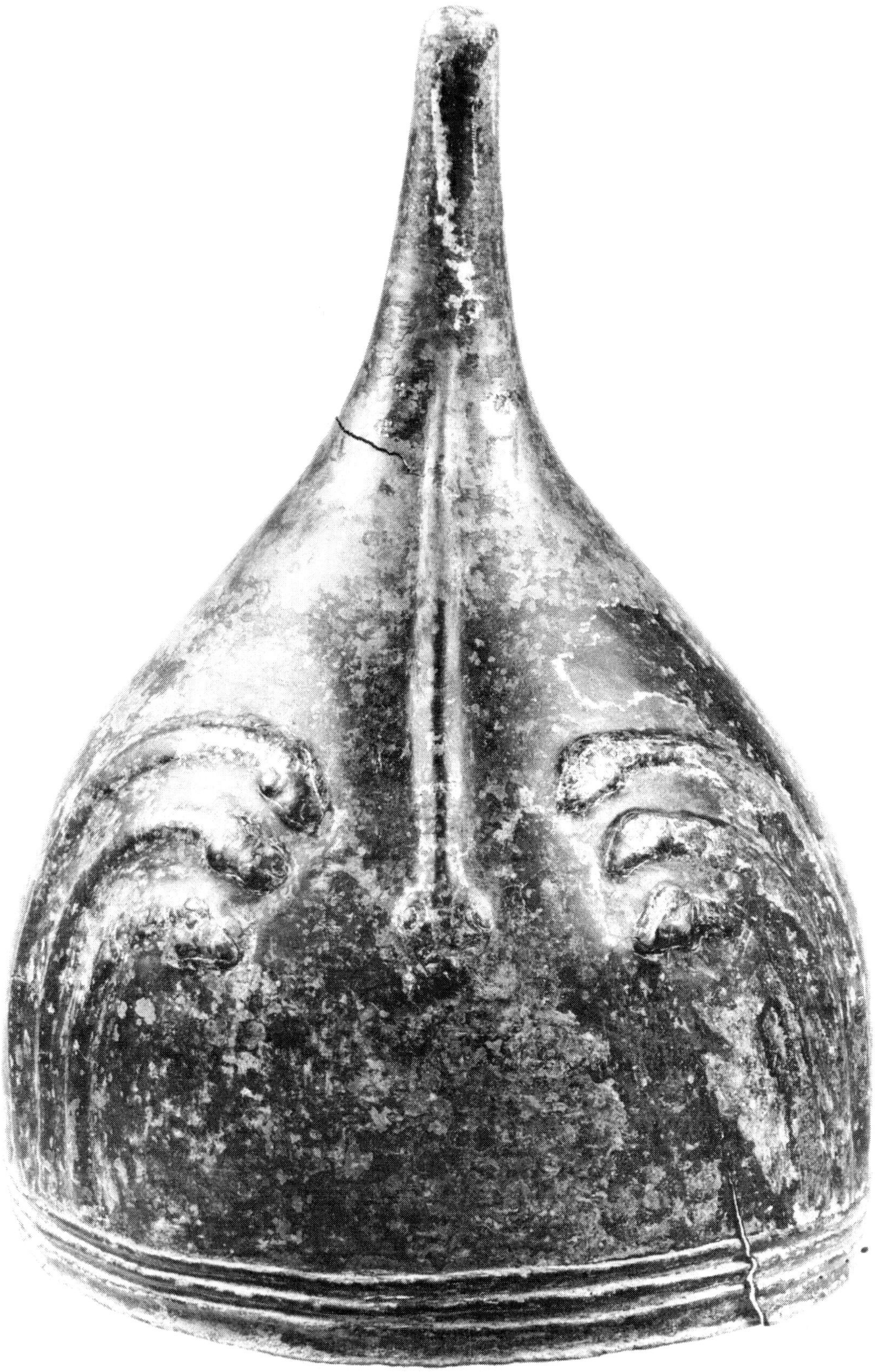

Cat. no. 42. Copenhagen, National Museum, Department of Near Eastern and Classical Antiquities, 14495.

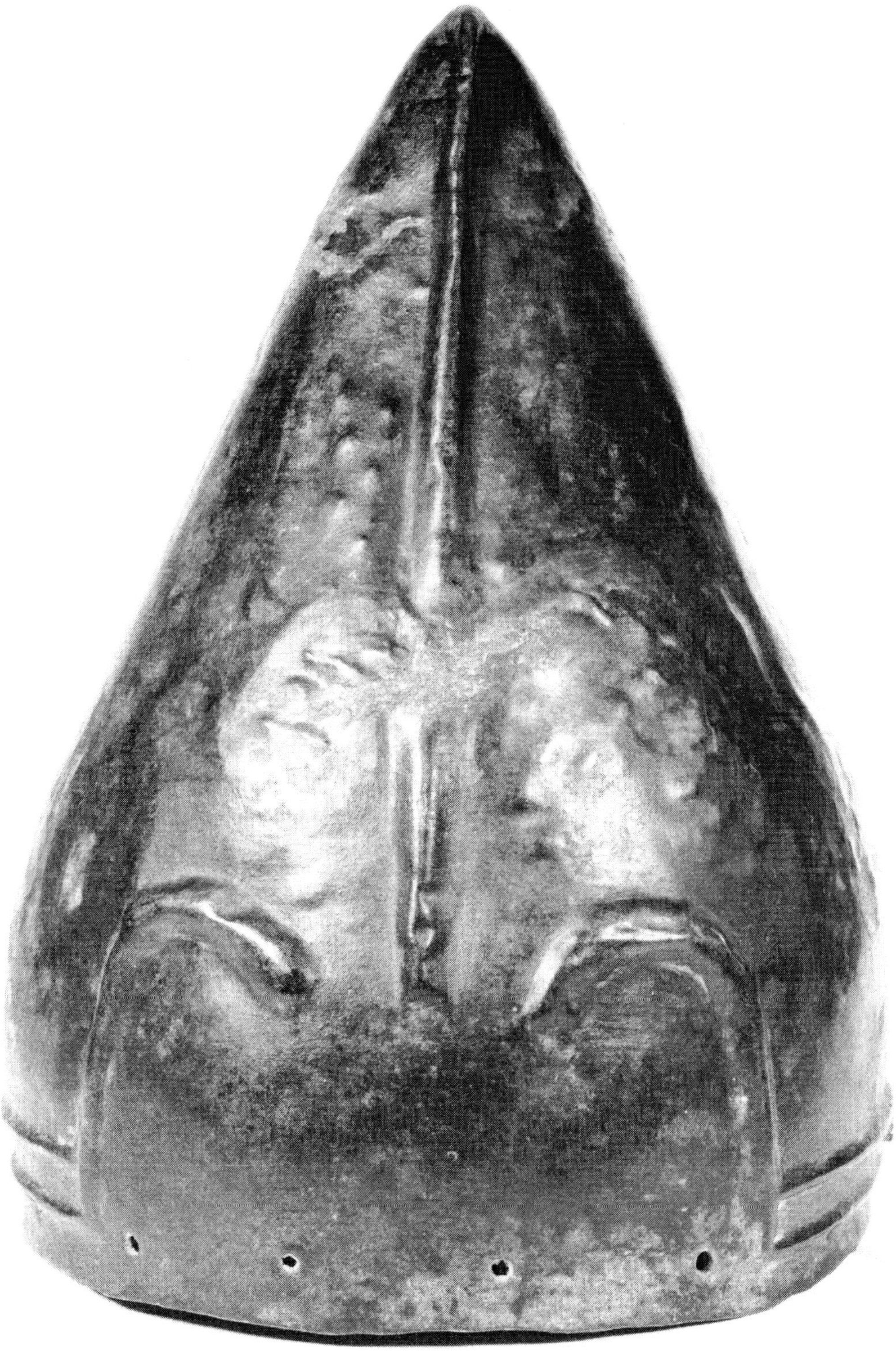

Cat. no. 45. Mainz, Römisch-Germanisches Zentralmuseum, O.39591.

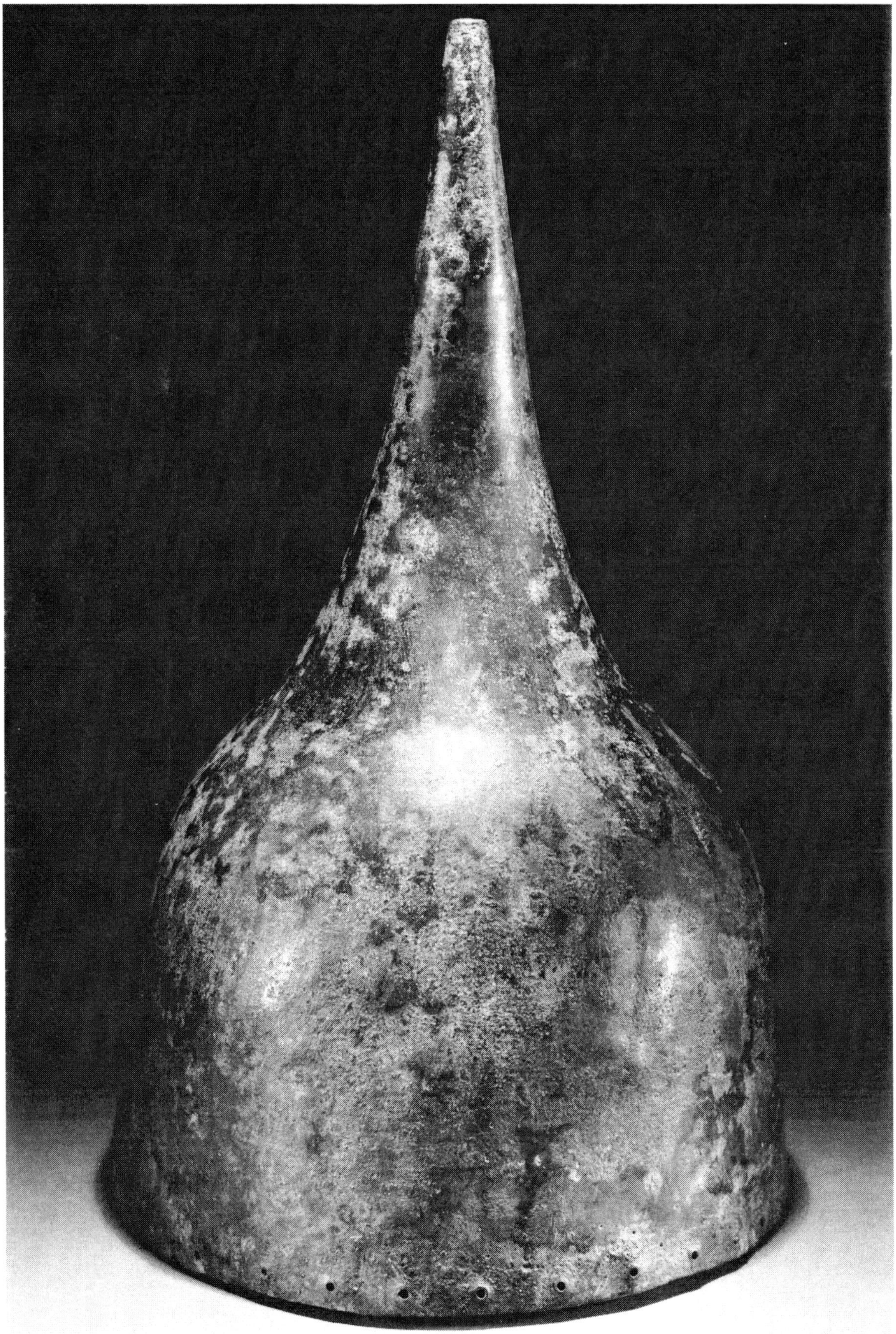

Cat. no. 47. London, Mahboubian Family Collection, 237.

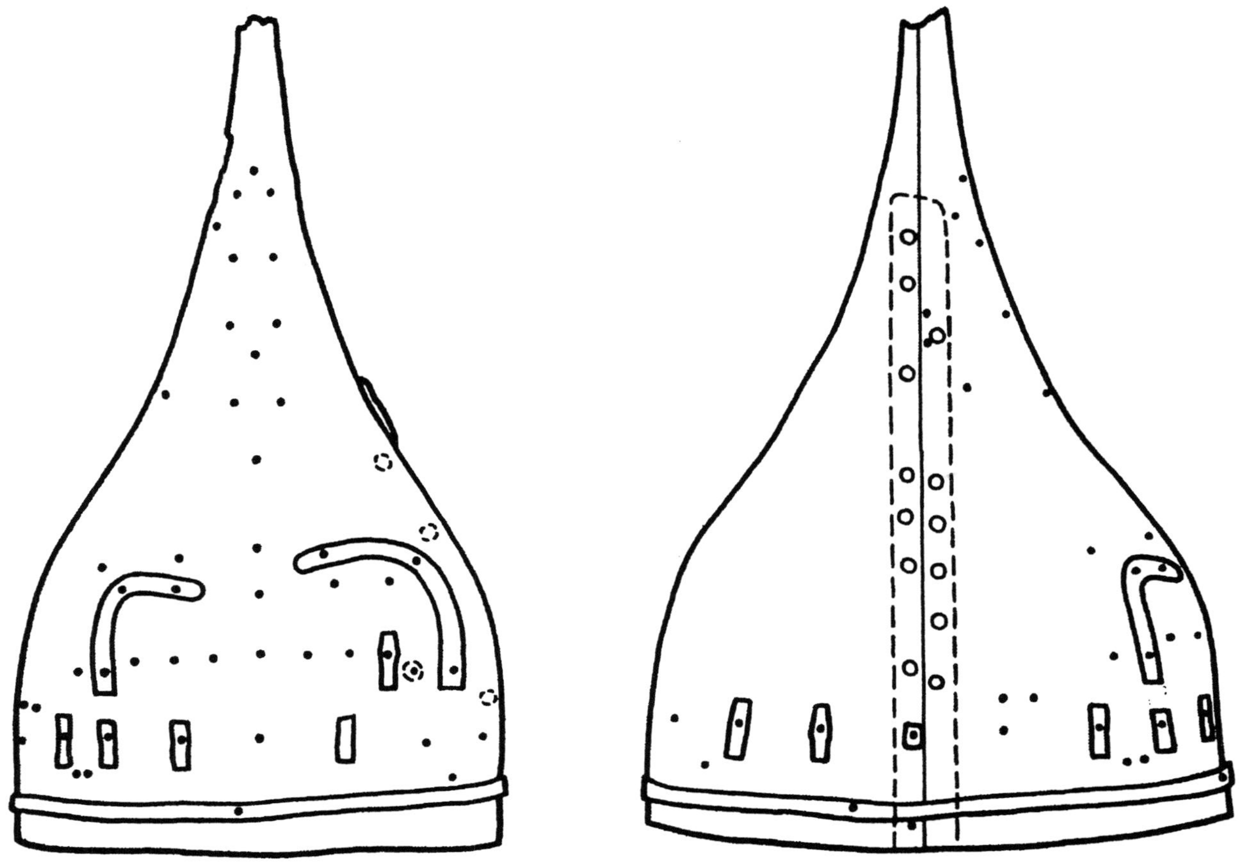

Cat. no. 48. Bruxelles, Musées Royaux d'Art et d'Histoire, IR 1241.
Drawing of helmet construction.

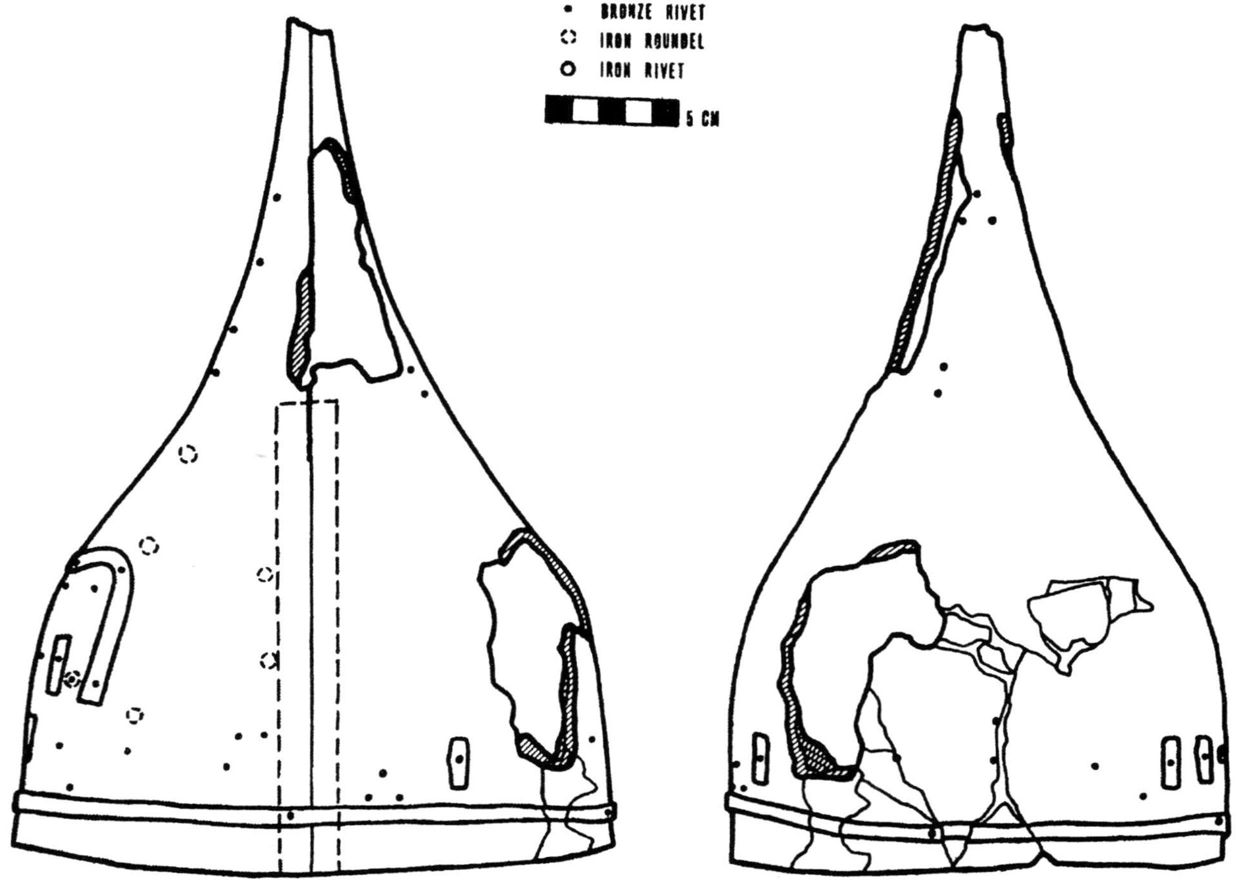

Cat. no. 48. Bruxelles, Musées Royaux d'Art et d'Histoire, IR 1241.
Drawing of helmet construction.

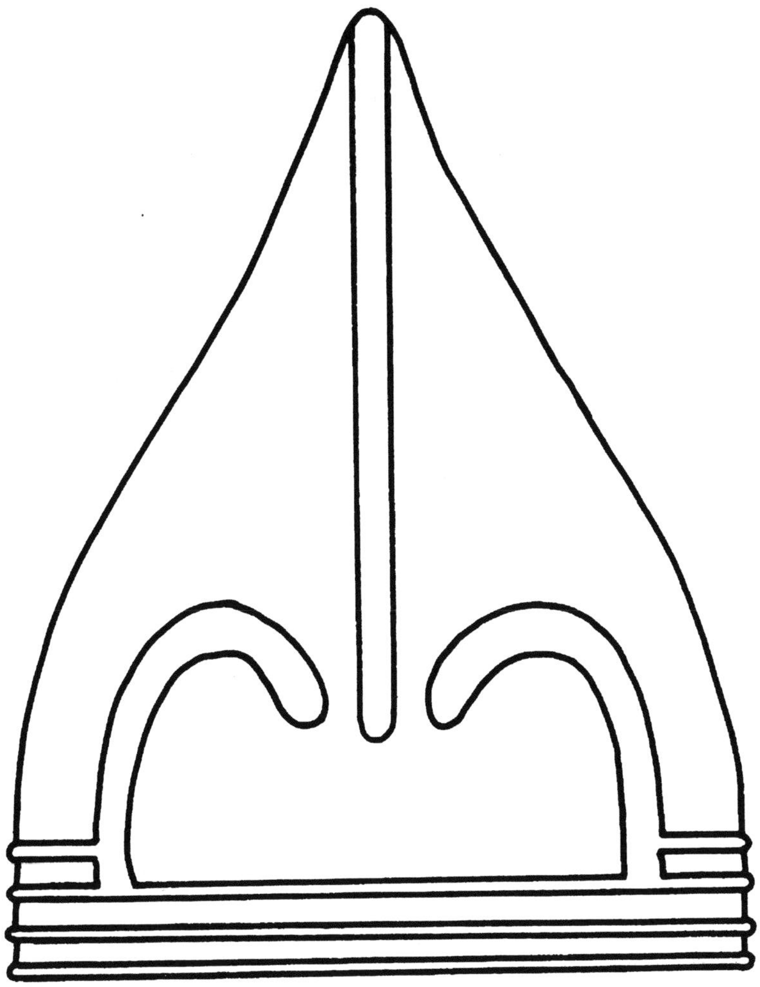

Cat. no. 49. Tehran, Musée Iran Bastan, HAS 60-883.
Drawing of helmet.

Pl. 68

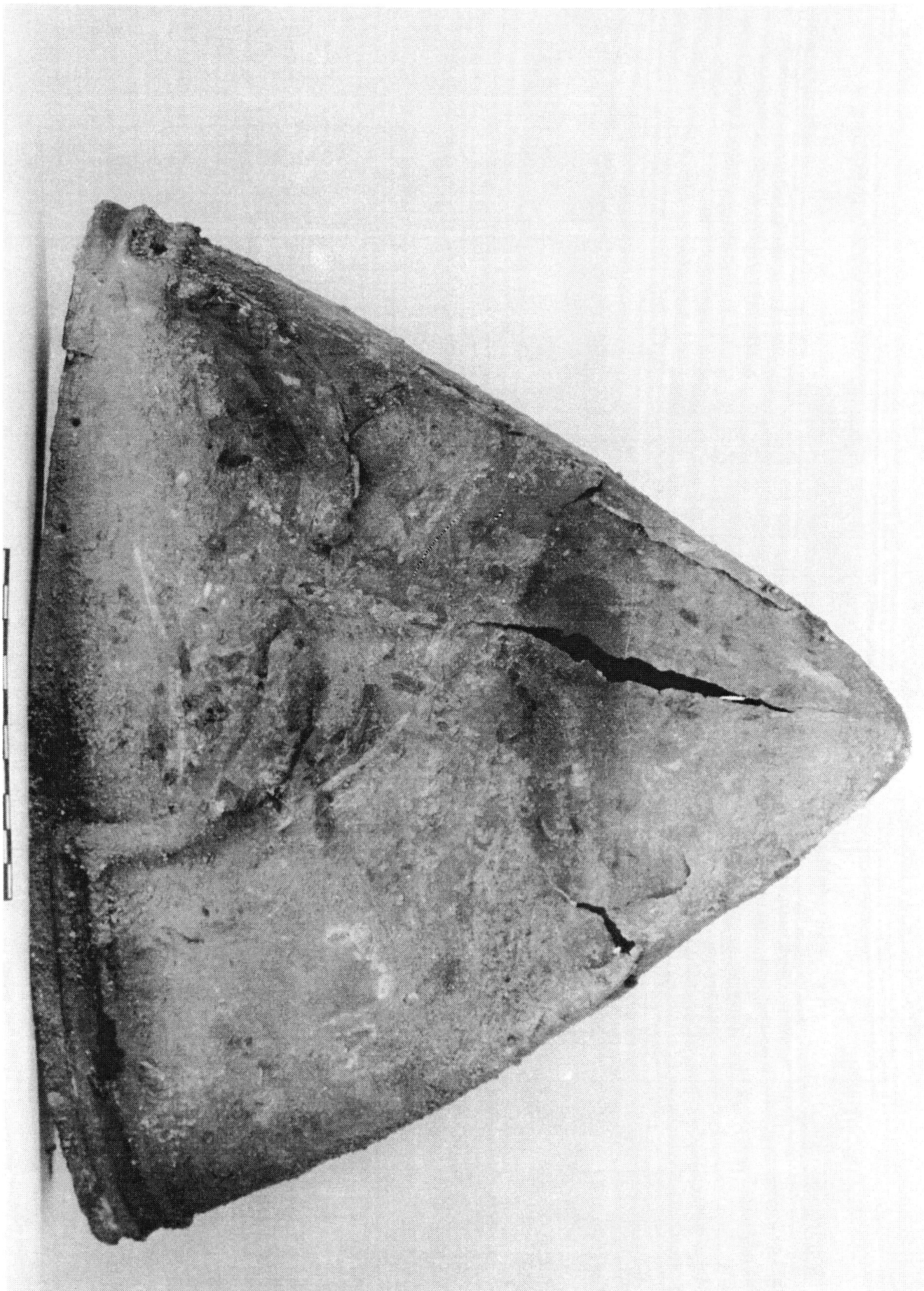

Cat. no. 50. Philadelphia, University of Pennsylvania Museum of Archaeology and Anthropology, UM 61-5-352 (HAS 60-1030).

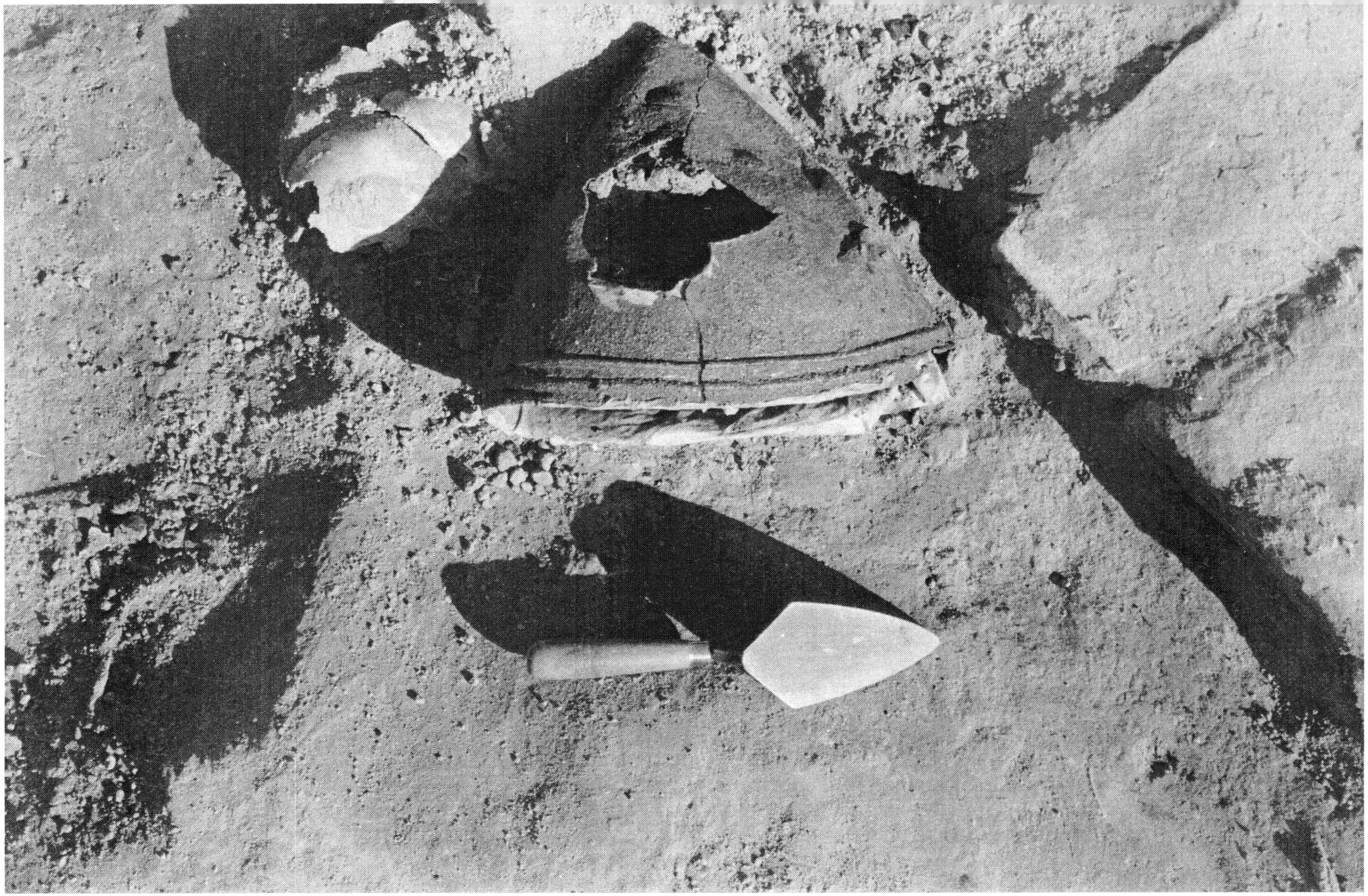

Cat. no. 50. Philadelphia, University of Pennsylvania Museum of Archaeology and Anthropology,
UM 61-5-352 (HAS 60-1030).
In situ.

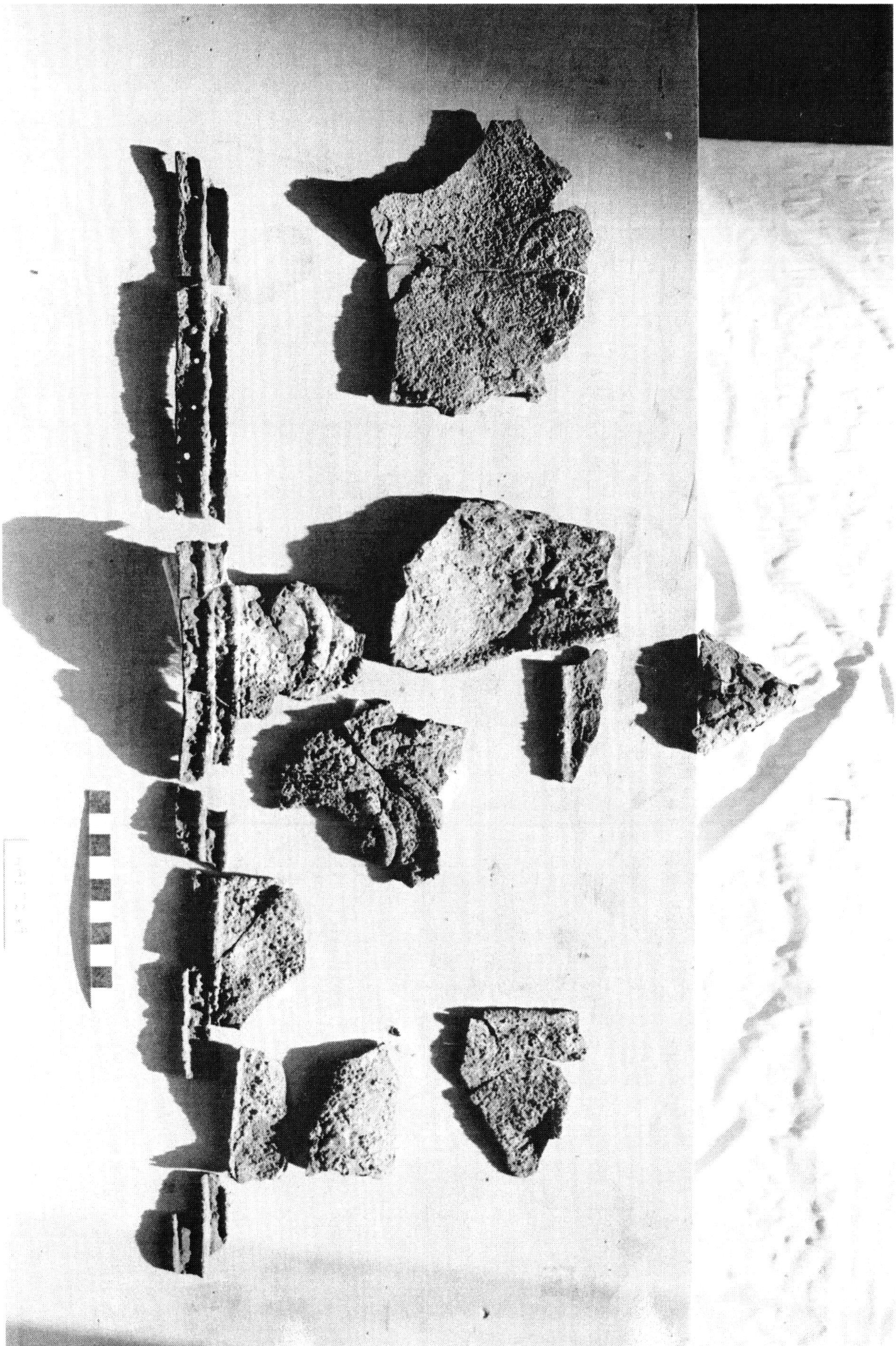

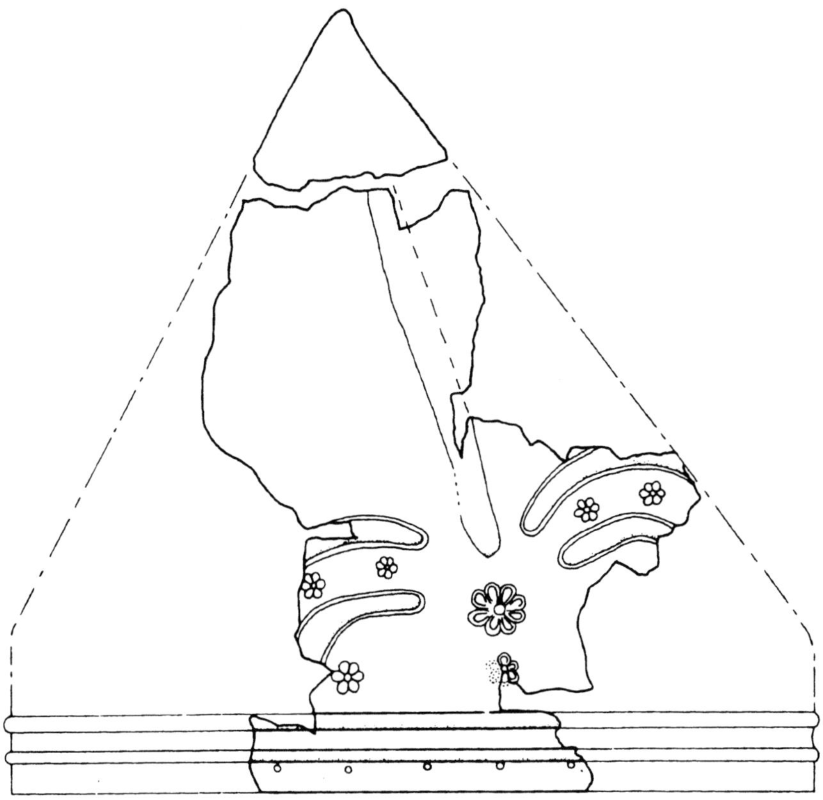

Cat. no. 51. Tehran, Musée Iran Bastan, HAS 72-69.
Reconstruction.

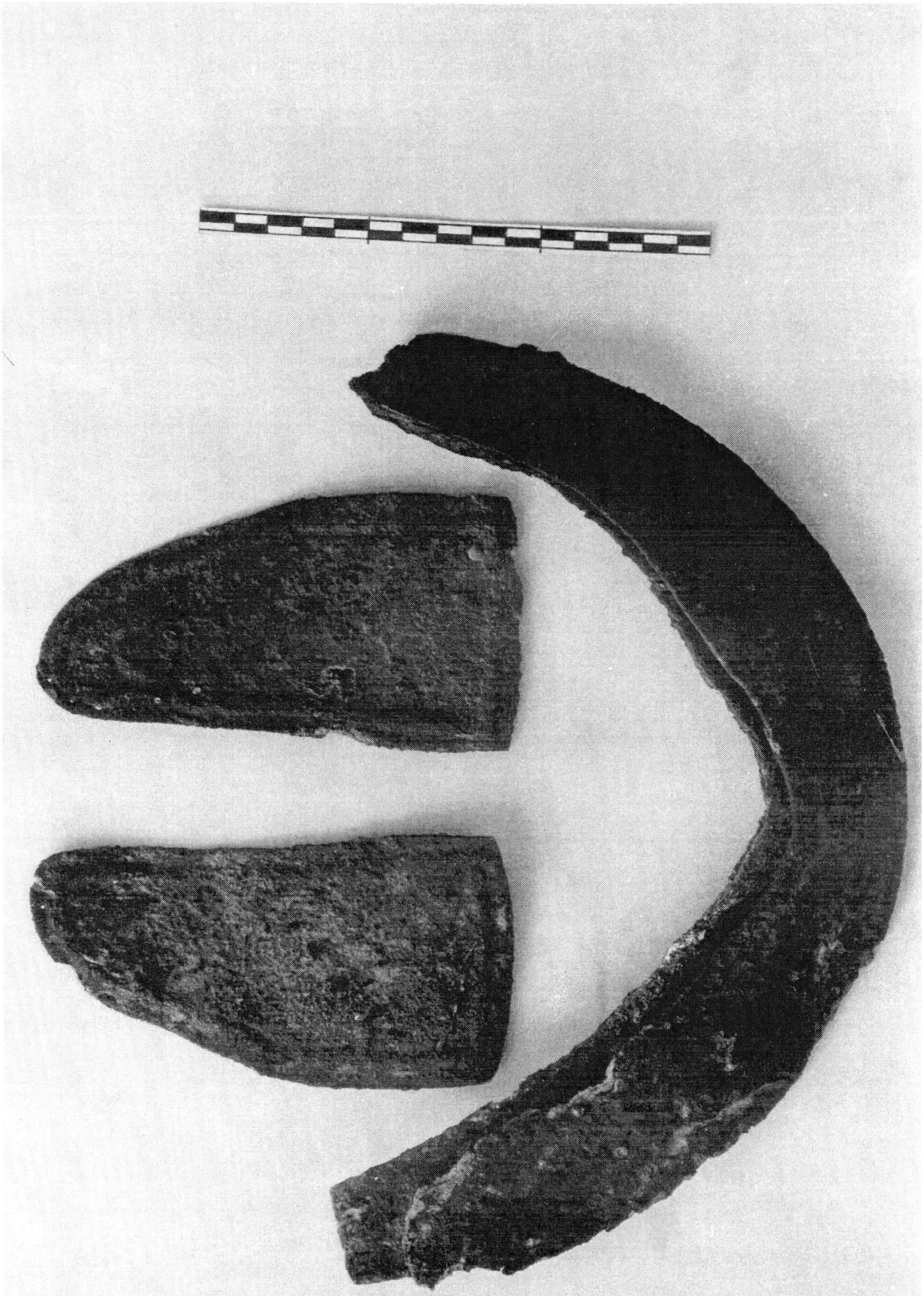

Cat. no. 52. New York, Metropolitan Museum of Art, MMA 61.100.39a-m (HAS 60-620).

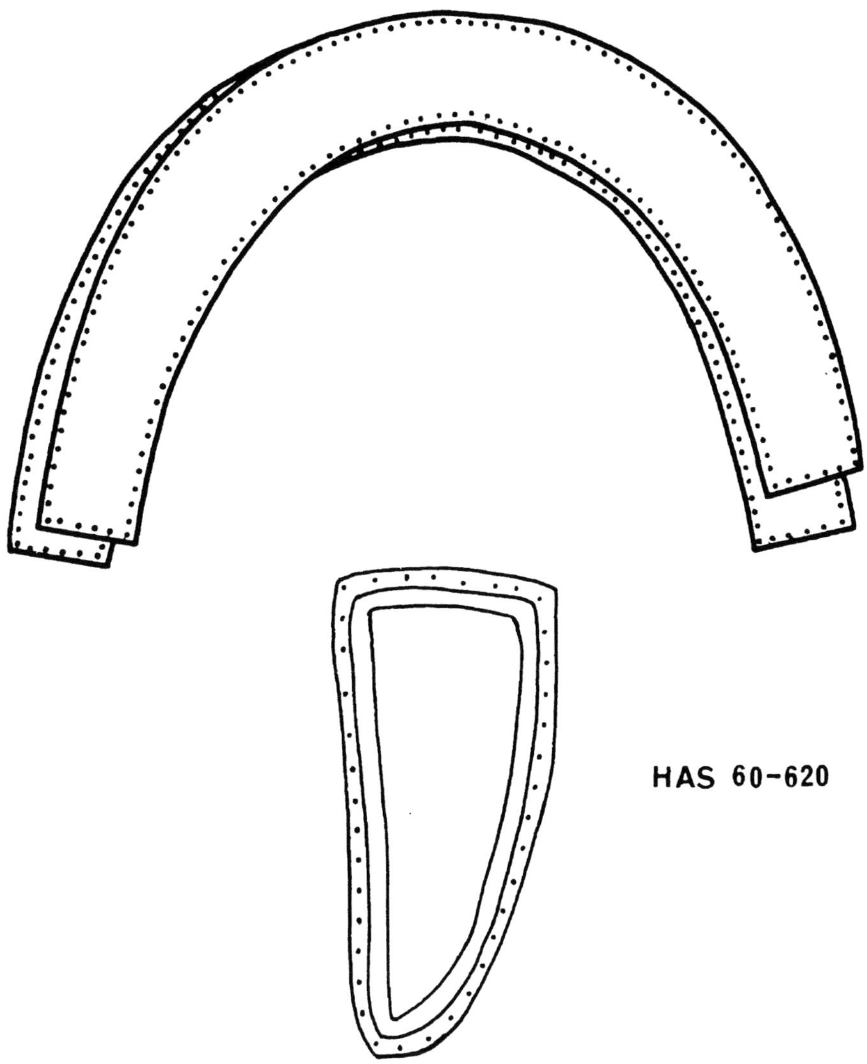

Cat. no. 52. New York, Metropolitan Museum of Art, MMA 61.100.39a-m (HAS 60-620).
Drawing of helmet.

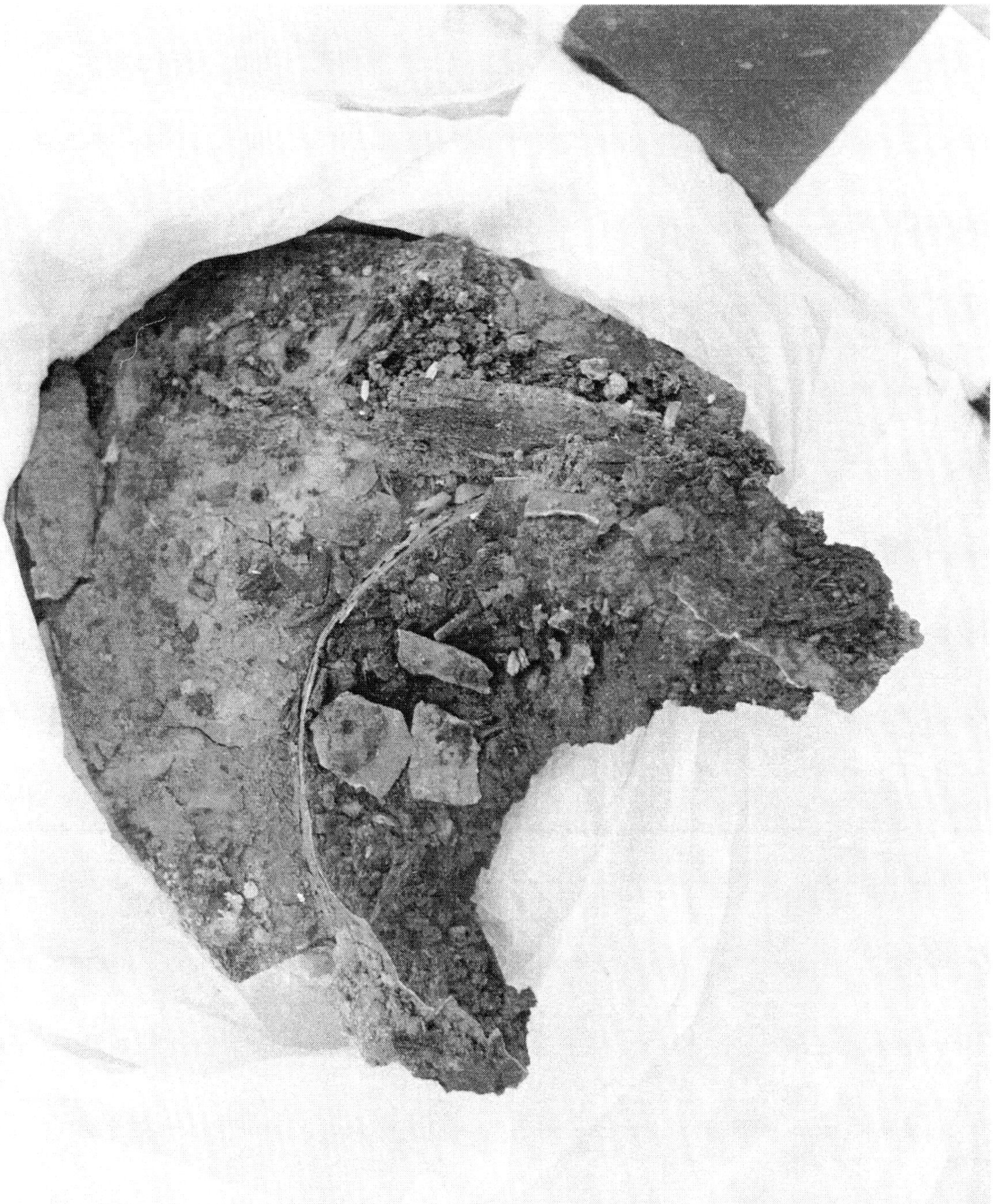

Cat. no. 53. Tehran, Musée Iran Bastan, HAS 60-528.

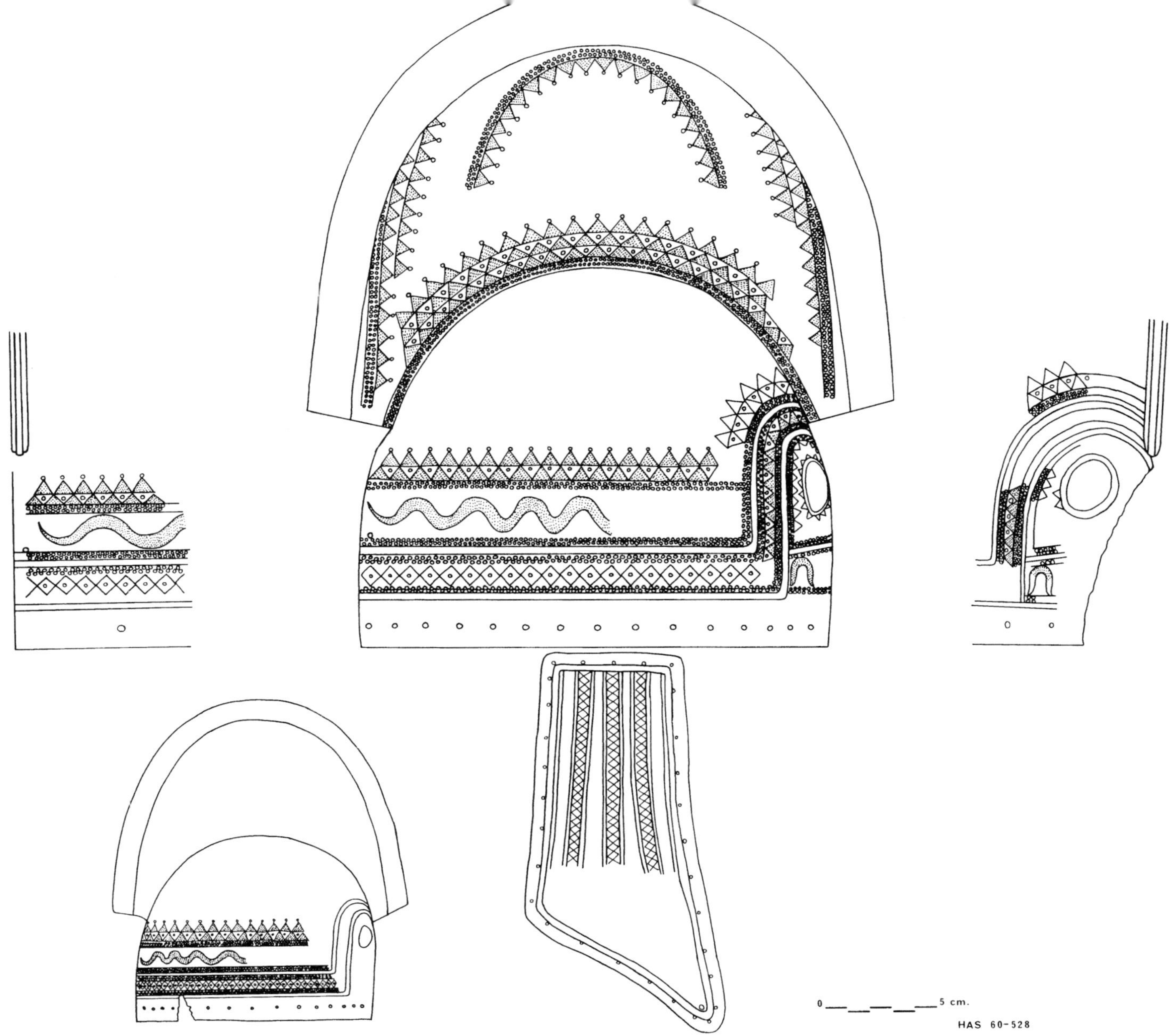

Cat. no. 53. Tehran, Musée Iran Bastan, HAS 60-528.
Drawing of decoration.

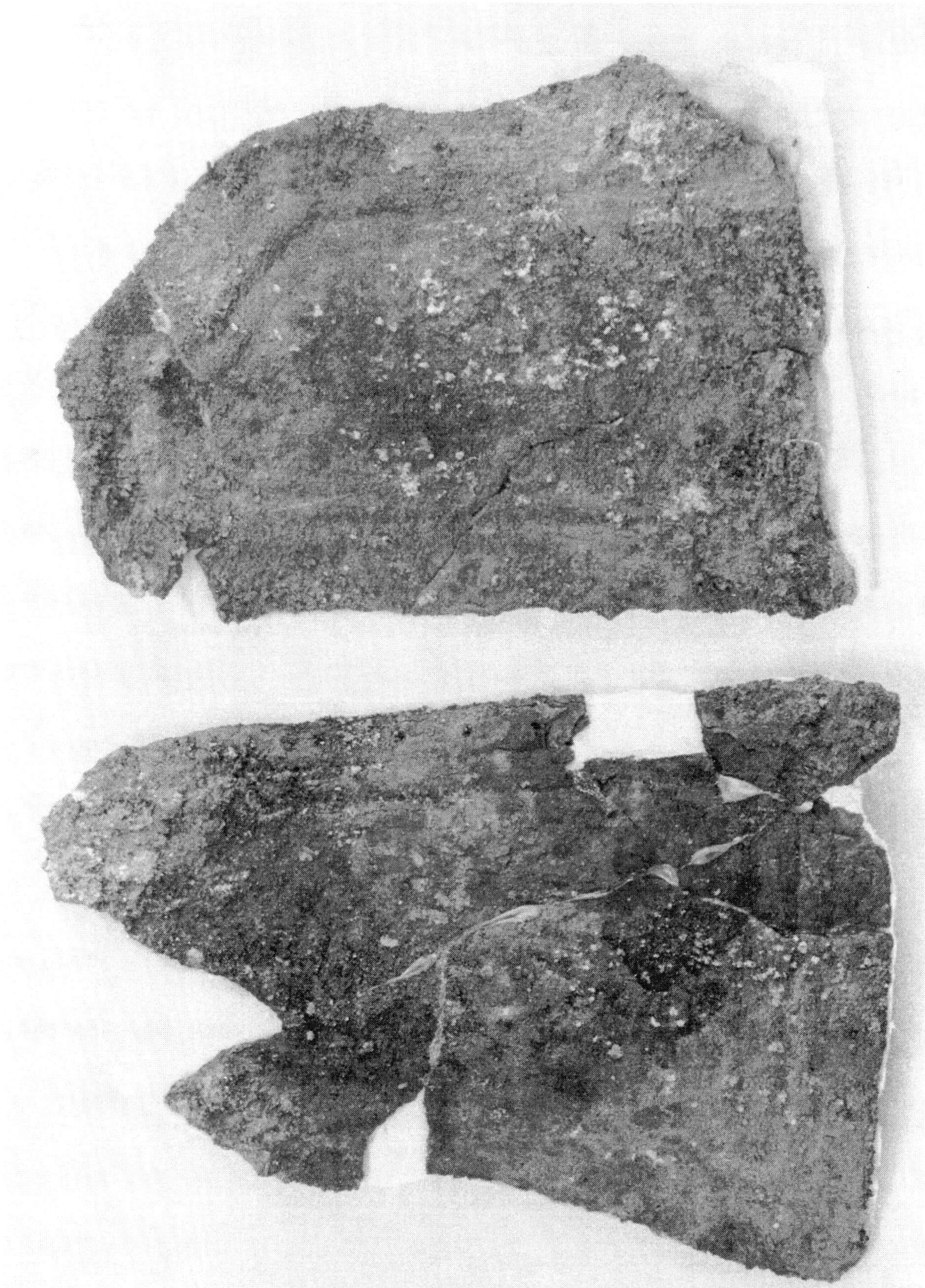

Cat. no. 54. Discarded in field, HAS 60-906.

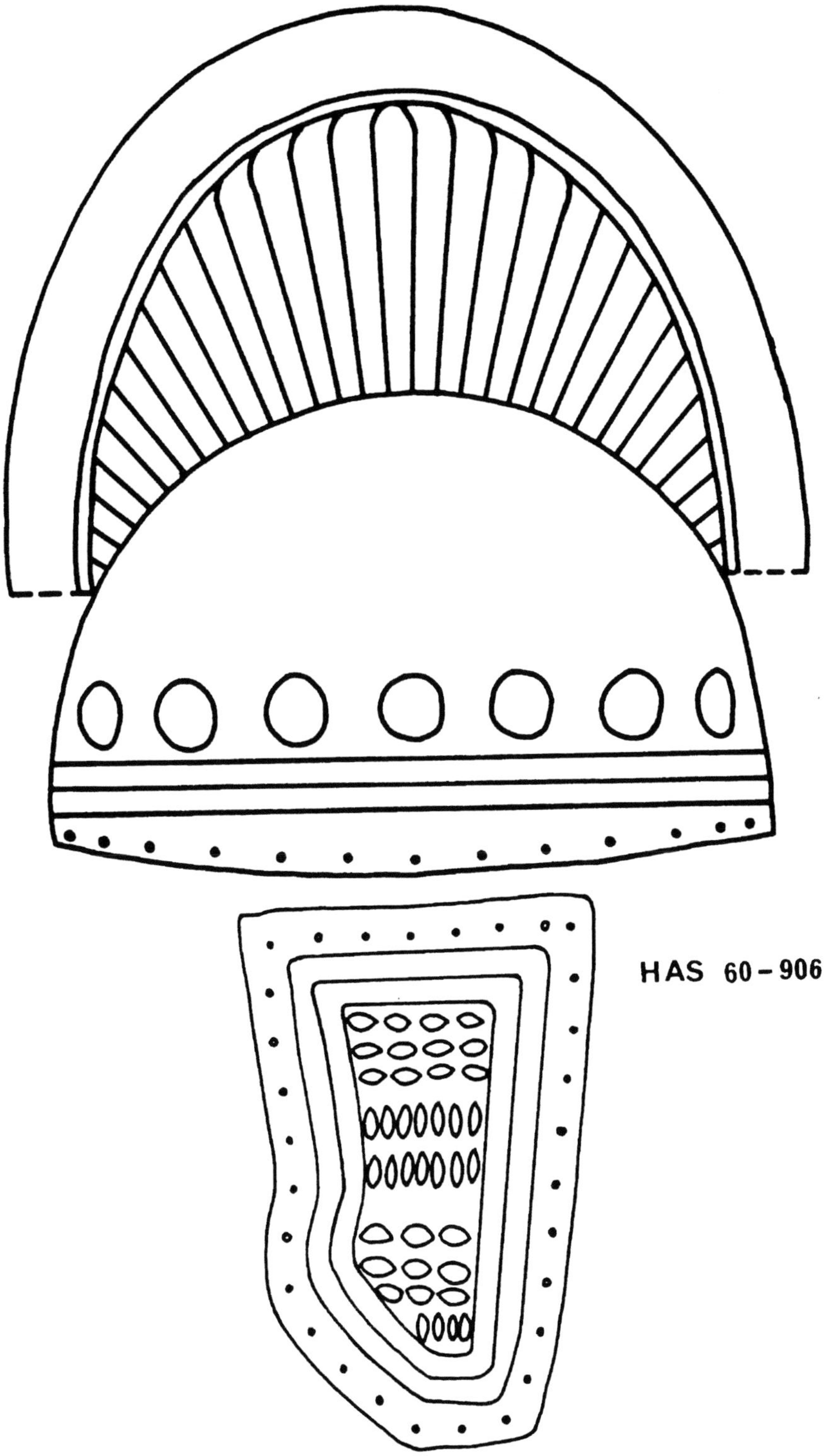

Cat. no. 54. Discarded in field, HAS 60-906.
Drawing of helmet.

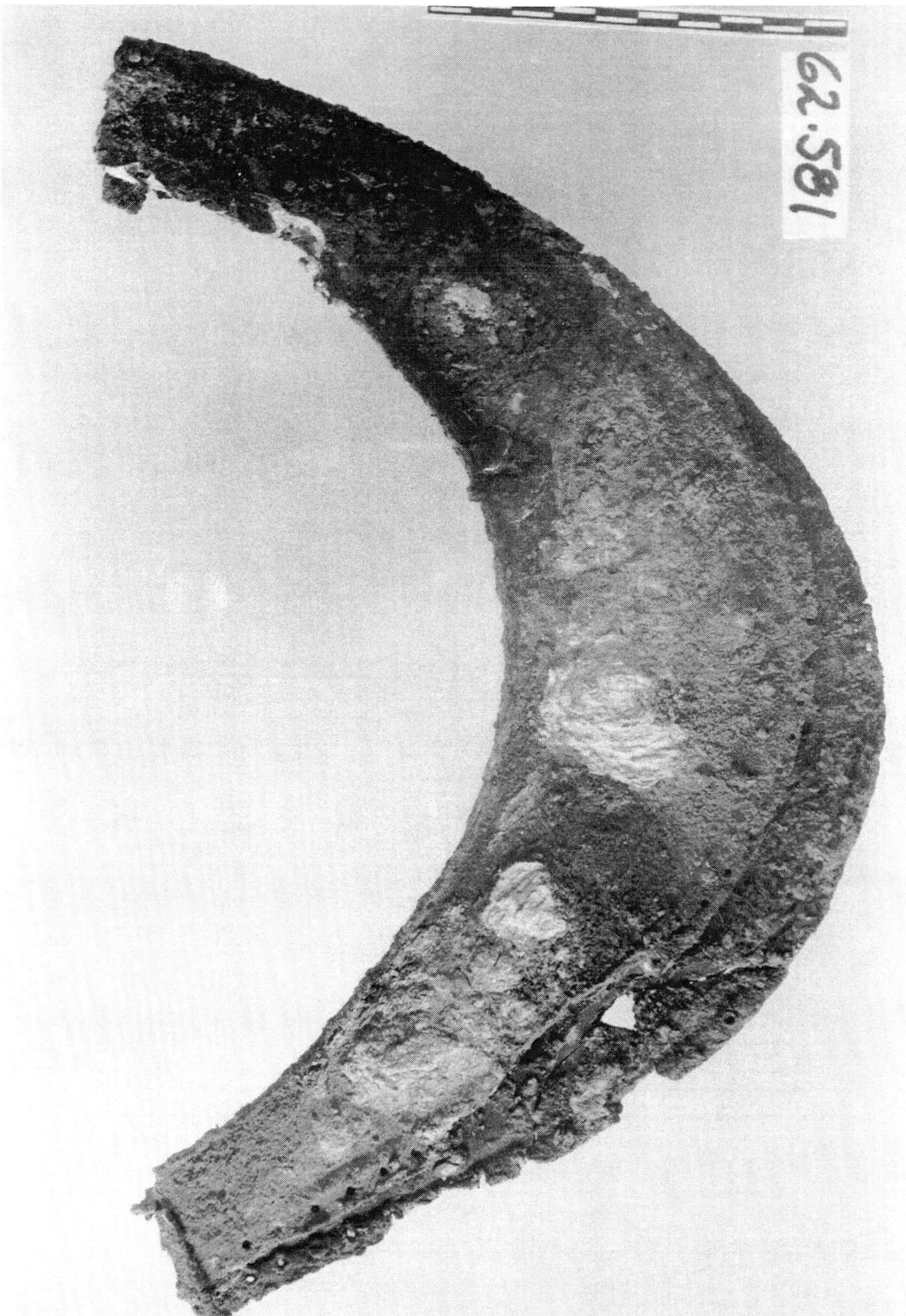

Cat. no. 55. Philadelphia, University of Pennsylvania Museum of Archaeology and Anthropology, UM 63-5-247 (HAS 62-581).

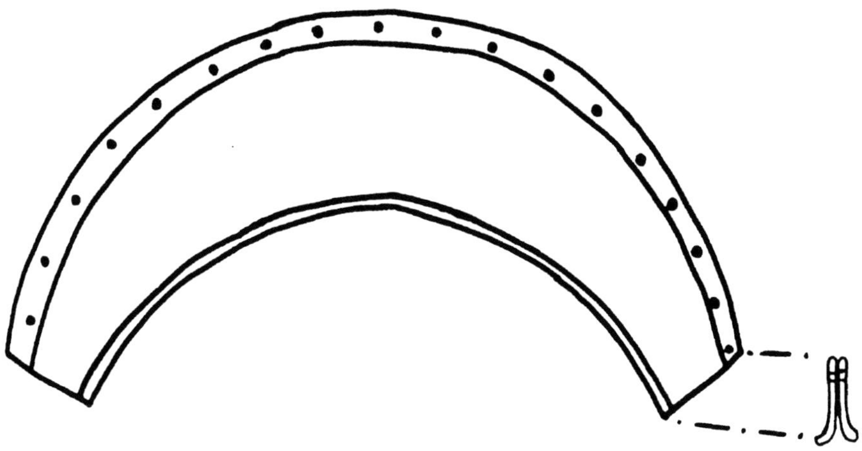

Cat. no. 56. Tehran, Musée Iran Bastan, HAS 60-1078.
Drawing of crest.

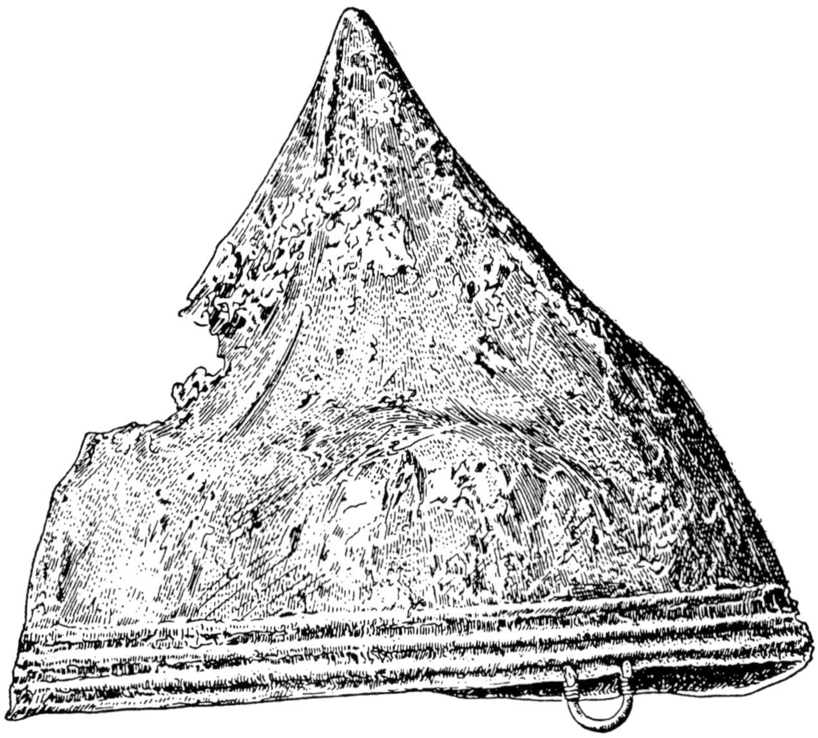

Cat. no. 57. Erevan, Historical Museum of Armenia.
Drawing of helmet (Piotrovsky 1959, fig. 27).

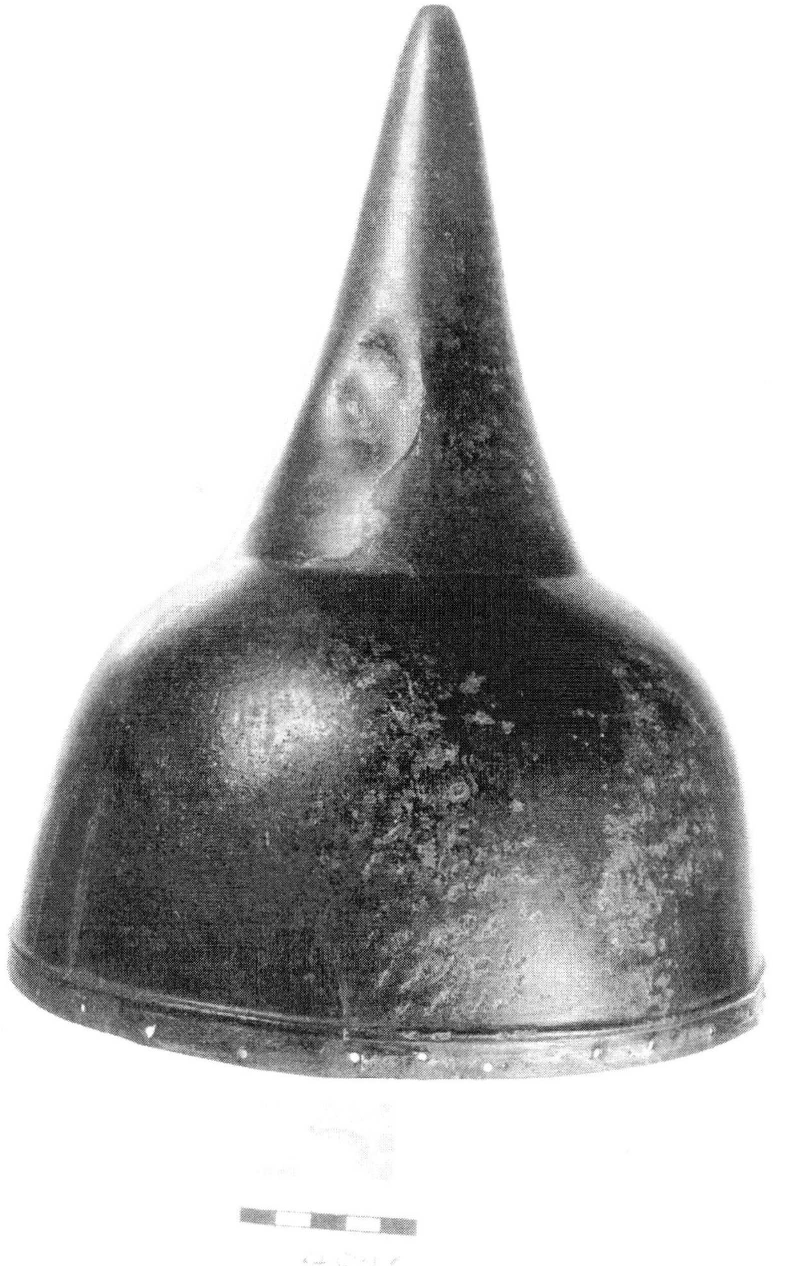

Cat. no. 58. Berlin, Staatliche Museen zu Berlin, Museum für Vor- und Frühgeschichte, XIc 4826.

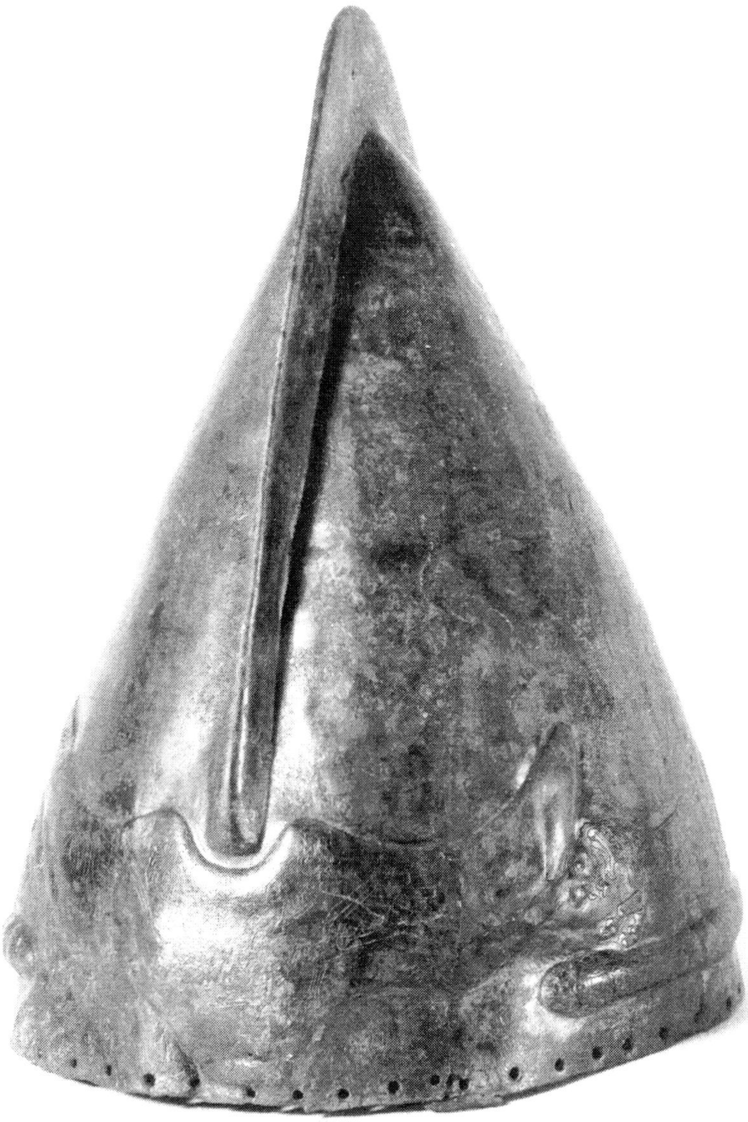

Cat. no. 59. Tokio, Ancient Orient Museum.
(Galvano Copy: Munich, Archäologische Staatssammlung, Museum für Vor- und Frühgeschichte, PS 1975/3176).

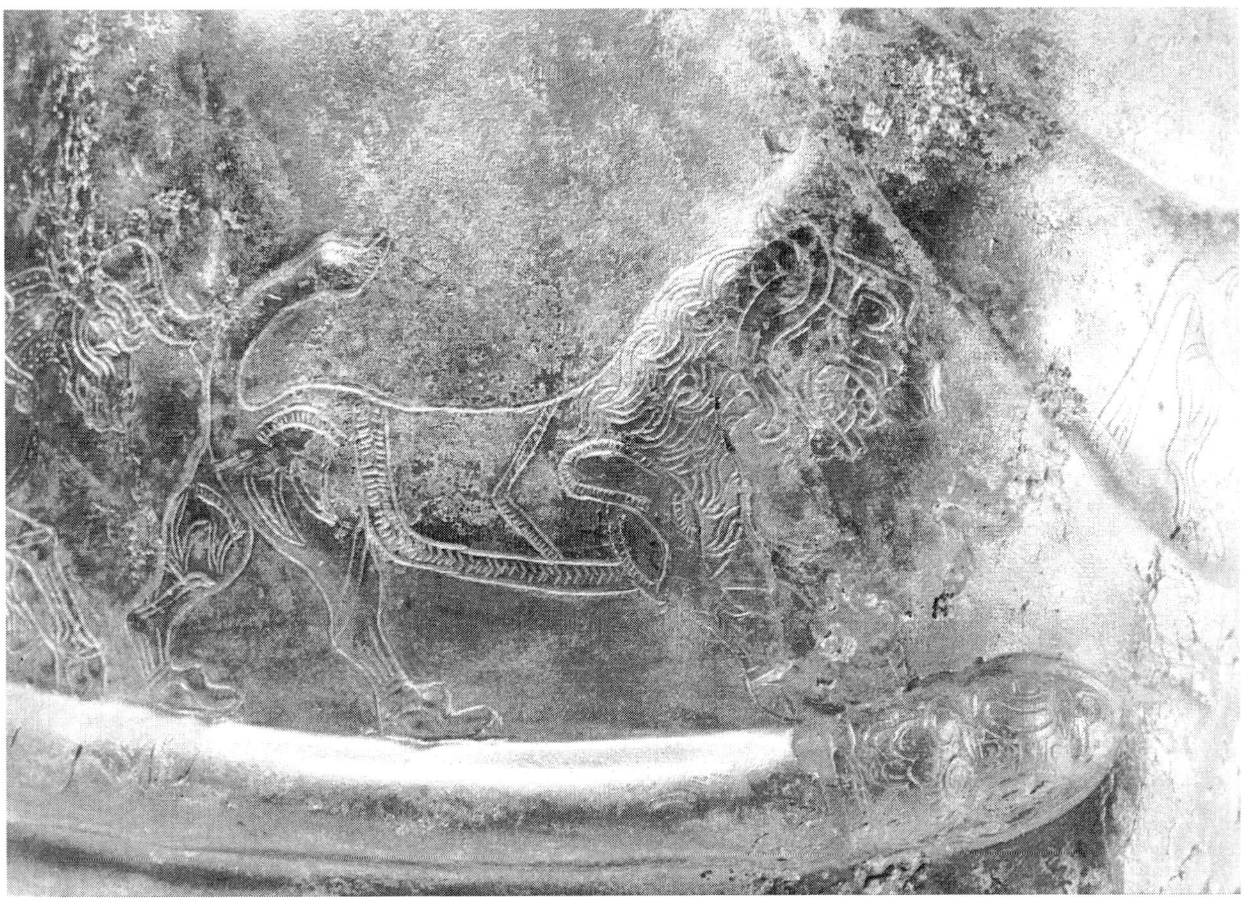

Cat. no. 59. Tokio, Ancient Orient Museum.
(Galvano Copy: Munich, Archäologische Staatssammlung, Museum für Vor- und Frühgeschichte, PS 1975/3176).
Detail.

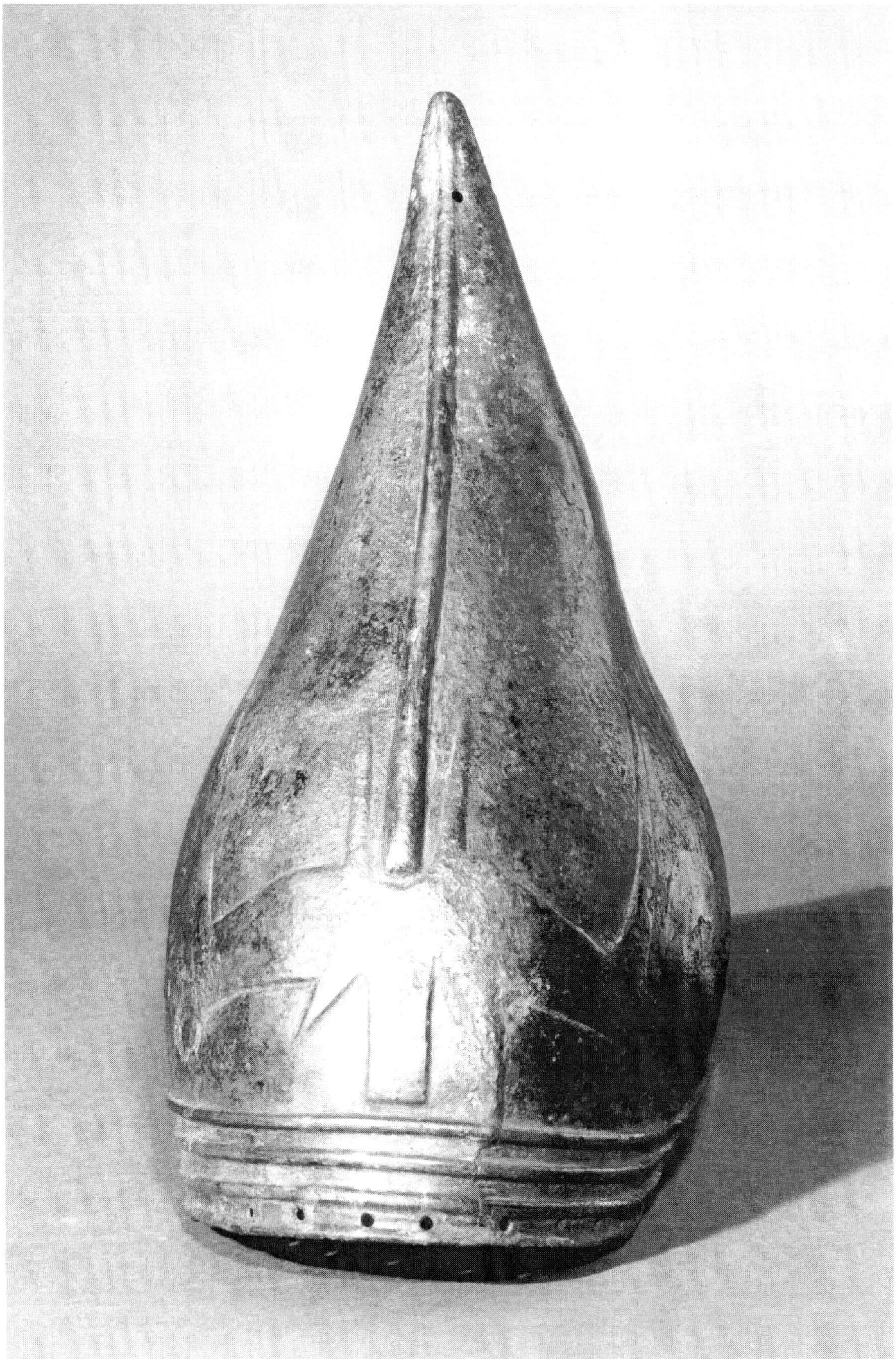

Cat. no. 62. London, British Museum, WA 135061.

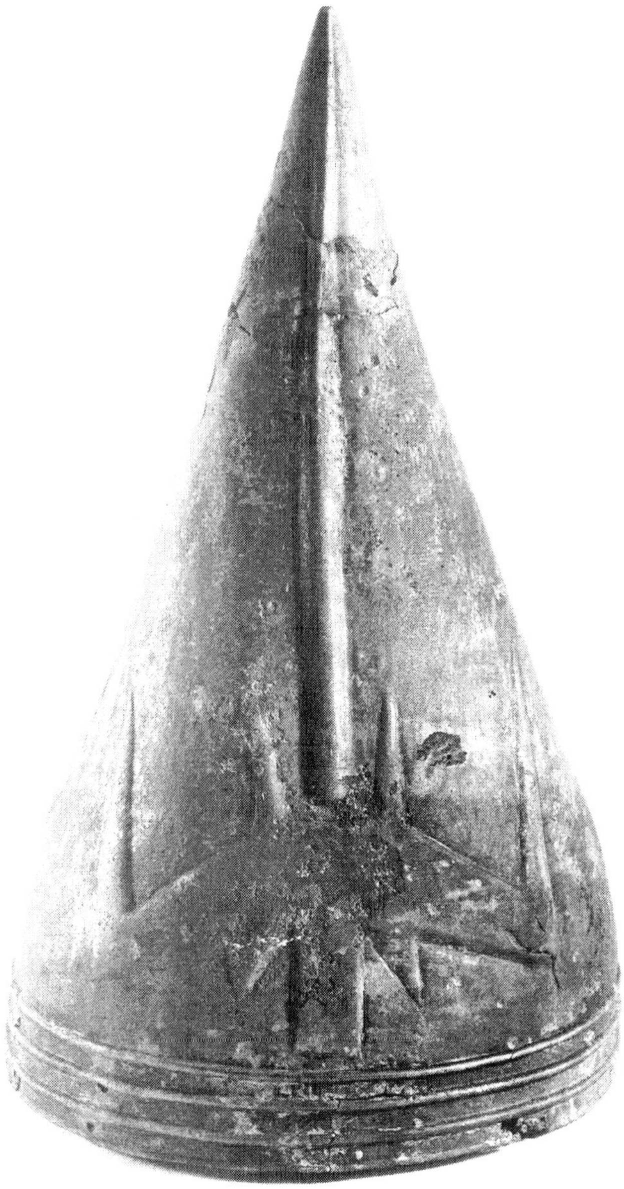

Cat. no. 68. Munich, Archäologische Staatssammlung, Museum für Vor- und Frühgeschichte, PS 1985/521.

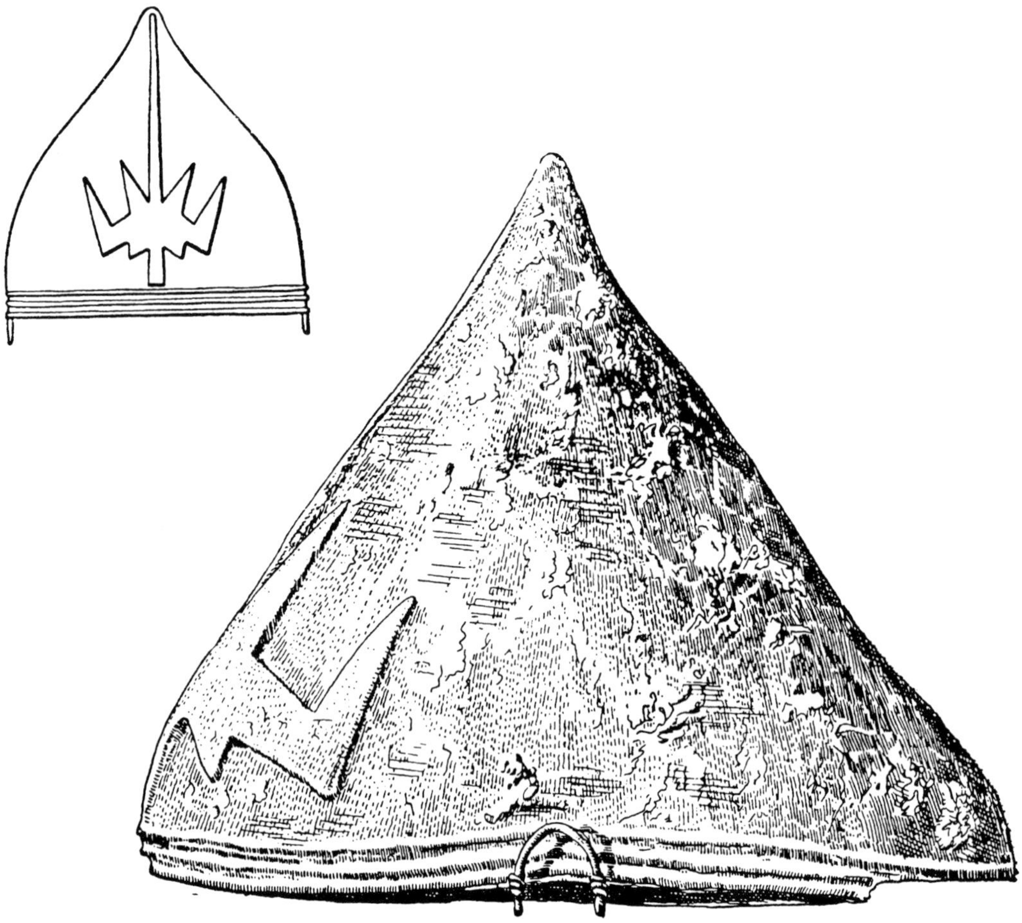

Cat. no. 69. Erevan, Historical Museum of Armenia.
Drawing of helmet (Piotrovsky 1959, fig. 26).

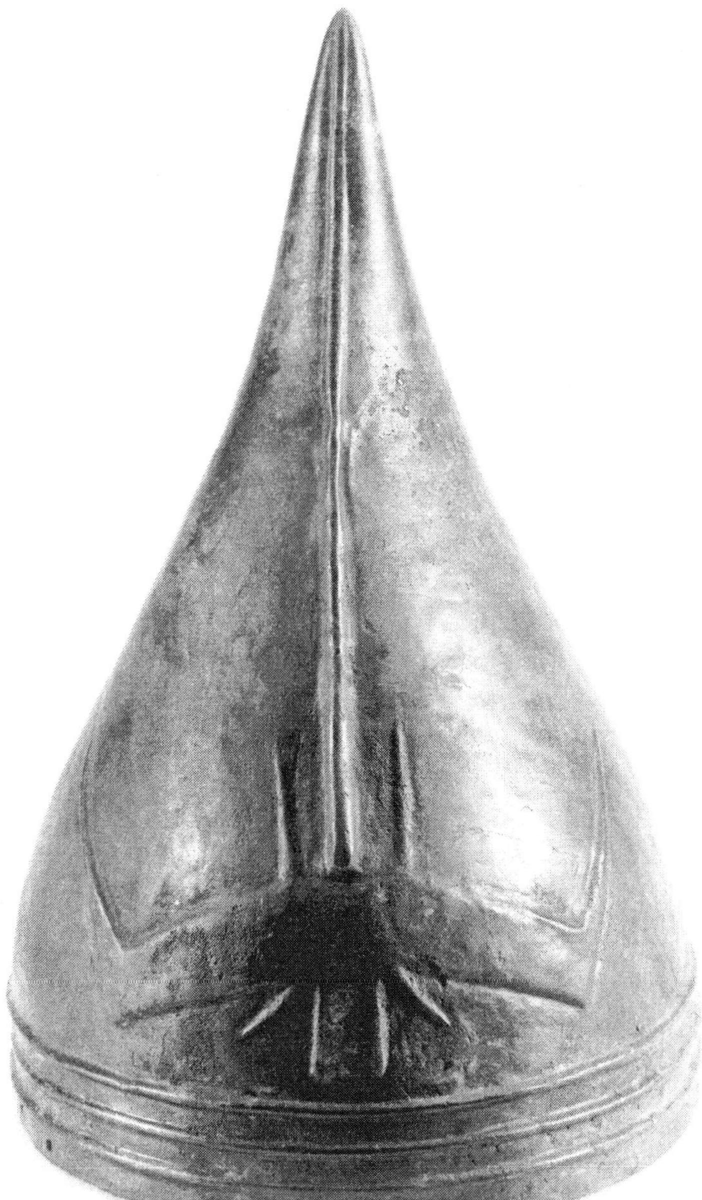

Cat. no. 73. Karlsruhe, Badisches Landesmuseum, 89/13.

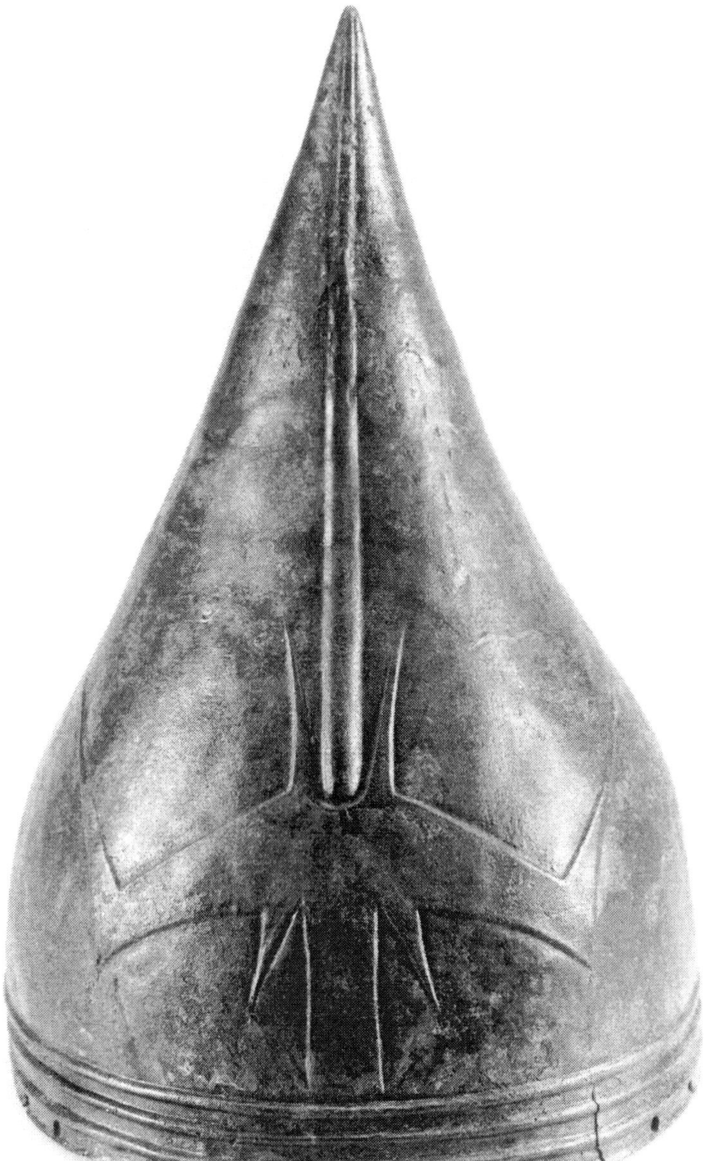

Cat. no. 74. Karlsruhe, Badisches Landesmuseum, 89/14.

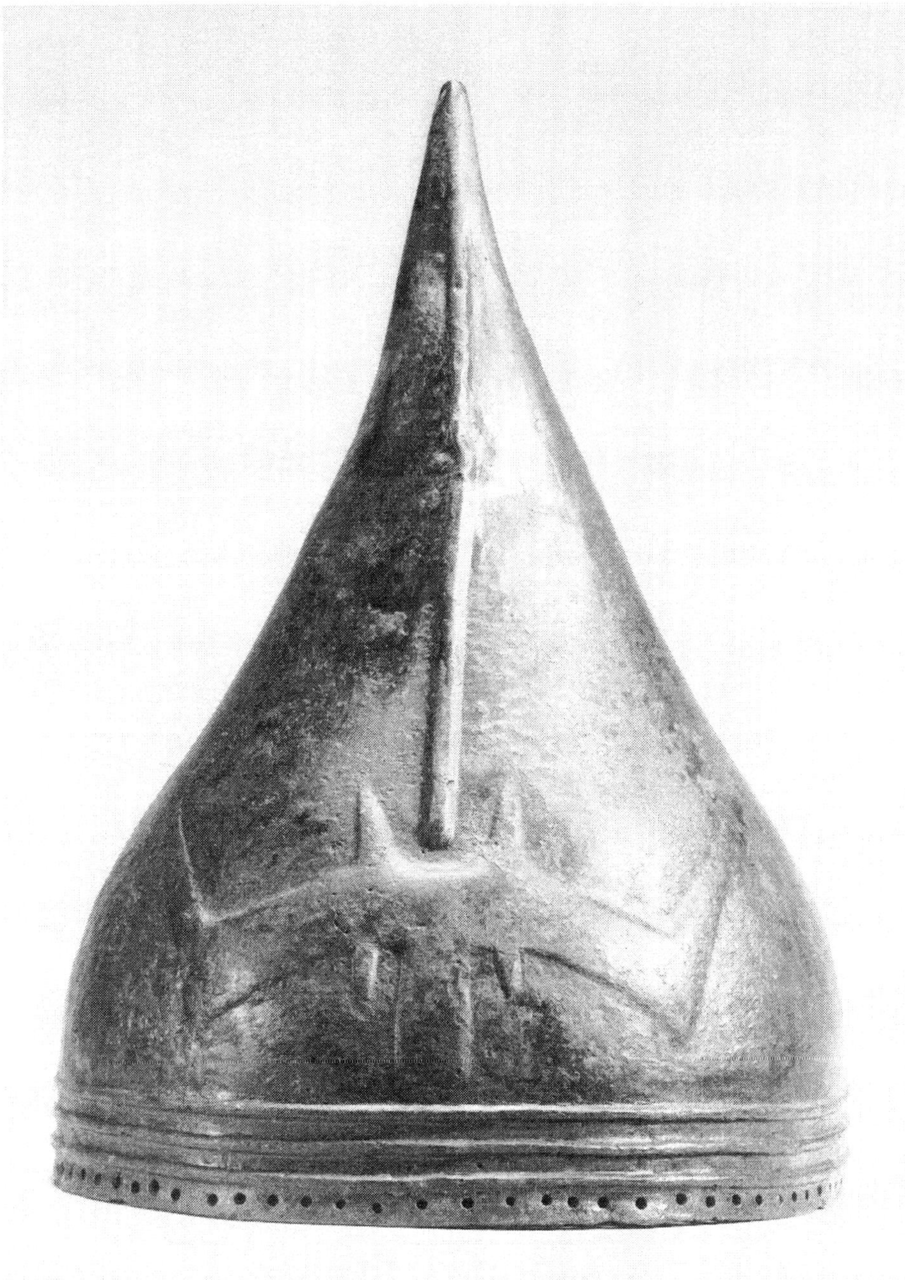

Cat. no. 75. Karlsruhe, Badisches Landesmuseum, 84/69.

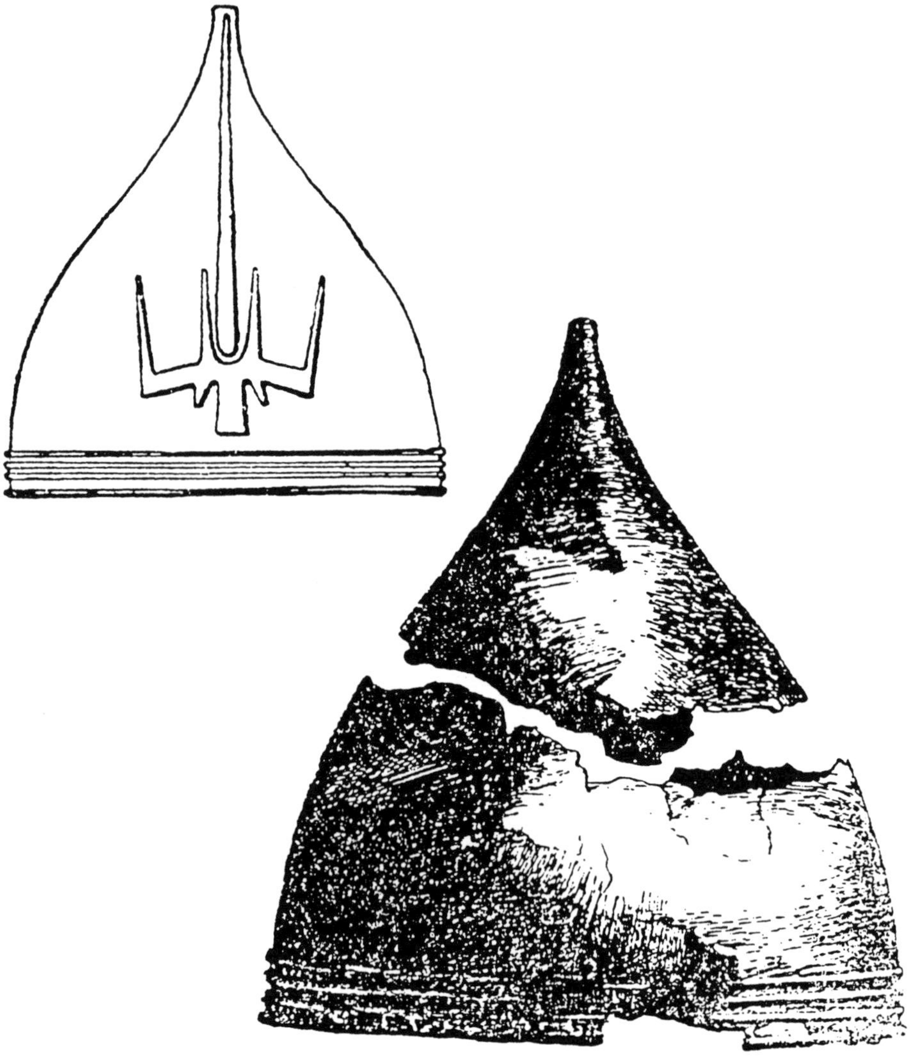

Cat. no. 76. Erevan, Historical Museum of Armenia.
Drawing of helmet (Piotrovsky 1955, fig. 16).

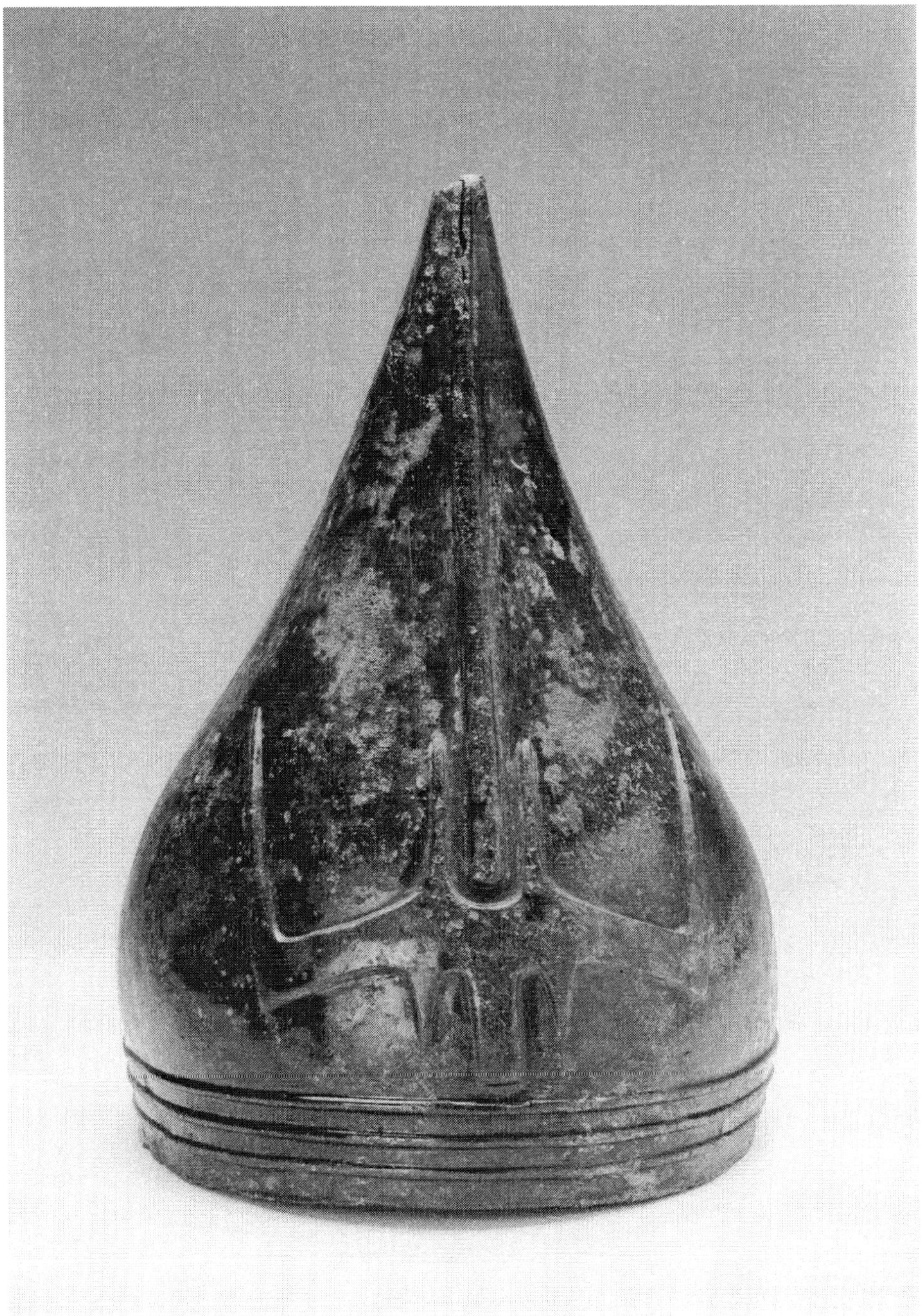

Cat. no. 78. Berlin, Staatliche Museen zu Berlin, Museum für Vor- und Frühgeschichte, IIId 4882.

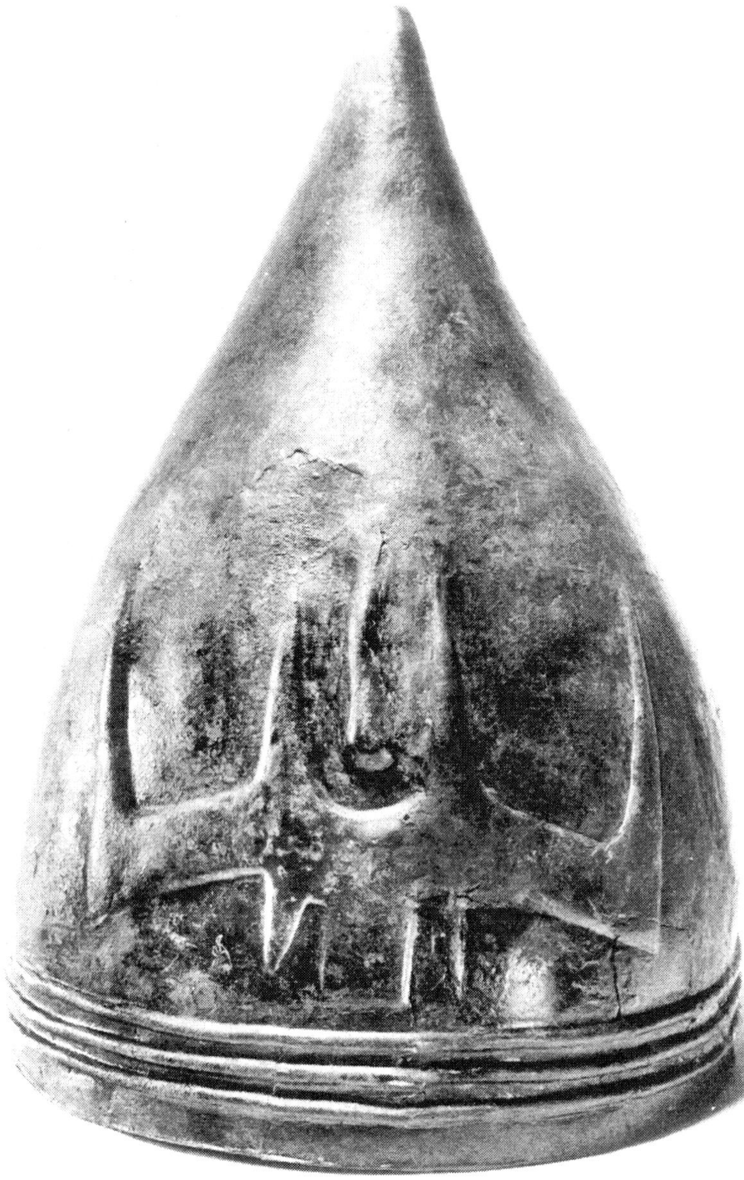

Cat. no. 82. Munich, Archäologische Staatssammlung, Museum für Vor- und Frühgeschichte, PS 1971/1767.

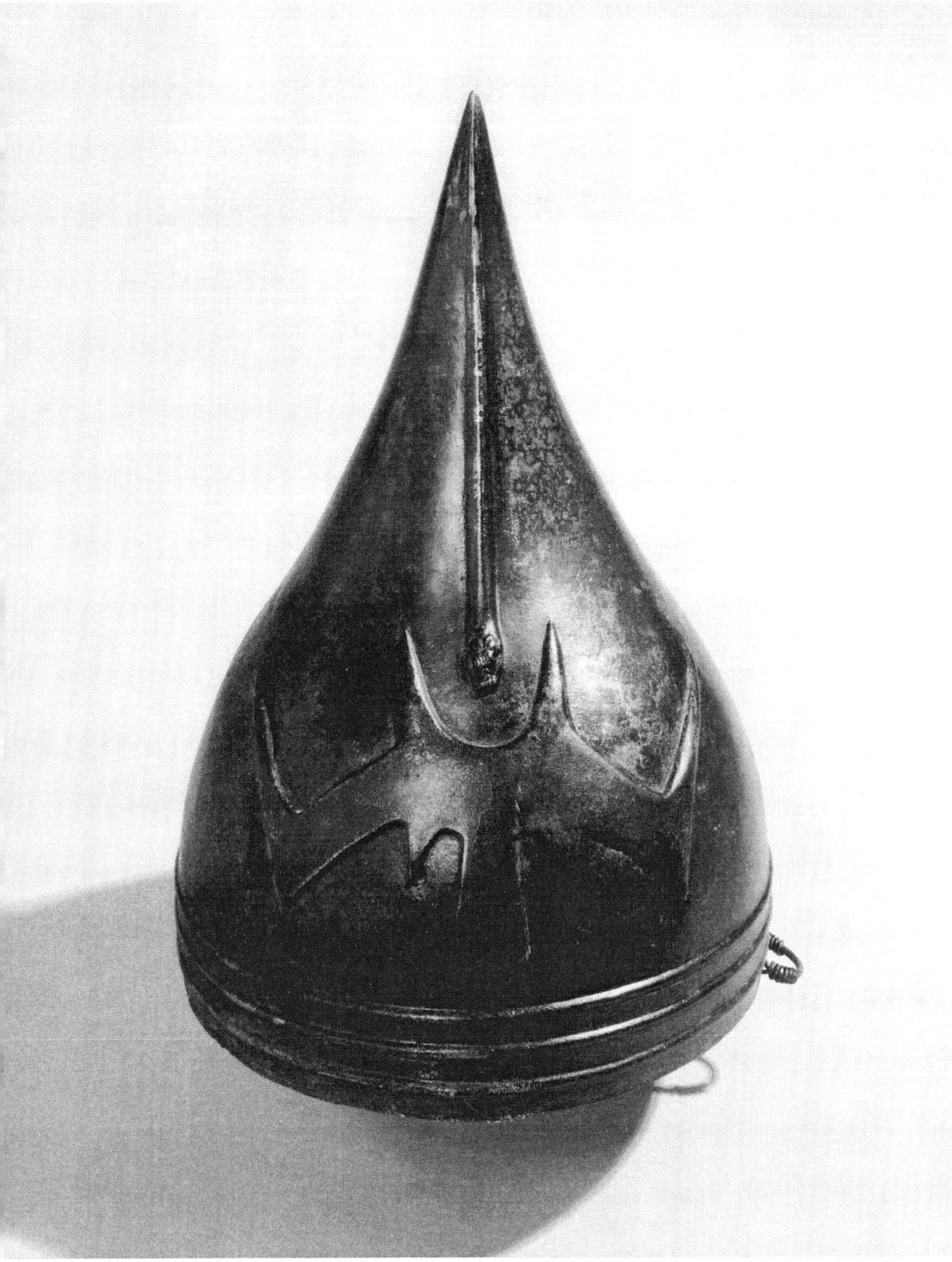

Cat. no. 86. Boston, Museum of Fine Arts, 1981.25.
Egyptian Curator's Fund with money given by Mrs. Horace L. Mayer, 1981.

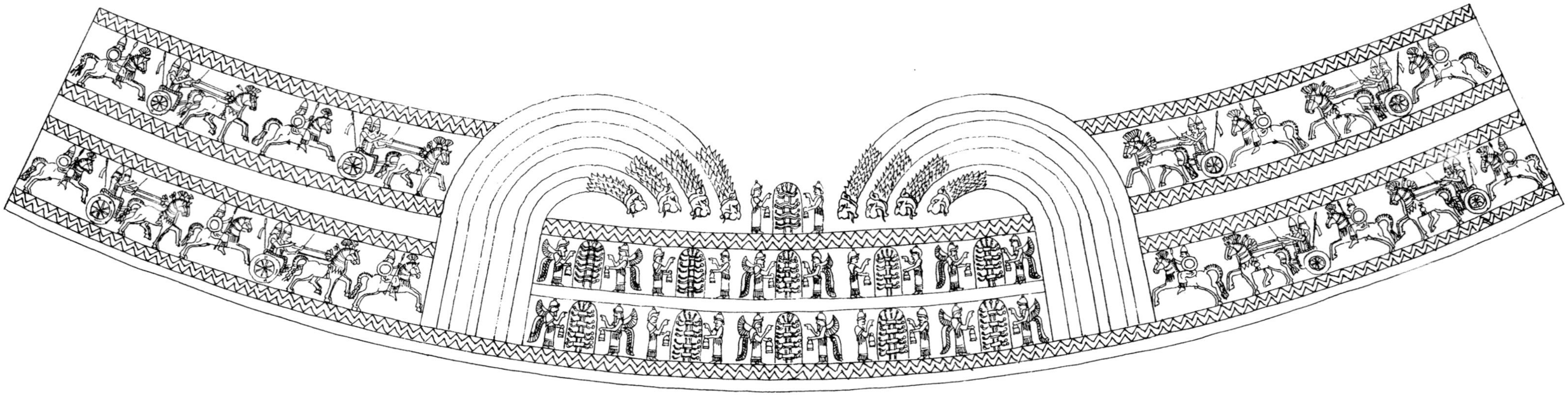

Cat. nos. 91-97. Erevan, Historical Museum of Armenia; St. Petersburg, Hermitage.
Drawing of decoration.

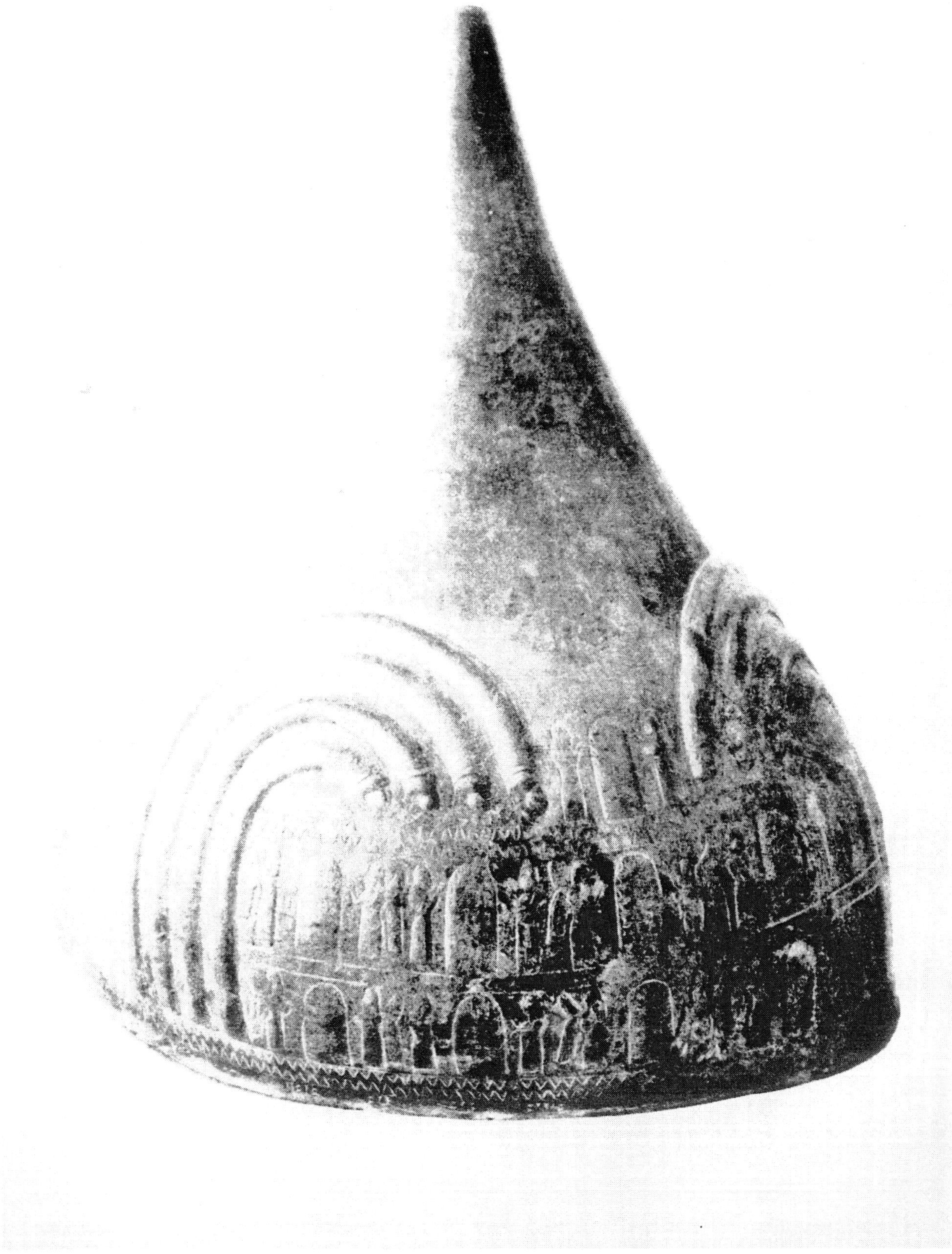

Cat. no. 91. Erevan, Historical Museum of Armenia.

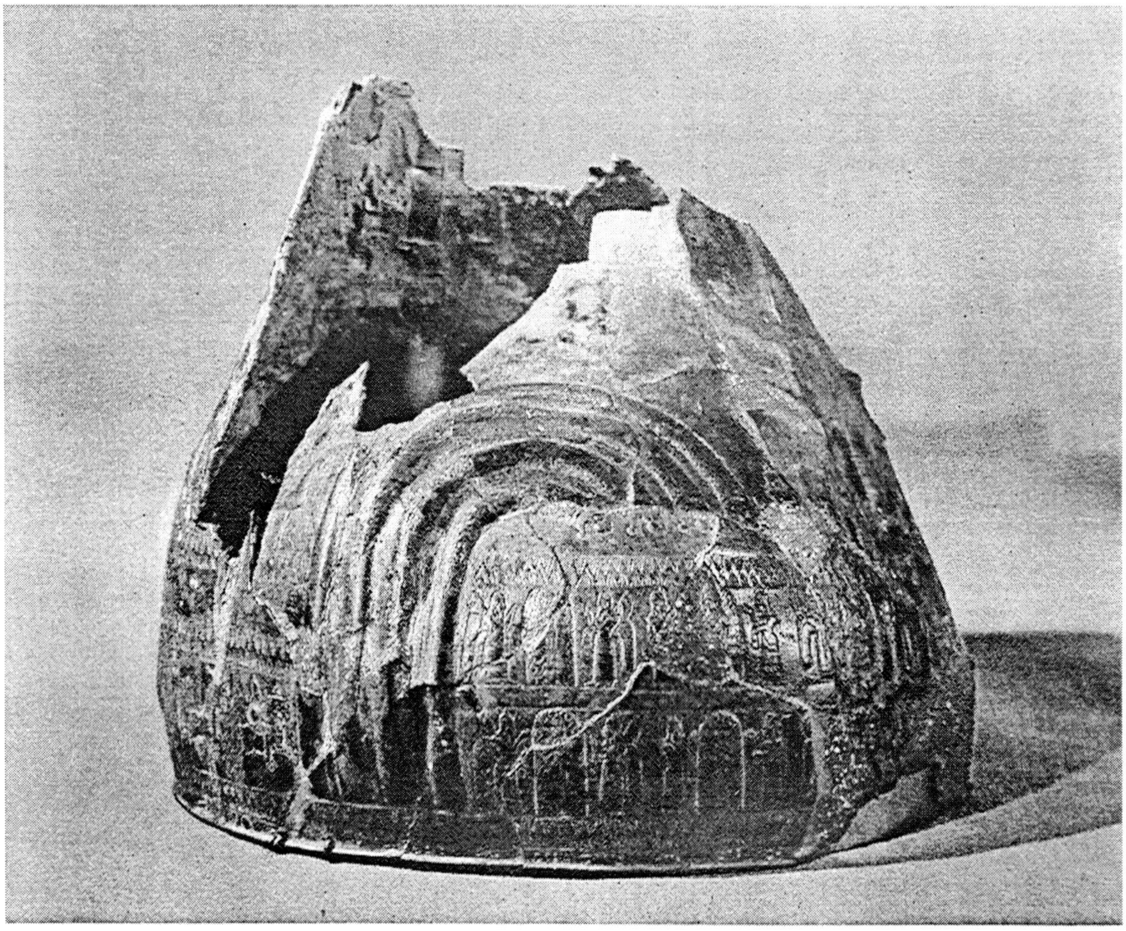

Cat. no. 93. St. Petersburg, Hermitage.

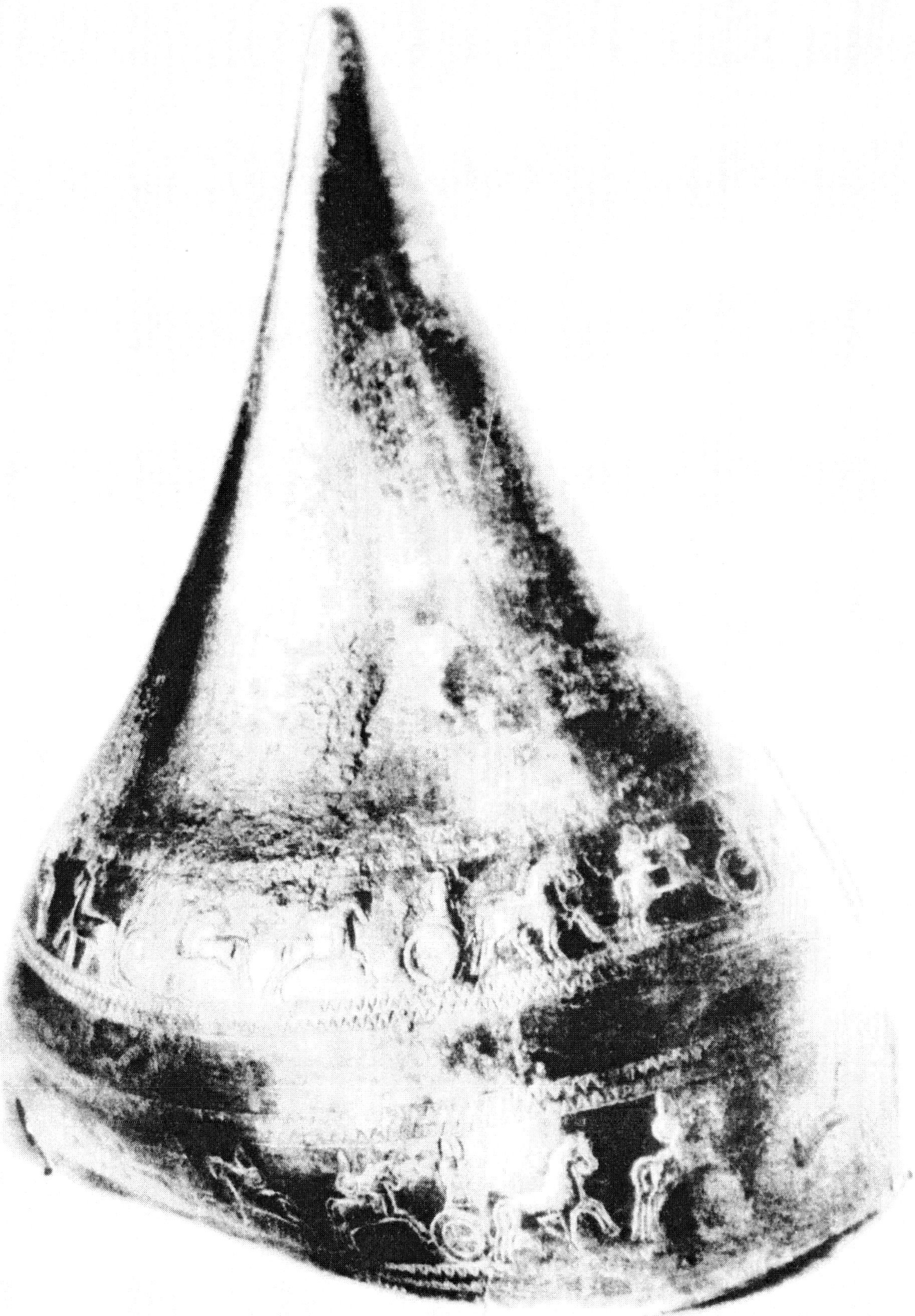

Cat. no. 94. Erevan, Historical Museum of Armenia.

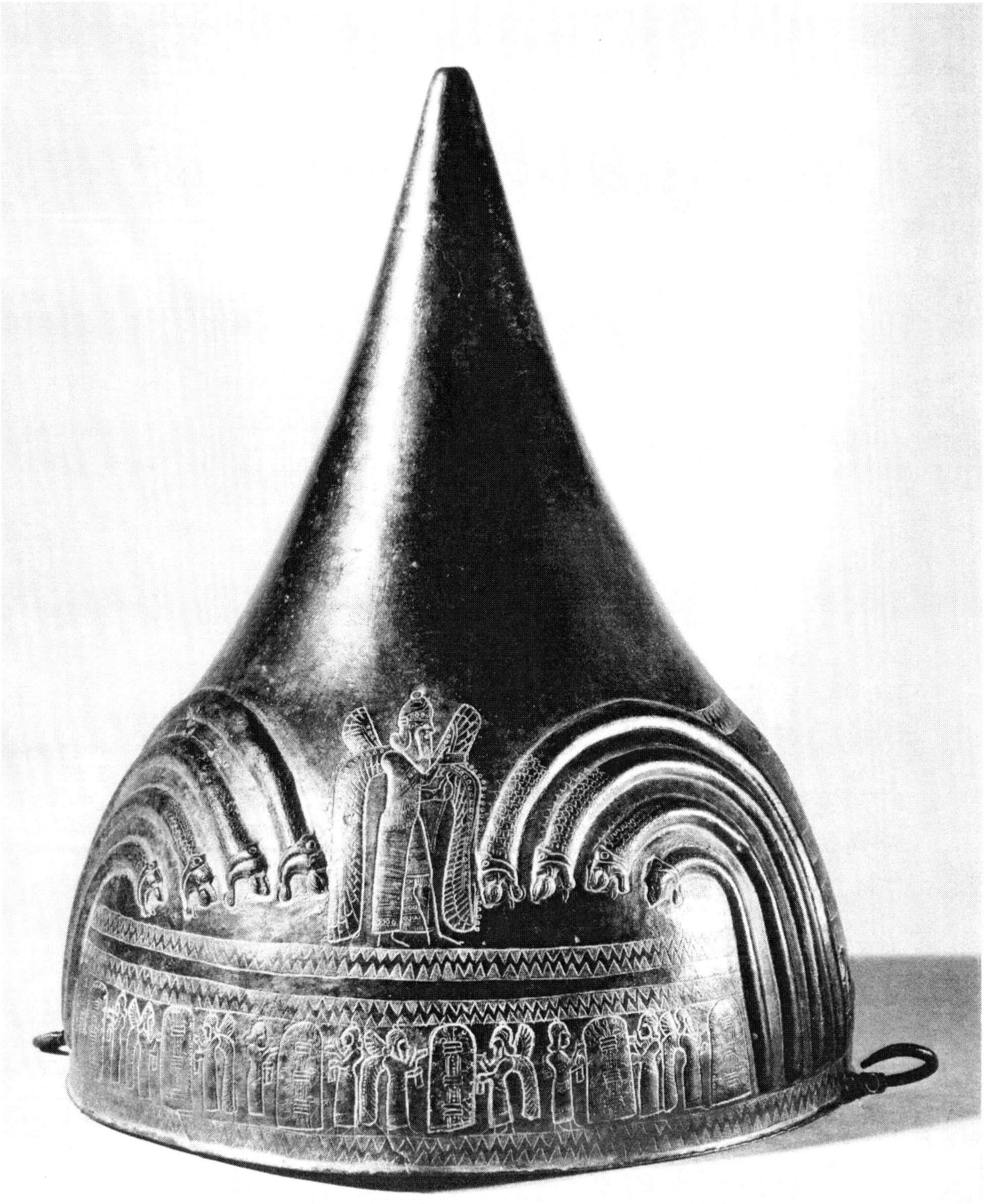

Cat. no. 98. Karlsruhe, Badisches Landesmuseum, 89/1.
Front view.

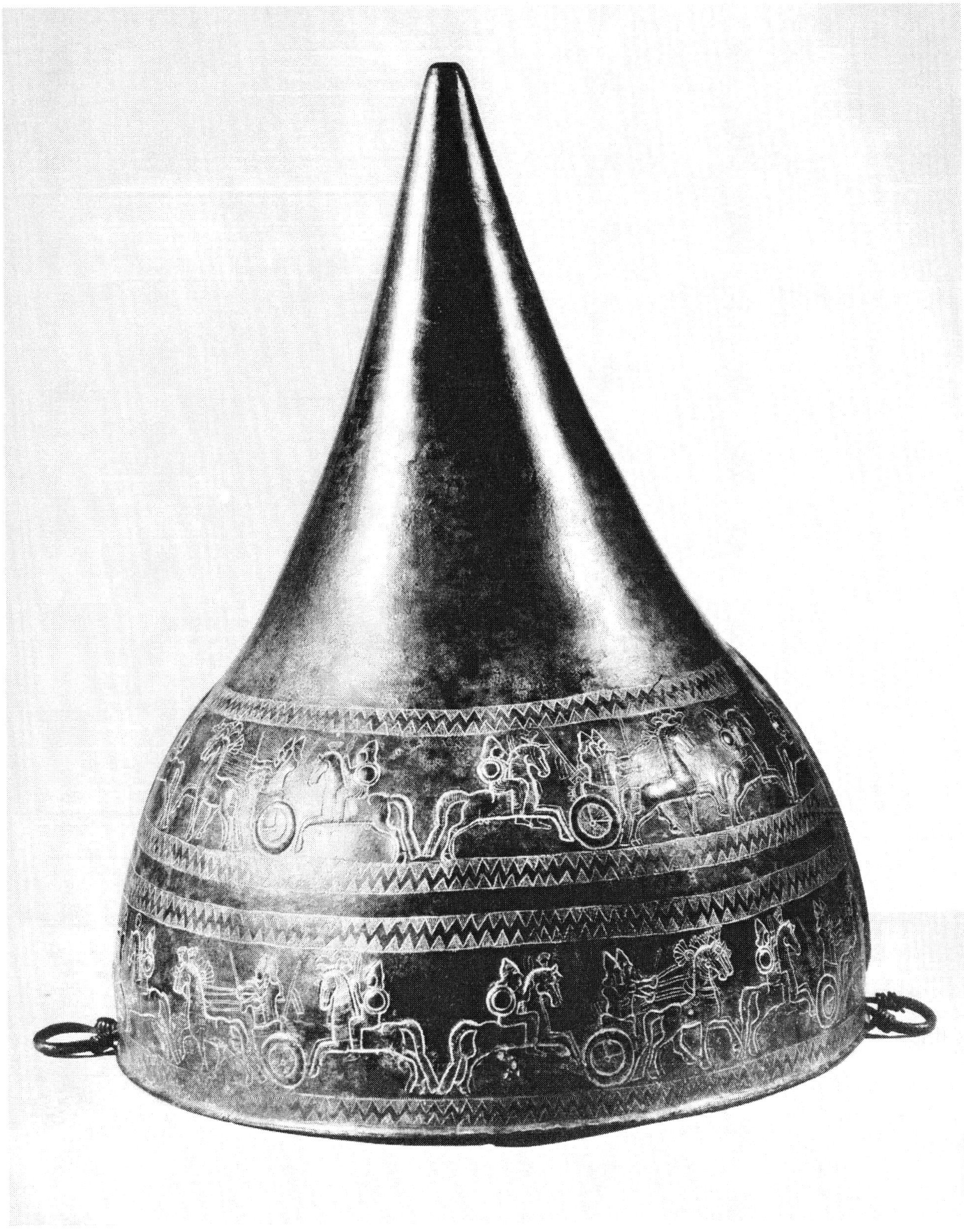

Cat. no. 98. Karlsruhe, Badisches Landesmuseum, 89/1.
Back view.

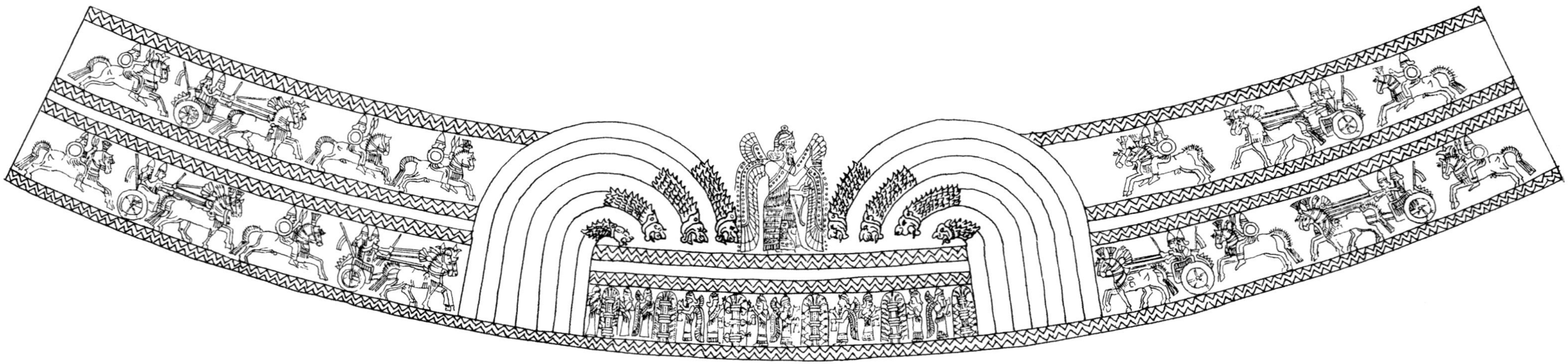

Cat. no. 98. Karlsruhe, Badisches Landesmuseum, 89/1.
Drawing of decoration.

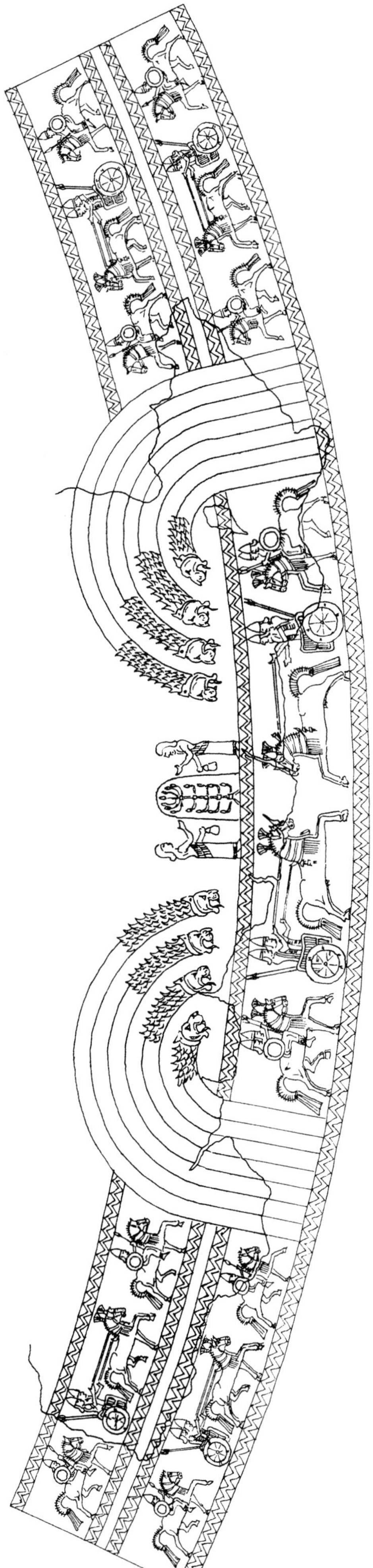

Cat. no. 99. Gaziantep, Gaziantep Museum, 64.3.1973.

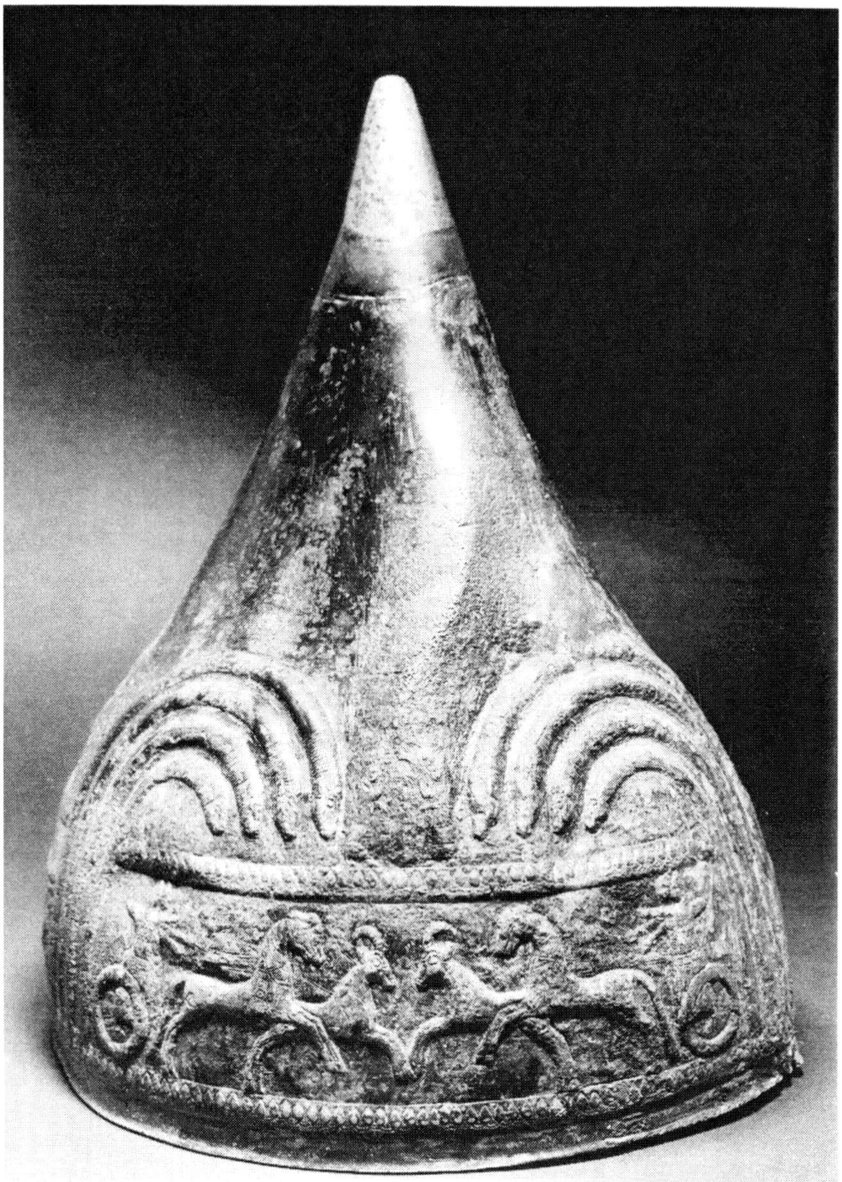

Cat. no. 100. Private collection.
Galvano Copy: Munich, Archäologische Staatssammlung, Museum für Vor- und Frühgeschichte, PS 1971/1782a.

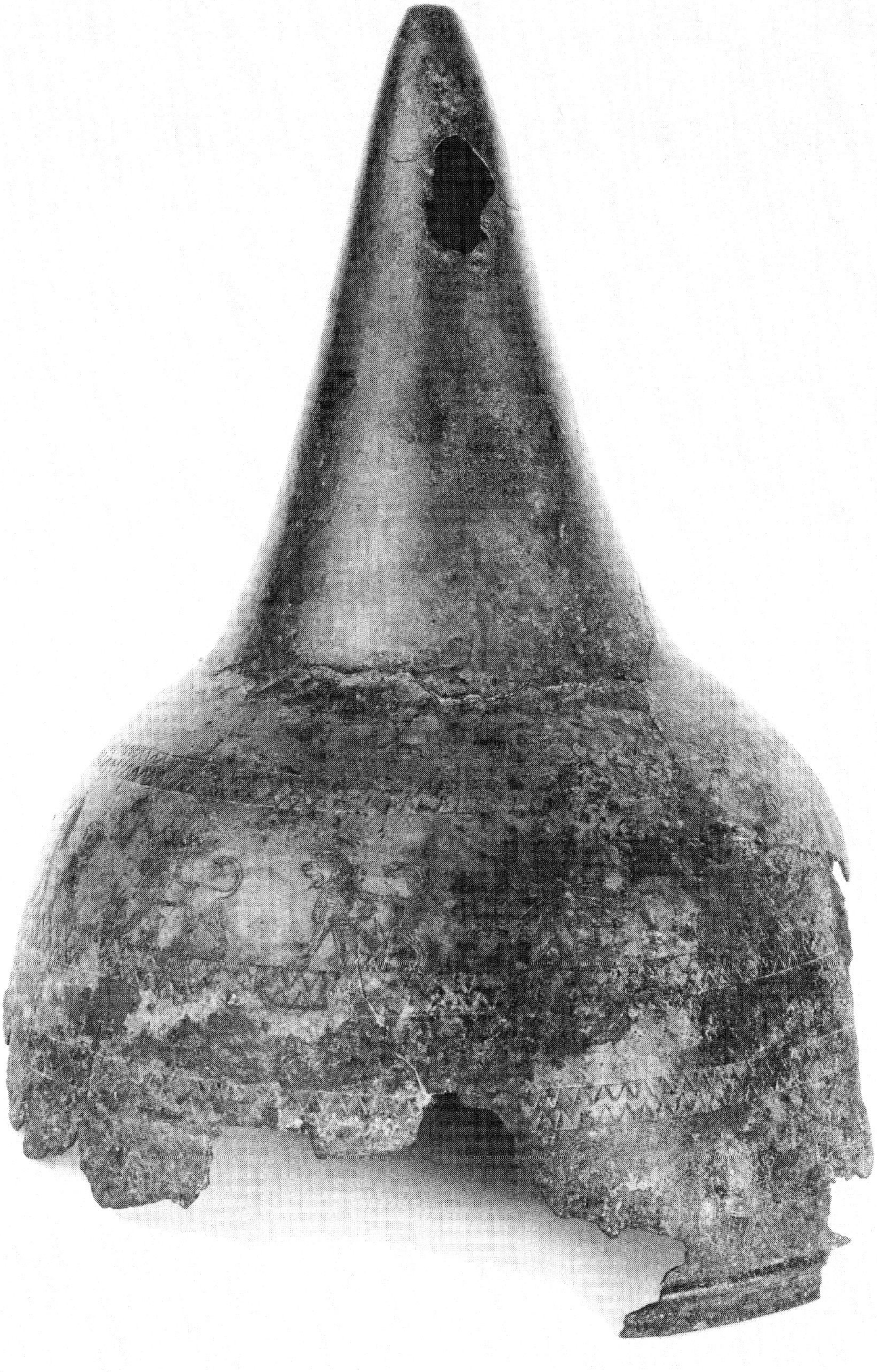

Cat. no. 102. Philadelphia, University of Pennsylvania Museum of Archaeology and Anthropology, UM 67-39-21.

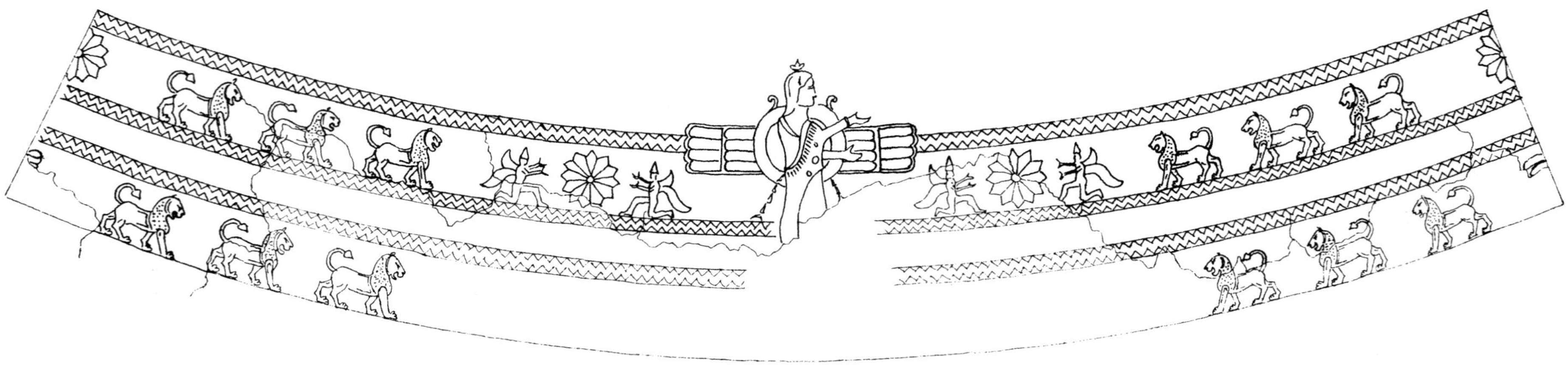

Cat. no. 102. Philadelphia, University of Pennsylvania Museum of Archaeology and Anthropology, UM 67-39-21.
Drawing of decoration.

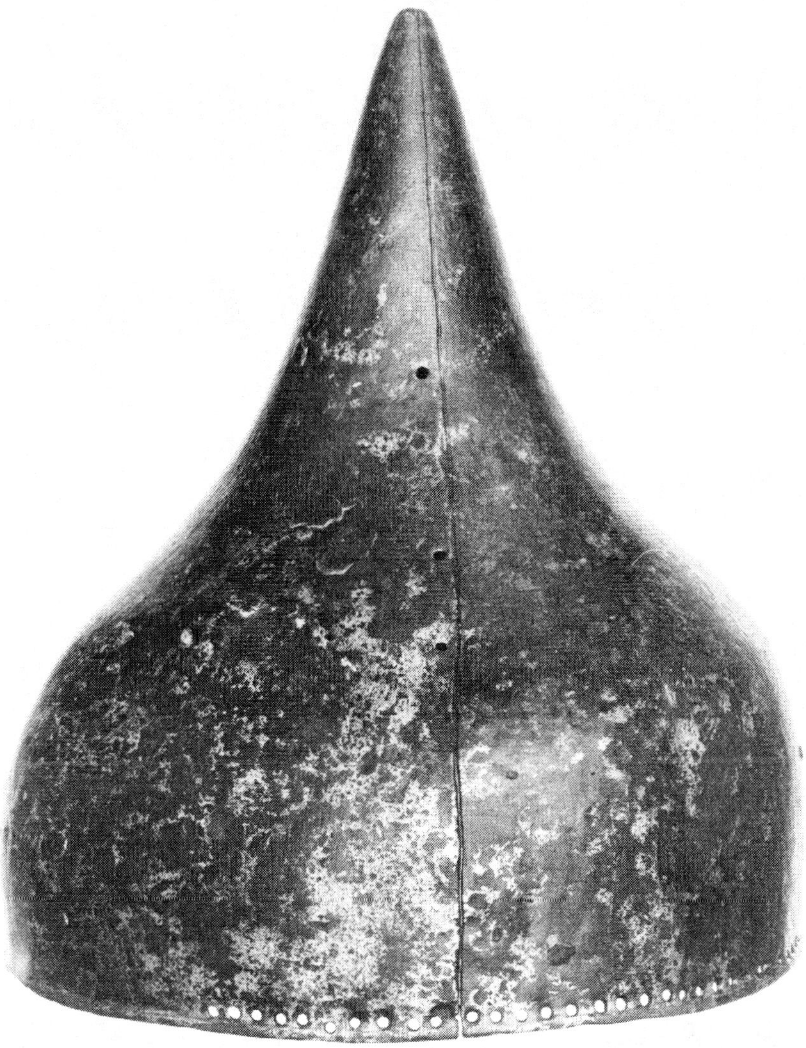

Cat. no. 104. Munich, Archäologische Staatssammlung, Museum für Vor- und Frühgeschichte, PS 1971/2022.

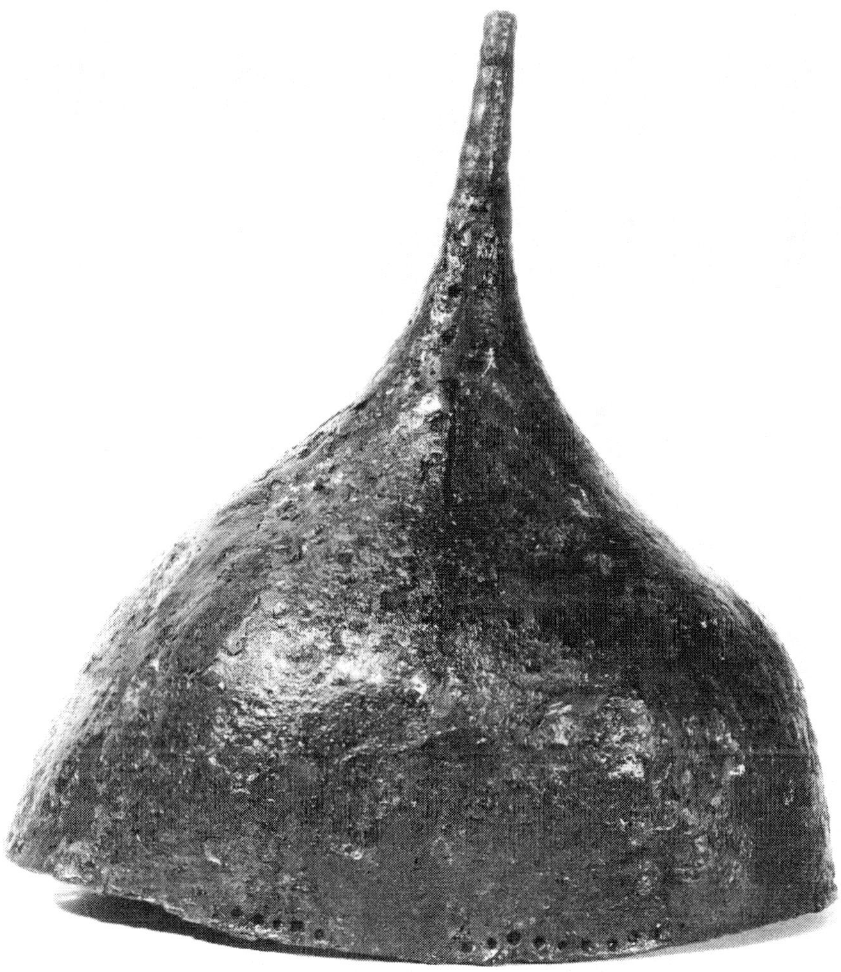

Cat. no. 109. Munich, Archäologische Staatssammlung, Museum für Vor- und Frühgeschichte, PS 1971/1997.

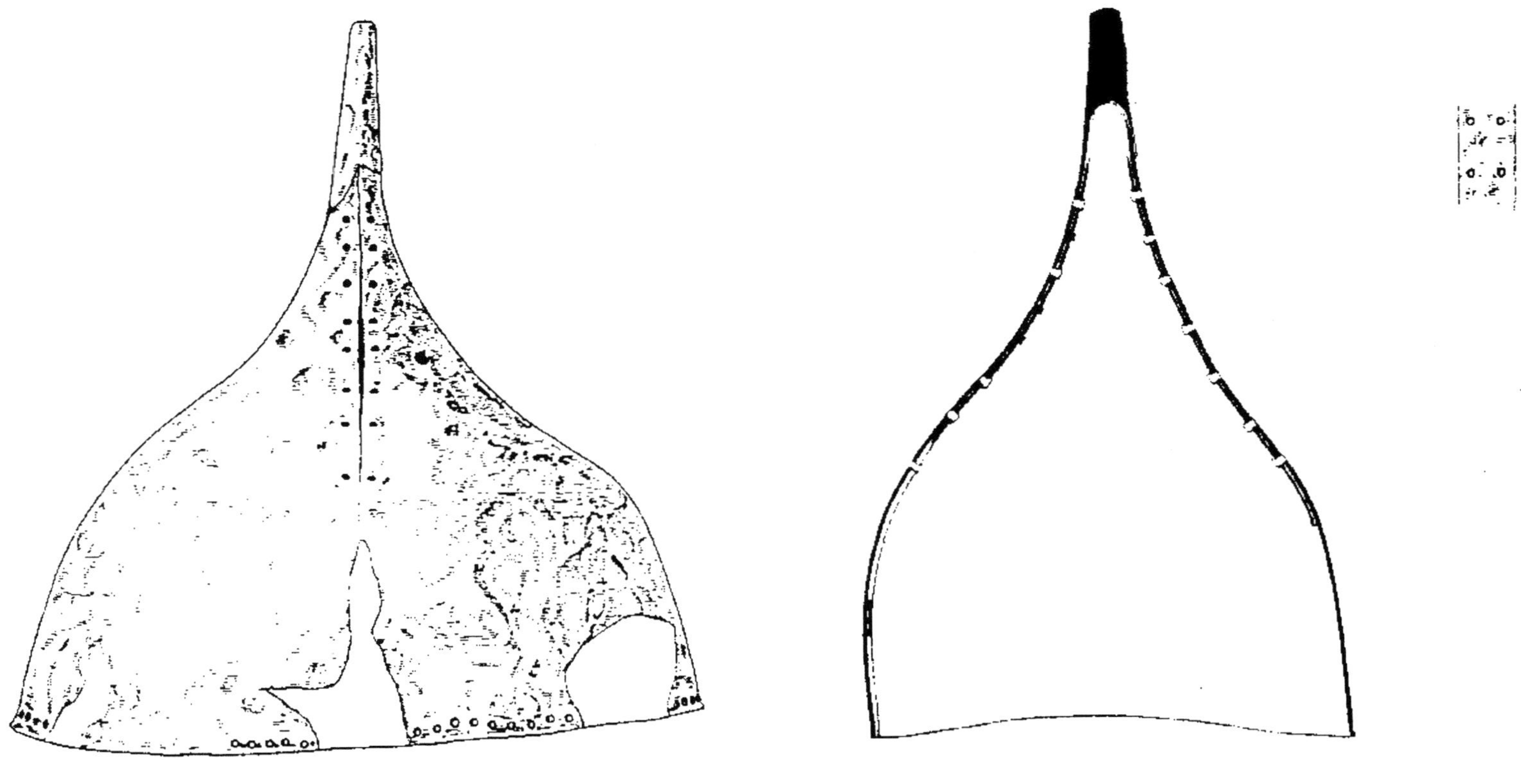

Cat. no. 107. Munich, Archäologische Staatssammlung, Museum für Vor- und Frühgeschichte, PS 1971/1997.
Drawing of construction.

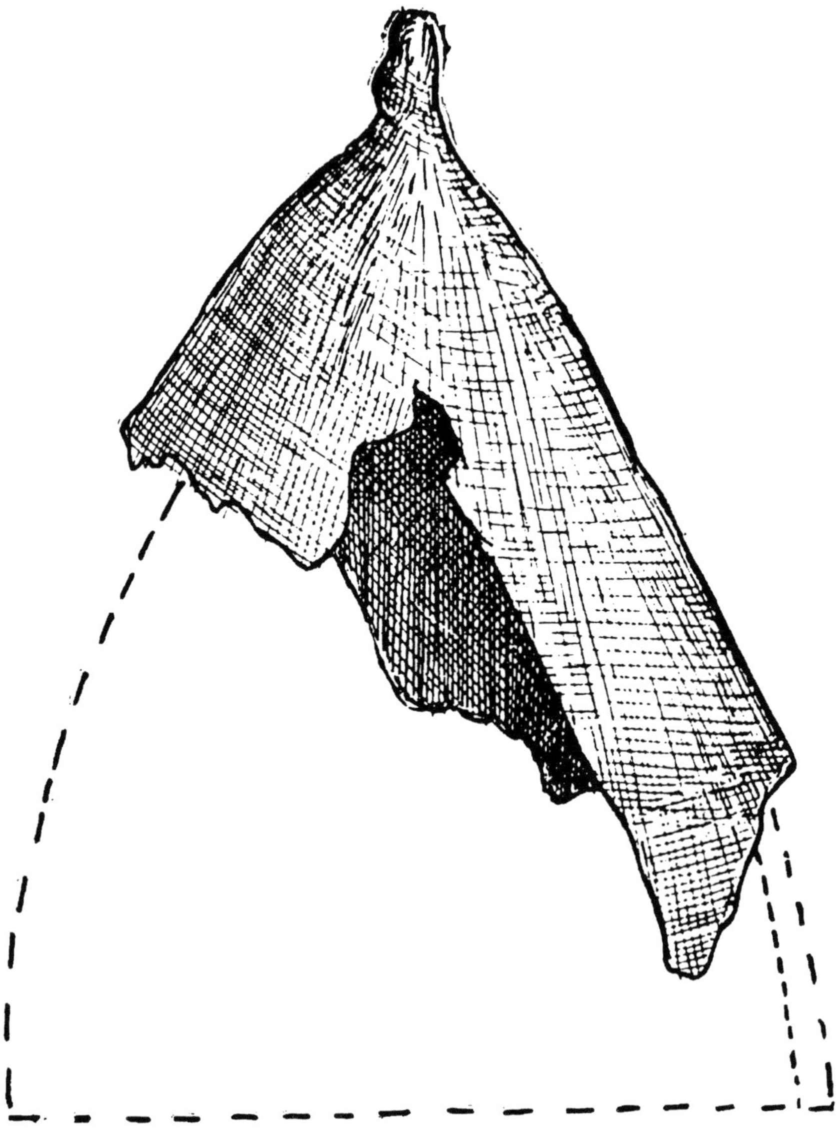

Cat. no. 108. Erevan, Historical Museum of Armenia.
Drawing of helmet.

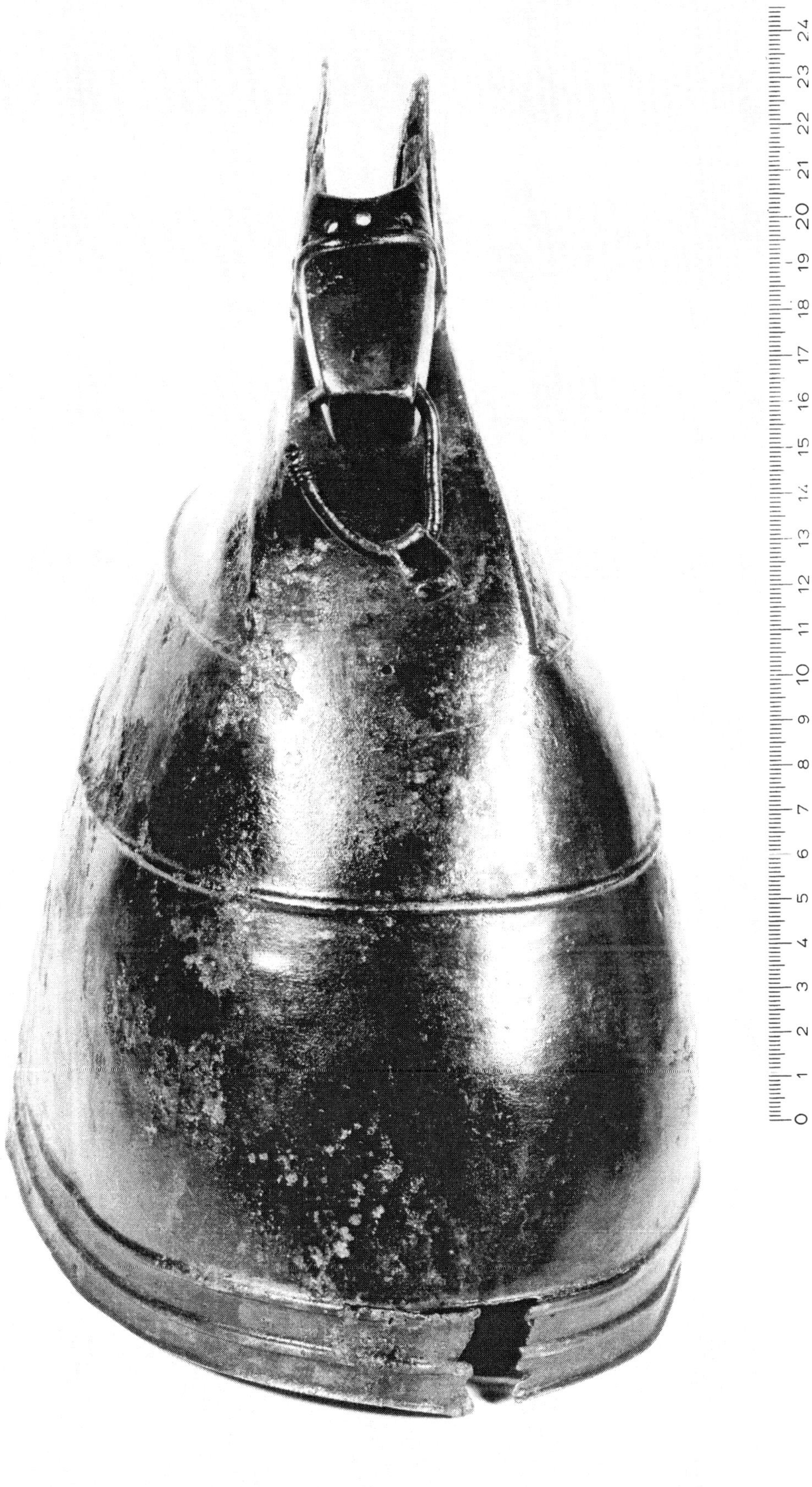

Cat. no. 109. Mainz, Römisch-Germanisches Zentralmuseum, O.39702.
Front view.

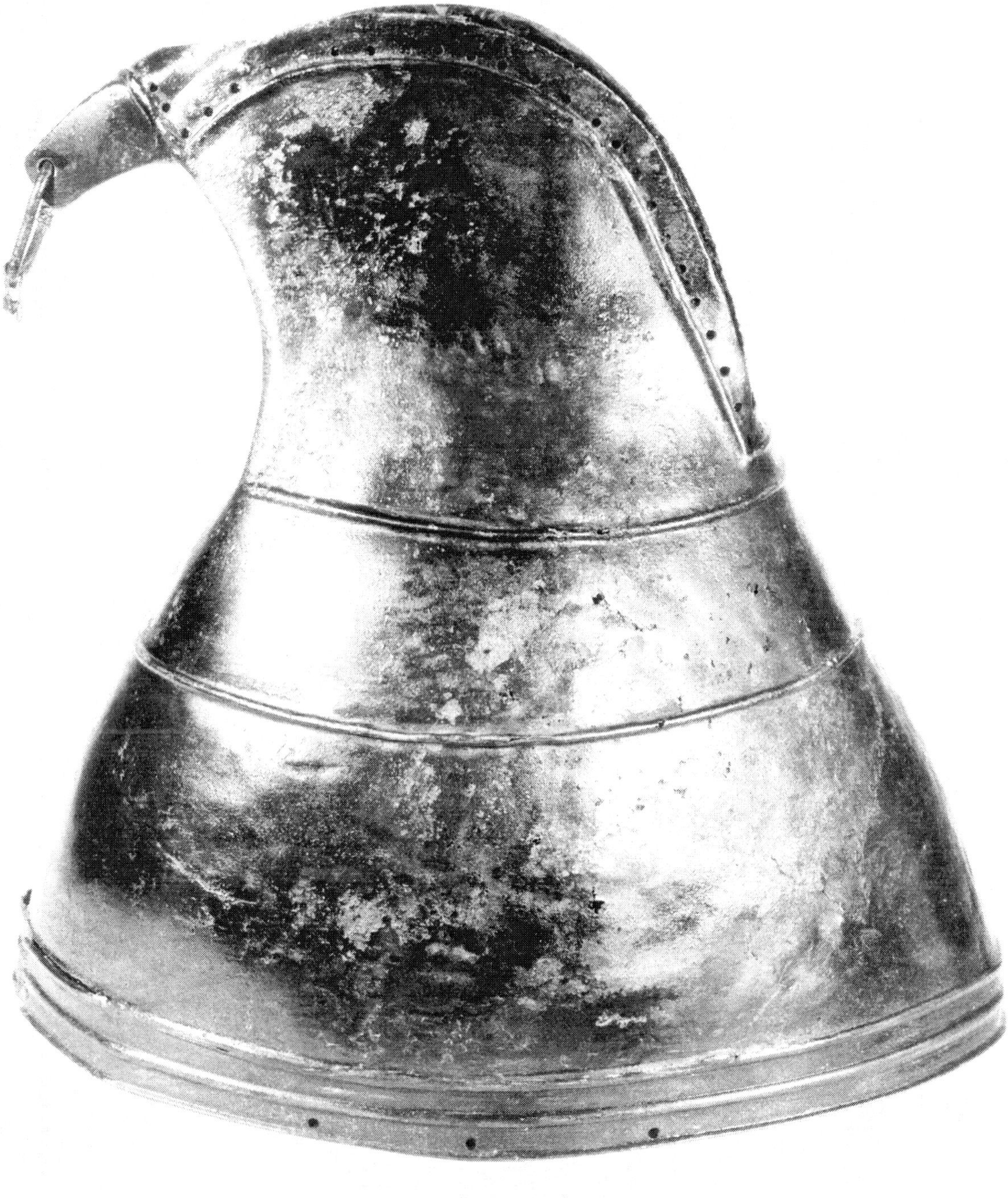

Cat. no. 109. Mainz, Römisch-Germanisches Zentralmuseum, O.39702.
Side view.

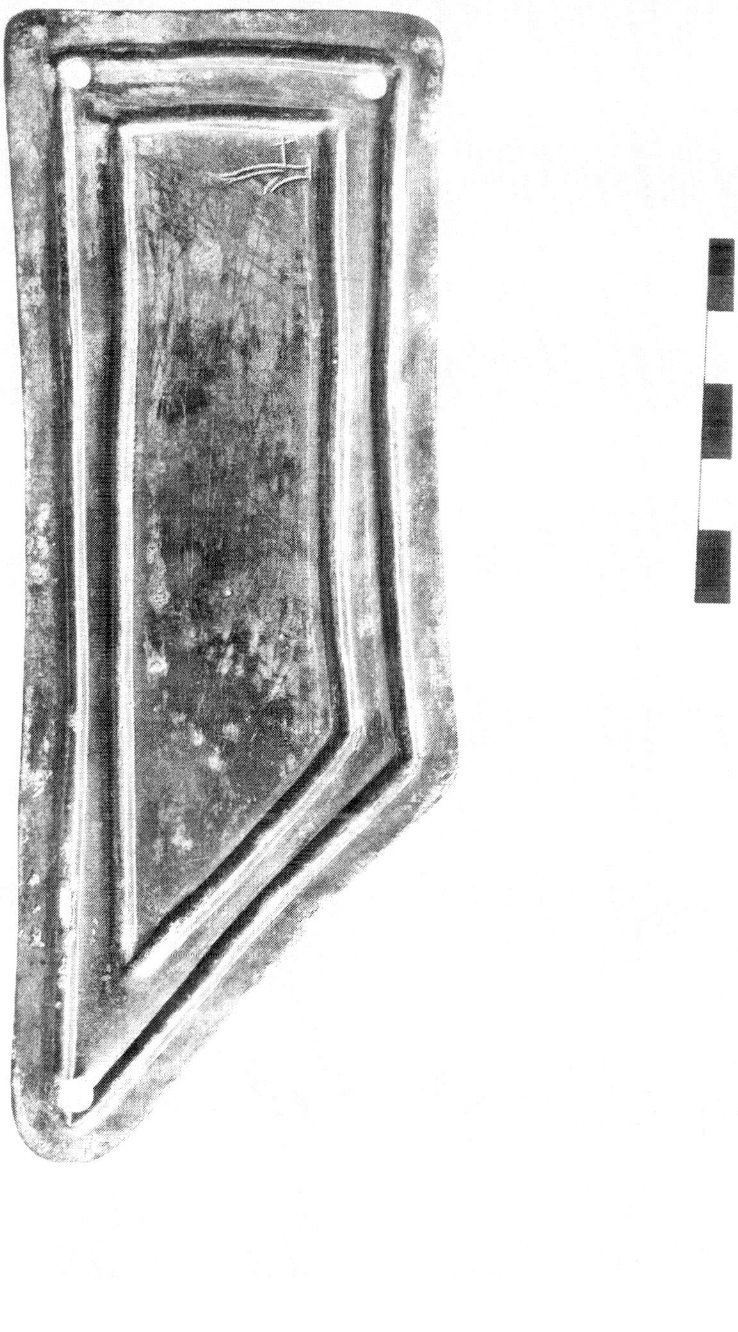

Cat. no. 112. Munich, Archäologische Staatssammlung, Museum für Vor- und Frühgeschichte, PS 1971/1829.

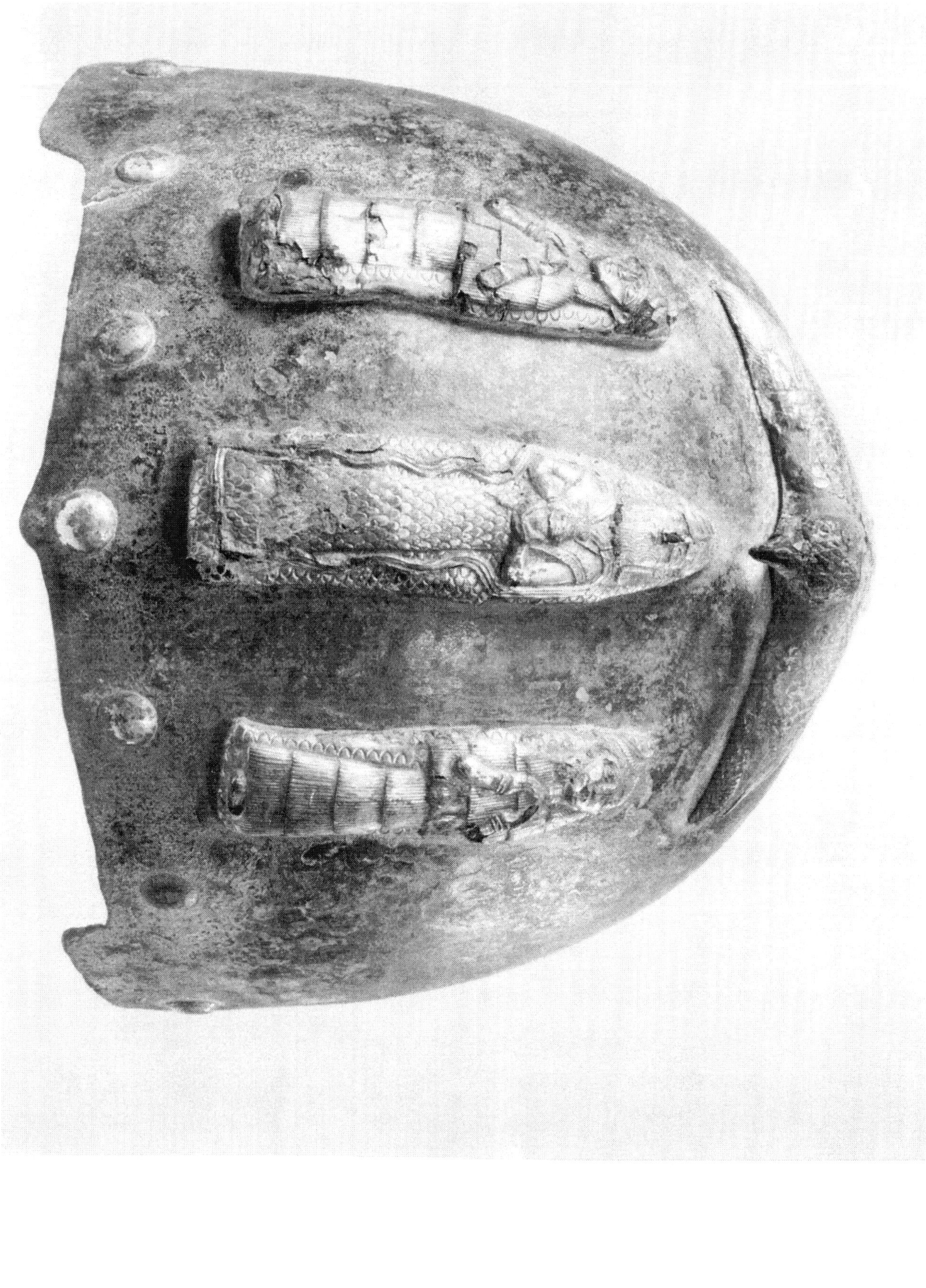

Cat. no. 122. New York, Metropolitan Museum of Art, MMA 63.74.

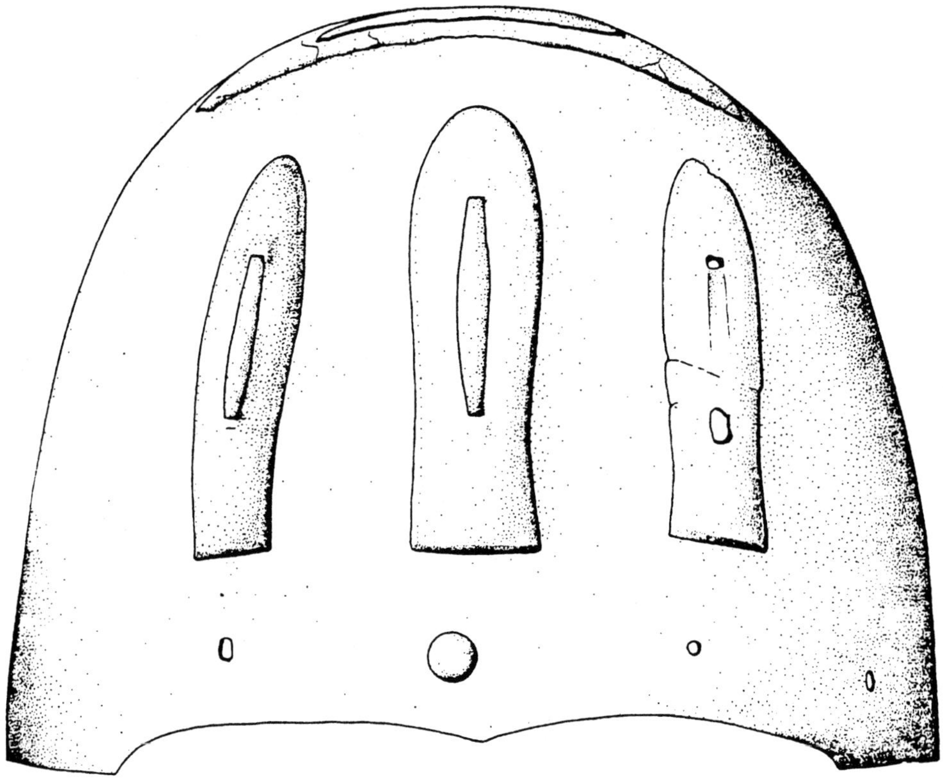

Cat. no. 124. Mainz, Römisch-Germanisches Zentralmuseum, O.39704.
Drawing of helmet (Schauer 1979-1980, fig. 11).

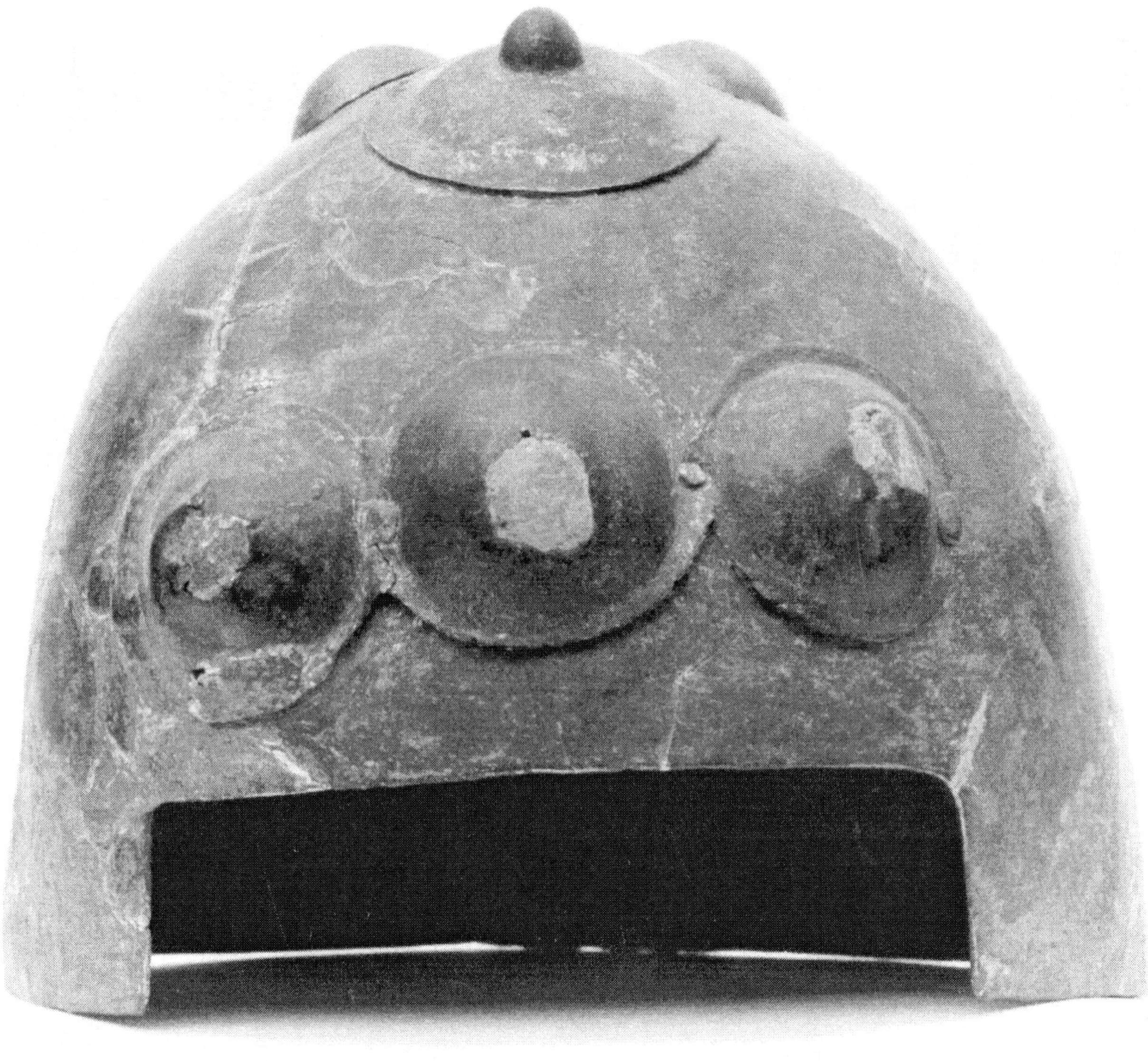

Cat. no. 126. Kibbutz Hazorea, Wilfrid Israel Museum of Oriental Art and Studies, P-259.

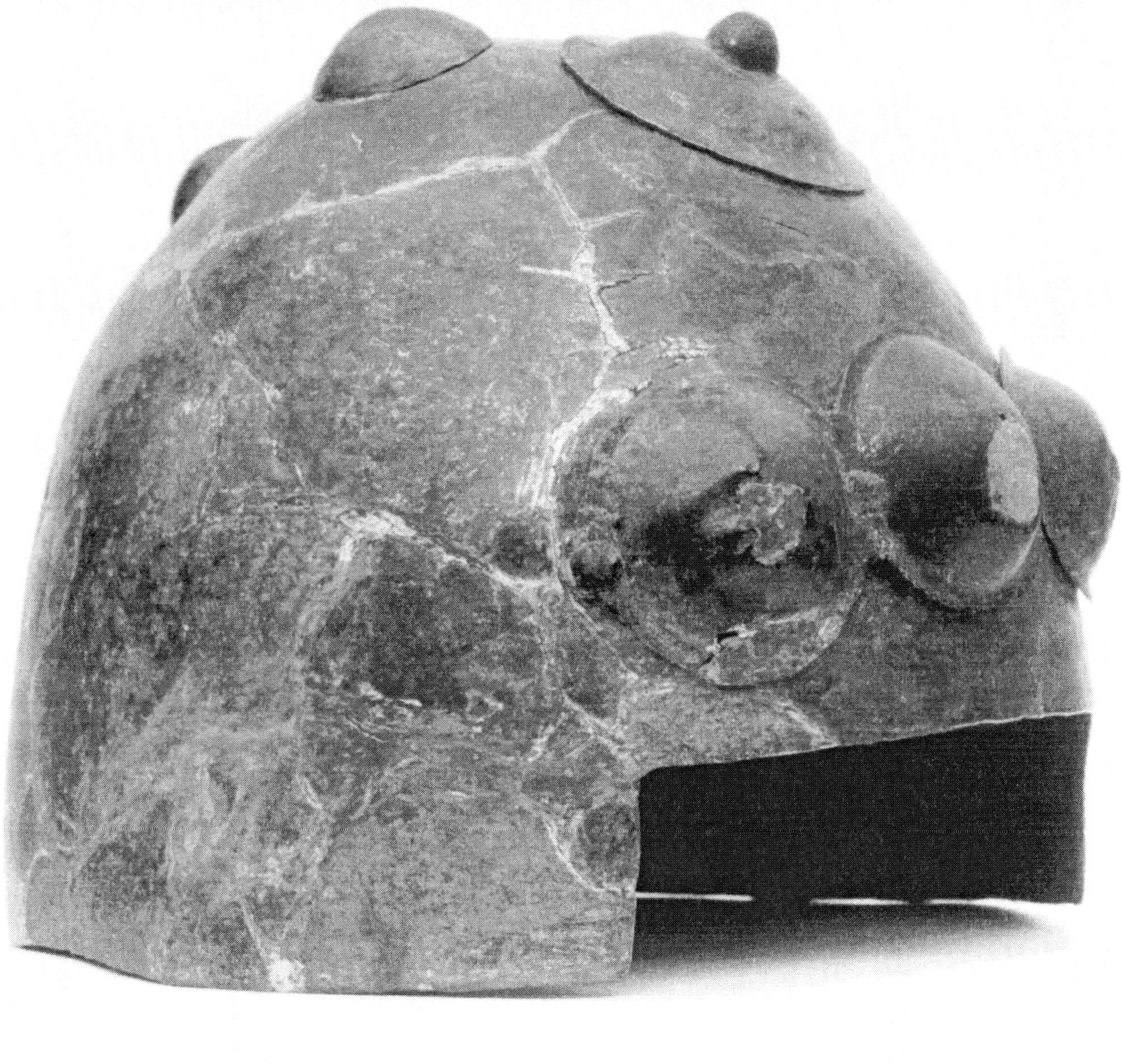

Cat. no. 126. Kibbutz Hazorea, Wilfrid Israel Museum of Oriental Art and Studies, P-259.

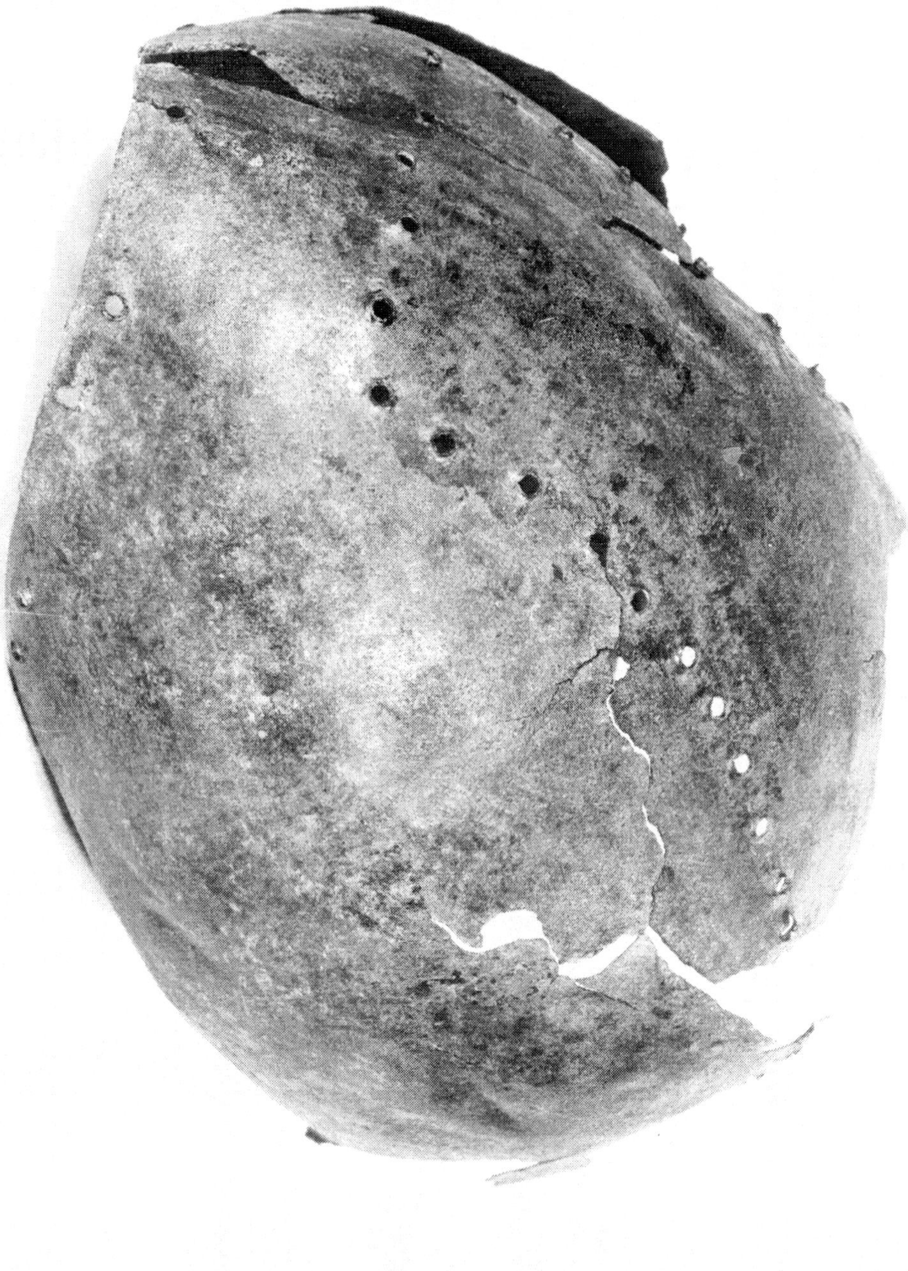

Cat. no. 128. Oxford, Ashmolean Museum, 1965.859.

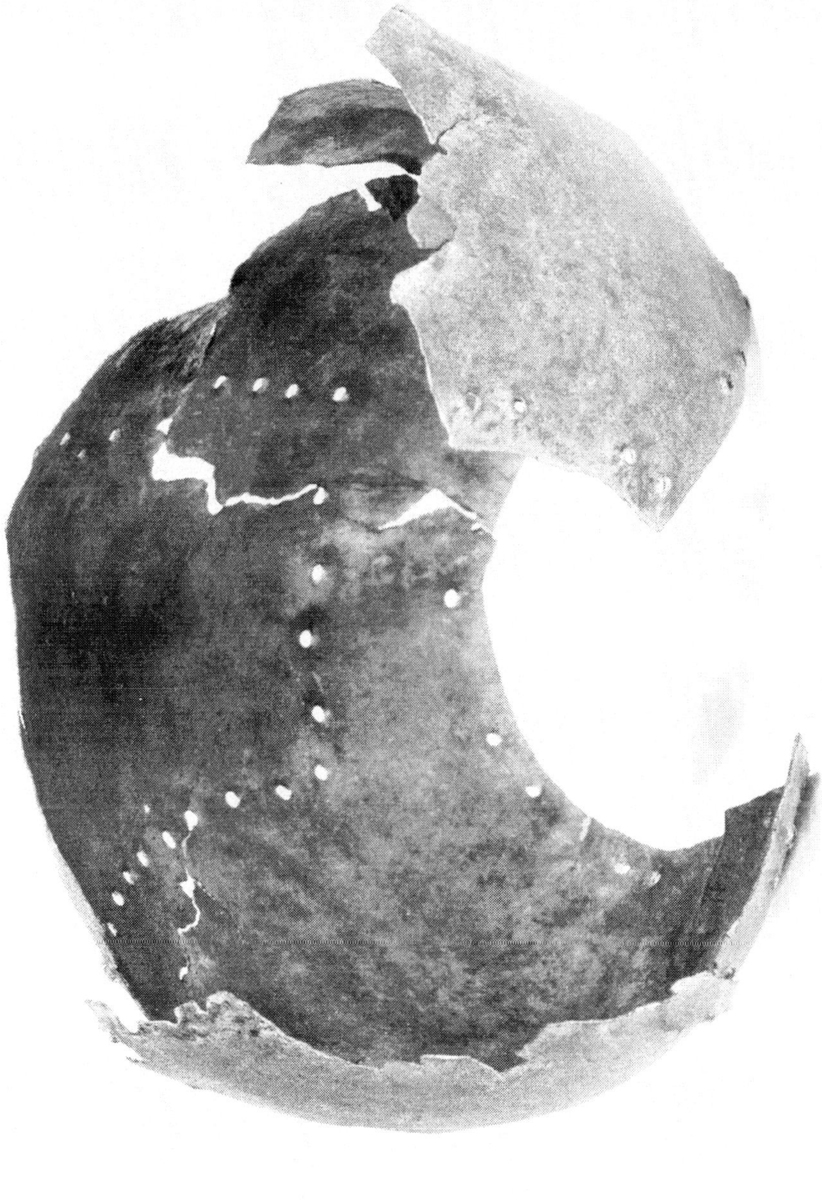

Cat. no. 128. Oxford, Ashmolean Museum, 1965.859.

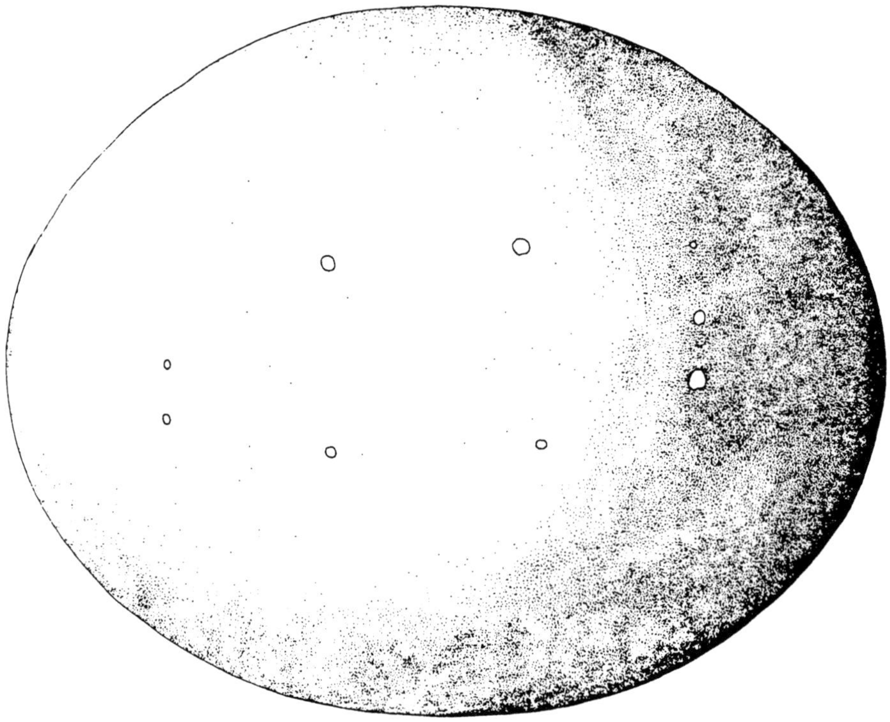

b

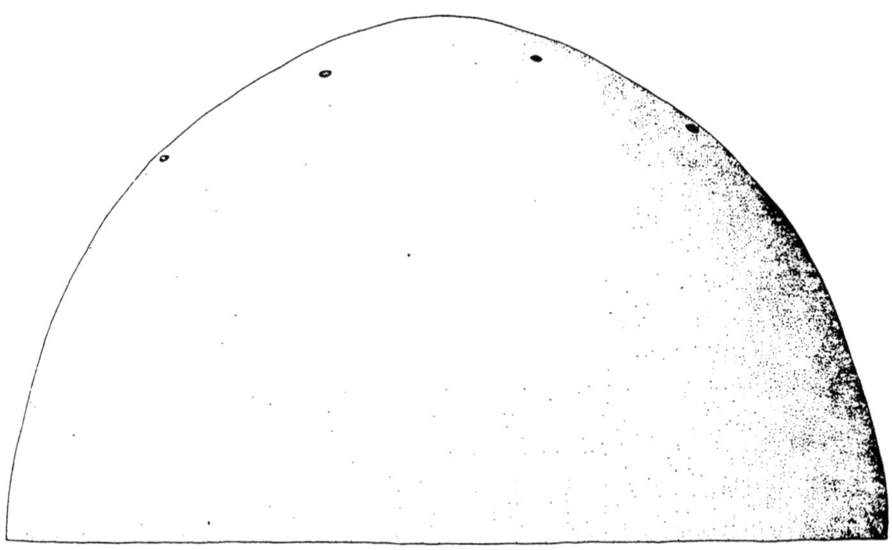

a

Cat. no. 129. Mainz, Römisch-Germanisches Zentralmuseum, O.39671.
Drawing of helmet (Schauer 1979-1980, fig. 9).
a) top, b) side view.

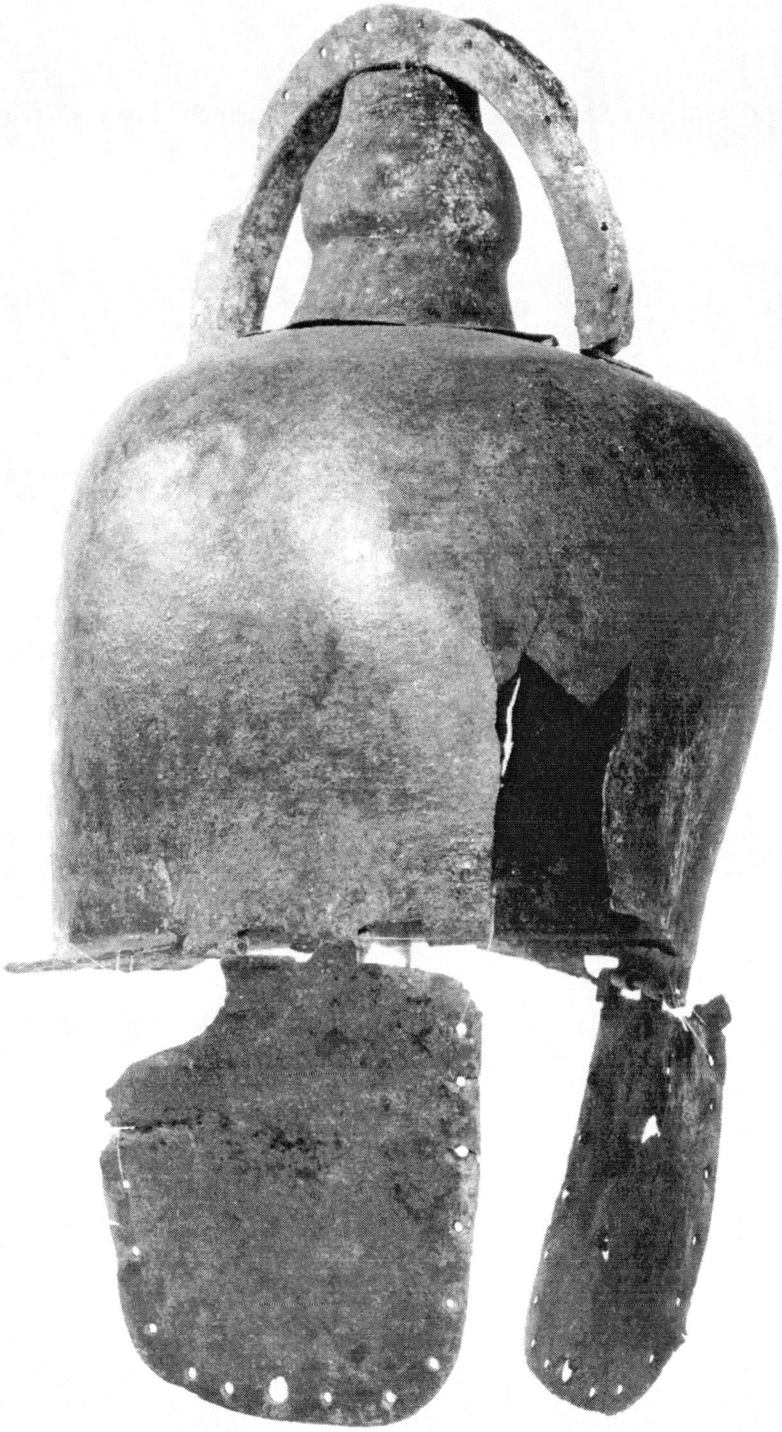

Cat. no. 133. London, British Museum, WA 134705.

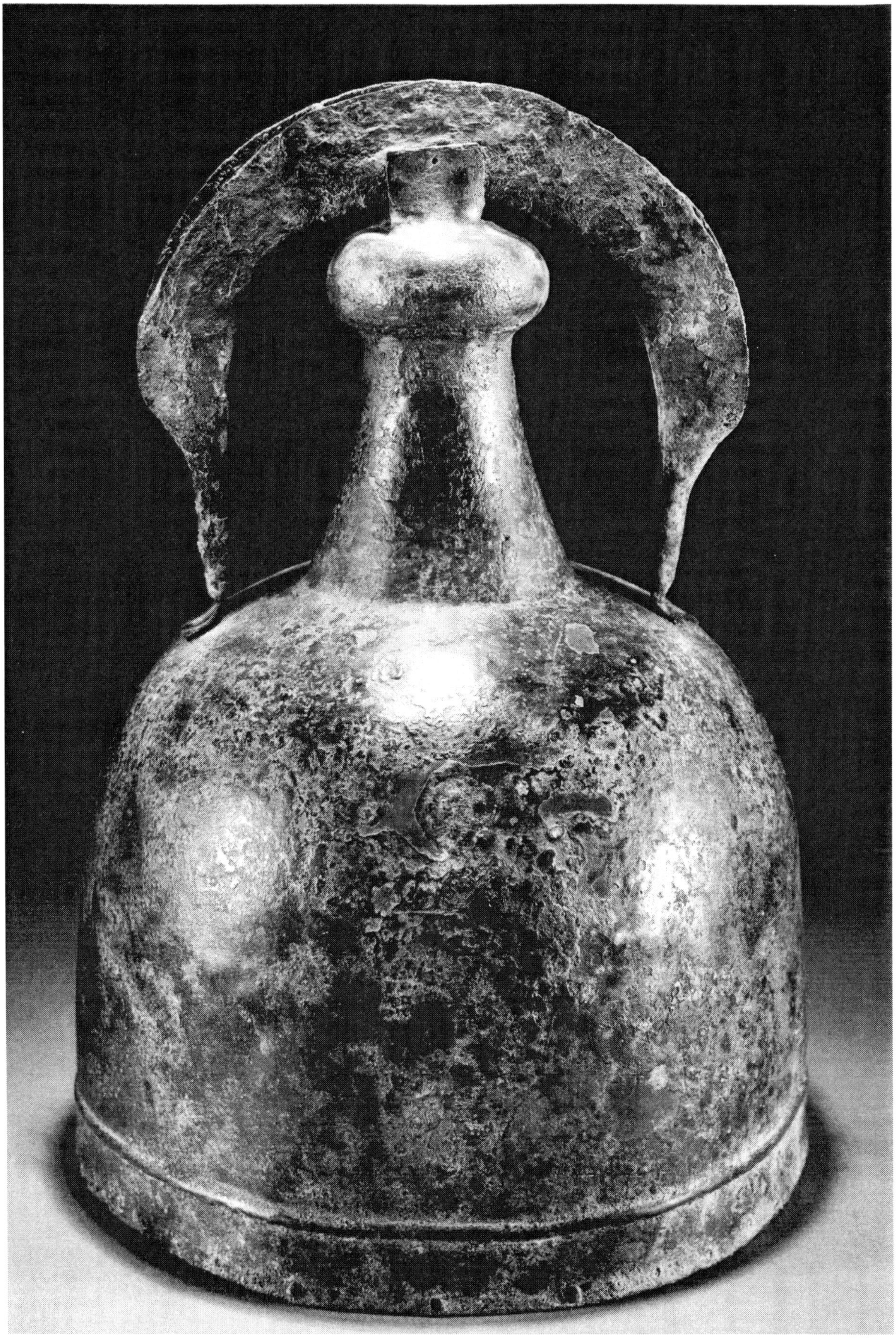

Cat. no. 134. London, Mahboubian Family Collection.

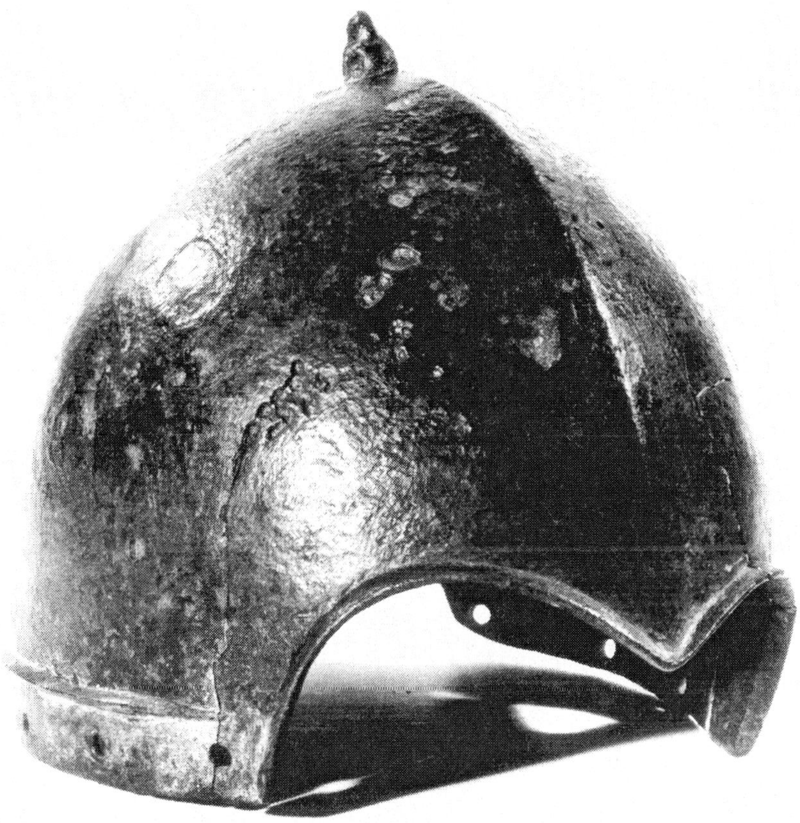

St. Petersburg, Hermitage; Kelermes, Kurgan number 2.

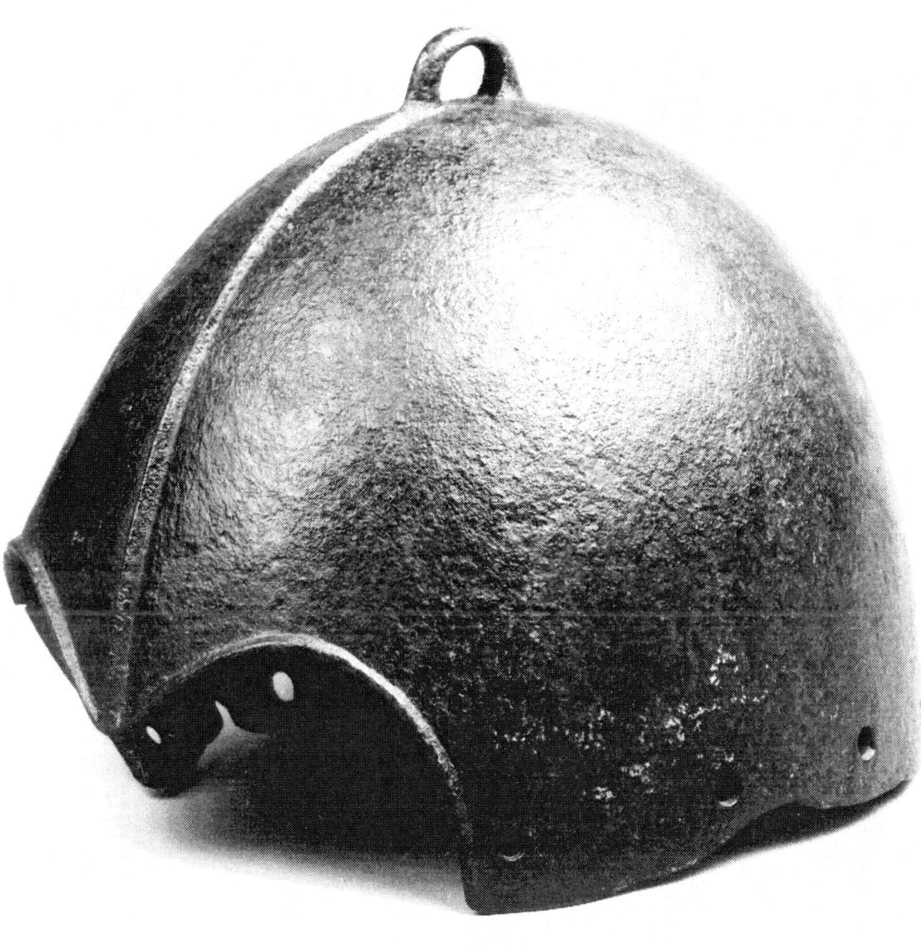

St. Petersburg, Hermitage; Kelermes, Kurgan number 2.

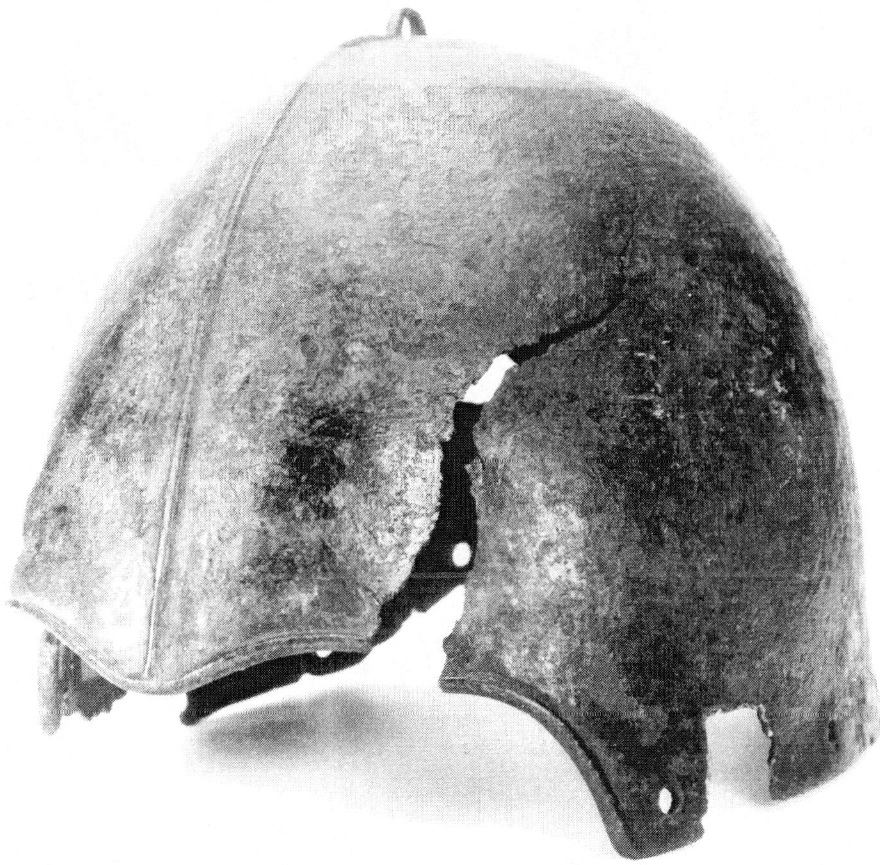

St. Petersburg, Hermitage; Kelermes, Kurgan number 1.

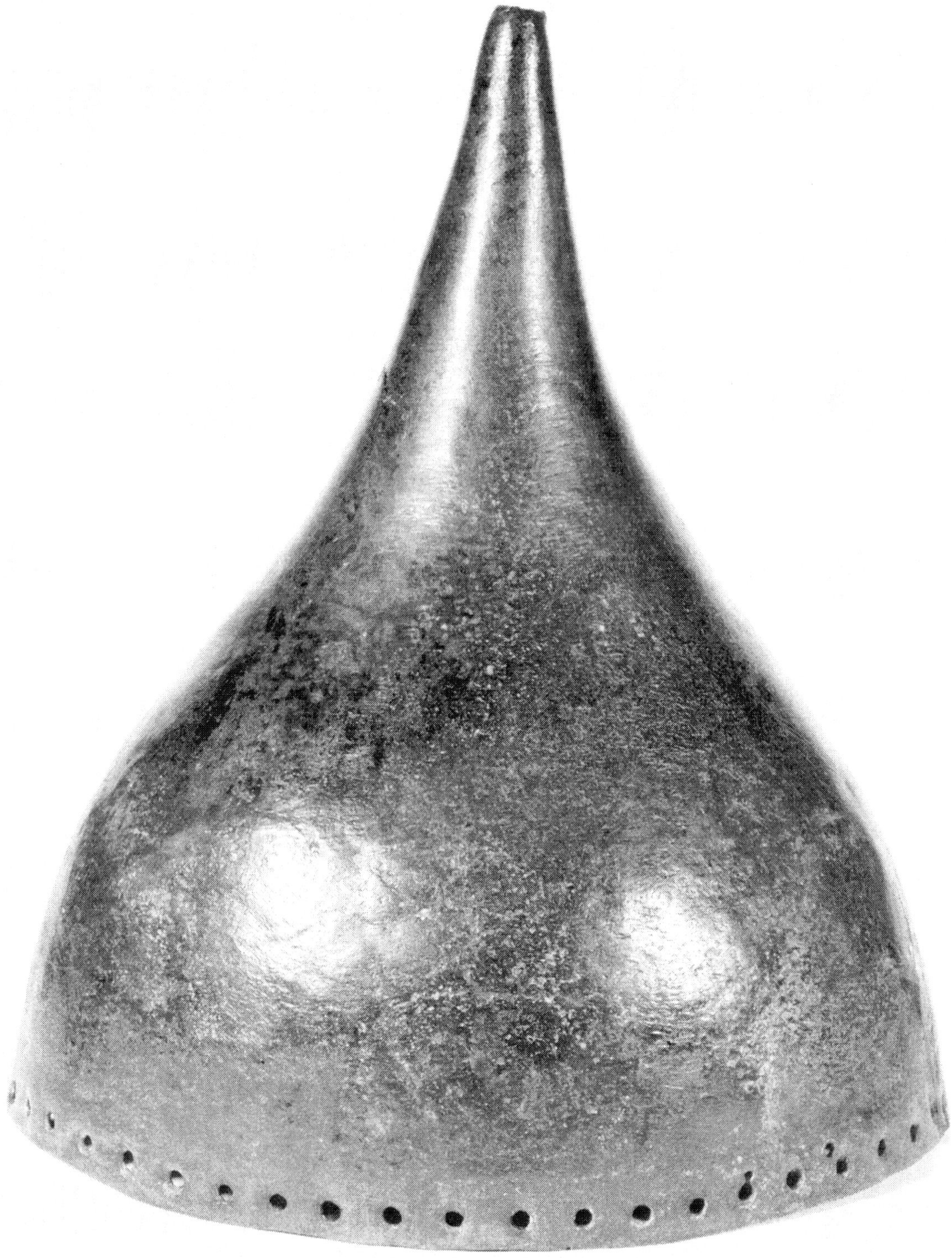

Cat. no. 139. Mainz, Römisch-Germanisches Zentralmuseum, O.39606.

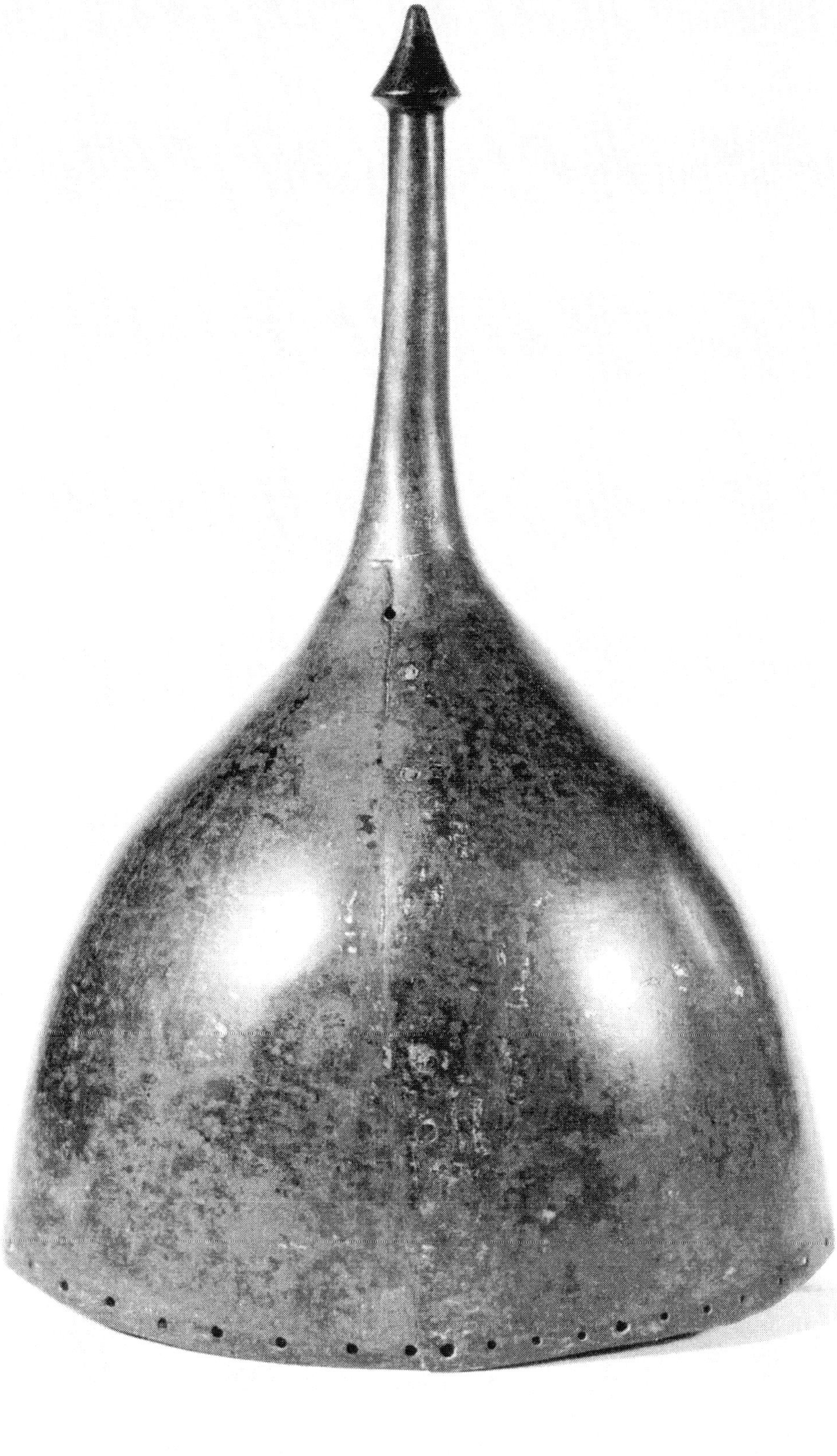

Cat. no. 140. Mainz, Römisch-Germanisches Zentralmuseum, O.39585.

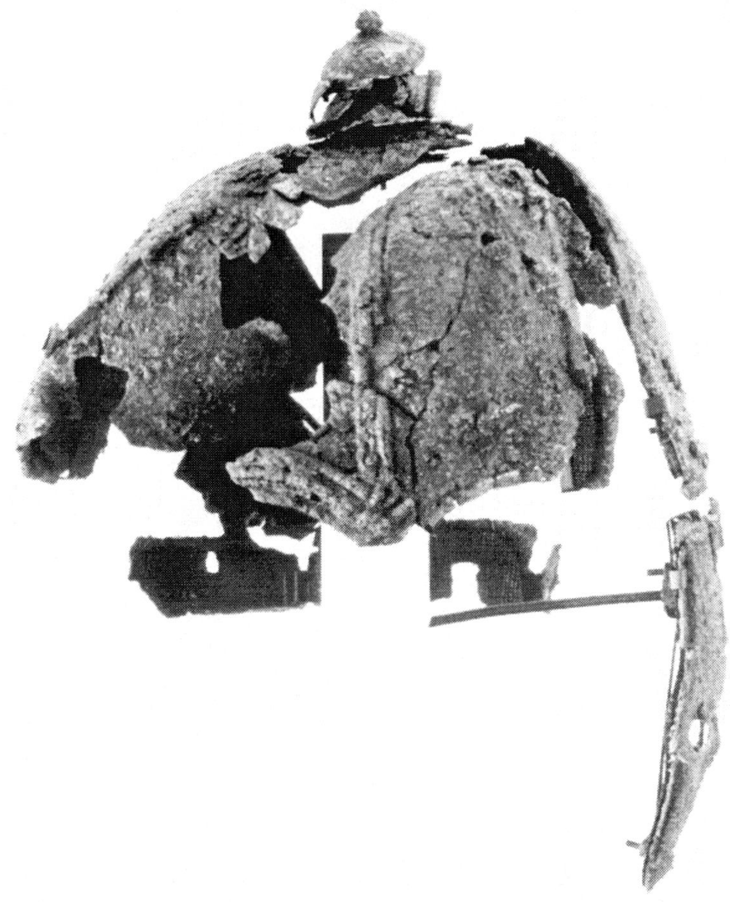

Cat. no. 142. Sardis, M87.7/9444.
Front view.

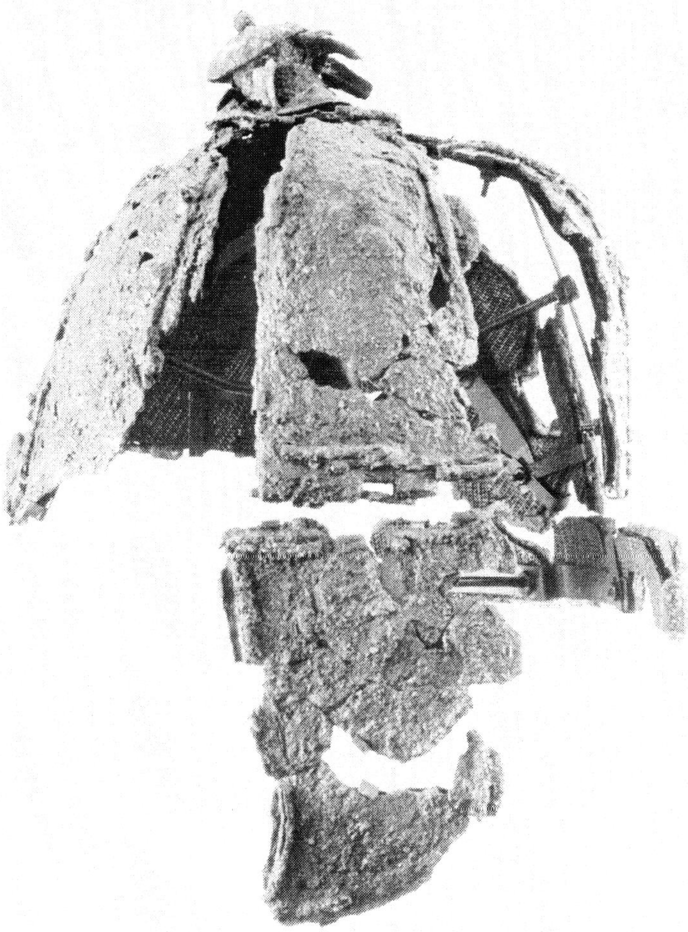

Cat. no. 142. Sardis, M87.7/9444.
Side view.

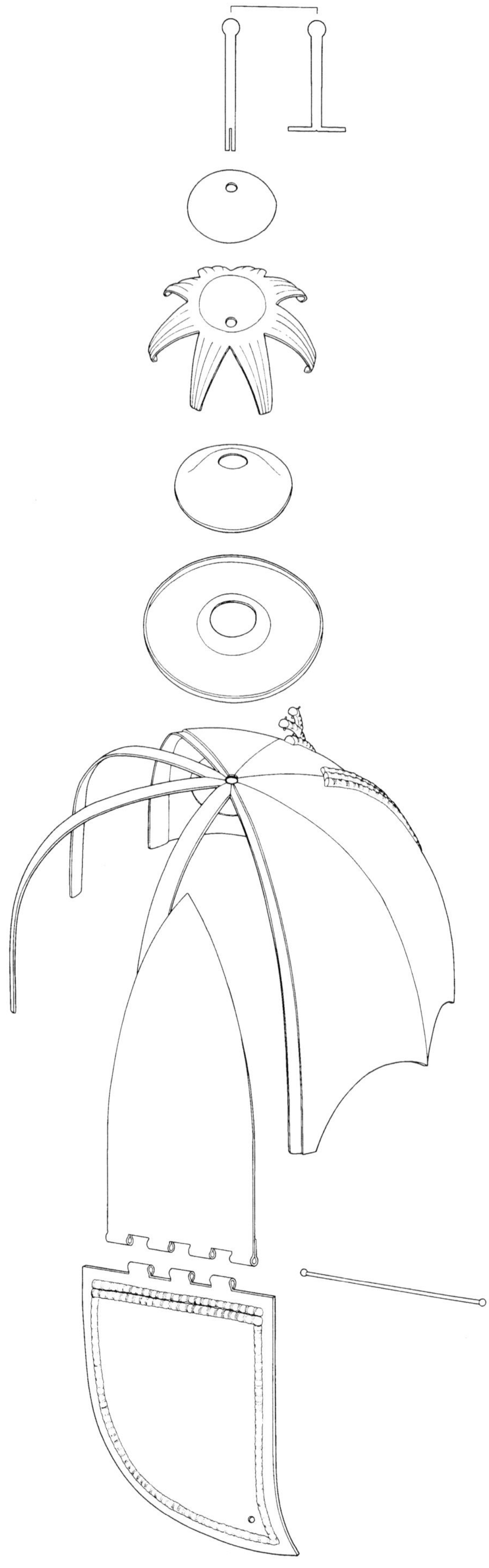

Cat. no. 142. Sardis, M87.7/9444.
Drawing of construction.

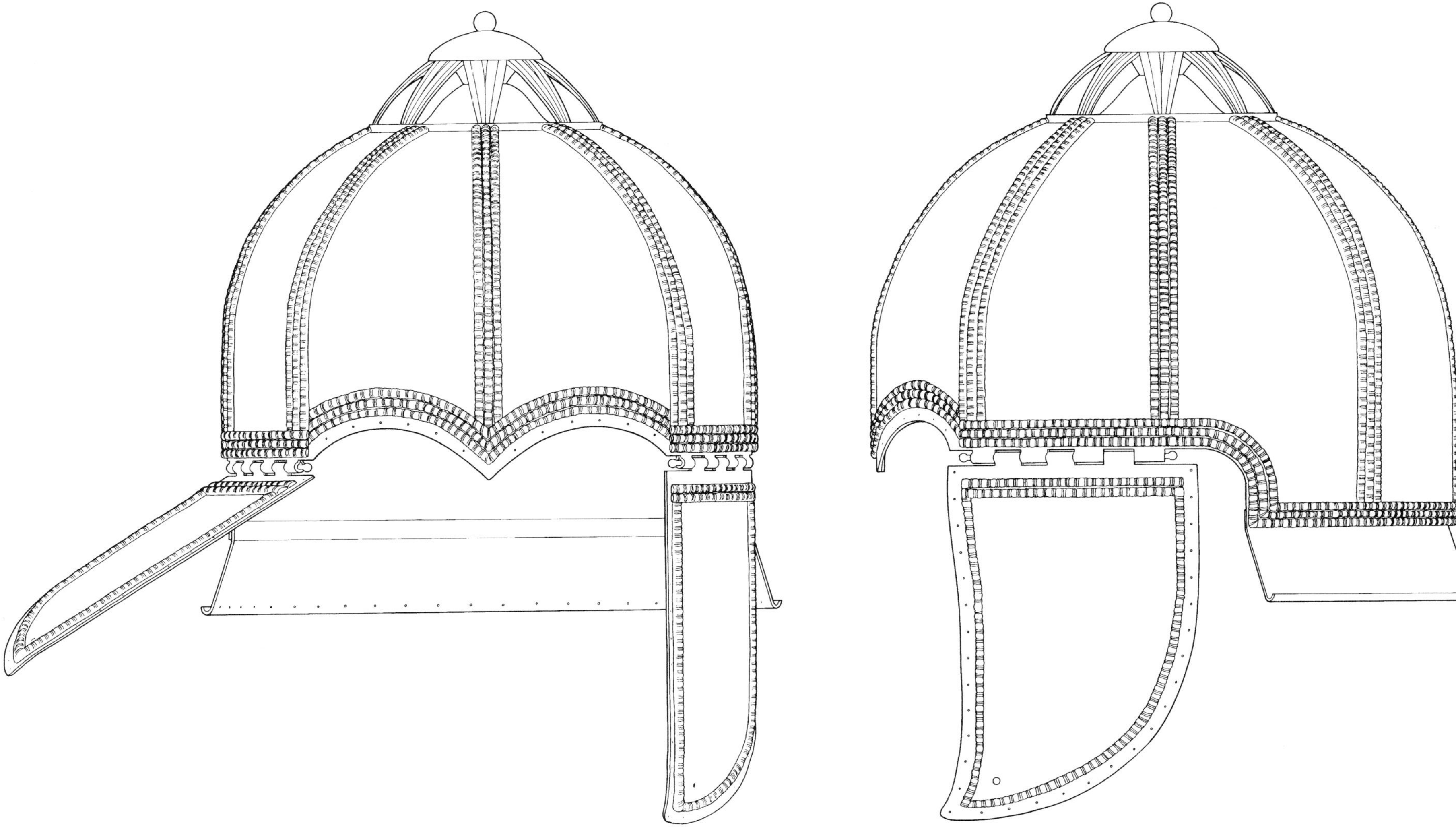

Cat. no. 142. Sardis, M87.7/9444.
Drawing of construction.

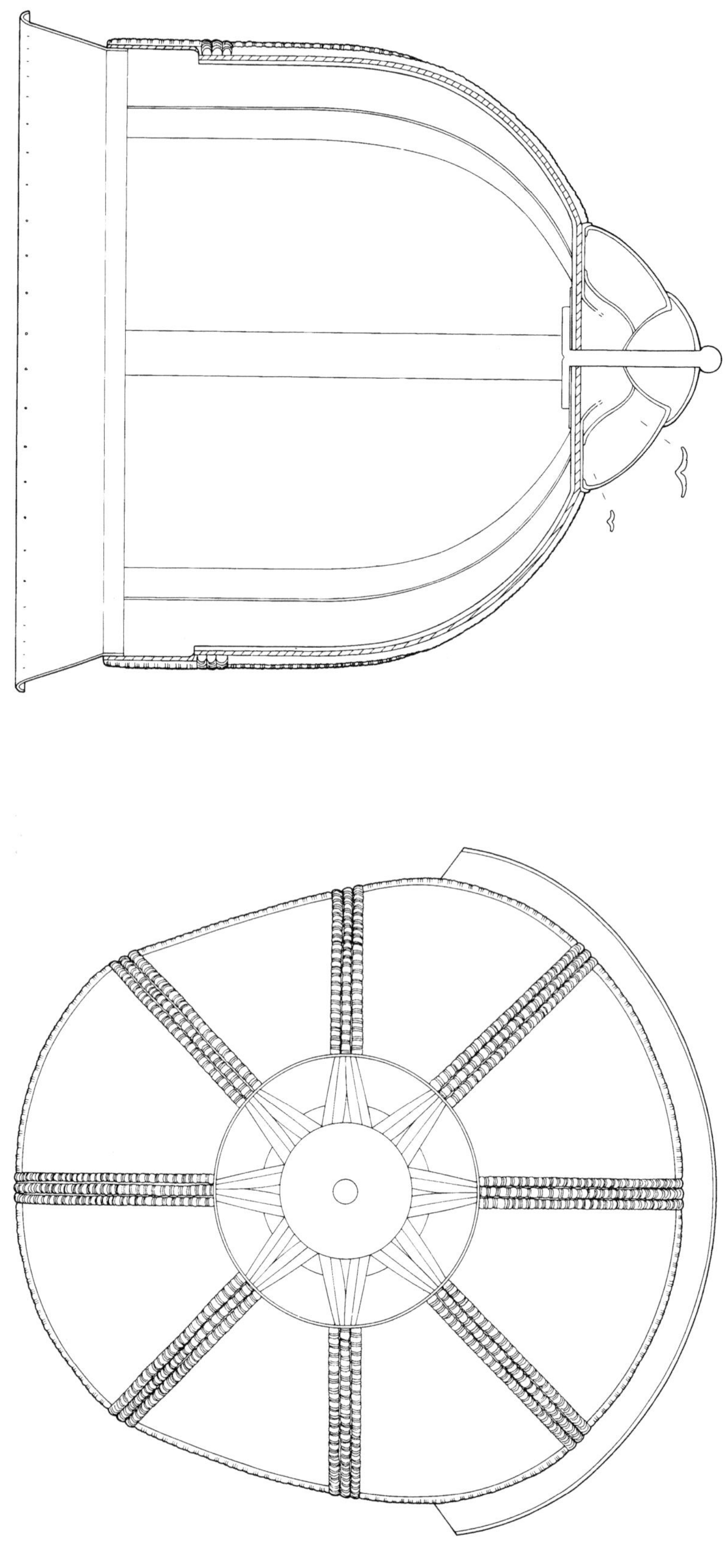

Cat. no. 142. Sardis, M87.7/9444.
Drawing of construction.

DURHAM UNIVERSITY LIBRARY
3 0104 00996027 3